TM 9-755

WAR DEPARTMENT TECHNICAL MANUAL

76-MM GUN MOTOR CARRIAGE M18 HELLCAT AND ARMORED UTILITY VEHICLE M39

By *WAR DEPARTMENT* • *APRIL 1945*

RESTRICTED

©2013 Periscope Film LLC
All Rights Reserved
ISBN#978-1-937684-46-4
www.PeriscopeFilm.com

DISCLAIMER:

This manual is sold for historic research purposes only, as an entertainment. It contains obsolete information and is not intended to be used as part of an actual operation or maintenance training program. No book can substitute for proper training by an authorized instructor.

This document reproduces the text of a manual first published by the Department of the Army, Washington DC. All source material contained in the reproduced document has been approved for public release and unlimited distribution by an agency of the U.S. Government. Any U.S. Government markings in this reproduction that indicate limited distribution or classified material have been superseded by downgrading instructions that were promulgated by an agency of the U.S. government after the original publication of the document. No U.S. government agency is associated with the publishing of this reproduction.

©2013 Periscope Film LLC
All Rights Reserved
ISBN#978-1-937684-46-4
www.PeriscopeFilm.com

WARNING

Authority for release of this document to a foreign government must be secured from the Assistant Chief of Staff, G-2.

When this document is released to a foreign government, it is released subject to the following conditions: This information is furnished with the understanding that it will not be released to another nation without specific approval of the United States of America, Department of the Army; that it will not be used for other than military purposes; that individual or corporation rights originating in the information whether patented or not will be respected; and that the information will be afforded substantially the same degree of security as afforded by the United States of America, Department of the Army.

RESTRICTED

WAR DEPARTMENT TECHNICAL MANUAL
TM 9-755

This TM supersedes TM 9-755, dated 15 Jul 43; OFSTB 755-1, dated 22 Nov 43; OFSTB 755-2, dated 22 Nov 43; OFSTB 755-3, dated 18 Dec 43; WDTB 9-755-4, dated 23 Dec 43; WDTB 9-755-5, dated 5 Jan 44; WDTB 9-755-6, dated 6 Jan 44; WDTB 9-755-7, dated 11 Jan 44; WDTB 9-755-8, dated 11 Jan 44; WDTB 9-755-9, dated 11 Jan 44; WDTB 9-755-10, dated 11 Jan 44; WDTB 9-755-11, dated 13 Jan 44; WDTB 9-755-12, dated 23 Jun 44; and WDTB 9-755-13, dated 2 Aug 44. This TM supersedes portions of WDTB ORD 29, dated 24 Jan 44; WDTB ORD 75, dated 10 Apr 44; WDTB ORD 88, dated 4 May 44; WDTB ORD 126, dated 19 Jul 44; WDTB ORD 130, dated 1 Aug 44; WDTB ORD 210, dated 10 Oct 44; WDTB ORD 217, dated 8 Nov 44; WDTB ORD 235, dated 19 Dec 44; WDTB ORD 254, dated 17 Feb 45; and OFSTB 700-29 and TB ORD 196, dated 2 Jan 43; which apply to the materiel covered in this TM; however, these TB's remain in force until incorporated in all other affected TM's or specifically rescinded.

76-MM GUN MOTOR CARRIAGE M18 AND ARMORED UTILITY VEHICLE M39

WAR DEPARTMENT • *APRIL 1945*

RESTRICTED

WAR DEPARTMENT
Washington 25, D. C., 25 April 1945

TM 9-755, 76-mm Gun Motor Carriage M18 and Armored Utility Vehicle M39, is published for the information and guidance of all concerned.

[A.G. 300.7 (2 Oct 43)
O.O.M. 461/Rar. Ar. (4-27-45)]

BY ORDER OF THE SECRETARY OF WAR:

G. C. MARSHALL,
Chief of Staff.

OFFICIAL:

J. A. ULIO,
*Major General,
The Adjutant General.*

DISTRIBUTION: AAF (10); AGF (10); ASF (2); S Div ASF (1); Dept (10); AAF Comd (2); Arm & Sv Bd (2); Tech Sv (2); SvC (10); PC&S (1); PE (Ord O) (5); Dist O, 9 (5); Dist Br O, 9 (3); Reg O, 9 (3); Establishments, 9 (5); Decentralized Sub-O, 9 (3); Gen & Sp Sv Sch (10); USMA (20); A (10); CHQ (10); D (2); Bn 18 (2); AF (2); T/O & E: 9-7 (3); 9-9 (3); 9-65 (2); 9-67 (3); 9-76 (2); 9-127 (3); 9-197 (3); 9-318 (3); 9-325 (2); 9-327 (3); 9-328 (3); 9-377 (3); 18-27 (3); 18-28 (3); 18-37 (3).

(For explanation of symbols, see FM 21-6.)

CONTENTS

PART ONE—INTRODUCTION

		Paragraphs	Pages
Section	I. General	1– 2	1– 6
	II. Description and data	3– 4	7– 11
	III. Tools, parts, and accessories	5– 8	11– 25

PART TWO—OPERATING INSTRUCTIONS

		Paragraphs	Pages
Section	IV. General	9	26
	V. Service upon receipt of equipment	10– 12	26– 31
	VI. Controls and instruments	13– 16	31– 43
	VII. Operation under ordinary conditions	17– 20	43– 52
	VIII. Turret controls and operation—M18	21– 22	52– 60
	IX. Operation of accessory equipment	23– 26	60– 68
	X. Operation under unusual conditions	27– 32	68– 74
	XI. Demolition to prevent enemy use	33– 34	74– 75

PART THREE—MAINTENANCE INSTRUCTIONS

		Paragraphs	Pages
Section	XII. General	35	76
	XIII. Special organizational tools and equipment	36	76– 78
	XIV. Lubrication	37– 38	79– 95
	XV. Preventive maintenance services	39– 45	95–125
	XVI. Trouble shooting	46– 66	126–176
	XVII. Engine description and maintenance	67– 73	177–191
	XVIII. Engine removal and installation	74– 76	192–201
	XIX. Engine oiling system	77– 85	201–220
	XX. Ignition system	86– 90	220–231
	XXI. Fuel and air intake and exhaust systems	91–100	231–262
	XXII. Transfer case assembly (rear)	101–104	262–268

CONTENTS—Contd.

PART THREE—MAINTENANCE INSTRUCTIONS—Contd.

		Paragraphs	Pages
Section	XXIII. Propeller shaft and universal joint	105–107	268–276
	XXIV. Torqmatic transmission assembly	108–112	276–286
	XXV. Controlled differential assembly	113–117	286–294
	XXVI. Transmission, differential, and transfer case lubrication system	118–125	294–315
	XXVII. Final drive	126–129	315–321
	XXVIII. Tracks and suspension	130–140	322–354
	XXIX. Batteries and charging system	141–145	354–371
	XXX. Starter system	146–149	371–376
	XXXI. Lighting system	150–154	376–384
	XXXII. Instrument panel assembly	155–162	384–394
	XXXIII. Electrical equipment	163–174	394–429
	XXXIV. Radio interference suppression	175–177	429–431
	XXXV. Hull	178–189	431–447
	XXXVI. Turret—M18	190–200	447–464
	XXXVII. Fixed fire extinguisher system	201–204	464–471

PART FOUR—AUXILIARY EQUIPMENT

		Paragraphs	Pages
Section	XXXVIII. General	205	472
	XXXIX. Armament	206–216	472–491
	XXXX. Sighting and fire control equipment	217–221	491–497
	XXXXI. Ammunition	222–223	497–500
	XXXXII. Radio and interphone equipment	224–228	500–507

APPENDIX

		Paragraphs	Pages
Section	XXXXIII. Shipment and limited storage	229–231	508–512
	XXXXIV. References	232–234	513–517
Index			518–528

TM 9-755
1-2

RESTRICTED

This TM supersedes TM 9-755, dated 15 Jul 43; OFSTB 755-1, dated 22 Nov 43; OFSTB 755-2, dated 22 Nov 43; OFSTB 755-3, dated 18 Dec 43; WDTB 9-755-4, dated 23 Dec 43; WDTB 9-755-5, dated 5 Jan 44; WDTB 9-755-6, dated 6 Jan 44; WDTB 9-755-7, dated 11 Jan 44; WDTB 9-755-8, dated 11 Jan 44; WDTB 9-755-9, dated 11 Jan 44; WDTB 9-755-10, dated 11 Jan 44; WDTB 9-755-11, dated 13 Jan 44; WDTB 9-755-12, dated 23 Jun 44; and WDTB 9-755-13, dated 2 Aug 44. This TM supersedes portions of WDTB ORD 20, dated 24 Jan 44; WDTB ORD 75, dated 10 Apr 44; WDTB ORD 88, dated 4 May 44; WDTB ORD 126, dated 19 Jul 44; WDTB ORD 130, dated 1 Aug 44; WDTB ORD 210, dated 10 Oct 44; WDTB ORD 217, dated 8 Nov 44; WDTB ORD 233, dated 19 Dec 44; WDTB ORD 254, dated 17 Feb 45; and OFSTB 700-29 and TB ORD 196, dated 2 Jan 43, which apply to the material covered in this TM; however, these TB's remain in force until incorporated in all other affected TM's or specifically rescinded.

PART ONE—INTRODUCTION

Section I
GENERAL

1. SCOPE.

a. These instructions are published for information and guidance of all concerned. They contain information on operation and maintenance of equipment as well as descriptions of major units and their functions in relation to other components of the vehicles. They apply only to the 76-mm Gun Motor Carriage M18 and the Armored Utility Vehicle M39. These instructions are arranged in four parts; Part One, Introduction; Part Two, Operating Instructions; Part Three, Maintenance Instructions; Part Four, Auxiliary Equipment.

b. The appendix at the end of the manual contains instructions for shipment and limited storage, and a list of references including standard nomenclature list, technical manuals, and other publications applicable to the vehicles.

c. The stock and part numbers which appear throughout the manual are extracted from ORD 7, SNL G-163, 15 November 1944.

2. RECORDS.

a. Forms and records applicable for use in performing prescribed operations are listed below with a brief explanation of each.

(1) W.D., A.G.O. Form No. 7360. Army Motor Vehicle Operator's Permit. This form will be issued by commanding officers of posts, camps, stations, or organizations, to all operators of military vehicles who have passed the driver's examination (TM 21-300) and are qualified to drive the particular vehicles noted on the permit.

(2) War Department Lubrication Order. War Department Lubrication Order L.O. 9-755 prescribes lubrication maintenance for these vehicles. A Lubrication Order is issued with each vehicle and is to be carried with it at all times.

(3) Standard Form. No. 26, Driver's Report—Accident, Motor Transportation. One copy of this form will be kept with

RESTRICTED

the vehicle at all times. In case of an accident resulting in injury or property damage, it will be filled out by the driver on the spot, or as promptly as practical thereafter.

(4) WAR DEPARTMENT FORM NO. 48, DRIVER'S TRIP TICKET AND PREVENTIVE MAINTENANCE SERVICE RECORD. This form, properly executed, will be furnished to the driver when his vehicle is dispatched on nontactical missions. The driver and the official user of the vehicle will complete in detail appropriate parts of this form. These forms need not be issued for vehicles in convoy or on tactical missions. The reverse side of this form contains the driver's daily and weekly preventive maintenance service reminder schedule.

(5) W.D., A.G.O. FORM NO. 478, MWO AND MAJOR UNIT ASSEMBLY REPLACEMENT RECORD. This form, carried with the vehicle, will be used by all personnel completing a modification or major unit assembly (engine, transmission, transfer case, and tracks) replacement to record clearly the description of work completed, date, vehicle hours, and/or mileage, and MWO number or nomenclature of unit assembly. Personnel performing the operation will initial in the column provided. Minor repairs, parts, and accessory replacements will not be recorded.

(6) W.D., A.G.O. FORM NO. 460, PREVENTIVE MAINTENANCE ROSTER. This form will be used for scheduling and maintaining a record of motor vehicle maintenance operations.

(7) W.D., A.G.O. FORM NO. 6, DUTY ROSTER. This form slightly modified, can be used for scheduling and maintaining a record of vehicle maintenance operations if W.D., A.G.O. Form No. 460 is not available. It may be used for lubrication records.

(8) W.D., A.G.O. FORM NO. 462, PREVENTIVE MAINTENANCE SERVICE AND TECHNICAL INSPECTION WORK SHEET FOR FULL-TRACK AND TANK-LIKE WHEELED VEHICLES. This form will be used for all 50-hour (500-mile) or 100-hour (1,000-mile) services, and for technical inspections of these vehicles.

(9) W.D., O.O. FORM NO. 9-70, SPOT-CHECK INSPECTION REPORT FOR ALL MOTOR VEHICLES. This form may be used by all commanding officers or their staff representatives in making spot-check inspections on all vehicles.

(10) W.D., A.G.O. FORM NO. 468, UNSATISFACTORY EQUIPMENT REPORT. This form will be used for reporting manufacturing, design, or operational defects in materiel with a view to improving and correcting such defects, and for use in recommending modifications of materiel. This form will not be used for reporting failures, isolated materiel defects, or malfunctions of materiel resulting from fair wear and tear or accidental damage; nor for the replacement, repair, or the issue of parts and equipment. It does not replace currently authorized operational or performance records.

(11) W.D., O.O. FORM NO. 9-81 EXCHANGE PART OR UNIT IDENTIFICATION TAG. This tag, properly executed, may be used when exchanging unserviceable items for like serviceable assemblies, parts, vehicles, and tools.

GENERAL

Figure 1—76-mm Gun Motor Carriage M18—Right Front

Figure 2 — 76-mm Gun Motor Carriage M18 — Right Rear

GENERAL

Figure 3 — Armored Utility Tractor M39 — Right Front

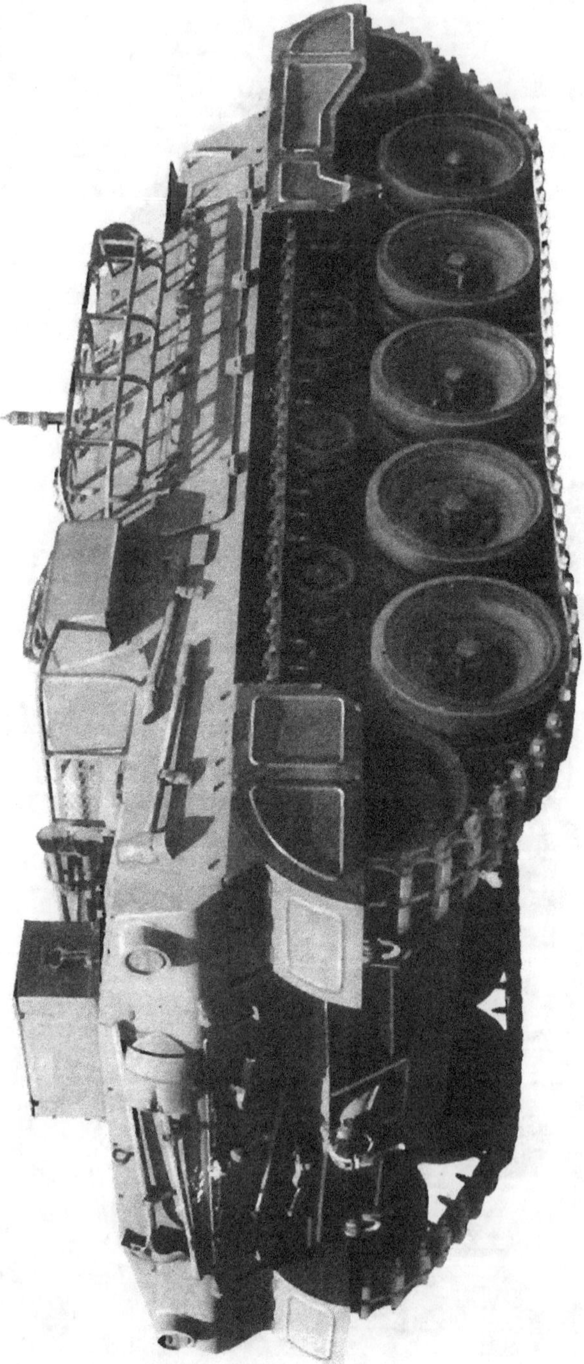

RA PD 344543

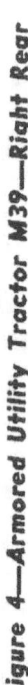

Figure 4—Armored Utility Tractor M39—Right Rear

Section II

DESCRIPTION AND DATA

3. DESCRIPTION.

a. **General.** The 76-mm Gun Motor Carriage M18 and Armored Utility Tractor M39 are armored, full-track laying vehicles. The M18 vehicle has a 360-degree traversing turret with a 76-mm gun mounted over the middle compartment of the hull. The M39 utility tractor does not have the turret and 76-mm gun; the middle of the hull has an open compartment for cargo or personnel. The power train, track, and suspension are identical in both vehicles. The descriptions, data, and maintenance instructions contained in this manual apply equally to both vehicles except where a difference between models is specified.

b. **Power Train** (fig. 5). Power is provided by a 9-cylinder, radial, air-cooled, gasoline engine mounted in rear compartment of hull. Power from engine is transmitted through a universal joint to rear transfer case which is mounted on a transverse bulkhead between engine and middle compartments. A propeller shaft, provided with a universal joint at each end, transmits power from rear transfer case to front transfer case which is assembled on the Torqmatic transmission. This arrangement of transfer cases permits location of propeller shaft under subfloor (M18) or seats (M39) in middle compartment of hull. The Torqmatic transmission is bolted to rear end of controlled differential to form a compact unit assembly, which is mounted in front, or driving compartment of the hull. Power from controlled differential is transmitted through universal joints to right and left final drive units which are mounted on front ends of the hull side plates. The final drive units transmit power through dual sprockets to endless steel-link tracks which provide necessary traction to propel the vehicle.

c. **Tracks and Suspension.** The tracks and suspension are completely described in Section XXVIII.

d. **Hull and Turret.** The hulls are completely described in Section XXXV. The turret for the M18 is described in Section XXXVI.

e. **Crew.**

(1) The crew of the M18 vehicle consists of five men. The driver sits to the left of the transmission and assistant driver sits to the right of the transmission in the front compartment of the vehicle. The commander is stationed at the antiaircraft gun on the left side of the turret; the gunner sits to the left of the 76-mm gun; the loader sits to the right of the 76-mm gun.

(2) The crew of the M39 vehicle consists of 10 men. The driver sits to the left of the transmission and assistant driver sits to the right of transmission in front compartment of vehicle. All

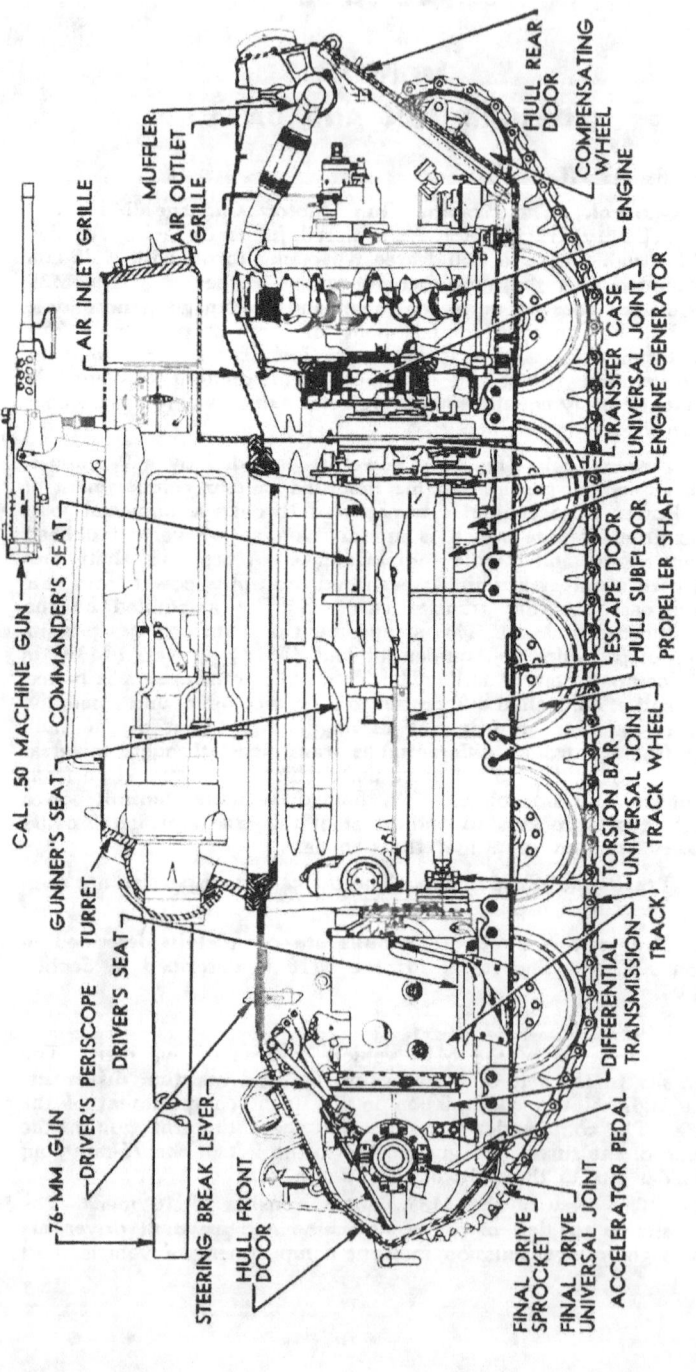

Figure 5—76-mm Gun Motor Carriage M18—Longitudinal Section

Description and Data

other members of crew occupy seats in middle compartment of vehicle.

f. Communications.

(1) Vehicles are equipped with radio for inter-tank, and telephone for intra-tank, communication.

g. Vehicle Serial Numbers. Serial numbers of the vehicle and unit assemblies are located in the following places:

Unit	Serial Number Location
Vehicle	Left sponson bottom plate in front of instrument panel.
Engine	Plate on front crankcase, also stamped on rear crankcase just forward of starter.
Carburetor	Top of housing on left side.
Magneto	Right side of housing.
Starter	Right side of housing.
Engine generator	Top of frame.
Auxiliary generator	Front side of flywheel housing.
Battery	Cell connector at top of battery.
Transmission	Stamped on boss on case to right of shift lever.
Oil cooler blower	Rear side of housing.
Turret traversing electric motor	Rear side of frame.
Turret traversing hydraulic pump	Top of pump head.
Turret traversing hydraulic motor	Top of housing.
Turret hand traversing mechanism	Stamped on traverse case cover.

4. TABULATED DATA.

a. Vehicle Specifications.

Length, over-all
 M18 .... 21 ft 10 in.
 M39 .... 17 ft 10 in.
Width, over-all .... 9 ft 5 in.
Height, over-all, A.A. gun horizontal
 M18 .... 8 ft 5 in.
 M39 .... 6 ft 8½ in.
Height, lowest operable
 M18 .... 7 ft 9¼ in.
 M39 .... 6 ft 5½ in.
Track width .... 12 in.
Tread (center-to-center) .... 7 ft 10⅝ in.
Crew
 M18 .... 5 men
 M39 .... 10 men
Weight, fighting (less crew)
 M18 .... 36510 lb
 M39 .... 33500 lb
Weight, gross
 M18 .... 37557 lb

Part One—Introduction

M39	35500 lb
Ground pressure	
M18	11.9 lb psi
M39	11.25 lb psi
Ground contact area	3156 sq in.
Ground clearance	
M18	14¼ in.
M39	14¼ in.
Pintle height (loaded)	
M18	25½ in.
M39	25½ in.
Kind and grade of fuel	80 octane gasoline
Approach angle	
M18	28 deg
M39	28 deg
Departure angle	
M18	26½ deg
M39	26½ deg

b. **Performance.**

Vehicle cruising speeds	
1st range	12 mph
2nd range	25 mph
3rd range	45 mph
Reverse	20 mph
Minimum turning radius	33 ft
Fording depth	
M18	48 in.
M39	48 in.
Towing facilities, front	
M18	2 shackles
M39	2 shackles, 1 pintle
Towing facilities, rear	2 shackles, 1 pintle
Maximum draw-bar pull	24,500 lb
Maximum grade ascending ability	60% @ 3 mph
Engine cruising speed	2100 rpm
Maximum allowable engine speed	
R 975-C1	350 @ 2,400 rpm
R 975-C2	400 @ 2,400 rpm

c. **Capacities.**

Transmission, only	44 qt
Transmission and cooler	48 qt
Differential and transfer case, including cooler	20 qt
Final drives (each)	5 qt
Fuel tank, left	75 gal
Fuel tank, right	90 gal
Engine oil tank	11 gal

Tools, Parts, and Accessories

Auxiliary generator fuel tank		5 gal

   d. Communications.

Radio	Model SCR 610
Interphone	Model RC-99

	M18	M39
Gun, 76-mm, M1A1, M1A1C, or M1A2	1	0
Gun, Machine, Cal. .50, M2H.B	1	1
Carbine, Cal. .30, M1	5	2

	M18	M39
76-mm, rounds	45	0
Cal. .50, rounds	800	900
Cal. .30, rounds	450	1620
*3 in., rounds	0	42*
Grenades, smoke, WP M50	6	6
Grenades, fragmentation Mk. II	6	6
Smoke pots	4	4

Section III
TOOLS, PARTS, AND ACCESSORIES

5. PURPOSE.

   a. The lists in this section are for information only and must not be used as a basis for requisition.

6. ON-VEHICLE TOOLS.

   a. Pioneer Tools. Items listed are the same for both vehicles except when model is shown in quantity column.

Quantity Per Vehicle	Item, Name and Stock Number	Stowage Location
1	AX, chopping, 4-lb (41-A-1277)	Rear outside hull
1	BAR, crow, 5-ft, pinch point (41-B-175)	Rear outside hull
1	CUTTER, wire (M1938)	Issued 1 per company (22 vehicles) M18 Rear Sponson extension M39
1	HANDLE, mattock (41-H-1286)	Rear outside hull
1	MATTOCK, pick, M1 (W/O Handle) (41-W-722)	Rear outside hull
1 (M18)	SHOVEL, short handled (41-S-3170)	Rear outside hull

*When used as prime mover for gun, 3-in., M6.

Part One—Introduction

Quantity Per Vehicle	Item, Name and Stock Number	Stowage Location
2 (M39)	SHOVEL, short handled (41-S-3170)	Rear outside hull
2 (M39)	SHOVEL, long handle (41-S-3220)	Rear outside hull
1 (M18)	SLEDGE, blacksmith, double face, 10-lb (41-S-3726)	Rear outside hull
2 (M39)	PICK	Rear outside hull

b. **Vehicular Tools.** Vehicular tools are stowed in tool stowage box located at rear end of turret on M18 vehicle, or at rear end of right sponson extension on M39 vehicle. Some tools are placed in tool bag (41-B-9-250), tool box (41-B-1624-500), or socket box (41-B-1642-500), which are stowed in tool stowage box. Exact location of each tool is given below:

Quantity Per Vehicle	Item, Name and Stock Number	Stowage Location
1	ADAPTER, 1/2-in. sq-drive, female 3/8-in. sq-drive, male, snap on (41-A-20-220)	Tool bag
1	BAG, tool (41-B-9-250)	Tool stowage box
1	BAR, 1/4-in., hex, 12 in. long (G163-7050459)	Tool bag
2	BAR, utility (41-B-2-53)	Tool bag
1	BAR, track pin removing (41-B-327)	Tool stowage box
1	BAR, cross (41-B-154)	Tool bag
1	BAR, 1/2-in. sq-drive, 10 in. (41-B-309)	Tool bag
1	BAR, handy grip, 1/2-in. sq-drive, 5 in. long (41-B-311-200)	Tool bag
1	BAR, 3/4-in., sq-drive, 8 in. long (7050457)	Tool bag
1	BAR, 3/4-in. sq-drive, 16 in. long (B187272)	Tool bag
1	BOX, tool (41-B-1624-500)	Tool stowage box
1	BOX, socket (41-B-1620-700)	Tool stowage box
1	CHISEL, cold, 3/4-in. (41-C-1124)	Tool bag
1	FILE, 3-in. sq smooth, 6-in. (41-F-1572)	Tool bag
1	FILE, hand, smooth, 8-in. (41-F-1028)	Tool bag
1	HAMMER, machinist ball peen, 32 oz (41-H-527)	Tool bag
1	HANDLE, combination tee, 1/2-in. sq-drive, 11 in. long (41-H-1509-55)	Tool bag
1	HANDLE, combination tee, 3/4-in. sq-drive, 17 in. long (41-H-1509-60)	Tool bag
1	HANDLE, flexible, 1/2-in. sq-drive, 12 in. (41-H-1502)	Tool bag
1	HANDLE, track adjusting wrench (41-H-1498-15)	Tool bag
1	HANDLE, speeder, 1/2-in. sq-drive, 17 in. long (41-H-1508)	Tool bag
1	JOINT, universal, 1/2-in. sq-drive (41-J-380)	Tool bag

Quantity Per Vehicle	Item, Name and Stock Number	Stowage Location
1	PLIERS, combination, slip-joint, 8-in. (41-F-1652)	Tool bag
1	PLIERS (side cutting, 8-in. (41-P-1977)	Tool bag
1	RATCHET, reversible, ½-in. sq-drive, 9 in. (41-H-1505)	Tool bag
1	SCREWDRIVER, machinist's, 5-in. blade (41-S-1385)	Tool bag
1	SCREWDRIVER, special purpose, 1¾-in. blade (41-S-1063)	Tool bag
1	SCREWDRIVER, special purpose, 1½-in. blade (41-S-1062-90)	Tool bag
1	SCREWDRIVER, non-magnetic, for compass (41S-1067-700)	On compass mount
2	SLING, ⅜-in. dia wire rope (41-S-3829-75)	Tool stowage box
1	WRENCH, adjustable, single-end, 8 in. (41-W-486)	Tool bag
1	WRENCH, adjustable, single-end, 12 in. (41-W-488)	Tool bag
1	WRENCH, auto, 11 in. (41-W-448)	Tool bag
1	WRENCH, box end offset, 1¼-in.(shock absorber link adjusting) (41-W-576)	Tool bag
1	WRENCH, engr, single-hd, 1⅞ in. (41-W-1314)	Tool bag
1	WRENCH, track adjust, open-end, 2$^{1}/_{16}$ in.	Tool bag
1	WRENCH, engr, dble-hd, $^{5}/_{16}$ in. x ⅜ in. (41-W-990)	Tool box
1	WRENCH, engr, dble-hd, $^{7}/_{16}$ in. x ½ in. (41-W-1000)	Tool box
1	WRENCH, engr, dble-hd, $^{9}/_{16}$ in. x $^{11}/_{16}$ in. (41-W-1005-5)	Tool box
1	WRENCH, engr, dble-hd, ⅝ in. x ¾ in. (41-W-1008)	Tool box
1	WRENCH, engr, dble-hd, $^{13}/_{16}$ in. x ⅞ in. (41-W-1015)	Tool box
1	WRENCH, engr, dble-hd, $^{15}/_{16}$ in. x 1 in. (41-W-1021)	Tool box
1	WRENCH, engr, dble-hd, 1⅛ in. x 1⅜ in. (41-W-1028-10)	Tool box
1	WRENCH, plug, ⅝-in. hex(41-W-1961)	Tool bag
1	WRENCH, plug, 1-in. hex (41-W-1961-50)	Tool bag
1	WRENCH, socket head set screw, $^{5}/_{32}$-in. hex (41-W-2449)	Tool bag
1	WRENCH, socket head set screw, 1⅛-in. hex (41-W-2450)	Tool bag
1	WRENCH, socket head set screw, $^{3}/_{16}$-in. hex (41-W-2452)	Tool bag

TM 9-755

Part One—Introduction

Quantity Per Vehicle	Item, Name and Stock Number	Stowage Location
1	WRENCH, socket head set screw, ¼-in. hex (41-W-2454)	Tool bag
1	WRENCH, socket head set screw, 5/16-in. hex (41-W-2455)	Tool bag
1	WRENCH, socket head set screw, ⅜-in. hex (41-W-2456)	Tool bag
1	WRENCH, socket head set screw, ½-in. hex (41-W-2457)	Tool bag
1	WRENCH, socket, ½-in. sq-drive, ⅜-in. hex (41-W-3004)	Socket box
1	WRENCH, socket, ½-in. sq-drive, ⅜-in. sq (41-W-3001-200)	Socket box
1	WRENCH, socket, ½-in. sq-drive, 7/16-in. hex (41W-3005)	Socket box
1	WRENCH, socket, ½-in. sq-drive, ½-in. hex (41-W-3007)	Socket box
1	WRENCH, socket, ½-in. sq-drive, 9/16-in. hex (41-W-3009)	Socket box
1	WRENCH, socket, ½-in. sq-drive, ⅝-in. hex (41-W-3013)	Socket box
1	WRENCH, socket, ½-in. sq-drive, ¾-in. hex (41-W-3017)	Socket box
2	WRENCH, socket, ½-in. sq-drive, ⅞-in. hex (41-W-3023)	Socket box
1	WRENCH, socket, ½-in. sq-drive, 1-in. hex (41-W-3027)	Socket box
1	WRENCH, socket, ½-in. sq-drive, 1 1/16-in. hex (41-W-3029)	Socket box
2	WRENCH, socket, ¾-in. sq-drive, 15/16-in. hex (41-W-3033-25)	Socket box
1	WRENCH, socket, ¾-in. sq-drive, 1-in. hex (41-W-3033-50)	Socket box
1	WRENCH, socket, ¾-in. sq-drive, 1⅛-in. hex (41-W-3036)	Socket box
1	WRENCH, socket, ¾-in. sq-drive, 1 5/16-in. hex (41-W-3037)	Socket box
1	WRENCH, socket, track adjusting (41-W-2574-240)	Tool bag

c. **Gun Tools.** Gun tools are placed in the gun parts box (D41197), which is stowed in the stowage box located at rear end of turret on M18 vehicle, or at rear end of the right sponson extension on the M39 vehicle.

Quantity Per Vehicle	Item, Name and Stock Number	Stowage Location
1	EYEBOLT, breechblock removing, 76-mm (41-E-3135)	Gun parts box
1	ROD, push, 76-mm gun (41-R-2619)	Gun parts box
1	TOOL, breechblock removing, 76-mm gun (41-T-3076-800)	Gun parts box

TM 9-755
6-7

Tools, Parts, and Accessories

Quantity Per Vehicle	Item, Name and Stock Number	Stowage Location
1	WRENCH, socket head set screw, recoil oil check, 76-mm gun (41-W-2410-25)	Gun parts box
1	WRENCH, combination, cal. .50, M2 (41-W-3349-50)	Gun parts box

7. ON-VEHICLE EQUIPMENT.

a. Ammunition, M18 Vehicle.

Quantity Per Vehicle	Name	Stowage Location
450	ROUNDS, cal. .30 (for model M1 carbine)	Periscope head box behind driver
1000	ROUNDS, cal. .50 (in box M2, D73913)	700 in right turret box; 200 under turret floor; 100 on gun
45	ROUNDS, 76-mm (75 pct AP; 25 pct He)	Shell racks
6	GRENADES, smoke, WP, M50	In cal. .50 M2 box in right turret box
6	GRENADES, fragmentation, Mk. II	In cal. .50 M2 box in right turret box
4	SMOKE POTS	In turret rear box

b. Ammunition, M39 Vehicle.

1620	ROUNDS, cal. .30 (for model M1 carbine)	L. H. upper sponson
900	ROUNDS, cal. .50 (in box M2, D-73913)	400 L. H. upper front sponson; 400 top of battery case; 100 on gun
*42	ROUNDS, 3-in.	Right and left sponsons
6	GRENADES, fragmentation, Mk. II	In cal. .50 M2 box top of battery case
6	GRENADES, smoke, WP, M50	In cal. .50 M2 box top of battery case
4	SMOKE POTS	In right rear sponson extension

c. Armament—M18 Vehicle.

1	GUN, 76-mm, M1A1, M1A2, or M1A1C	Turret
1	GUN, machine, cal. .50, M2, HB	Turret

*When used as prime mover for: Gun, 3-in., M6.

TM 9-755

Part One—Introduction

Quantity Per Vehicle	Name	Stowage Location
5	CARBINE, M1	(1) Left hull side; (2) Right hull side; (1) Right side of recoil guard; (1) Left side turret
1	BRAKE, muzzle, M2 (when M1A2 or M1A1C gun is used)	On gun

d. **Armament—M39 Vehicle.**

1	GUN, machine, cal. .50, M2 HB	Front of crew compartment
2	CARBINE, M1	Front side of bulkhead

e. **Accessories, 76-mm Gun, M18 Vehicle.** The 76-mm gun spare parts box is stowed in the turret rear extension stowage box.

Quantity Per Vehicle	Item, Name and Stock Number	Stowage Location
1	BOOK, artillery gun, O.O. Form 5825	Behind driver's seat
1	BRUSH, bore, M15, w/staff, complete (38-B-992-855)	Staff on left sponson; brush in turret extension box
1	COVER, bore brush, M516 (24-C-1059-100)	Turret extension box
1	COVER, breech (G163-01-40082)	On gun
1	COVER, empty shell case (G163-01-40080)	On gun mount
1	COVER, muzzle (7069075)	On gun
1	GUN, lubricating oil (2 oz) (41-G-1362-500)	On rear turret extension box
1	HOSE, assembly (33-H-581-450)	Turret extension box
1	OIL, recoil, special (spec. AXS 808), in container 1 qt, type 1, class D spec. (100-13).	
1	RAMMER, cleaning and unloading, M3 (C021-03-00106)	Right sponson
1	ROLL, spare parts, M13 (41-R-2690)	Spare parts box
1	SETTER fuse, M14 (F002-02-00873)	Spare parts box
1	SIGHT, bore, complete (muzzle, RF11AD; breech, RF11GA) (15-2A-11)	Spare parts box
1	TABLE, firing (76-A-4)	Spare parts box
1	TARGET, testing (set of 4)	Spare parts box

f. **Accessories, Cal. .30 Carbine.**

5	COVER, cal. .30 carbine (G163-5653357)	On gun

TM 9-755

Tools, Parts, and Accessories

g. **Accessories, Cal. .50 Machine Gun, HB, M2.** Items listed are identical for both vehicles except where model is shown in quantity column unless otherwise specified, items are carried in the cal. .50 gun spare parts box which is stowed in turret rear extension box (M18) or sponson rear extension (M39).

Quantity Per Vehicle	Item, Name and Stock Number	Stowage Location
1	BAG, empty cartridge (D90999)	On gun
1	BAG, metallic belt link (A037-03-00005)	On gun (M18); spare parts box (M39)
10 (M18)	BOX, ammunition, M2 (A037-0300002)	Right turret box (M18)
9 (M39)	BOX, ammunition, M2 (A037-0300002)	Over battery box
4	BRUSH, cleaning, cal. .50, M4 (38-B-992-27)	Spare parts box
1	CASE, cleaning rod, M15 (41-C-454-200)	Spare parts box
1	CHUTE, metallic belt link, M1 (A019-01-00540)	Spare parts box
1	COVER, spare barrel, M13, 45 in. (A039-88-00260)	On barrel (M18); spare parts box (M39)
1	COVER, gun cradle, cal. .50 (G163-568120)	On gun
1	COVER, tripod mount (A039-0300140)	On tripod (M18); spare parts box (M39)
2	ENVELOPE, spare parts, M1 (w/o contents) (M003-01-04590)	Spare parts box (M39)
1	EXTRACTOR, ruptured cartridge (41-E-557-75)	Spare parts box (M39)
1	MOUNT, tripod, cal. .50, M3 (A001-0305801)	Right rear hull roof (M18) rack on rear roof
1	OILER, filling, oil buffer (M003-01-09960)	Spare parts box
1	ROD, cleaning, jointed, M7, cal. .50 (41-R-2567-75)	Spare parts box

17

TM 9-755
7

Part One—Introduction

b. **Communications.** Items listed are identical for both vehicles except where model is shown in quantity column.

Quantity Per Vehicle	Name	Stowage Location
18 (M18)	CARTRIDGES, pyrotechnic	(2) w/pistol; (8) right turret box; (8) box on left air cleaner
10 (M39)	CARTRIDGES, pyrotechnic	(2) w/pistol; (8) box on left air cleaner
1	FLAG SET, M238, composed of: 1 case, CS-90; 1 flag, M-6-273, (red); 1 flag, MC-274 (orange); 1 flag, MC-275 (green); 3 flag-staffs, MC-270	Right turret rack (M18) or right sponson rack (M39)
1	INTERPHONE SYSTEM, Rc-99	Turret extension (M18), or right driving compartment ceiling (M39)
1	PISTOL, pyrotechnic, M2	Right turret box (M18) or bulkhead (M39)
1	RADIO SET, SCR-610, composed of:	
	2 ANTENNA, 3 sect.	1 mounted; 1 spare right turret rack (M18) or right sponson rack (M39)
	1 CASE, CS-79 (dry batteries)	Right rear turret wall (M18), or bulkhead (M39)
	1 CHEST, CH-72-A (access.)	Chest, less crystals, picked up by second echelon maintenance
	1 POWER UNIT, PE-117C	Turret extension (M18) or right front upper sponson (M39)

TM 9-755

Tools, Parts, and Accessories

Quantity Per Vehicle	Name	Stowage Location
1	RADIO, Rec. and Trans., BC-659-A	Turret extension (M18) or right front upper sponson (M39)
1	REEL, interphone, assembly, RL-108/VI	Turret left wall (M18)

i. Fire Extinguishers.

1	EXTINGUISHER, fire, 4 lb, CO_2, portable	Right front sponson

j. Rations.

30 cans	(M18) Type "K", 2-day rations for 5 men	In bags under subfloor
2 cans	(M18) Type "D", 1-day rations for 5 men	In bags under subfloor
54 cans	(M39) Type "C", 2-day rations for 9 men	Rear sponson extension

Quantity Per Vehicle	Name	Stowage Location
	k. Sighting Equipment—M18 Vehicle.	
2	BINOCULAR, M3, complete, composed of:	(1) Left turret wall;
	1 BINOCULAR, M3; 1 case, carrying, M17	(1) Right turret wall
1	CASE, carrying, gunner's quadrant, M1	Turret left wall
2	COVER, spare telescope and aux. right	On spare telescope
9	HEAD, extra for periscope, M4 and M6	(4) Right front sponson; (1) Left front subfloor; (4) Turret floor behind driver
1	HEADREST, telescope	On turret
10	LAMP (for elevation quadrant and azimuth indicator)	In spare battery and lamp box C101039

Part One—Introduction

Quantity Per Vehicle	Name	Stowage Location
2	LAMP, electric (for telescope reticle lights)	In spare battery and lamp box C101039
1	LIGHT, instrument, M30 (for elev. quad)	On gun mount
1	LIGHT, instrument, M33 (for telescope)	On telescope mount
1	MOUNT, telescope, M55	On gun mount
2	PERISCOPE, M4, w/telescope, M47	(1) Installed; (1) spare in right turret box
4	PERISCOPE, M6	(2) Installed; (2) spares inside left sponson
1	QUADRANT, elevation, M9	On gun mount
1	QUADRANT, gunner's M1	In case on turret left wall
1	SIGHT, open metal, aux.	
2	TELESCOPE, direct, M76C or 793	(1) Installed; (1) spare under right turret box

l. Sighting Equipment—T41.

Quantity Per Vehicle	Name	Stowage Location
4	HEAD, extra for periscope, M6	(4) right front sponson
4	PERISCOPE, M6, w/o telescope	(2) installed; (2) spares inside right front sponson

m. Accessories and Equipment, Misc. Items listed are identical for both vehicles except where vehicle model is shown in quantity column.

Quantity Per Vehicle	Name	Stowage Location
2	APPARATUS, decontaminating, 1½ qt M2 (spec. 197-54-113)	Assistant driver's compartment
1	ASSEMBLY, pistol grip and cable	Right turret wall (M18); Right sponson wall (M39)
5 (M18)	BAG, canvas, field, O.D., M1936	(3) Under right turret floor; (2) Under left turret floor
9 (M39)	BAG, canvas, field, O.D., M1936	
1	BAG, grease gun	Tool bag
1 (M18)	BAG, ration, type D	Under right turret floor

Tools, Parts, and Accessories

Quantity Per Vehicle	Name	Stowage Location
2 (M18)	BAG, ration, type K	Under right turret floor
1	BAG, tool	Turret extension box (17718); rear sponson extension (M39)
26 (M18)	BATTERY, flashlight	8 in flashlights; 6 in instrument lights; 12 spares in box C101039 in turret extension box
12 (M39)	BATTERY, flashlight, spares	In box C101039, in sponson extension
5 (M18)	BELT, safety	On seats
2 (M39)	BELT, safety	On seats
1	BOX, assembly, battery and lamp stowage	Turret extension box (M18); sponson extension (M39)
1	BUCKET, canvas, folding, 18 qt	Turret extension box (M18); rear sponson extension (M39)
1	CABLE, towing, 1 in. x 20 ft	On left hull roof
5 (M18)	CANTEEN, M1910, with cup and cover, M1910	1 at each crew position
1 (M18)	CONTAINER, water, 5 gal (QMC st'd)	Under left turret floor
1 (M39)	CONTAINER, water, 5 gal (QMC st'd)	Outside rear of sponson extension
1 (M18)	COVER, azimuth indicator	On indicator
1	COVER, air intake, engine	Right turret rack (M18) right rack (M39)
1	COVER, air outlet, engine	Right turret rack (M18); right rack (M39)
1 (M18)	COVER, mantlet, 76-mm gun mount	On turret and gun
1	COVER, oil cooler outlet, trans.	On outlet
1	CRANK, engine	Hull rear door
1	EXTENSION, hose	Tool bag
1	EXTINGUISHER, fire, carbon dioxide 4 lb, portable	Right front sponson
4 (M18)	FLASHLIGHT	1 each crew position

TM 9-755

Part One—Introduction

Quantity Per Vehicle	Name	Stowage Location
2 (M39)	FLASHLIGHT	Driver's position
1	GUN, lubricating, hand type	Tool bag
2	HOOD, hatch, driver's	Turret sides (M18); sponson extension sides (M39)
1	KIT, first-aid (24 unit) (spec. 1553)	Turret extension box (M18) sponson extension (M39)
4 (M18)	LAMP (spare for flashlight)	
2 (M39)	LAMP (spare for flashlight)	
1	LIST, Organizational Spare Parts and Equipment SNL C-163	Manual tray, escape hatch door (M18); manual box in crew compartment (M39)
1	MANUAL, Field, for cal. .30 Carbine, M1, FM 23-7	Back of ass't or driver's seat
1	MANUAL, Field, for cal. .50 M.G., M2, FM 23-65	Back of ass't or driver's seat
1	MANUAL, Technical, for 76-mm Gun Motor Carriage M18 and Armored Utility Vehicle (M39) TM 9-755	Manual tray, escape hatch door (M18); manual box in crew compartment (M39)
2 (M18)	MITTENS, asbestos, pairs	Right turret box
1 (M18)	NET, camouflage, 45 ft x 45 ft T1534	Turret racks
1 (M39)	NET, camouflage, cotton shrimp, 29 ft x 29 ft	Racks on rear roof
1	OILER (trigger type, 1 pt)	Front left turret floor (M18); oil cooler outlet (M39)
1	ORDER, Lubrication, L.O. 9-755	Holder B193728 in ass't driver's compartment, or in Technical Manual
1	PAULIN, 12 ft x 12 ft	Left turret rack (M18); left sponson rack (M39)

Tools, Parts, and Accessories

Quantity Per Vehicle	Name	Stowage Location
5 (M18)	ROLL, blanket	Left turret rack
9 (M39)	ROLL, blanket	Sponson extension racks
1 (M18)	SHEET, instruction, for compass	On indicator
1	SHIELD, spotlight	On spotlight
1	SPOTLIGHT	Under right turret box (M18); left front sponson box (M39)
1 (M18)	STOVE, cooking, gasoline, M1941, 1 burner, consists of: Coleman Military Burner No. 520 w/accessory cups	Rear turret box
2 (M39)	STOVE, cooking, gasoline, M1941, 1 burner, consists of: Coleman Military Burner No. 520 w/accessory cups	Rear sponson extension
4	STRAPS, canvas, 1½ x 50 in.	Tool bag
1	TAPE, adhesive, 4 in. wide (O.D.) 15 yds long	Tool bag
1	TAPE, friction ¼ in. wide, 30 ft roll	Tool bag
1 (M18)	TOP, turret, canvas, assembly	Right turret rack
1 (M39)	TOP, crew compartment, canvas	Right sponson rack
2	TUBE, flexible nozzle	Turret extension box (M18); rear sponson extension (M39)
1	WIRE, soft iron, 14 ga, 10-ft roll	Tool bag

8. ON-VEHICLE SPARE PARTS.

a. **76-mm Gun, M18 Vehicle.** The 76-mm gun spare parts box is stowed in the turret rear extension stowage box.

Quantity Per Vehicle	Name	Stowage Location
1	FORK, firing pin cocking	Spare parts box
2	GASKET, recoil cylinder filling plug	Spare parts box
1	MECHANISM, percussion, assembly (composed of: 1 Guide, firing pin; 1 pin, firing; 1 pin, straight, $\frac{3}{32}$ x ⅝ in. (firing pin guide); 1 spring (firing pin retracting); 1 stop, firing spring	Spare parts box
3	PIN, cotter, ⅛ x 1¾ in.	Spare parts box
1	PIN, firing	Spare parts box
2	PLUG, filling recoil cylinder (rear)	Spare parts box
1	PLUNGER, cocking fork	Spare parts box

TM 9-755

Part One—Introduction

Quantity Per Vehicle	Name	Stowage Location
1	RETAINER, sear	Spare parts box
1	SPRING, cocking fork plunger	Spare parts box
1	SPRING, firing pin retracting	Spare parts box
1	SPRING, firing	Spare parts box
1	SPRING, sear	Spare parts box

b. Cal. .50 Machine Gun, HB, M2. Parts are identical in both vehicles. Unless otherwise specified, parts are carried in cal. .50 gun spare parts box which is stowed in turret rear extension box (M18) or sponson rear extension (M39).

1	BARREL, assembly	Hull left side (M18); rear upper wall (M39)
1	BOX, spare parts	Turret extension box (M18); sponson rear extension (M39)
1	DISK, buffer	Spare parts box
1	EXTENSION, firing pin assembly	Spare parts box
1	EXTRACTOR, assembly	Spare parts box
1	LEVER, cocking	Spare parts box
1	PIN, cotter, belt feed lever pivot stud	Spare parts box
2	PIN, cotter, $1/16$ x $3/4$ in. (switch pivot)	Spare parts box
1	PIN, cotter, $1/8$ x $5/8$ in. (cover pin)	Spare parts box
1	PIN, cotter, $3/32$ x $3/4$ in. (belt feed lever pivot stud)	Spare parts box
1	Pin, firing	Spare parts box
1	PLUNGER, belt feed lever	Spare parts box
1	ROD, driving spring w/spring assembly	Spare parts box
1	SLIDE, belt feed group, consisting of: 1 ARM, belt feed pawl, B8914 1 PAWL, feed belt, assembly, B8961 1 PIN, belt feed pawl, assembly B8962 1 SLIDE, belt feed assembly, B261110 1 SPRING, belt feed pawl, A9351	Spare parts box
1	SLIDE, sear	Spare parts box
1	SPRING, belt holding pawl	Spare parts box
1	SPRING, belt feed lever plunger	Spare parts box
1	SPRING, cover extractor	Spare parts box
1	SPRING, locking barrel	Spare parts box
1	SPRING, sear	Spare parts box
1	STUD, bolt	Spare parts box

c. Vehicular. Parts are identical for both vehicles except where model is shown in quantity column. Tool bag is stowed in turret rear extension box (M18) or sponson rear extension (M39).

Tools, Parts, and Accessories

Quantity Per Vehicle	Name	Stowage Location
3	FITTINGS, lubr, straight, 1/8 in.	Turret extension box (M18); sponson rear extension (M39)
3	FITTINGS, relief, 1/8-27 N.P.T., male	Turret extension box (M18); sponson rear extension (M39)
4	LAMP, 3 cp, 24-28 V	Turret extension box (M18); sponson rear extension (M39)
1	LINK, track, assembly, complete, consisting of: 6 LINK w/bushing (C121476) 7 PIN (G163-03-38587) 6 KEY (G163-02-58753) 6 NUT (G163-03-14842) 6 WASHER, lock (H001-15-17013)	Rack on rear of turret (M18); rack on hull front door (M39)
1	PLUG, drain, differential	Tool bag
1	PLUG, drain, eng. oil tank	Tool bag
1	PLUG, drain, eng. oil tank, hull	Tool bag
1	PLUG, drain, final drive	Tool bag
1	PLUG, drain, torque converter	Tool bag
1	PLUG, drain, transmission	Tool bag

TM 9-755
9-11

Part Two—Operating Instructions

PART TWO—OPERATING INSTRUCTIONS

Section IV
GENERAL

9. SCOPE.

a. Part two contains information for the guidance of the personnel responsible for the operation of the equipment. It contains information on the operation of the equipment with the description and location of controls and instruments.

Section V
SERVICE UPON RECEIPT OF EQUIPMENT

10. PURPOSE.

a. When a new or reconditioned vehicle is first received by the using organization, it is necessary for second echelon personnel to determine whether the vehicle has been properly prepared for service by the supplying organization, and to be sure it is in condition to perform any mission to which it may be assigned when placed in service. For this purpose, inspect all assemblies, subassemblies, and accessories to be sure they are properly assembled, secure, clean, and correctly adjusted and/or lubricated. Check all tools and equipment against Section III (Tools, Parts and Accessories), to be sure every item is present, in good condition, clean, and properly mounted or stowed.

b. In addition, the using organization will perform a run-in test of at least 50 miles as directed in AR 850-15, according to procedures in paragraph 11 which follows.

c. Whenever practicable, the first echelon personnel (crew) will assist in the performance of these services.

11. CORRECTION OF DEFICIENCIES.

a. Deficiencies disclosed during the course of these services will be treated as follows:

(1) Correct any deficiencies within the scope of the maintenance echelons of using organization before vehicle is placed in service.

(2) Refer deficiencies beyond the scope of the maintenance echelons of the using organization to a higher echelon for correction.

(3) Bring deficiencies of a serious nature to the attention of supplying organization through proper channels.

Service Upon Receipt of Equipment

12. RUN-IN TEST PROCEDURES.

a. Preliminary Service. Before vehicle is moved to make actual run-in test, certain inspections and services will be performed as follows:

(1) FIRE EXTINGUISHER. See that portable and fixed cylinders are fully charged, that fixed cylinders, lines, and nozzles are securely mounted, and that all nozzles are properly aimed, and not clogged. The contents of the cylinders can only be determined by weighing, see paragraphs 24 a and 202 c.

(2) FUEL AND OIL. Check fuel in main and auxiliary (M18) tanks, and see that oil in main supply tank is at proper level. CAUTION: *Mix ⅛ pint SAE 30 engine oil with each gallon of gasoline for Homelite auxiliary generator engine, according to instructions in lubrication order, par. 38 d.* If a tag attached to main oil tank filler cap concerning contents, follow instructions on tag before starting engine, when step (18) is reached.

(3) FUEL STRAINER. On Homelite auxiliary generator engine only, remove and clean fuel strainer bowl and screen. Clean carburetor air filter.

(4) BATTERIES. Make hydrometer test of batteries and, if needed, add clean water to ½ inch above plates, or as specified on battery filler caps. Inspect terminal connections and bolts to be sure they are clean, secure and lightly greased.

(5) AIR CLEANERS AND BREATHER CAPS. Examine engine breather cap and carburetor air cleaners to see if in good condition and secure. Remove oil cups, wash cups and elements in dry-cleaning solvent, and refill reservoirs to proper depth with fresh engine oil, (par. 38) and reassemble, using new rubber seals. Be sure air ducts and air horn connections are tight.

(6) ACCESSORIES AND BELTS. See that accessories such as carburetor, magnetos, starter, generator, auxiliary engine and generator (M18), and filters are securely mounted. See that engine generator belts and oil cooler blower belts are adjusted to ½-inch deflection under finger pressure. Turn oil filter cleaner handle clockwise several complete turns.

(7) ELECTRICAL WIRING. Examine all accessible wiring and conduits to see if in good condition, securely connected and properly supported.

(8) TRACKS (LINKS, LINK PINS, AND LOCK KEYS). See that these items are in good condition, correctly assembled and secure. Check track adjustment, paragraph 132 h.

(9) SPROCKET AND TRACK WHEEL NUTS AND SUPPORT ROLLER HUB BOLTS. Examine these items to see that they are in good condition, and that all assembly and mounting nuts or screws are secure.

(10) TRACK GUARDS. Examine track guards to see if in good condition and secure.

(11) TOWING CONNECTIONS. Inspect pintle hook, tow hooks or shackles, cables, and all connections for good condition and proper

operation. On M39, examine trailer receptacle for towed vehicle electric brakes and lights to see that it is in good condition and securely mounted.

(12) SIDE AND BOTTOM ARMOR PLATE (ENTRANCE AND ESCAPE HATCHES, DRAINS, PAINT AND MARKING). Inspect these items to see that armor plate is securely welded, that drivers entrance and escape doors and stop and lock latches operate properly, and are well lubricated. CAUTION: *Do not fully open escape door latch during inspection.* See that hull drain valves open and close freely; that there are no bright spots on exterior finish to cause glare or rust, and that markings are legible, unless covered for tactical reasons. See that all tape or other water-proofing or corrosion-preventive materials are removed from hull or turret, and hatches, and that silica-gel bags are removed.

(13) VISION DEVICES. See that periscopes are secure in holders, holders are properly mounted, and spares are in good condition and properly stowed. See that the traversing, elevating, and locking devices and cleaning blades operate properly. CAUTION: *Prisms should be cleaned only with a soft cloth or brush.* Be sure all tape and corrosion preventive materials are removed.

(14) LUBRICATE. Perform a complete lubrication of the vehicle, covering all intervals according to instructions on lubrication order, paragraph 38, except gear case oil levels. Fill traversing hydraulic motor adapter with ½ pint of hydraulic oil of the specified grade (par. 38). Add oil to gear cases as necessary to bring to correct levels. Change only if condition of oil indicates the necessity, or if oil is not of proper grade for existing atmospheric temperature. NOTE: *Perform step (15) to (17) during lubrication.*

(15) TRACK SUSPENSION. Inspect track wheels, support rollers, and wheel tires for looseness and damage. Inspect shock absorbers, and see that above items are in good condition, correctly assembled and securely mounted. See that oil seals and gaskets are not leaking excessively.

(16) STEERING BRAKES. Examine steering brake controls for good condition and security. Check brake controls for free action and positive application of brakes by applying driver's hand levers independently and together; then repeating test with auxiliary hand levers. Controls must operate freely and brake shoes must be fully applied when hand levers are slightly back of vertical position, with both levers of each set having equal travel. When all hand levers are forward in non-operating position, control levers on differential must be held up against stop pins on differential carrier by the return springs.

(17) PROPELLER SHAFT AND UNIVERSAL JOINTS. Inspect propeller shaft and universal joints to see that they are in proper alinement and securely mounted.

(18) ENGINE WARM-UP. Start and warm up engine as described in paragraph 17 b and c, observing if starter has satisfactory speed and engages and disengages properly. On M18, start and test Homelite auxiliary generator engine to see if it operates properly.

TM 9-755

12

Service Upon Receipt of Equipment

(19) PRIMER. While starting main engine, observe if primer action is satisfactory, and look for leaks at pump or connections.

(20) INSTRUMENTS.

(a) Engine Oil Pressure Gage. Engine oil pressure must be indicated when engine is idling. Oil pressure must be 50 to 90 pounds at operating speeds. Stop engine immediately when pressure drops below minimum.

(b) Ammeter. After starting, ammeter may show high charge until current used in starting is restored to batteries, then if battery is full and lights and accessories turned off, a zero, or slight charge, reading is normal.

(c) Engine Oil Temperature Gage. Reading should rise gradually, during warm-up period, to normal range, which is 150° F to 190° F. CAUTION: *Do not move vehicle until temperature is over 100° F as indicated by gage.*

(d) Tachometer. Tachometer should register engine speed in revolutions per minute, and record accumulating revolutions.

(e) Fuel Gage. Operate fuel gage switch in right and left positions; gage should indicate approximate amount of fuel in each tank. Ordinarily, tanks will have been filled and gage should register "F."

(21) ENGINE CONTROLS. Observe whether or not engine responds to hand throttle and accelerator promptly and properly, and if there is excessive looseness or binding in controls.

(22) SIREN AND WINDSHIELD WIPERS. Test siren for proper tone and operation and if driver's hoods are in use, operate windshield wipers to see that blades press windshield firmly through full stroke.

(23) LAMPS (LIGHTS) AND REFLECTORS. Observe whether or not all lights, including blackout lights, respond to the switches in both the "OFF" and "ON" positions. Stop lights must just start to burn when locking pawls on driver's brake hand levers are engaged in fourth or fifth notches from front ends of quadrants.

(24) LEAKS (GENERAL). Examine the inside of engine compartment and under vehicle, and check all accessible lines and seals, for fuel and oil leaks.

(25) TOOLS AND EQUIPMENT. Check tools and equipment, Section III, (Tools, Parts and Accessories) to be sure all items are present, and see that they are serviceable and properly mounted or stowed.

b. Run-in Test. Perform the following procedures, (1) to (9) inclusive, during the road test of the vehicle. On vehicles which have been driven 50 miles or more in the course of delivery from the supplying to the using organization, reduce the length of the road test to the least mileage necessary to make the observations following: CAUTION: *During the road test of the vehicle, continuous operation of vehicle or engine at speeds beyond those recommended on the caution plates must be avoided.*

(1) INSTRUMENTS AND GAGES. Do not move vehicle until engine oil temperature reaches 100° F. Observe readings of oil pressure gage, ammeter, fuel gage, and tachometer to see if they register proper

function of the units to which they apply. With vehicle in motion, the speedometer should register vehicle speed, and the odometer should register both trip and accumulating mileage.

(2) BRAKES: STEERING AND PARKING. Steering brakes must stop vehicle effectively with one-third of quadrant travel in reserve. With vehicle on incline, pull back on brake hand levers and trip parking brake locking levers. Parking brake must hold vehicle with one-third of quadrant travel in reserve and levers remain in applied position. Apply steering brakes independently and notice whether or not they steer the vehicle properly.

(3) TRANSMISSION. Mechanism should operate and shift easily without unusual noise or vibration.

(4) ENGINE. The engine must respond to controls and have maximum pulling power without unusual noise, stalling, overheating or exhaust smoke.

(5) UNUSUAL NOISES. Be on the alert continually for unusual noises that would indicate damage or looseness of tracks, sprockets, support rollers, track or compensator wheels, and suspension. NOTE: *Halt vehicle at 10-mile intervals or less for steps (6) to (9) below.*

(6) TEMPERATURES. Place hand cautiously on each track wheel and support roller hub, to feel if abnormally hot. Shock absorbers should feel warm when the vehicle is being operated. If shock absorbers do not become warm during operation, it is an indication that the fluid is low or that shock absorbers are not functioning. Check transmission and final drives for overheating or excessive oil leaks.

(7) LEAKS, FUEL AND OIL. Inspect within engine and fighting compartments, underneath vehicle and all visible lines, for fuel or oil leaks. Trace any leaks to source and remedy or report them.

(8) GUNS: ELEVATING AND TRAVERSING MECHANISM (M18 ONLY). Place vehicle on a 10 degree lateral incline (tilted sideways). Traverse the turret through its full 360-degree range by both hand and power controls, and observe if they operate properly, and if there is any indication of looseness or of binding. With the gun pointed forward or rearward, elevate it through its entire range with the hand controls to see if there is binding, excessive lash, or erratic action. See that all preservative oils have been cleaned from gun mechanism, and that proper lubricant has been applied.

(9) TRACK TENSION. Inspect tracks for satisfactory tension. CAUTION: *Tracks should not be adjusted too tightly.* Be sure adjustment locking devices are secure, see paragraph 132 h.

c. Vehicle Publications and Reports.

(1) PUBLICATIONS. See that Vehicle Operator's Manual, Lubrication Order, Standard Form No. 26 (Drivers Report-Accident, Motor Transportation) and W.D., A.G.O. Form No. 478 (Major Unit Assembly Replacement Record), are in the vehicle, legible, and properly stowed. NOTE: *U.S.A. registration number and vehicle nomenclature must be filled in on Form No. 478, for new vehicles.*

(2) REPORTS. Upon completion of run-in test, correct or report any deficiencies found. Report general condition to the designated individual in authority.

Section VI
CONTROLS AND INSTRUMENTS

13. CONTROLS.

   a. **Drivers' Doors.** Two double-section doors are hinged to roof of hull to provide entrances to the driving compartment (fig. 6). The doors are water tight with rubber seals and are anchored in the closed position by two latch handles on each outer section. Each section is anchored in the open position by a stop latch knob and pin located in bosses welded to hull roof. CAUTION: *Lock periscope in lower position, and lock housing so that periscope is at right angle to straight-ahead position while opening and closing drivers' doors.*

   b. **Drivers' Hoods.** A detachable drivers' hood, incorporating a windshield, a windshield wiper, and an electric defroster is provided for installation over each door hatch during inclement weather (fig. 7). Each hood is anchored in position over hatch by two hook rods which engage bosses welded to hull roof and are tightened by wing nuts. The assembly contains a windshield wiper switch and a defroster switch. Current is supplied through a shielded cable, which is plugged into instrument panel outlet socket for left hood and into socket in dome light junction box for right hood.

   c. **Drivers' Periscope.** A periscope housing mounted in outer section of each driver's door (fig. 6) supports periscope (fig. 263). The periscope is secured in either upper or lower position by a knurled locking nut and stud on periscope, and a sliding latch on housing prevents periscope from dropping out of housing. A wiper blade attached to housing is provided to wipe the periscope window when the periscope is moved up and down. The top of housing is closed by a spring-loaded door when periscope is removed. The periscope housing may be rotated and tilted to point periscope in any direction, and may be locked in any position by two knurled nuts.

   d. **Adjustment of Drivers' Seats.** (fig. 8). Both drivers' seats are adjustable to four horizontal positions and four vertical positions. To move seat forward or backward, lift horizontal release knob on left side of seat, slide seat to desired position, release knob and slide seat until plunger enters nearest hole in latch plate. To lower seat, take weight off seat sufficiently to permit lifting vertical release handle on right side, push seat down almost to desired position and release handle, then push seat down until lock engages. A heavy spring will raise seat when weight of occupant is removed and release handle is pulled upward. The seat back may be removed by lifting it up out of sockets in seat frame. The early production seat back is hinged to

TM 9-755
13-14

Part Two—Operating Instructions

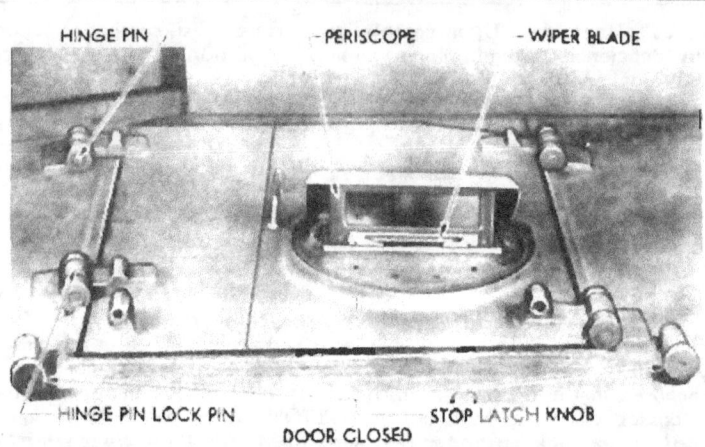

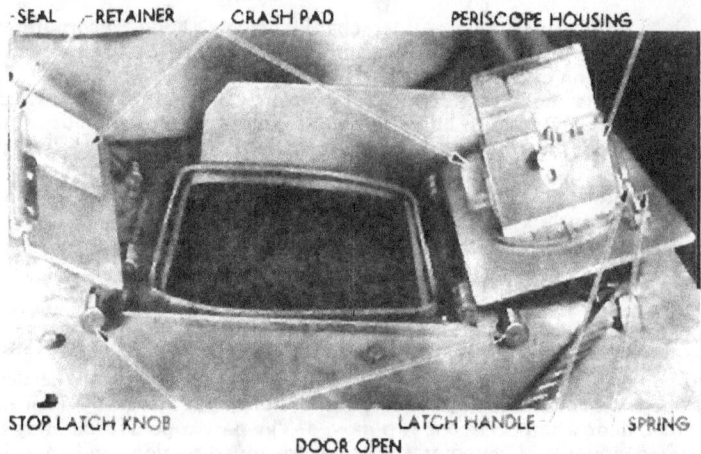

Figure 6—Driver's Doors

seat frame so that it may be folded down over seat by pulling upward until hinge is out of socket in seat frame, then pulling forward to horizontal position. When seat back is raised to vertical position, springs will draw the hinges down into the tubular frame to hold the back in vertical position.

14. OPERATING CONTROLS.

　a. Fuel Valve Control Handles (fig. 9). Three fuel valve control handles are mounted on a plate located on front side of bulkhead.

TM 9-755
14

Controls and Instruments

Figure 7—Driver's Hood—Installed

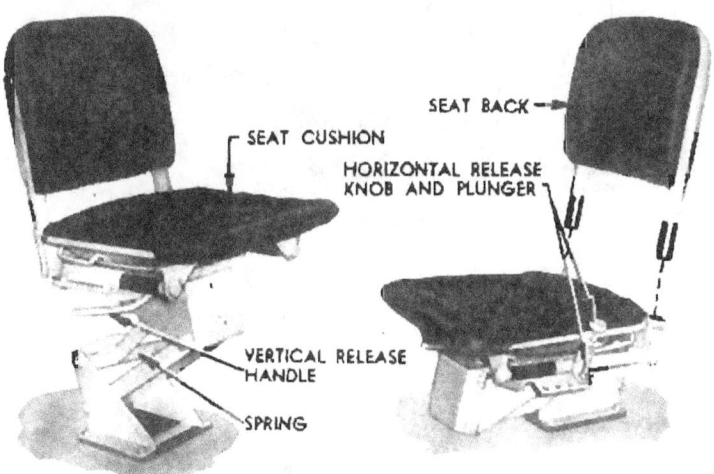

Figure 8—Driver's Seat Adjustments

Two of these handles operate the fuel shut-off valves at fuel tanks, and the third handle operates shut-off valve in the balance pipe which connects to both tanks. The plate identifies each handle and indicates the "ON" (open) and "OFF" (closed) positions. To open fuel shut-off valves, pull control handles outward and turn them one-quarter turn

TM 9-755
14

Part Two—Operating Instructions

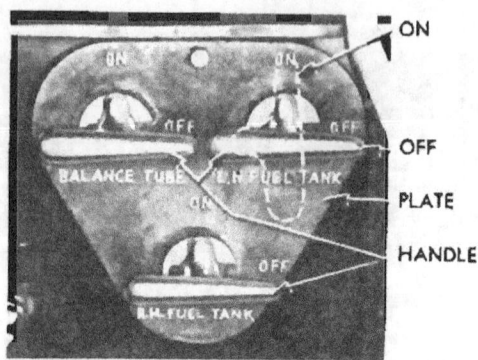

Figure 9—Fuel Valve Control Handles

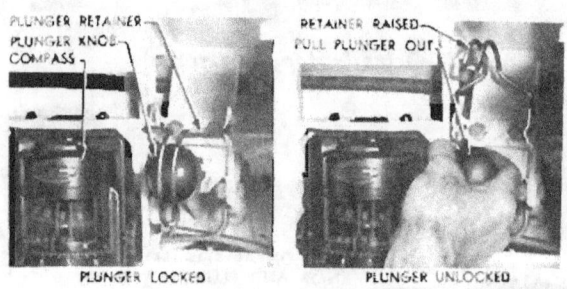

Figure 10—Engine Primer Pump

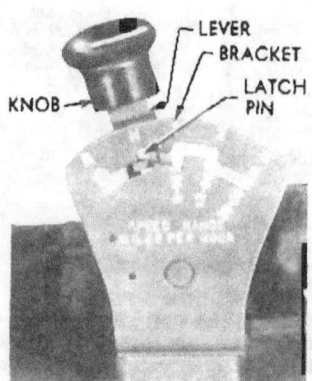

Figure 11—Manual Shift Lever and Bracket

TM 9-755
14

Controls and Instruments

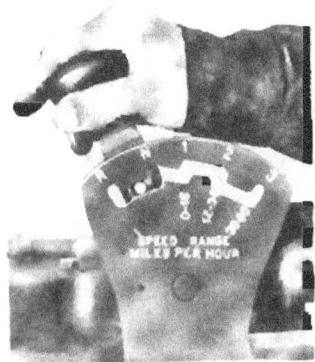

Figure 12—Operating Manual Shift Lever

counterclockwise, making sure that lock pins in handles engage holes in plate; turn handles one-quarter turn clockwise to close valves. In normal operation, these three control handles must be turned to "ON" position before starting engine, and must always be turned to "OFF" position when engine is stopped, and when filling fuel tanks. If vehicle, with tanks nearly full, is operated for some time over terrain which causes one tank to be much lower than the other, close balance pipe shut-off valve to prevent fuel in higher tank from draining into lower tank and causing it to overflow. If one fuel tank pump becomes inoperative, leave balance pipe shut-off valve open. If one tank is punctured in combat, close shut-off valves in balance pipe and at punctured tank.

b. **Engine Primer Pump** (fig. 10). A primer pump, located in front of the driver to the right of the compass, provides a means of injecting a spray of fuel into the engine intake pipes to facilitate starting engine. A retainer which swings down over the primer knob holds plunger in forward position. To use primer pump, push plunger retainer upward to clear knob, pull primer plunger out slowly and push it in quickly to thoroughly atomize the priming charge. Push plunger retainer down over plunger knob after priming is completed.

c. **Clutch Pedal.** A clutch pedal is not provided since the conventional type of clutch is not required with the Torqmatic transmission. When it is necessary to disconnect the engine from the power train for any reason, use the rear transfer case shifter lever (subpar. g below).

d. **Transmission Manual Shift Lever** (fig. 11). A shift lever for manually changing speed range in the transmission is mounted on top of the transmission where it may be conveniently reached by driver and assistant driver. The lever has five positions: reverse,

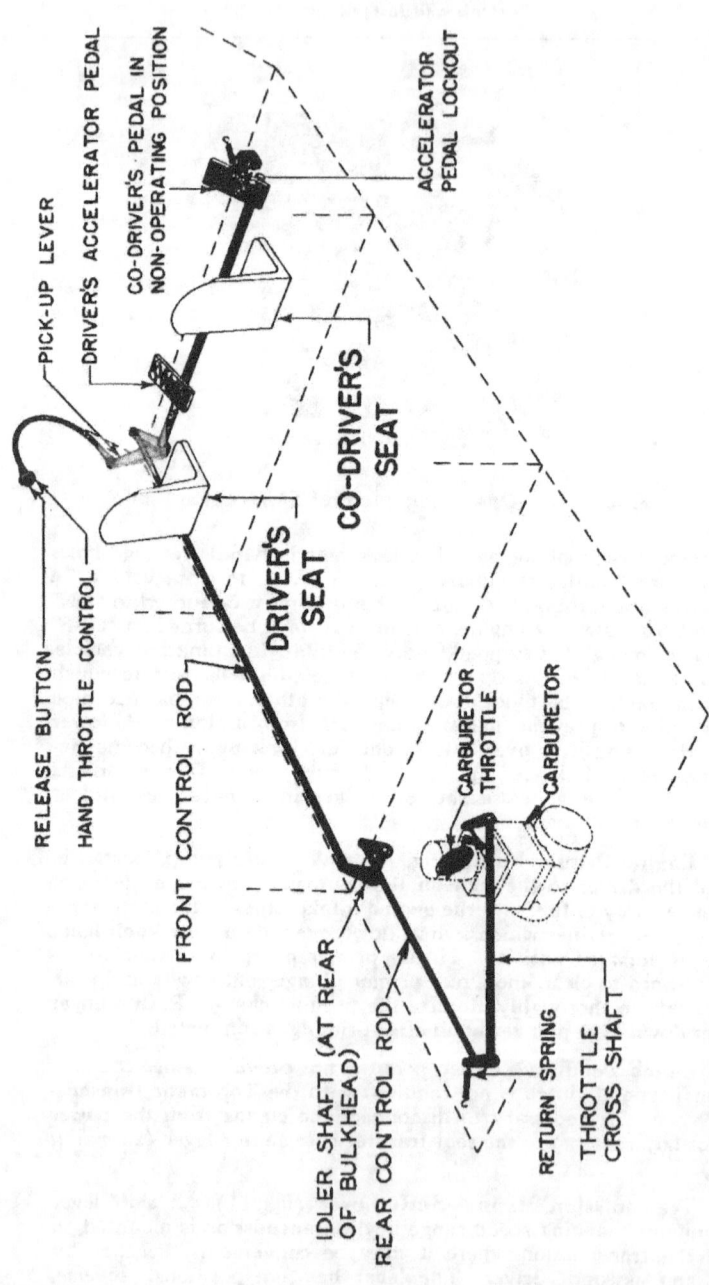

Figure 13 — Throttle Control System

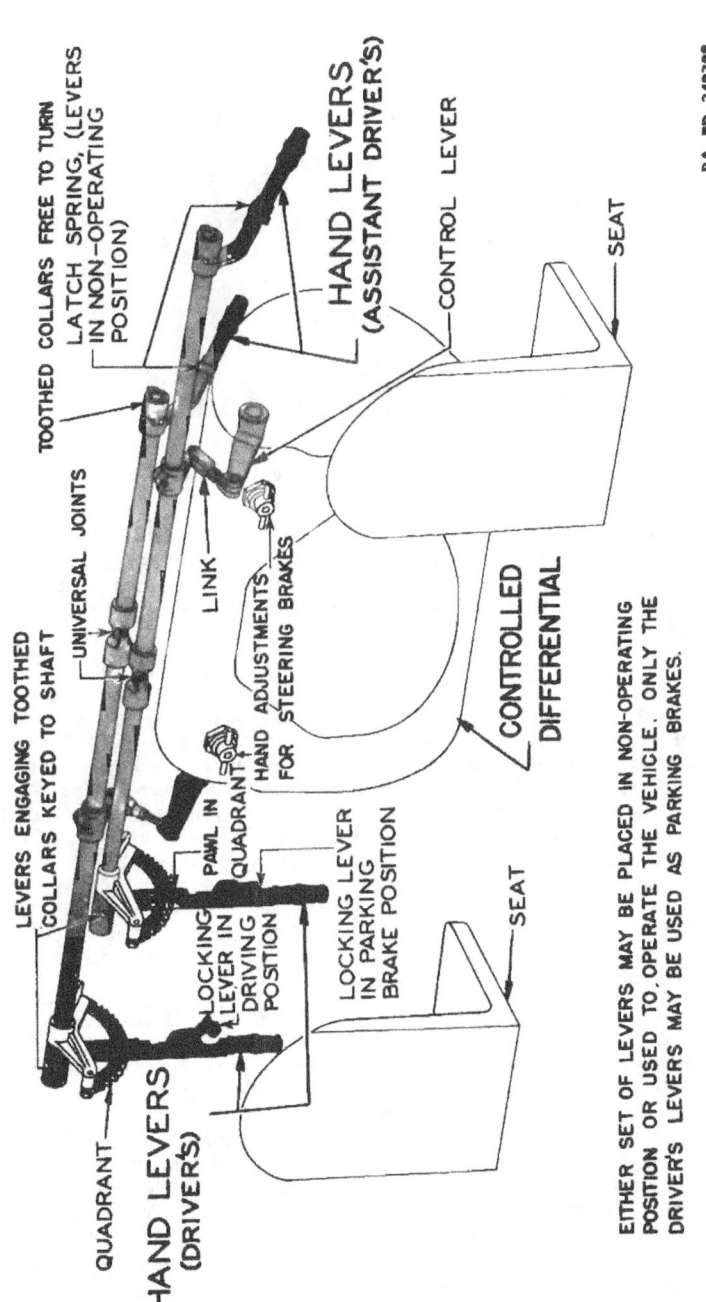

Figure 14 — Steering and Brake Controls

TM 9-755
14

Part Two—Operating Instructions

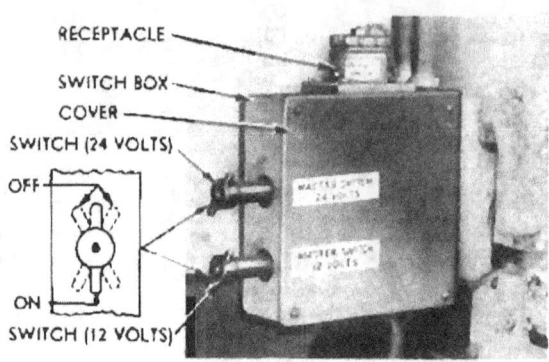

Figure 15—Master Switch Box

neutral, first, second, and third. A speed range is selected by moving the lever to desired position as indicated by the markings on the shift lever bracket (fig. 12). Refer to paragraph 17 f for operating instructions. The shift lever also actuates starter neutral safety switch (par. 15 c).

e. **Accelerator Pedals and Hand Throttle Control** (fig. 13). Individual accelerator pedals for controlling engine speed are provided for driver and assistant driver. A lock out device on right side of assistant driver's accelerator pedal is provided to hold this pedal out of contact with accelerator pedal cross shaft when assistant is not driving vehicle. Push lock out away from pedal when this pedal is to be used. A hand throttle control, located forward of the driver's left steering brake lever, may be used to set engine at any speed above idle. This control connects to the accelerator cross shaft through a pick-up lever which permits independent operation of the pedals. Pull the control knob rearward to increase engine speed; a self-locking device will hold the control in desired position. Push inward on the button in the center of control knob to release the lock and return the control to slow idle position.

f. **Steering Brake Hand Levers** (fig. 14). Two steering brake hand levers are located forward of each driver's seat. These levers swing from cross shafts mounted on ceiling of driving compartment and are positioned so that driver (or the assistant driver) can conveniently operate one lever with each hand. Use of these levers for steering and braking is explained in paragraph 17 e. The left hand (driver's) set of levers is equipped with quadrants and locking pawls so that these levers can be used for parking brakes; the right (assistant driver's) set of levers cannot be used for parking. To set the parking brakes, pull rearward on both levers until brakes are firmly applied and push locking levers inward against braking levers, making sure that locking pawls engage notches in quadrants. To release parking brakes, pull slightly rearward on brake hand levers, push locking levers away from brake levers with thumbs and ease brake levers forward. Either set of levers may be used to operate the vehicle while the other set is placed in non-operating position. The driver's brake hand levers are held forward in non-operating position by engaging the locking pawls in the first notches in quadrants. The auxiliary assistant driver's brake hand levers are held forward in non-operating position by pushing levers into engagement with latch springs on hull.

g. *Transfer Case Shifter Lever.* A shifter lever, which actuates a sliding clutch in the rear transfer case, provides a means of disconnecting the engine from the power train when required for maintenance operations. The shifter lever is stowed in the tool stowage box. Installation and use are described in paragraph 102.

15. SWITCHES, OUTSIDE OF INSTRUMENT PANEL.

a. **Master Switch Box Switches** (fig. 15). The master switch box is located in the driving compartment behind the driver's seat. Two spring-loaded switches mounted in the right side of this box control all current between the battery and the vehicle electrical systems. The lower switch controls 12-volt current to radio and interphone circuits. The upper switch controls 24-volt current to all other circuits in vehicle. The switches are in the "ON" position when switch handles are vertical. Pulling the handles out and turning them in either direction locks the switches in the "OFF" position. Turning the handles to the vertical position allows the springs to force them inward, thereby closing the circuits. A receptacle covered with a screw type cap is mounted on top of the switch box. This receptacle provides a means of connecting an outside or slave battery to the vehicle.

b. **Heater Control Box Switch** (fig. 28). A heater control box is mounted on the ceiling of driving compartment to left of driver's hatch. It is used only when vehicle has winterization equipment installed. The switches in this box are explained in paragraph 26 a (1).

c. **Starter Neutral Safety Switch** (fig. 125). A safety switch, mounted on transmission shift lever bracket and actuated by manual shift lever, prevents accidental starting of engine when lever is not in neutral.

d. **Dome Light Switches.** Individual switches are mounted in each of the dome lights. Two dome lights are located in the driving compartment. In M18 vehicle only, one dome light is located in the turret at the radio and one dome light is located above the turret wiring switch box.

e. **Siren Switch.** A switch for operating siren is located in front of driver's seat where it can be conveniently depressed with left foot.

f. **Stop Light Switch.** Two stop light switches are mounted above the transmission in position to be actuated by cams on brake

TM 9-755
15

Part Two—Operating Instructions

A	BOOSTER COIL AND STARTER CONTROL CIRCUIT BREAKER
B	TEST LIGHT SWITCH (FOR WARNING LIGHT C)
C	TEMPERATURE WARNING LAMP CAP (CONVERTER)
D	FUEL PUMPS SWITCH
E	BOOSTER SWITCH
F	MAGNETOS SWITCH
G	STARTER SWITCH
H	FUEL CUT-OFF SWITCH (CARBURETOR IDLE)
I	HEADLIGHT, BLACKOUT LIGHT AND TAILLIGHT SWITCH
J	LOCK OUT BUTTON (PART OF SWITCH I)
K	OUTLET PLUG
L	SIREN CIRCUIT BREAKER.
M	OUTSIDE LIGHTS CIRCUIT BREAKER
N	INSIDE LIGHTS CIRCUIT BREAKER
O	INSTRUMENT PANEL GAGES CIRCUIT BREAKER
P	PANEL LIGHT SWITCH
Q	BATTERY AMMETER
R	ENGINE OIL PRESSURE GAGE
S	INSTRUMENT PANEL LAMP CAP
T	SPEEDOMETER
U	ENGINE TACHOMETER
V	ENGINE OIL TEMPERATURE THERMO GAGE
W	FUEL GAGE SWITCH
X	FUEL GAGE
Y	AUXILIARY FUEL PUMP CIRCUIT BREAKER

RA PD 340378

Figure 16—Instrument Panel

TM 9-755
15-16

Controls and Instruments

cross shafts. They are connected in series so stop lights will burn only when both steering brake hand levers of either set are pulled rearward together.

16. **INSTRUMENT PANEL.**

a. **General.** The instrument panel is located in the sponson to left of driver's seat. It contains the instruments, switches, circuit breakers, etc. described in this paragraph. All key letters refer to figure 16.

b. **Circuit Breakers.** Circuit breakers having manual reset buttons are installed in instrument panel, instead of fuses, to protect wiring and electrical units in case an overload condition, which would cause damage or fire, develops in a circuit. The two circuits breakers on left end of the panel protect circuits of booster coil, starter, carburetor idle cut-off (A), and fuel tank pumps (Y). Four circuit breakers on the right end of the panel protect circuits of the siren (L), outside lights (M), inside lights (N), and gages (O). Each circuit breaker reset button is clearly identified on instrument panel. An overload in a circuit will cause the breaker reset button to snap outward with an audible click. When a button snaps out, wait about 2 minutes to allow breaker to cool, then push button in. The button will stay in if overload was temporary; however, if the overload remains circuit breaker will continue to cut out until circuit is corrected.

c. **Converter Oil Temperature Warning Light and Test Switch.** This light (C) on the instrument panel burns only when oil in the transmission torque converter becomes overheated. When this light burns, the vehicle should be stopped and measures should be taken to determine and correct cause of overheating. The push button located to left of the converter oil temperature warning light (B) is provided for testing the lamp in warning light. When button is depressed, warning lamp should burn.

d. **Auxiliary Fuel Pumps Switch.** The auxiliary fuel pumps switch (D) on the instrument panel operates the pumps in both fuel tanks. The pumps operate continuously when switch is in "ON" position. This switch must always be turned to "OFF" position when engine is stopped, otherwise the carburetor and engine may be flooded with fuel.

e. **Magneto, Booster and Starter Switch.** The magneto, booster and starter switch assembly in the middle of instrument panel contains individual switches for controlling the booster coil, the magnetos, and the starter. The booster (E) and starter (G) switch levers are spring-loaded so that they must be held in "ON" position; when released they automatically return to the "OFF" position. Turn the magneto switch lever (F) to "BOTH" position to connect both magnetos for normal operation of the engine. Turn the lever to the "R" or "L" position to test operation of magnetos separately.

f. **Carburetor Idle Fuel Cut Off Switch.** This push button switch (H) is provided for stopping the engine. Pushing in on switch

TM 9-755
16

Part Two—Operating Instructions

button causes a solenoid on carburetor to cut off the fuel supply to the idle system of carburetor.

g. **Headlight, Blackout Light and Taillight Switch.** The light switch (I) on the instrument panel controls all outside lights for both blackout and bright beams. The switch has four positions besides "OFF" and contains a button operated lock-out device to prevent accidental turning on of bright lights. The free positions of the switch lever are "OFF" and "BO-MK"; all other positions require release of lock-out by depressing the button (J) above the lever. The switch positions and the lights controlled by each position are as follows:

Switch Position	Lights On
BO-MK (Free position)	Blackout marker lights (RH and LH)
	Blackout taillights (RH and LH)
	Blackout stop light (RH)
	Trailer blackout taillight and stop light (M39)
HD-LT (Press lock-out button)	Both headlights, bright
	Left stop light, bright
	Left taillight, bright
STOP-LT (Press lock-out button)	Left stop light, bright
BO-DR (Press lock-out button)	Left blackout headlight
	Blackout marker lights (RH and LH)
	Blackout taillights (RH and LH)
	Blackout stop light (RH)
	Trailer blackout taillight and stop light (M39)

h. **Instrument Panel Outlet.** An outlet socket located in the upper right corner of instrument panel is closed by a removable plug (K) attached to panel by a chain. Remove the plug by turning it counterclockwise and pulling it out of socket. The plug should be installed when outlet is not in use. This outlet provides a connection to the 24-volt lighting circuit for the driver's hood windshield wiper (par. 13 h), trouble light, etc.

i. **Instrument Panel Light Switch.** The panel light switch (P) in lower right corner of instrument panel controls lights in panel and compass. The switch is off when switch knob is turned clockwise as far as it will go. Turning knob counterclockwise from this position turns on panel and compass lights which become brighter as knob is turned farther in counterclockwise direction.

j. **Battery Ammeter.** The ammeter (Q) on instrument panel indicates total charge entering the batteries with either generator, or both generators, in operation. It does not indicate current leaving the batteries, nor current used by electrical units; therefore, it should never indicate discharge, left side of dial. The ammeter should register a high charging rate for a few minutes after engine is started until the generator restores to batteries the current used by the starter.

Thereafter, the charging rate will depend upon state of charge of batteries, being high if batteries are low or very low if batteries are fully charged.

k. **Engine Oil Pressure Gage.** The engine oil pressure gage (R) on instrument panel indicates oil pressure in the lubrication system of the engine. Normal oil pressure is 50 to 90 pounds at operating speeds.

l. **Speedometer.** The speedometer (T) on instrument panel is driven by a flexible shaft which is connected to a drive gear in the transmission. The speedometer indicates vehicle speed in miles per hour. It also registers accumulated mileage and trip mileage. The trip mileage may be set at zero by turning the small knob located below instrument panel. The vehicle speed should never be permitted to exceed 55 miles per hour.

m. **Engine Tachometer.** The tachometer (U) on instrument panel is driven by a flexible shaft which is connected to a drive gear in engine. This instrument indicates the engine speed in revolutions per minute. It also registers the accumulated revolutions of the engine. The engine speed should never be permitted to drop below 1,600 revolutions per minute under operating load.

n. **Engine Oil Temperature Gage.** The engine oil temperature gage (V) records temperature of engine lubricating oil at point where it enters the engine from oil tank. Oil temperature must not exceed 190° F.

o. **Fuel Gage Switch.** This switch (W) in the lower left corner of instrument panel operates fuel gage just above it. Turning switch lever to "LEFT" position causes fuel gage to indicate amount of fuel in left tank; turning switch lever to "RIGHT" position causes gage to indicate fuel in right tank. NOTE: *Each position has two detent notches for switch lever; fuel gage will read same in either notch.*

p. **Fuel Gage.** The fuel gage (X) on instrument panel registers fuel level in right or left fuel tank when fuel gage switch below it (W) is turned to the "RIGHT" or "LEFT" position. Since gage shows fractions of a full tank, instead of gallons, it is necessary to know that the capacity of the right tank is 90 gallons and capacity of left tank is 75 gallons.

Section VII
OPERATION UNDER ORDINARY CONDITIONS

17. **USE OF INSTRUMENTS AND CONTROLS IN VEHICULAR OPERATION.**

a. **Service Upon Receipt of Equipment.** Before a new or reconditioned vehicle is placed in service, be sure that services described in paragraphs 10, 11, and 12 have been performed.

b. **Before-operation Service.** Perform services in items 1 to 6 in paragraph 41 b before attempting to start engine. CAUTION:

TM 9-755
17

Part Two—Operating Instructions

If water is present in engine compartment floor it may be drawn in through carburetor and damage engine. Open drain valves and allow water to drain before starting engine. Start and warmup engine (subpar. c below) and complete the before-operation services.

c. **Starting the Engine.** The following instructions apply when starting the engine at temperatures above 0° F. If temperature is below 0° F., refer to paragraph 27 for instructions on starting the engine.

(1) Firmly apply and lock brakes (par. 14 f) and place transmission manual shift lever in neutral position (fig. 11).

(2) CAUTION: *Before attempting to start engine, crank engine at least 50 turns with the hand crank to make certain it turns freely and is free of hydrostatic lock (par. 19).*

(3) Turn three fuel valve control handles to the "ON" position (fig. 9) if vehicle is to be operated on reasonably level terrain. However, if tanks are nearly full and vehicle will be operated where one side will be lower than other, keep "BALANCE TUBE" valve closed to prevent fuel from overflowing from low tank.

(4) Close the upper switch in the master switch box (fig. 15), place auxiliary fuel pumps switch in the "ON" position (O, fig. 16), and pull hand throttle control knob (fig. 13) out about ½ inch.

(5) If the engine is already warmed up to 72° F or hotter, prime it with two strokes of primer pump (fig. 10) by pulling pump plunger out slowly and pushing it in quickly. Progressively increase number of strokes as engine temperature falls below 72° F, being guided by experience with the particular equipment. Over-priming before starting is to be avoided as it will cause hard starting and will also wash oil off cylinder walls, tending to cause scoring of cylinders and pistons. *Do not pump accelerator pedal.*

(6) Place the magneto switch lever (F, fig. 16) in the "BOTH" position, press the starter and booster switch levers (E, G, fig. 16) together to "ON" positions and hold them there until the engine starts.

(7) If engine fails to fire within 30 seconds, release starter and booster switch levers and allow starter to cool for several minutes. Then, if engine is cold, prime it with several more strokes of primer pump and attempt to start engine again. If engine is hot, however, it may have been over-primed and is flooded; therefore, it should not be primed again. To clear flooded engine, hold throttle wide open, turn magneto switch off, crank engine several revolutions and then attempt to start engine without priming. If engine still does not start, refer to paragraph 47. NOTE: *The engine cannot be cranked for starting by towing the vehicle.*

(8) As soon as the engine fires evenly, push hand throttle control all the way in. CAUTION: *Pumping the accelerator pedal may flood air inlet ducts and cause a fire.* Check oil pressure; if oil pressure gage (R, fig. 16) does not show an increasing pressure within 10 seconds at 700 revolutions per minute, turn magneto switch off and correct difficulty (par. 48) before starting again.

Operation Under Ordinary Conditions

(9) Warm the engine up at 700 revolutions per minute for 5 minutes or more until oil temperature gage hand (V, fig. 16) starts to raise. The oil pressure should be between 50 and 90 pounds at this speed.

(10) When engine has been completely warmed up, increase engine speed to 1,800 revolutions per minute and turn magneto switch to "L" position. A maximum drop of 100 revolutions per minute is permissible at this speed. Do not leave switch in this position for more than 30 seconds, then turn switch to "BOTH" position for a short time to clean inoperative spark plugs which may have fouled. Then repeat test by turning magneto switch to "R" position; finally, turn magneto switch to "BOTH" position.

(11) Set engine speed at 700 revolutions per minute, by means of hand throttle control, for normal operation of vehicle.

d. Placing Vehicle in Motion. Before attempting to place vehicle in motion, driver should adjust his seat (par. 13 d) so that he has maximum visibility and all operating controls can be reached conveniently.

(1) With engine warmed up (subpar. c above) and running at idle speed, move transmission manual shift lever rearward, without depressing knob, until latch pin is at figure "1" on bracket (fig. 11).

(2) Fully release brakes and press slowly and firmly on accelerator pedal to move vehicle forward in first range. For shifts to other ranges refer to subparagraph f below.

(3) When starting forward up on incline, vehicle may roll down grade when brakes are released. To prevent this increase engine speed to approximately 2,100 revolutions per minute and then release brakes, to provide sufficient power to move vehicle forward.

(4) The vehicle must not be held stationary by use of accelerator pedal when headed up an incline in first range because this will cause overheating of oil in torque converter.

(5) To drive vehicle in reverse, same procedure should be followed as for forward motion except that transmission shift lever knob must be depressed and lever moved forward to position marked "R" on bracket (fig. 11). The assistance of an observer outside the vehicle is desirable when driving in reverse.

e. Steering the Vehicle. Except when turning vehicle, steering brake hand levers should be kept in forward position so that both brakes are free. Do not hold levers back to keep brakes partially applied in anticipation of making a turn or stopping vehicle because this will cause unnecessary wear of brake shoes and loss of performance.

(1) To steer vehicle to right, pull rearward on right hand lever while holding left lever forward.

(2) To steer vehicle to left, pull rearward on left hand lever while holding right hand lever forward.

(3) Operate steering brake hand levers smoothly and firmly, to avoid sudden application of brake shoes, which will unnecessarily

TM 9-755
17

Part Two—Operating Instructions

strain the parts of the differential, final drive and track, and may cause skidding of vehicle on hard pavement.

f. **Operating Transmission Manual Shift Lever** (fig. 12). After vehicle is in motion in first range (subpar. d above) it is necessary to move the manual shift lever to higher range positions as the vehicle speed is increased. Running at top engine speed in low range will cause overheating of transmission and abnormal fuel consumption.

(1) Select proper speed range of the transmission to correspond with speed at which vehicle is to be driven. The correct speed ranges as shown in shift lever bracket are:

Transmission Range	Vehicle Speed
First	Up to 16 mph
Second	12 to 34 mph
Third	30 to 60 mph

(2) Move shift lever to position required, depressing knob when necessary, while holding accelerator pedal steady at position when shift is made. Do not release nor depress accelerator pedal in an attempt to help shift as this may cause backlash and rough shifting. Shifts to higher or lower range can be made even though engine is running at maximum speed.

(3) When decreasing speed of vehicle, keep shift lever in third gear position down to 30 miles per hour, and in second gear position down to 12 miles per hour.

(4) When vehicle is moving forward at a speed above 5 miles per hour do not shift transmission into reverse. When moving in reverse at a speed above 5 miles per hour do not shift transmission into first range. Such shifts will severely strain the entire power train, and may cause injury to personnel. The shift lever bracket is designed to prevent accidental shifting from first to reverse and from reverse to first ranges.

g. **Use of Reverse Range as a Brake on Steep Grade.** Occasion will arise in operation of the vehicle when it will be desirable to use reverse range as a brake when descending a steep grade or very long grade. Continued use of steering brakes for braking under these conditions may cause abnormal wear or damage to brake shoes and drums, and does not leave any reserve braking power to meet an emergency. Under such conditions, descending grades in reverse and using engine power to control forward speed of the vehicle is desirable because this permits full use of the regular brakes for steering or meeting some unexpected emergency. The following procedure is to be used to prevent damage to the transmission:

(1) Stop vehicle at crest of down grade and release both brake levers. Do not go over crest and then try to hold vehicle with brakes while shifting into reverse.

(2) Set engine idle speed at 700 to 900 revolutions per minute by means of hand throttle control to prevent engine stalling while descending grade. CAUTION: *This is essential as any attempt by the operator to set engine idle by use of accelerator pedal will result*

Operation Under Ordinary Conditions

in loss of vehicle control, should the operators foot slip from the accelerator.

(3) Shift transmission manual shift lever to first range position, start down grade and before vehicle has reached a speed of 5 miles per hour, shift lever to reverse position. CAUTION: *Do not shift into reverse above 5 miles per hour.*

(4) Accelerate engine momentarily to 1,200-1,400 revolutions per minute to insure proper engagement within the transmission.

(5) While vehicle is descending, continue to apply power from engine so that the vehicle does not exceed 10 miles per hour at any time while proceeding down hill. This is accomplished by the use of the accelerator pedal, as speeding up engine increases braking effort. CAUTION: *Stop using reverse for a brake if converter oil temperature warning light burns red.*

(6) Shifting transmission into forward range after completion of descent may be accomplished without harm, provided operator has maintained vehicle speed below 10 miles per hour. However, attempts to shift into forward range at any higher vehicle speed will be attended with considerable shock and damage to the transmission.

h. **Stopping the Vehicle.** Except in an emergency, the brakes should not be applied at high vehicle speed as this will produce considerable strain on differential, final drives and tracks, and may cause vehicle to get out of control on slippery ground. Stop the vehicle by following procedure:

(1) Fully release accelerator pedal, which will cut off driving power and allow vehicle to lose speed and come to a stop.

(2) For a more rapid stop, pull rearward equally and smoothly on both steering brake hand levers to apply the brakes just firmly enough to accomplish the stop in required distance. Avoid harsh or jerky application of brakes.

(3) During the braking action some steering may be necessary to control direction of vehicle.

(4) After vehicle is stopped, apply and lock brakes (par. 14 f) and move transmission shift lever to neutral position.

i. **Stopping the Engine.** When a cold engine has just been started, it should not be stopped until it has been thoroughly warmed up, or until the oil temperature gage (V, fig. 16) shows a rise of at least 10° F. Repeated starting and stopping of a cold engine will result in lack of lubrication in some parts of engine. A warmed-up engine, or one that has been operating under load, must be idled at 700 revolutions per minute for 5 minutes before it is stopped, in order to allow oil vapor to liquefy and drain into oil sump and be removed from sump by the scavenge pump. This is very important, as hydrostatic lock will result if an engine is stopped without idling as specified. After idling, stop engine as follows:

(1) Depress carburetor idle fuel cut-off switch button (H, fig. 16) on instrument panel and hold it depressed until engine stops.

(2) After engine stops, turn magneto switch off.
(3) Turn auxiliary fuel pumps switch off.
(4) Turn fuel shut-off valves to "OFF" position.

j. **After-operation Services.** After vehicle and engine are stopped and before leaving vehicle, make sure that the following conditions exist:

(1) Parking brakes are firmly applied.
(2) Transmission manual shift lever is in neutral position.
(3) Hand throttle control is in released or idle position.
(4) Fuel pumps switch is in "OFF" position.
(5) Magneto switch is in "OFF" position.
(6) Three fuel valve control handles are in "OFF" position.
(7) Both master switch box switches are turned off.

18. DRIVING PRECAUTIONS AND INSTRUCTIONS.

a. **High Speed Operation.** This vehicle, with its all-steel track, may be driven safely at high speeds; however, until the driver becomes thoroughly familiar with the individual vehicle being driven, he must use every precaution not to "over drive" and allow vehicle to go out of control. Careless handling at high speed on hard pavement or slippery ground may result in loss of steering control and possible injury to personnel and damage to vehicle.

b. **Engine Speed.** Do not allow engine to idle at less than 700 revolutions per minute. When operating under load, use a transmission range which will allow engine to run above 1,600 revolutions per minute. The most desirable engine cruising speed is 2,100 revolutions per minute.

c. **Attention to Instruments and Gages.** Frequently observe instruments and gages on instrument panel during operation of vehicle. If any instrument or gage shows an abnormal reading, make an investigation to determine and correct the cause (Section XVI) before continuing to operate vehicle. Normal readings are as follows:

(1) ENGINE OIL PRESSURE GAGE. The engine oil pressure gage must register between 50 and 90 pounds at operating speeds.

(2) ENGINE OIL TEMPERATURE GAGE. The engine oil temperature gage must not register above 190° F.

(3) CONVERTER OIL TEMPERATURE WARNING LIGHT. The converter oil temperature warning light must not show red except when lamp is tested by pushing test switch button.

(4) TACHOMETER. The tachometer must register not less than 700 revolutions per minute with engine idling, and must register 1,600 revolutions per minute or over under load.

(5) SPEEDOMETER. The speedometer must register within the vehicle speed limits specified on shift lever bracket for the transmission range with which vehicle is being operated (par. 17 f (1)).

(6) FUEL GAGE. The fuel gage must indicate an ample supply of fuel.

Operation Under Ordinary Conditions

d. **Attention to Unusual Noises.** Give immediate attention to any unusual noises in engine, power train, or other parts of the vehicle. Stop the vehicle as soon as possible for investigation and correction. Safety usually warrants spending time for necessary corrections.

e. **Use of Brakes.** Insure safety and conserve brake shoe linings by observing the following rules:

(1) Do not apply brakes suddenly at high speeds, or on hard pavement or slippery ground.

(2) Avoid harsh or jerky application; pull brake hand levers back smoothly and firmly.

(3) Avoid excessive use of brakes; anticipate stops and allow vehicle to coast to low speed before applying brakes, whenever possible.

(4) Do not hold brakes partially applied while driving straight ahead.

(5) Use reverse range as a brake when descending a steep or very long grade (par. 17 g).

f. **Selection of Transmission Gears.** The Torqmatic transmission automatically compensates for variable load and speed conditions when the vehicle is being driven in the correct speed range, and the shift from one range to another is made under full torque, without any change in engine speed.

(1) Always use the range specified on the shift lever bracket for the speed at which vehicle is moving. Operation in the wrong range will cause unnecessary fuel consumption, excessive heating of transmission oil, and poor over-all performance.

(2) When vehicle speed drops below minimum specified for the range in which it is being operated, immediately shift to next lower range. Do not continue in a higher range than first after engine speed drops below 1,600 revolutions per minute with wide open throttle.

(3) First range provides a starting and an emergency range for conditions when maximum pulling power is needed at speeds below 16 miles per hour.

(4) Second range provides the greatest tractive effort for operation at speeds between 12 and 34 miles per hour.

(5) Third range should be used at speeds above 34 miles per hour whenever the engine speed can be maintained above 1,600 revolutions per minute.

(6) When driving in deep mud, keep the engine at maximum speed. When the vehicle begins to slow down, immediately shift to a lower range without easing off on accelerator pedal.

g. **Making Turns.** Making a turn requires additional power, which necessitates keeping engine speed and power up proportionately; otherwise, a shift to a lower range will be necessary before turn is completed. Avoid turning while climbing; however, if turning is necessary, shift to a lower range before starting turn, to insure ample power for both turning and climbing.

h. **Starting Forward Up a Steep Incline.** When starting forward up an incline, hold both steering brake hand levers back and depress accelerator until engine reaches 2,100 revolutions per minute. Release brake hand levers suddenly and sharply.

i. **Going Down Grades.** Use reverse range as a brake when descending steep or long grades (par. 17 g).

j. **Crossing a Gully.** When entering a gully, release accelerator pedal and shift to gear required for climbing out; allow vehicle to settle to bottom of ditch and then apply full power to climb out; release accelerator pedal upon reaching top of bank and allow vehicle to roll over edge before applying full power again.

k. **Driving Over an Obstacle.** When driving over an obstacle, first shift to required range then apply sufficient power to negotiate the climb; release accelerator pedal upon reaching the crest and allow vehicle to roll over obstacle before applying full power again.

l. **Driving Over Rough Roads.** When driving on rough roads or cross country, it is better to drive fast and float over the bumps rather than drive slow and drive into each depression.

m. **Driving in Deep Mud or Sand.** When driving in deep mud or sand, use a transmission range low enough to provide steady pulling with a reserve of power to avoid stalling the vehicle. When starting in deep mud or sand, use first range and accelerate slowly and smoothly to avoid slipping the track and digging in.

19. HYDROSTATIC LOCK.

a. **Description.** All radial engines are subject to hydrostatic lock in the inverted cylinders, under certain conditions. Hydrostatic lock results when oil, gasoline, or water accumulates in a cylinder combustion chamber in sufficient quantity to block piston before it moves over top center on the compression stroke (both valves closed). If engine is cranked by hand, hydrostatic lock will be apparent in a definite lock-up of the engine. If engine is cranked with the starter, however, the turning power will be great enough to force crank over top center when piston strikes liquid, with the result that connecting rod will be bent and piston and cylinder may be badly damaged. A locked engine thus cranked may start and run for some time before failure of parts become apparent. Conditions which cause hydrostatic lock are described in the following subparagraphs; correction procedures are described in paragraph 47.

b. **Failure to Idle Engine Before Stopping It.** When engine is operating at high speed, a considerable quantity of liquid and vaporized oil is in suspension in the crankcase and other sections of the engine. While engine is running, the surplus oil drains into the oil sump, from which it is removed and returned to the oil tank by the scavenge pump. If engine is stopped while the oil is in suspension in large quantity, the oil is not removed, but accumulates in the crankcase and drains into inverted cylinders. It seeps past the pistons and rings to fill the combustion chamber of any inverted cylinder which has both valves closed, and hydrostatic lock results. NOTE: *This condition can be prevented by idling the engine at 700 revolutions per minute for 5 minutes, as specified in paragraph 17 i.* While idling, suspended oil vapor becomes liquid and all surplus oil drains to the sump and is removed while the scavenge pump is operating.

c. Leaking Oil Tank Check Valve. The function of oil tank check valve is to prevent oil in oil tank from draining by gravity through pressure pump and flooding crankcase when engine is not running. If the check valve leaks due to faulty seat, or foreign matter holding the seat open, flooding and hydrostatic lock will result while the engine is stopped. If the engine is properly idled before stopping (subpar. b above), but hydrostatic lock from oil occurs each time engine stands idle for some time, the check valve is probably leaking. See paragraph 84 d for correction procedure.

d. Leaking Carburetor Float Needle Valve. If carburetor float needle valve is held off its seat by dirt or gum while the engine is stopped, fuel will drain from tanks by gravity, pass through the carburetor and flood crankcase, resulting in hydrostatic lock. NOTE: *This condition can be prevented by always closing fuel shut-off valves when engine is stopped as specified in paragraph 17 i.* If hydrostatic lock from fuel does occur, use the following method of stopping engine until condition can be corrected by replacement of carburetor (par. 93): Close fuel shut-off valves and allow engine to run until fuel in carburetor is exhausted, without depressing fuel cut-off switch button.

e. Water Drawn In Through Carburetor. If vehicle has been fording streams, or has been standing in rain with air inlet and outlet grilles uncovered, sufficient water may collect in engine compartment to submerge the carburetor air scoop. A leak at air scoop flexible connector seals will permit water to enter the air scoop and be drawn through carburetor when engine is started. If water is drawn into engine in any quantity, serious damage and probably hydrostatic lock will result. NOTE: *This condition can be prevented by inspecting engine compartment and draining water before starting engine, as specified in paragraph 17 b.*

20. TOWING THE VEHICLE.

a. General. Towing shackles are installed on tow blocks welded to the hull, at front and rear ends, and a towing cable is carried on the vehicle. Tow bars may be attached to the tow blocks by removing shackles. If towing cable is used, a driver must be in towed vehicle to steer and stop it as required. If a driver is not available, or vehicle is in such condition that the steering brakes cannot be used to control it, two bars must be used. In an emergency, the cable "short hitch" (subpar. b below), may be used if two bars are not available.

b. Attaching Towing Cable. When towing a vehicle that will be controlled by a driver, attach towing cable with shackles passed

TM 9-755
20-21

Part Two—Operating Instructions

through eyes of cable. Avoid doubling the cable, as this will cause failure of the strands and leave cable extremely hazardous to use and handle. If vehicle cannot be controlled by a driver and tow bars are not available, attach cable with a "short hitch," which causes a minimum of bending and movement at shackles and also furnishes clearance between cable and tracks. To make a "short hitch," pass cable through shackles of vehicle to be towed, then cross the ends and attach eyes to shackles on tow vehicle.

c. **Preparation of Vehicle for Towing.** When the vehicle is in condition to be towed by a cable and will be controlled by a driver at the steering brakes, place transmission manual shift lever in neutral position and check lubricant in differential and transmission (par. 38). It is very important to have ample lubricant in these units because lubricant will not be circulated through the coolers while vehicle is being towed. If vehicle is to be towed a long distance during which considerable steering and braking will be required, it is advisable to use tow bars to avoid use of brakes. If tracks are off, steering brakes cannot be used and vehicle must be controlled by tow bars. If final drives are disabled, remove tracks and run on track wheels, using tow bars. If differential or transmission is disabled, remove tracks or the differential to final drive universal joints, and use tow bars.

d. **Towing Procedures.**

(1) Start the tow vehicle in first gear and accelerate until both vehicles are moving freely then, and then only, shift to higher speeds.

(2) Do not tow a disabled vehicle at a speed greater than 10 miles per hour.

(3) The two vehicles must make changes in direction by a series of slight turns so that the vehicle being towed is, as nearly as possible, directly behind the tow vehicle.

(4) If tow cable is used, the driver in vehicle being towed must control speed of vehicle by brakes so as to keep towing cable taut at all times, particularly on down grades.

Section VIII

TURRET CONTROLS AND OPERATION—M18

21. DESCRIPTION OF CONTROLS.

a. **Armament.** Refer to Part Four of this manual for description of controls pertaining to armament.

b. **Turret Platforms and Seats.** A commander's platform, gunner's platform and seats for vehicle commander, gunner, and gun loader are attached to turret and rotate with it. The cushion which provides a seat for the commander may be lifted out of commander's platform so that it is not necessary to stand on the cushion. The gunner's seat (AI, fig. 40) is mounted on a threaded stud and may be adjusted

TM 9-755
21

Turret Controls and Operation—M18

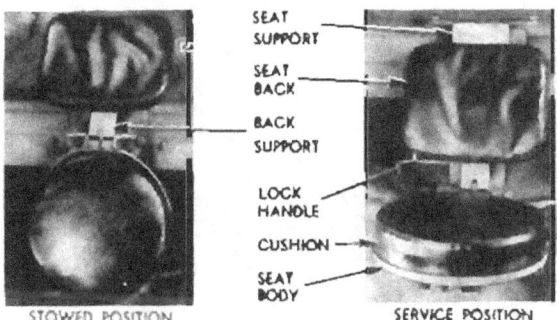

Figure 17—Gun Loader's Seat

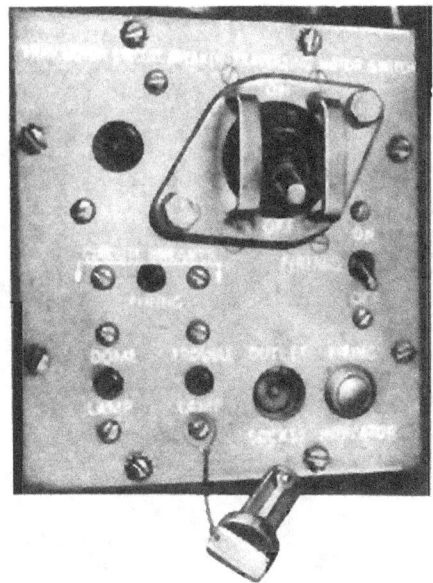

Figure 18—Turret Wiring Switch Box

vertically by turning it. Gun firing foot switch stirrup mounted on gunner's platform may be adjusted both for height and for distance from switch to suit gunner. Gun loader's seat may be installed in upper or lower channel of support, either in a stowed or a service position (fig. 17). Pull outward on lock handle when removing or installing seat, and make certain seat is firmly locked when installed.

TM 9-755
21

Part Two—Operating Instructions

Figure 19—Turret Lock

Figure 20—Hand Traversing Mechanism and Hydraulic Motor

Turret Controls and Operation—M18

c. **Turret Wiring Switch Box** (fig. 18). Turret wiring switch box, which is located to right of turret traversing electric motor, is control center of turret electrical system. It contains following components:

(1) TRAVERSE MOTOR MASTER SWITCH. This switch controls turret traversing electric motor and hydraulic pump. CAUTION: *Before this switch is turned "ON" make sure that hydraulic pump control handle (fig. 21) is in neutral (vertical) position.*

(2) FIRING CIRCUIT SWITCH. This switch controls current to gun firing foot switch and solenoid. Turn switch to "ON" position only when 76-mm gun is to be fired.

(3) FIRING INDICATOR. Firing indicator burns red when firing circuit switch is turned "ON" to indicate circuit is closed and ready for firing.

(4) OUTLET SOCKET. Outlet socket provides a connection to 24-volt current for plugging in trouble light or other electrical accessories. When not in use, socket is closed by a removable plug which is attached to switch box by a chain. Turn plug counterclockwise and pull out to remove it from socket.

(5) CIRCUIT BREAKERS. Circuit breakers having manual reset buttons are installed in switch box, instead of fuses, to protect the turret electrical units and wiring in case an overload condition develops which would damage units or wiring. Each circuit breaker reset button is clearly marked on panel to indicate circuit it protects. An overload in a circuit will cause breaker button to snap outward with audible click. When a button snaps out, wait about 2 minutes to allow breaker to cool, then push button in. Button will stay in if overload was temporary; however, if the overload remains, circuit breaker will continue to cut out until circuit is corrected.

d. **Turret Lock** (fig. 19). Turret lock, which is mounted on turret upper race ring to left of traversing electric motor, contains a toothed pawl which is actuated by an eccentric. Turning eccentric handle rearward to "LOCK" position until it enters retaining clip moves pawl into engagement with teeth in turret lower race ring and locks turret. Turning eccentric handle forward to "FREE" position moves pawl out of engagement with lower race ring so turret is unlocked. Turret lock must be kept in "LOCK" position except when turret is traversed.

e. **Hand Traversing Mechanism** (fig. 20). Turret may be traversed 360 degrees by means of hand traversing mechanism assembled on traverse case (gear box), which is mounted on turret to left of gunner's seat. This mechanism is engaged for manual operation or disengaged for hydraulic (power) operation by means of shifting lever on the lower front side of traverse case. Push shifting lever down for manual operation and pull it up for hydraulic operation. When hand traversing mechanism is engaged and turret lock is in "FREE" position, turret is locked by a spring-applied brake incorporated in the traverse mechanism. Brake is released by pressing brake release lever against the brake handle on break cover, which

TM 9-755

Part Two—Operating Instructions

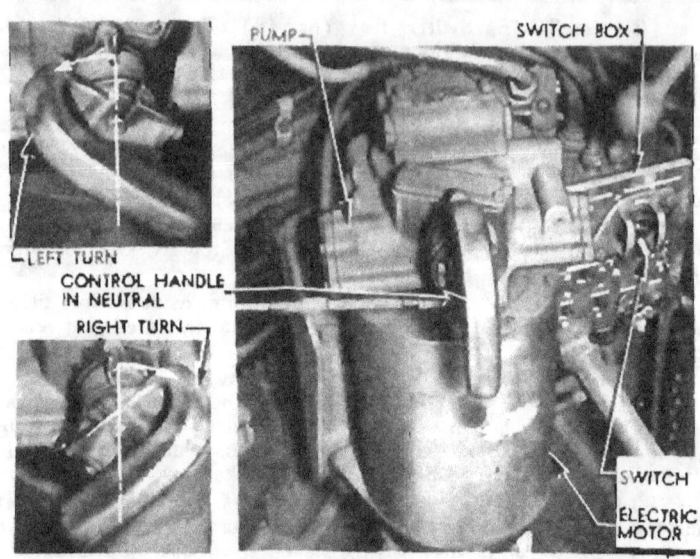

Figure 21—Control Handle, Pump, and Electric Motor

then may be operated as a crank to traverse turret through reduction gears and pinion contained in traverse case. Turret traverses in same direction as brake handle is cranked, viewed from above. When brake lever is released, brake is applied by spring pressure to lock turret.

f. **Hydraulic Traversing Mechanism.** Normally, the turret is power traversed by means of hydraulic traversing mechanism which enables gunner, or vehicle commander, to traverse turret quickly and accurately with minimum of effort. Hydraulic traversing mechanism includes the following components:

(1) HYDRAULIC PUMP AND ELECTRIC MOTOR (fig. 21). A hydraulic pump driven by an electric motor is located in front of gunner. This pump actuates hydraulic motor (step (4) below) through medium of oil delivered under high pressure through tubes (step (5) below). Oil delivered to the hydraulic motor is regulated by a control handle (step (2) below) assembled on the rear side of the hydraulic pump.

(2) HYDRAULIC PUMP CONTROL HANDLE (fig. 21). Control handle on hydraulic pump is provided to control movement and speed of turret in either direction. Turret remains stationary when the handle is in a vertical (neutral) position. Turning control handle clockwise causes turret to traverse to right; turning handle counterclockwise causes turret to traverse to left. Turret speed is increased

TM 9-755
21

Turret Controls and Operation—M18

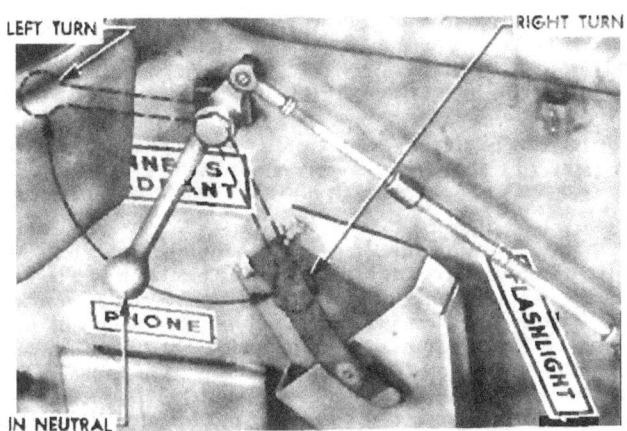

Figure 22—Remote Control Lever

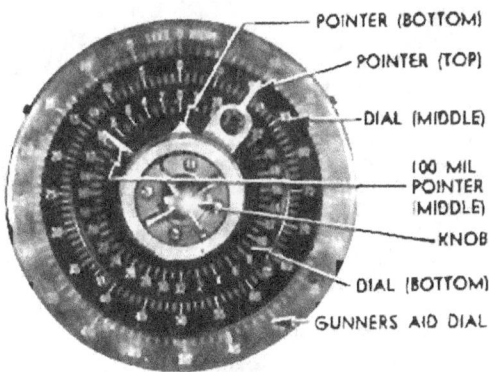

Figure 23—Dials and Pointers—Azimuth Indicator M20

as handle is rotated farther from vertical position. Turret may be reversed instantly by turning control handle to opposite side of vertical position.

(3) REMOTE CONTROL LEVER (fig. 22). A remote control lever located on turret wall just forward of caliber .50 machine gun mount is connected to control handle on hydraulic pump. This lever enables vehicle commander to traverse turret while standing on his platform.

(4) HYDRAULIC MOTOR (fig. 20). The oilgear hydraulic motor mounted on top of traverse case is actuated by oil delivered under

57

TM 9-755
21-22

Part Two—Operating Instructions

high pressure from the hydraulic pump (step (1) above). When shifting lever is in upper position and control handle (fig. 21) is turned away from vertical position, this motor rotates main drive pinion to traverse turret through reduction gears in traverse case.

(5) RESERVOIR AND TUBES, OIL (fig. 274). An oil reservoir located to right of turret wiring switch box supplies oil to hydraulic pump (step (1) above) and receives oil discharged from hydraulic motor (step (4) above) through oil tubes which connect these three units.

g. **Azimuth Indicator** (fig. 23). Azimuth Indicator, M20 is mounted on turret to rear of hand traversing mechanism. The azimuth indicator enables gunner to set off horizontal angles for indirect laying. This is a dialed instrument with two systems of pointers. The first system, bottom pointer indicates number of mils which turret has been traversed from longitudinal axis of tank. When gun points straight forward, pointer reads zero. The second system, middle and top pointers, can be set at zero for any position of gun and thereafter reads number of mils which gun has been traversed. The top pointer provides a fine reading supplementing coarse reading obtained from either bottom or middle pointer. To set middle pointer, press down on knob and turn it. To set top pointer, turn knob without pressing down.

h. **Gun-sighting Periscope, Housing, and Drag Link.** Gun-sighting Periscope M4A1 containing Telescope M47A2, is located in a housing mounted in the upper left corner of turret in front of gunner's seat. The periscope housing is connected to 76-mm gun mount rotor by a drag link so housing and periscope follow movement of gun as it is elevated and depressed. Periscope can be raised to using position, or lowered for complete protection by loosening knurled locking nut and moving periscope to desired position. Top of housing is closed by a spring-loaded door when periscope is lowered. A wiper blade attached to housing is provided to wipe periscope window when periscope is moved up and down.

22. OPERATION OF TURRET.

a. **Armament.** Refer to Part Four of this manual for instructions pertaining to operation of armament.

b. **Before-operation Precautions.** Before operation of turret make certain that sub-floor doors are closed and securely latched. Also make sure turret lock is in "LOCK" position (fig. 19), shifting lever (fig. 20) is in down position, and 76-mm gun is elevated sufficiently to clear all equipment.

c. **Operating Turret by Hand Traversing Mechanism** (fig. 20). Before operating turret observe instructions in subparagraph b above.

(1) Grasp brake handle and press brake release lever against handle to free brake. Push shifting lever to lower position, moving brake handle back and forth as required to secure engagement of gears in traverse case.

Turret Controls and Operation—M18

(2) Release the brake release lever to apply brake and turn turret lock handle to "FREE" position (fig. 19) to unlock turret.

(3) Traverse turret in desired direction by pressing brake release lever and cranking brake handle in same direction, viewed from above. To lock turret temporarily during traversing operation, release brake lever to apply brake.

(4) When traversing operation is completed, turn turret so that 76-mm gun points straight forward on vehicle and turn turret lock handle to "LOCK" position until it enters retaining clip.

d. **Operating Turret by Hydraulic Traversing Mechanism.** Before operating turret, observe instructions in subparagraph b above.

(1) Turn on the master switch box upper switch (fig. 15).

(2) Make sure hydraulic pump control handle (fig. 21) is in vertical position, then turn traverse motor master switch (fig. 18) to "ON" position.

(3) Move shifting lever (fig. 20) to upper position. It may be necessary to release hand brake and move brake handle slightly to secure engagement of gears in traverse case.

(4) Turn turret lock handle to "FREE" position (fig. 19).

(5) Turn pump control handle (fig. 21) in clockwise direction to rotate turret to right, or in counterclockwise direction to rotate turret to left. A slight turn of control handle will give slow speed and a turn to limit of travel will give maximum speed. NOTE: *When using remote control lever (fig. 22), move lever forward to traverse to right and rearward to traverse to left.*

(6) To stop turret, release control handle or remote control lever which will be returned to neutral position by spring on control handle.

(7) While traversing in one direction, turret rotation can be reversed instantly by turning control handle to opposite side of neutral position. This can be done without damage to the traversing mechanism as automatic braking is provided.

e. **After-operation Service.** When traversing operation is completed, set controls in the following sequence:

(1) Turn turret so 76-mm gun points straight forward in vehicle and lock turret by turning turret lock handle to "LOCK" position until it enters retaining clip.

(2) Move shifting lever to lower or manual position to secure additional locking of turret by means of hand brake.

(3) Turn traverse motor master switch to "OFF" position, and turn off master switch box upper switch, if no other electrical units are in operation.

f. **Precautions About Traversing Operation.**

(1) When hydraulic pump is running and turret is locked, either by turret lock or by having shifting lever in lower or manual position, do not turn pump control handle from neutral position, as this will place an unnecessary strain on all parts of traversing mechanism.

TM 9-755
22-23

Part Two—Operating Instructions

(2) If vehicle is on an incline, do not turn traversing motor master switch "OFF" unless turret lock is engaged and shifting lever is in lower or manual position, because weight of 76-mm gun will cause turret to rotate rapidly until gun reaches lowest point.

(3) Do not get out of turret after vehicle is parked without first making sure turret lock is in "LOCK" position and shifting lever is down.

(4) Lock 76-mm gun by applying gun traveling lock (fig. 284 or 285).

Section IX
OPERATION OF ACCESSORY EQUIPMENT

23. AUXILIARY GENERATOR—M18 VEHICLE.

a. *Description.* The Homelite auxiliary generator in the M18 vehicle is mounted on the hull subfloor in the right front corner of the fighting (turret) compartment. This generator is used to charge the batteries when the engine generator is not operating or to supplement the engine generator when the various electrical units impose a heavy load on the battery. The generator assembly incorporates a small gasoline engine which is supplied with fuel from 5-gallon tank mounted in the sponson to right of assistant driver's seat (fig. 25). A separate fuel tank is provided because lubricating oil is mixed with fuel, as described in paragraph 38 d (4). The engine is connected by a flexible coupling to a muffler which discharges through exhaust pipe welded to outside of sponson. A control box mounted on generator contains ammeter which shows generator charging rate, circuit breaker for protecting generator and wiring against overload, and switch for starting engine (fig. 24).

b. *Operation* (fig. 24). The generator engine usually is started by using current from batteries; however, if battery current is not sufficient to crank engine, it may be cranked by means of starting rope wound in a counterclockwise direction in groove on magneto rotor. Procedure for starting is otherwise same with either method of cranking.

(1) Fully open the fuel shut-off cocks on fuel tank (fig. 25) and fuel strainer by turning cocks in counterclockwise direction.

(2) Close choke as far as required for starting by moving choke lever on carburetor in clockwise direction. If weather is warm, or engine is warm from recent running, little or no choke may be required. CAUTION: *Over-choking will cause flooding and hard starting.*

(3) Crank engine by depressing switch button on control box; release button as soon as engine fires.

(4) As soon as engine starts firing, open choke as far as possible without stalling engine. As engine warms continue opening choke until fully open with engine firing evenly. NOTE: *Engine speed is regulated by automatic governor; do not use choke as a throttle.*

Operation of Accessory Equipment

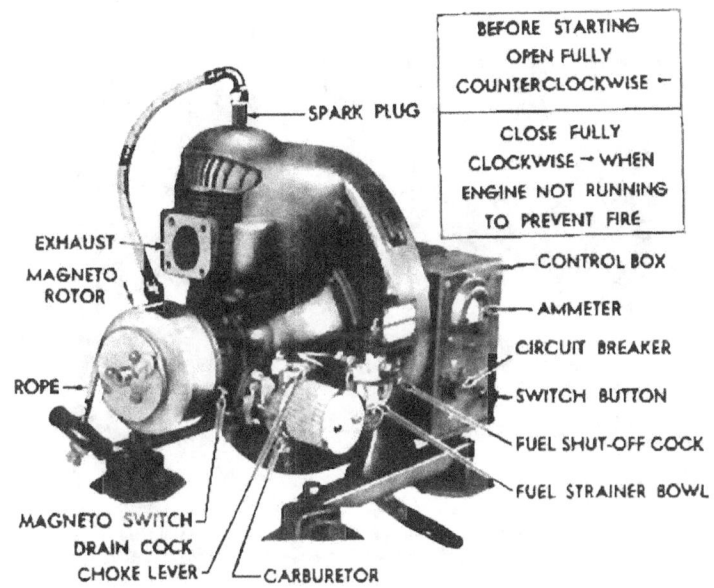

Figure 24—Auxiliary Generator

(5) If engine fails to start because of flooding from over-choking, release starter button, open choke, open drain cock on bottom of crankcase, and allow excess fuel to drain out while turning engine over several times. Close drain cock and start engine again, avoiding use of choke.

(6) If ammeter in control box does not indicate charge with engine running, push in red button marked "CHARGING CIRCUIT RESET." This button resets charging circuit breaker; if button will not stay in, investigate and correct cause of circuit overload.

(7) To stop engine, press red magneto switch button at edge of magneto rotor and hold depressed until engine stops.

(8) Always shut off fuel after stopping engine by turning fuel shut-off cocks on fuel tank and fuel strainer in clockwise direction until tight.

24. FIRE EXTINGUISHERS.

a. Portable Fire Extinguisher (fig. 25). A portable carbon dioxide fire extinguisher is mounted in a stowing bracket in sponson beside assistant driver's seat in both vehicles. It may be removed for use by lifting latch on stowing bracket. The portable fire extinguisher must be recharged when weight drops to 6 ounces below total weight (includes horn) stamped on valve body.

Part Two—Operating Instructions

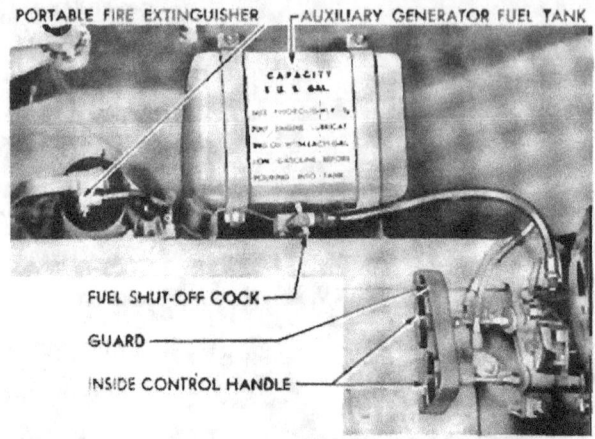

Figure 25—Auxiliary Generator Fuel Tank, Portable Fire Extinguisher, and Fixed Extinguisher Control Handles

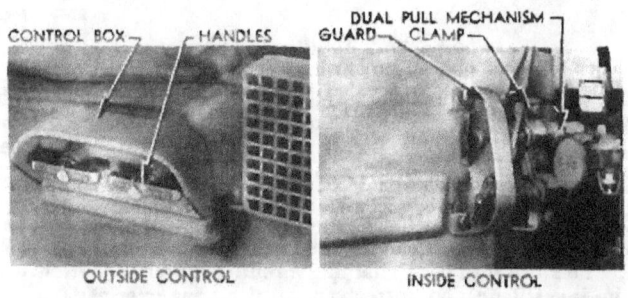

Figure 26—Fixed Fire Extinguisher Controls—M18

(1) OPERATION OF PORTABLE FIRE EXTINGUISHER. To operate portable fire extinguisher after removal from stowing bracket, swing horn one-half turn and press trigger to control the rate of discharge while directing horn close to base of flame. CAUTION: *The white discharge is dry ice which will cause frost-bite; do not permit extended contact with skin.*

b. Fixed Fire Extinguisher System. The fixed fire extinguisher system is provided solely for extinguishing fires in engine compart-

Operation of Accessory Equipment

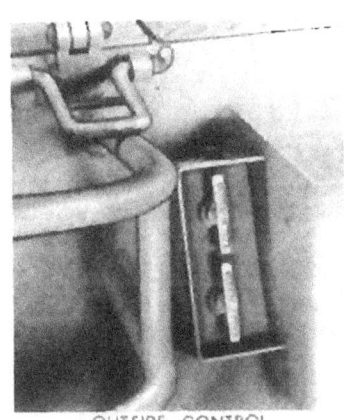

Figure 27—Fixed Fire Extinguisher Controls—M39

ment by use of carbon-dioxide gas. This system is described in paragraph 201.

(1) LOCATION OF CONTROLS AND CYLINDERS—M18 (fig. 26). The outside control handles on M18 vehicle are located in control box on hull roof adjacent to auxiliary generator air outlet cover. The inside control handles are located on hull wall behind assistant driver's seat. Cylinders are located under the right rear subfloor door (fig. 276).

(2) LOCATION OF CONTROLS AND CYLINDERS—M39 (fig. 27). Outside control handles on M39 vehicle are located in an open box on left sponson between stowage rack and windshield stowage box. Inside control handles are located under bulkhead ledge at rear of crew compartment. Cylinders are located under rear seat center cover (fig. 277).

(3) OPERATION OF CONTROLS. Fixed fire extinguisher cylinders may be discharged by pulling outward on either outside or inside control handles (fig. 26 or 27). At each location, a separate control handle is provided for each cylinder so that one cylinder, or both, may be discharged as required. Each cylinder may be discharged also by raising right rear subfloor door (M18) or rear seat center cover (M39), and moving lever on local control (fig. 276 or 277).

c. Precautions in Handling Fire Extinguishers.

(1) Any cylinder containing gas under high pressure is as dangerous as a loaded shell. The cylinder should never be dropped, struck, roughly handled, or exposed to unnecessary heat.

Part Two—Operating Instructions

(2) Make sure extinguisher cylinders are always securely fastened and nothing interferes with accessibility of the portable extinguisher or controls of fixed extinguisher system.

(3) After use, immediately exchange discharged cylinder, or portable extinguisher, for a fully charged one. Do not install a portable cylinder that weighs less than 3 pounds 12 ounces.

25. DECONTAMINATING APPARATUS.

a. Decontaminating apparatus consists of two 1½-quart, M2 (Spec. 197-54-113) portable units. In M18 vehicle, one unit is mounted in clamps in driving compartment in front of assistant driver's seat, and the other is mounted in clamps on roof support at front of fighting compartment. In M39 vehicle: one unit is mounted in front of assistant driver's seat, and the other is mounted on under side of front seat cover in crew compartment. Refer to paragraph 32 for use.

26. WINTERIZATION EQUIPMENT.

a. **Description.** M18 vehicles below serial No. 1700 do not have any winterization equipment components unless they have been installed in the field. M18 vehicles starting with serial No. 1701 have all components installed at factory except heater power and burner units, connecting air tube, fuel pipes and air intake shutters. The M39 vehicle has only the most inaccessible hot air distribution tubes installed at factory. Installed heater components have no use in vehicle operation except where equipment has been completed in field. Engine oil dilution components, when installed, can be used in sub-zero weather as described in paragraph 27. Winterization equipment components are as follows:

(1) HEATER CONTROL BOX (fig. 28). Heater control box is installed on driving compartment ceiling to left of driver in M18 vehicle. It is installed on ledge at rear of crew compartment in the M39 vehicle. The control box contains a circuit breaker, heater switch and clock mechanism, and a relay button. Circuit breaker connects 24-volt battery current to control box, but automatically snaps to "OFF" position if power unit circuit is overloaded. Heater switch when turned clockwise, connects current to power unit (step (2) below) and winds clock mechanism which automatically turns off current at end of time interval for which clock is set by switch handle. Switch may be set for any running time up to 10 minutes. Relay button, covered by a threaded cap, provides means of starting power unit manually if it fails to start when handle is turned clockwise.

(2) HEATER POWER UNIT. Heater power unit is located under left front subfloor door in the M18 vehicle (fig. 29). On the M39 vehicle, it is mounted over engine air inlet behind crew compartment. Power unit includes a motor generator unit, a fuel pump, a blower, and an ignition transformer. Motor generator drives fuel pump and blower and also supplies alternating current to ignition transformer. Fuel pump supplies fuel from vehicle left fuel tank to burner unit

Operation of Accessory Equipment

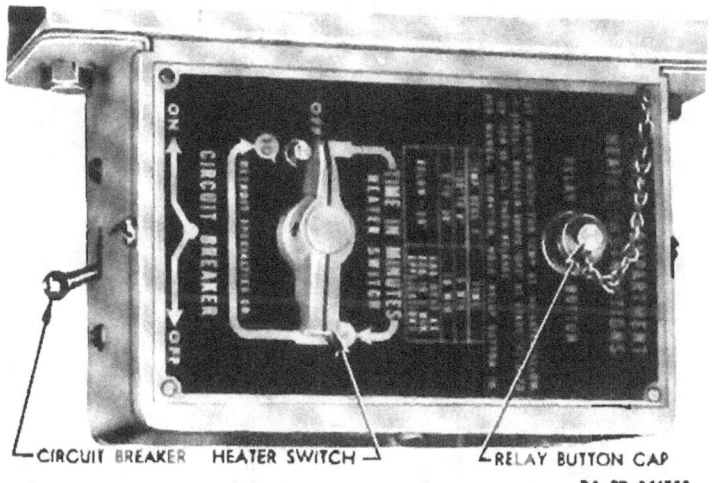

CIRCUIT BREAKER   HEATER SWITCH   RELAY BUTTON CAP

Figure 28—Heater Control Box

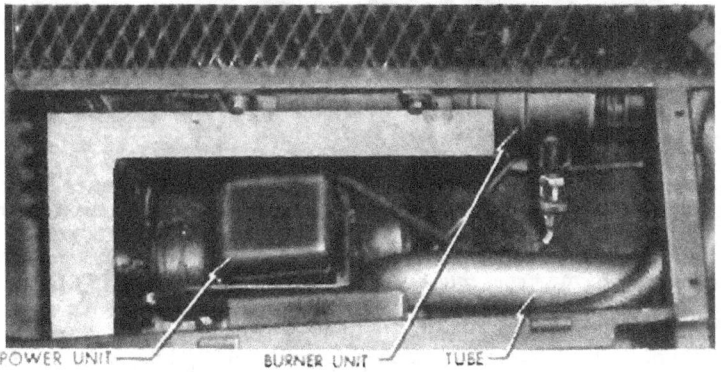

POWER UNIT   BURNER UNIT   TUBE

Figure 29—Heater Power and Burner Units—M18

(step (3) below), and blower supplies air to burner unit. Transformer provides automatic ignition of fuel in burner unit by supplying current to burner ignition electrodes.

(3) HEATER BURNER UNIT (fig. 29). Heater burner unit is located adjacent to power unit and connected to it by an air tube, fuel pipe, and an ignition cable in shielded conduit. Burner unit outlet is connected to hot air distribution tubes. Fuel from power unit fuel

TM 9-755
26

Part Two—Operating Instructions

INSTRUCTION PLATE — GAGE — VALVE HANDLE —

RA PD 344660

Figure 30—Dilution Valve Operating Handle, Pressure Gage, and Instruction Plate

pump is discharged through an atomizing nozzle in one end of burner ignition chamber to mix with air supplied by power unit blower. This mixture is ignited by an electric spark across electrode points in front of nozzle, and burns in ignition chamber to provide heat which is blown from burner into hot air distribution tubes (step (5) below).

(4) FUEL SUPPLY VALVE AND OPERATING LEVER. Fuel supply to heater power unit from left fuel tank is controlled by a valve in supply pipe. Valve is operated by a lever on hull wall to left of driver's seat, in M18 vehicle, or above engine near the power unit, on M39 vehicle. A spring actuated latch is provided to hold lever and valve in open position when latch is released lever will move by spring pressure to close the valve. Valve should be closed when heater is not in operation.

(5) HOT AIR DISTRIBUTION TUBES. Hot air from heater burner unit is blown through distribution tubes to transmission and differential, transmisson and differential oil cooler cores, battery, auxiliary generator (M18), and to rear side of engine. A valve in main distribution tube provides a means of directing hot air to either front or rear of vehicle, and this valve must be turned to heat first

Operation of Accessory Equipment

Figure 31—Air Bleed Plug on Power Unit Fuel Pump

one end and then other. Valve in M18 vehicle is reached by opening right rear subfloor door. On M39 vehicle it is operated by a control handle near heater power unit.

(6) ENGINE OIL DILUTION VALVE AND GAGE (fig. 30). Oil dilution valve is mounted on bulkhead in engine compartment and is actuated by an operating handle which projects through bulkhead. A gage to show pressure in dilution system is located to right of valve operating handle. An instruction plate above gage gives instructions for use of dilution system for temperatures below 0° F. The valve provides a means of diluting engine lubricating oil with fuel from vehicle tank for operation in sub-zero weather (par. 27).

(7) BLOWER TO BATTERY AIR TUBE. (fig. 142). A tube with shut off valve provides a means of supplying hot air from differential and transmission oil cooler blower housing to battery box.

(8) AIR INLET SHUTTER. A shutter installed in place of air inlet grille over front of engine provides a means of closing inlet to facilitate engine warm up. Shutter control lever projects through an opening in right side of bulkhead.

b. **Operating of Heater.** When heater is to be used (par. 27), perform following operations:

(1) On M18 vehicle, open right rear subfloor door and turn air distribution valve under propeller shaft to direct heat in desired direction. On M39 vehicle, this is accomplished by turning valve control handle near power unit.

(2) Open fuel supply valve by means of operating lever.

(3) Turn on the upper switch in master switch box (fig. 15).

(4) Move circuit breaker on control box to "ON" position and

TM 9-755
26-27

Part Two—Operating Instructions

turn heater switch to time interval specified on control box face plate for prevailing temperature (fig. 28). If circuit breaker snaps to "OFF" position at start, hold to "ON" position for 30 seconds, then release. NOTE: *If power unit fails to start remove cap and hold relay button in.*

(5) When power unit stops, turn air distribution valve to opposite position and repeat step (4) above.

(6) It is necessary to bleed trapped air out of fuel pump and line before initial operation of heater, or whenever the fuel line has been disconnected, or if fuel tank has been allowed to run dry. Loosen plug at side of pressure regulator on fuel pump (fig. 31), start heater (steps (2), (3), and (4) above), and when fuel flows out around plug, tighten securely.

c. Operation of Engine Oil Dilution Valve (fig. 30). When engine is running at 1,000 revolutions per minute and oil dilution valve is opened, fuel is forced into engine oil tank under pressure from left fuel tank fuel pump. The desired ratio of diluent to engine oil is obtained by holding dilution valve open for specified interval of time. Operation of valve for dilution of engine lubricating oil in cold weather is described in paragraph 27 e.

Section X

OPERATION UNDER UNUSUAL CONDITIONS

27. OPERATION AT SUZ-ZERO TEMPERATURES.

a. General. The operation and maintenance of these vehicles at temperatures of 0° F. to −65° F. involves factors which do not exist at normal temperatures; therefore, operators and maintenance personnel must spend more time in protective maintenance. Failure to give this extra service will result in actual damage, unnecessary and unwarranted expense and failure to start. Instructions contained in this paragraph must be carefully observed when operating vehicle in sub-zero temperatures.

b. Gasoline.

(1) GRADES. Winter grade of gasoline is designed to reduce cold weather starting difficulties; therefore winter grade of motor fuels procured under U. S. Army Specification 2-103, grade C, latest revision, will be used in M18 and M39 vehicles.

(2) STORAGE AND HANDLING. Due to condensation of moisture from air, water will accumulate in tanks, drums and containers. At low temperatures, this water will form ice crystals that will clog fuel lines and carburetor jets unless following precautions are taken:

(a) Be sure all containers are thoroughly clean and free from rust before storing fuel in them.

(b) If possible, after filling or moving a container, allow fuel to settle 24 hours before filling vehicle tank from it.

TM 9-755

Operation Under Unusual Conditions

(c) Keep all closures of containers tight to prevent snow, ice, dirt, and other foreign matter from entering.

(d) Wipe all snow or ice from dispensing equipment and from around fuel tank filler cap before removing cap to refuel vehicles. After filling tank replace cap securely.

(e) Add 1 quart of denatured alcohol, grade 3, to fuel tank at start of winter season, and ½ pint at each refueling. This will reduce the hazard of ice formation in the fuel.

(f) Strain fuel through any type of strainer that will prevent passage of water. CAUTION: *Gasoline flowing over a surface generates static electricity that will result in a spark unless means are provided to ground electricity. A metallic contact between container and tank will be provided to insure an effective ground.*

(g) Keep tank full, if possible. The more fuel there is in the tank, the smaller will be the volume of air from which moisture can be condensed.

c. **Keeping Engine Oil Fluid.**

(1) GENERAL. Engine lubrication system shall contain grade of engine oil prescribed in paragraph 38 for use between $+32°$ F. and $0°$ F. Following instructions for keeping engine oil fluid apply only for continuous temperatures below $0°$ F.

(2) VEHICLE EQUIPPED FOR DILUTION. If vehicle is equipped for dilution (par. 26) and engine is to be shut down for a period of 5 hours or more, engine oil must be diluted before it is shut down, while oil is hot. Use dilution valve (fig. 30) as follows:

(a) Check oil level in oil tank (AA, fig. 38) and add oil specified for use between $+32°$ F. to $0°$ F. (par. 38 d) to bring level to "FULL" mark on indicator.

(b) With engine running at 1,000 revolutions per minute, dilute oil in accordance with following table by lifting dilution valve handle for time specified. Pressure gage should show over 1 pound per square inch pressure drop when valve is open.

Temperature	Open Valve
$0°$ to $-10°$F	None
$-10°$ to $-20°$F	2 minutes
$-20°$ to $-30°$F	4 minutes
$-30°$ to $-40°$F	6 minutes
$-40°$ and below	8 minutes

CAUTION: *When maximum amount of diluent has been added, oil level in tank will be above filler neck; therefore, do not remove filler cap until after engine has been run long enough to cause oil level to drop.*

(c) Shut engine off immediately after diluting oil. Close air inlet shutter, if installed, and install other covers available; or, cover engine compartment with tarpaulin.

(3) VEHICLE NOT EQUIPPED FOR DILUTION. Several methods for keeping crankcase oil sufficiently fluid for proper lubrication are

TM 9-755
27

Part Two—Operating Instructions

listed below. Preference should be given to the different methods in the order listed, according to facilities available.

(a) Keep vehicle in a heated inclosure when it is not being operated.

(b) When engine is stopped, drain oil tank while oil is still hot and store in a warm place until vehicle is to be operated again. If warm storage is not available, heat oil before refilling tank. NOTE: *Do not get oil too hot; heat only to point where bare hand can be inserted without burning. Tag vehicle in a conspicuous place in driving compartment to warn personnel that oil tank is empty.*

(c) While engine is hot, dilute engine oil with gasoline in proportion shown in following table for prevailing temperature, using engine oil of grade specified for use at temperatures from +32° F to 0° F (par. 38 d). If oil level in tank is low enough, dilution may be accomplished without draining tank by adding specified amount of diluent and then filling tank with engine oil to the "FULL" mark on indicator. Run engine several minutes to mix oil and diluent, then shut engine off.

Temperature	Gasoline	Engine Oil
—10° F to —20° F	2 qt	41 qt
—20° F to —30° F	5 qt	39 qt
—30° F to —40° F	8 qt	36 qt
—40° F and below	11 qt	33 qt

(d) If vehicle is to be kept outdoors, and methods described above cannot be used, shelter engine compartment with a tarpaulin. About 3 hours before engine is to be started, place fire pots under tarpaulin. A Van Prag, Primus-type, or other type blowtorch or ordinary kerosene lanterns may be used, with due consideration to location to avoid a fire hazard.

(4) CHECKING LEVEL WHEN OIL IS DILUTED. The presence of a large percentage of light diluent will increase oil consumption and, for that reason, oil level should be checked frequently. Use grade of engine oil prescribed for use between +32° F to 0° F to maintain the oil level to manufacturer's "FULL" mark on oil level indicator during operation.

d. Lubrication.

(1) TRANSMISSIONS, TRANSFER CASE AND DIFFERENTIAL. Engine oil specified for transmission at +32° F to 0° F is satisfactory and no dilution of oil is required for operation to —40° F. Vehicle should be started and operated with transmission in first range at slow speed to warm up. In the differential, transfer case and final drives use SAE 30 engine oil to —10° F, and SAE 10 engine oil in temperatures between —10° F and —40° F.

(2) CHASSIS LUBRICANTS.

(a) If vehicle has been operated 1,000 miles using general purpose grease No. 0 for lubrication, no special precautions are necessary for wheels and track roller bearings. If quantities of general purpose grease, No. 1 are in these bearings, it will be necessary to disassemble

Operation Under Unusual Conditions

and wash in dry-cleaning solvent, dry and then relubricate with general purpose grease, No. 0 for satisfactory operation.

(b) All other places where general purpose greases are specified on the lubrication order for use between $\times 32°$ F to 0° F shall be lubricated with same lubricant below 0° F.

(c) When extremely low temperatures are encountered and No. 0 general purpose grease is not satisfactory where specified above, grease, O. D., No. 00, Ordnance Department Tentative Specification AXS-1169 may be used.

(d) For oil can points where engine oil is prescribed above 0° F use light preservative lubricating oil.

(3) AIR CLEANERS. Engine oil, SAE 10, diluted with gasoline or Diesel fuel in proportion of three parts oil to one part diluent will be used in servicing air cleaners for operation below 0° F. Oil and diluent should be mixed before adding to the air cleaner.

e. Electrical Systems.

(1) GENERATOR AND STARTER. Check brushes for wear and springs for tension and see that brushes and commutator are clean. Oil or grease on brushes or commutator will affect operation of generator and will prevent electrical contact required for large surge of current in starter required for good starting.

(2) WIRING. Check, clean and tighten all connections, especially battery terminals. Care must be taken that no short circuits are present.

(3) BOOSTER COIL. Check booster coil for proper functioning (par. 50 b).

(4) MAGNETOS. Check points more frequently and replace as necessary (par. 88).

(5) SPARK PLUGS. Test and replace spark plugs if necessary. If difficult to make engine fire, check to make sure plugs are Champion 63S, with gaps set at 0.018 inch to 0.020 inch.

(6) IGNITION TIMING. Check carefully, see paragraph 87 for detailed procedure.

(7) BATTERIES. The efficiency of battery drops sharply with decreasing temperatures and becomes practically nil at $-40°$ F. Do not try to start engine with battery when it has been exposed to temperatures below $-30°$ F without first warming up battery. Be sure battery is always fully charged with the hydrometer reading between 1.275 to 1.300. A fully charged battery will not freeze at temperatures usually encountered even in arctic climates, but a discharged battery will freeze at $+5°$ F.

(8) Do not add water to a battery when it has been exposed to sub-zero temperatures unless battery is to be charged immediately. If water is added and battery is not put on charge, layer of water will stay at top and freeze before it has a chance to mix with acid.

TM 9-755
27-28

Part Two—Operating Instructions

---

(9) STARTING. Before every start, be sure there is no ice or moisture on the spark plugs, wiring or other electrical equipment.

(10) LIGHTS. Inspect all lights carefully.

f. **Starting and Operating Engine.** In addition to instructions for starting engine described in paragraph 17 c, following instructions must be observed when starting engine in temperatures below 0° F.

(1) INSPECTION OF CARBURETOR AND FUEL PUMPS. The carburetor, which will give no appreciable trouble at normal temperatures may not operate satisfactorily at low temperatures. Be sure that carburetor is maintained in good condition and kept properly adjusted. Be sure that fuel pumps are in condition to deliver amount of fuel required to start and operate engine at low temperatures.

(2) HEATING ENGINE. If vehicle is equipped with heater, start and operate heater as described in paragraph 26 b, before starting engine. If vehicle is not equipped with heater, provide initial heat by use of available cold weather accessories described in paragraph 28.

(3) STARTING AND WARM-UP PROCEDURE.

(a) Turn engine over at least two revolutions by hand crank (20 turns of crank) to make certain that it turns freely and is free of hydrostatic lock.

(b) Flip booster coil switch on and off several times to remove any ice. Do not attempt to start until booster coil is working. If necessary, have another man listen for the buzz. This is important, because booster switch has a tendency to ice up and become inoperative. The same is true of magneto switch, and it should also be switched back and forth several times before attempting to start.

(c) Start engine as described in paragraph 17 c (3) through (8), holding booster switch on and continuing use of primer pump until engine firing is smooth; then continue warm up with steps (9) and (10).

(d) Adjust air inlet shutter, if installed, to maintain oil temperature of 120° F or above. CAUTION: *Overheating of engine will cause detonation.*

g. **Vehicle Inspection.** Inspect vehicle frequently. Shock resistance of metals, or resistance against breaking, is greatly reduced at extremely low temperatures. Operation of vehicles on hard, frozen ground causes strain and jolting which will result in screws breaking or nuts jarring loose.

h. **Parking Vehicle.** When a drop in temperature is anticipated, make sure that vehicle is parked on solid ground or footing to prevent tracks from being frozen in mud or water.

28. COLD WEATHER ACCESSORIES.

a. The following cold weather accessories are suggested for use at discretion of officers in charge of materiel.

Operation Under Unusual Conditions

(1) Winterization equipment and engine oil dilution valve, when installed on vehicle, may be used as described in paragraph 26 c.

(2) Vehicle covers listed in Section III provide convenient closures for all vehicle openings and armament.

(3) Tarpaulins, tents, or collapsible sheds are effective for covering vehicles.

(4) Firepots, Primus type or Van Prag blowtorches, ordinary blowtorches, oil stoves or kerosene lanterns can be used for heating vehicles. When used, particular attention must be given to their location to avoid fire hazard.

(5) Extra batteries and facilities for charging them quickly are aids in starting.

(6) Steel drums and suitable metal stands are helpful when heating engine oil.

(7) Small quantities of denatured alcohol, about ½ pint to each tank of fuel, will reduce crystallization of water in fuel.

29. OPERATION UNDER DUSTY CONDITIONS.

a. General. When operating under dusty or sandy conditions special precautions must be taken to prevent excessive wear and damage to the moving parts of power unit and suspension system.

b. Air Cleaners and Breathers. Under extremely dusty conditions the air cleaner oil cups and disks must be cleaned every 2 to 4 hours, or more frequently as required. Air cleaner prefilters must be cleaned when inspection reveals that any appreciable quantity of dirt has accumulated to restrict free flow of air or their capacity to trap dust has been reached. Continued operation of engine with dirty or saturated air cleaners will cause damage to engine which will continue and increase long after air cleaners have been cleaned. Transfer case and transmission breathers must be cleaned more frequently as required when operating under dusty conditions to prevent premature wear and damage to these units. Engine and differential oil filters will be cleaned more frequently as required to maintain its efficiency. Carefully examine all lubricating oil level indicators for evidence of discoloration or gritty substance that would indicate oil has become contaminated and must be changed.

c. Cooling Systems. Inspect engine, transmission and differential oil coolers frequently to make sure air passages are not restricted by accumulation of dirt. Clean cores by flushing with water under pressure or blowing out with compressed air. Cooling fins on engine cylinders must be inspected frequently, and accumulated dust must be blown out with compressed air.

d. Care of Track Suspension System Under Dusty Conditions. Lubricate track suspension system more frequently to cleanse the bearings of any sand or dirt that may have worked into hubs or housings. Inspect track suspension system units including track links for evidence of premature wear. Remove worn units promptly and install new ones to prevent ultimate failure.

TM 9-755
30-33

Part Two—Operating Instructions

30. CARE OF VEHICLE AFTER FORDING.

a. After fording, stop vehicle at once if tactical situation permits and operate hull drain valves to empty vehicle of any accumulated water. If suspension system has been submerged for even a few minutes, lubricate all suspension points to cleanse bearings of water or grit.

31. CARE OF BATTERY IN TORRID ZONES.

a. Water Level. In torrid zones, cell water level should be checked daily and replenished if necessary with pure distilled water. If this is not available, any water fit to drink may be used. However, continuous use of water with high mineral content will eventually cause damage to battery and should be avoided.

b. Specific Gravity. Batteries operating in torrid climates should have a weaker electrolyte than for temperate climates. Instead of 1.300 gravity, the electrolyte should be adjusted to a reading of from 1.210 to 1.230 for a fully charged battery. This will prolong the life of the negative plates and separators. Under this condition the battery should be recharged when reading drops to 1.160. Where freezing conditions do not prevail, there is no danger with gravities from 1.230 to 1.075.

c. Self-discharge. A battery will self-discharge at a greater rate at high temperatures if standing for long periods. This must be taken into consideration when operating in torrid zones. If necessary to park for several days, remove battery and store in a cool place.

32. DECONTAMINATION OF MATERIEL AFFECTED BY GAS.

a. Information on decontamination of materiel affected by gas is included in FM 17-59, Decontamination of Armored Force Vehicles.

Section XI
DEMOLITION TO PREVENT ENEMY USE

33. GENERAL.

a. Destruction of vehicle when subject to capture or abandonment in combat zone will be undertaken by using arm only when, in the judgment of the military commander concerned, such action is necessary.

b. The instructions which follow are for information only. Certain methods of destruction outlined require TNT and incendiary grenades which may not be normal items of issue. Issue of these materials, and conditions under which destruction will be effected *are command decisions in each case*, according to tactical situation.

Demolition To Prevent Enemy Use

c. If destruction is resorted to, vehicle must be so badly damaged that it cannot be restored to a usable condition in combat zone either by repair or cannibalization. Adequate destruction requires that all parts essential to operation of vehicle be destroyed or damaged beyond repair. Equally important, the same essential parts must be destroyed on all like vehicles so that the enemy cannot construct one complete operating unit from several partially damaged ones.

34. DETAILED INSTRUCTIONS.

a. Methods. Following instructions apply to 76-mm Gun Motor Carriage M18 and Armored Utility Vehicle M39.

b. Destruction of 76-mm Gun, M18 Vehicle.

(1) Insert four unfuzed incendiary grenades, M14, end to end halfway down the gun tube, with tube at 0 degrees elevation. Ignite these grenades with a fifth grenade equipped with a 15-second safety fuze. Elapsed time: 2 to 3 minutes.

(2) Metal from grenades will fuze with tube and fill grooves.

c. Destruction of Vehicle. Two methods of destroying vehicle are given below in their order of effectiveness.

(1) METHOD NO. 1—BY EXPLOSIVES.

(a) Remove and empty portable fire extinguishers. Discharge fixed fire extinguisher system. Puncture fuel tanks. Place a 3-pound TNT charge against left fuel tank, between engine and tank. Place a 2-pound TNT under transmission and differential as far forward as possible. Insert tetryl nonelectric caps with at least 5 feet of safety fuze in each charge. Ignite fuzes and take cover.

(b) If sufficient time and materials are available, additional destruction may be accomplished by placing a 2-pound TNT charge at about center of each track assembly. Detonate these charges in same manner as others.

(c) If charges are prepared beforehand and carried in vehicle, keep caps and fuses separated from charges until used.

(2) METHOD NO. 2—BY GUN FIRE.

(a) Remove and empty portable fire extinguishers. Discharge fixed fire extinguishers. Puncture fuel tanks. Open all vehicle doors and hatches if time is available. Fire on vehicle, using adjacent tanks, antitanks or other artillery, or antitank rockets or grenades. Aim at engine, suspension, and armament in order named. If a good fire is started, vehicle may be considered destroyed.

(b) Destroy the last remaining vehicle by the best means available.

TM 9-755
35-36

Part Three—Maintenance Instructions

PART THREE—MAINTENANCE INSTRUCTIONS

Section XII

GENERAL

35. SCOPE.

a. Part three contains information for guidance of personnel of using organizations responsible for maintenance (first and second echelon) of this equipment. It contains information needed for performance of scheduled lubrication and preventive maintenance services, as well as description and maintenance of major systems and units and their functions in relation to other components of equipment.

Section XIII

SPECIAL ORGANIZATIONAL TOOLS AND EQUIPMENT

36. PURPOSE.

a. The list of tools in this section is for information only. It is not to be used as a basis for requisition.

Name	Federal Stock Number	Mfr. Tool Number
ADAPTER, pressure gage, tube to transmission (use with 41-G-446)	45-A-198-435	
BAR, cross, extension, rail connecting	41-B-90	
BAR, socket wrench, extension, ⅜-in. square drive, 3 in. length	41-B-304-800	
BAR, socket wrench, sliding, 22-in. (use with 41-H-1779-50)	41-B-312-200	MTM-M3-16L
BRACKET, support, extension rail, L. H.	41-B-1926-250	
BRACKET, support, extension rail, R. H.	41-B-1926-255	
BOLT, eye, drive motor and armature lifting (transmission and differential)	41-B-1586-100	
DISC, timing engine	41-D-1266-35	
DRIVER, stud, ¼-in.-28	41-D-2984	MTM-M3-386
FIXTURE, lifting, final drive installing	41-F-2994-3	BMD-T-70-123
FIXTURE, lifting, track wheel sprocket assembly, and compensating wheel removing and replacing	41-F-2994-8	BMD-T-70-113

Special Organizational Tools and Equipment

Name	Federal Stock Number	Mfr. Tool Number
GAGE, engine installing alinement	41-G-13-300	
GAGE, thickness, special, 0.006-in. and 0.070-in., length 5 11/16 in.	41-G-412-77	MTM-M3-563
GAGE, transmission oil pressure (use with 41-A-198-435)	41-G-446	KM-J-1467-M6
HANDLE, socket wrench, speeder, brace type, 3/8-in. square drive, length 17 in.	41-H-1507-95	NBM-NB-85
HANDLE, socket wrench, T-sliding, 3/8-in. square drive, length 8 in.	41-H-1509-53	NBM-NB-70
HEAD, square, 1-in., male (use with 41-B-312-200)	41-H-1779-50-	MTM-M3-16E
HOLDER, coupling	41-H-2269-200	TSE-5275
HOOK, eye, extension rail supporting	41-H-2737	
HOOK, turnbuckle, rear door supporting	41-H-2742	
HOOK, twin turnbuckle, engine rear rail	41-H-2741	
INDICATOR, piston, top dead center, dial type	41-I-73-110	MTM-3-237
LIFTER, track wheel	41-L-1379	TEC-26827
POINTER, engine timing	41-P-2219-50	
PULLER, magneto gear, screw type	41-P-2941-800	MTM-M3-231
RAIL, extension differential carrier, L. H.	41-R-38	
RAIL, extension differential carrier, R. H.	41-R-38-10	
REMOVER, arm, wedge type, shock absorber	41-R-2366-975	BMD-T-70-114
REMOVER, track link pin	41-R-2372-565	BMD-T-70-109
REPLACER, bearing and seal, track compensating wheel	41-R-2383-950	BMD-T-70-103
REPLACER, grease retainer	41-R-2390-450	BMD-T-70-121
REPLACER, rubber, engine mounting	41-R-2397-150	BMD-T-70-205
REPLACER, rubber, engine mounting	41-R-2397-155	BMD-T-70-204
REPLACER, oil seal (axle shaft)	41-R-2391-150	
ROLLER, engine mounting auxiliary	41-R-2743	BMD-T-70-203
SCREWDRIVER, valve clearance adjusting	41-S-1725	MTM-M3-239
SLING, engine lifting, w/accessories	41-S-2831-835	BMD-T-70-202

TM 9-755

Part Three—Maintenance Instructions

Name	Federal Stock Number	Mfr. Tool Number
TONGS, lifting, torsion bar	41-T-2723	BMD-T-70-112
WRENCH, axle shaft plug	41-W-491-500	BMD-T-70-113
WRENCH, oil relief valve body, check nut and cap	41-W-636-620	MTM-M3-341
WRENCH, box (split) angle-end., dble-hex., flare nut, pipe and tubing, size of opening 1 3/8 in.	41-W-638-455	SN-RX-44
WRENCH, crowfoot, starter attaching, 11/16 in., special	41-W-871-45	MTM-M3-505
WRENCH, cylinder hold-down nut, 1/2-in. hex	41-W-871-37	MTM-M3-290
WRENCH, intake pipe packing nut	41-W-1537	MTM-M3-210
WRENCH, oil pump to crankcase rear section attaching nut, 7/16-in. hex	41-W-1577-500	MTM-M3-299
WRENCH, socket (detachable), 3/8-in. square drive, 12 point opening, size 7/16 in., universal joint (formerly 41-W-2610-15)	41-W-2990-90	IMC-101823
WRENCH, socket (detachable), 3/8-in. square drive, 12 point opening, size 1/2 in., universal joint (formerly 41-W-2610-20)	41-W-2990-120	IMC-101217
WRENCH, socket (detachable), 1-in. square drive, 6 point opening, size 2 3/16 in.	41-W-3058-415	
WRENCH, socket (detachable), 1-in. square drive, 2 5/8-in. hexagon (track wheel support arm spindle nut)	41-W-3058-480	MTM-M3-16K
WRENCH, socket, tubular, single end, 1 3/8-in. octagon opening, length 2 5/8 in.	41-W-3126	MTM-M3-558
WRENCH, spanner, pin, solid, circle diam. 2 5/8 in. pin size 7/32 in., length 13 3/16 in.	41-W-3255-350	BMD-T-70-104
WRENCHES, servo band adjusting	41-W-490-250	TEC-50-2
WRENCH, special, cylinder base screw, wide sweep, dble-hex., opening 1/2 in. sq.-drive, size of box opening 1/2 in., length overall 17 3/4 in.	41-W-3336-545	CWR-32101
WRENCH, torque indication, 3/4-in. square drive, capacity 0-300 lb-ft (final drive sprocket nuts)	41-W-3634	

Section XIV

LUBRICATION

37. LUBRICATION ORDER.

a. Reproduction of War Department Lubrication Order 9-755 (figs. 32 and 33) prescribes first and second echelon lubrication maintenance above 0° F. For lubrication below 0° F refer to paragraph 27. Lubrication to be performed by ordnance maintenance personnel is covered in paragraph 38 c.

b. A lubrication order is issued with each item of materiel and is to be carried with it at all times. If materiel is received without a copy, the using arm shall immediately requisition a replacement from closest Adjutant General Depot. See lists in FM 21-6.

c. Instructions on lubrication order are binding on all echelons of maintenance and there shall be no deviations.

d. Service intervals specified on lubrication order are for normal operating conditions during active service above 0° F. These intervals will be reduced under extreme conditions such as excessively high or low temperatures, prolonged periods of high-speed operation, continued operation in sand or dust, immersion in water, or exposure to moisture, any one of which may quickly destroy protective qualities of lubricant. Calendar intervals may be extended when materiel is not in use.

e. Lubricants are prescribed in "Key" in accordance with three temperature ranges, above +32° F, +32° F to 0° F, and below 0° F. When to change grades of lubricants is determined by maintaining a close check on operation of materiel during approach to change-over periods, especially during initial action. Sluggish starting is an indication of lubricants thickening, and signal to change to grades prescribed for next lower temperature range. Ordinarily it will be necessary to change grades of lubricants only when air temperatures are consistently in next higher or lower range.

38. DETAILED LUBRICATION INSTRUCTIONS.

a. Lubrication Equipment. Each piece of materiel is supplied with lubrication equipment adequate to maintain materiel. This equipment will be cleaned both before and after use. Lubrication guns will be operated carefully, and in such a manner as to insure a proper distribution of lubricant.

b. Points of Application.

(1) Lubrication fittings, grease cups, oilers, and oilholes are readily located by reference to the lubrication order. Wipe these devices and surrounding surfaces clean before lubricant is applied.

(2) When relief valves are provided, apply new lubricant until

WAR DEPARTMENT LUBRICATION ORDER L0 9-755

15 FEBRUARY 1945 (Supersedes WDLO No. 143, 1 April 1944)

CARRIAGE, MOTOR, 76-mm GUN, M18
VEHICLE, UTILITY, ARMORED, M39

References: ORD 7 SNL G-163; TM 9-755

Clean fittings before lubricating. Relubricate after washing and fording. Clean tank parts with SOLVENT, dry cleaning, or OIL, fuel Diesel. Dry before lubricating. Use CLEANER, rifle bore, or SOLVENT, dry cleaning, for artillery and small arms.

Intervals are based on average operation of 500 miles a month. Reduce to compensate for severe conditions and extended periods of operation. Intervals may be extended when the material is not in service. Lubricate dotted arrow points on both sides. Opposite points are shown by short arrows.

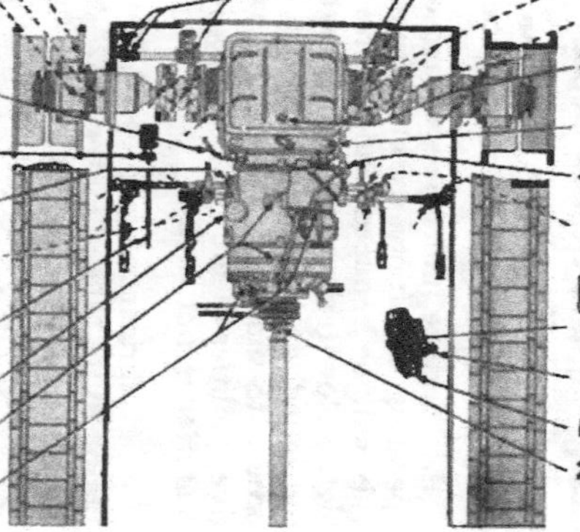

Lubricant • Interval

AV, FIG. 43	Final Drive Fill and Level — Check level
AW, FIG. 43	Final Drive Shaft — CG 2M
AV, FIG. 43	Final Drive Drain — Drain and refill Cap. 2½ qt. each (See Note B)

Serviced From Driver's Compartment

D, FIG. 34	Speedometer Cable — CG S (Remove and clean core, coat lightly with No. 0)
C, FIG. 34	Pedal Support Bracket Bushing — CG 2M (M39 only)
I, FIG. 35	Torqmatic Transmission Fill and Level — OE D Check level
A, FIG. 34	Steering Brake Shaft Bearings — CG 2M (Supported from hull casing)
C, FIG. 34	Brake Control Cable — CG S (Remove and clean core, coat lightly with No. 0) (M39 only)
J, FIG. 35	Transmission Oil Screen — 2M Remove and clean
	Transmission Drain — Drain and refill Cap. 44 qt. (See Note B) (Reached through hull floor)
H, FIG. 35	Transmission Vents — M Clean and inspect

Serviced From Driver's Compartment

L, M & N, FIG. 36	Accelerator Shaft Bearings — WOE (Open door to reach)
K, FIG. 36	Final Drive Universal Joint — 2M CG Remove plug, insert fitting
E, FIG. 35	Differential Breather — M OE Wash and oil (See Note I)
	Differential and Transfer Case Drain — 2M (Reached through hull floor) Drain and refill Cap. 20 qt. (See Note B)
G, FIG. 35	Differential and Transfer Case Fill and Level — D OE Check level
F, FIG. 35	Differential and Transfer Case Oil Screen — 2M Remove and clean
A, FIG. 34	Steering Brake Lever Bearings SAE 10 — WOE

Serviced From Fighting Compartment

O, FIG. 36	Auxiliary Generator Fill and Level — D Check level (See Note 2)
P, FIG. 36	Auxiliary Generator Air Cleaner — M OE Wash and oil (See Note 1)
Q, FIG. 37	Magneto Cam Follower — 1 or 2 drops on felt
R, FIG. 37	Universal Joint — 2M CG (M18 open sub-floor door, M39 remove seats) Remove plug, insert fitting

Figure 32— Lubrication

TM 9-755
38

Lubrication

RA PD 348753

Serviced From Engine Compartment

Ref	Item	Lubricant
W, FIG. 38	Tachometer Cable (Remove and clean core, coat lightly with No. 0 grease.) (Mark Type)	CG S
U, FIG. 37	Accelerator Shaft Bracket (Reached from fighting compartment)	OE W
AB, FIG. 38	Universal Joint (See Note 6)	CG
AD, FIG. 39	Engine Oil Pump Screen (Remove and clean (See Note 6))	OE W
Y & Z, FIG. 38	Accelerator Shaft Brgs. SAE 10	OE W
AE, FIG. 39	Engine Oil Filter Turn handle several times. Every 250 miles, drain sediment. (See Note 7)	D
AA, FIG. 39	Engine Oil Tank Vent (clean and inspect (See Note 3))	M
AA, FIG. 38	Engine Oil Tank Fill & Level Chart Level (Remove plug from rear of tank) Drain and refill Cap. 43 qt. (See Note 3)	OE D

Pintle CG 2M

Ref	Item	Lubricant
S, FIG. 37	Generator (Open sub-floor door) Turn cage until lubricant appears at relief hole	2M CG
V, FIG. 37	Air Cleaner Check level (See Note 1)	D OE
S, FIG. 37	Universal Joint (Open sub-floor door) Remove plug, insert fitting	2M CG
T, FIG. 37	Oil Filter (Differential and Transfer Case) Turn handle several times. Every 250 miles, drain sediment. Every 1,000 miles, clean element. (See Note 7)	D
T, FIG. 37	Transfer Case Breather Wash and oil (See Note 1)	M OE
AF, FIG. 39	Engine Breather Wash and oil (See Note 1)	M OE
X, FIG. 38	Scintilla Magneto SAE 30 Rotor lubricator 6 to 8 drops; Front bearing 20 to 25 drops. (For disassembly, see Note 5)	2M OE
	Pintle Spring Housing Fill to plug opening	RIG. 45

— NOTES and KEY —

1. AIR CLEANERS—[OIL bath Type] Fill engine air cleaner to bead level. Monthly, remove and wash all parts. (Mesh Type.) Wash and oil auxiliary generator cleaner or alternately. Change oil at inspection intervals on Lubrication Order. Used crankcase oil OE will be used in air cleaners above 0°F. Below 0°F. use OE (SAE 10).

2. AUXILIARY GENERATOR—Mix intermediately ½ pt. OE with each gallon of gasoline before pouring into tank.

3. ENGINE OIL TANK—Drain oil only when engine is hot. Keep oil tank vent well clean.

4. ENGINE OIL PUMP SCREEN—Monthly or when engine is removed for overhaul, remove oil pump screen and clean screen and oil pump.

5. GEAR CASES—Drain only after operation. [Transmission] CAUTION: To make accurate checking, if oil level in transmission, start engine and run five minutes to fill torque converter. Stop engine, check oil level and fill to "COLD" mark on dipstick. [Differential and Final Drive] Fill differential to "FULL" mark on dipstick and final drive cases to the plug level before operation and after draining.

6. UNIVERSAL JOINT (Located between engine flywheel and fan)—When engine is removed for inspection or overhaul, remove plug in universal, insert fitting and apply CG. Replace plug.

7. OIL FILTERS—After cleaning filter elements, drive vehicle a short distance and refill engine oil tank and differential case to "FULL" mark.

8. OIL CAN POINTS: Weekly, lubricate Tachometer Part Carrier Hinges, Driving Compartment Door Latches, Rear Door Hinges, Control Rod Clevises and Pins (except Throttle) with OE.

9. DO NOT LUBRICATE—OIL Cooler Blower Bearings, Belt Adjuster Bearings, Hydraulic Pump and Valve Bearings, Slip Joints, Hand Throttles, Driving Compartment Door Hinges, Periscope Mountings.

10. LUBRICATE ONLY BY HIGHER ECHELON—Turret Traversing Gears, Starter, Engine Magnetos (except other). (Refer to TM 9-75L.)

LUBRICANTS	EXPECTED TEMPERATURE			
OE—OIL engine	above +32°F.	+32°F. to 0°F.	below 0°F.	
Engine Oil Tank	OE 50	OE 30	See TM	
Transmission	OE 10	OE 10	OE 10	
Other Gear Cases	OE 30	OE 30	OE 30	
Other Points	OE 30	OE 10	PS	
CG—GREASE, general purpose	CG 1	CG 0	CG 0	
OG—GREASE O.D.	OG 0	OG 00	OG 00	

LUBRICANTS	INTERVALS
SAL—FLUID, shock-absorber, light	D—Daily
RS—OIL, recoil, special	W—Weekly
OH—OIL, hydraulic	M—Monthly
PS—OIL, lubricating, preservative, special	2M—2 Months
	S—Semi-
	Annually

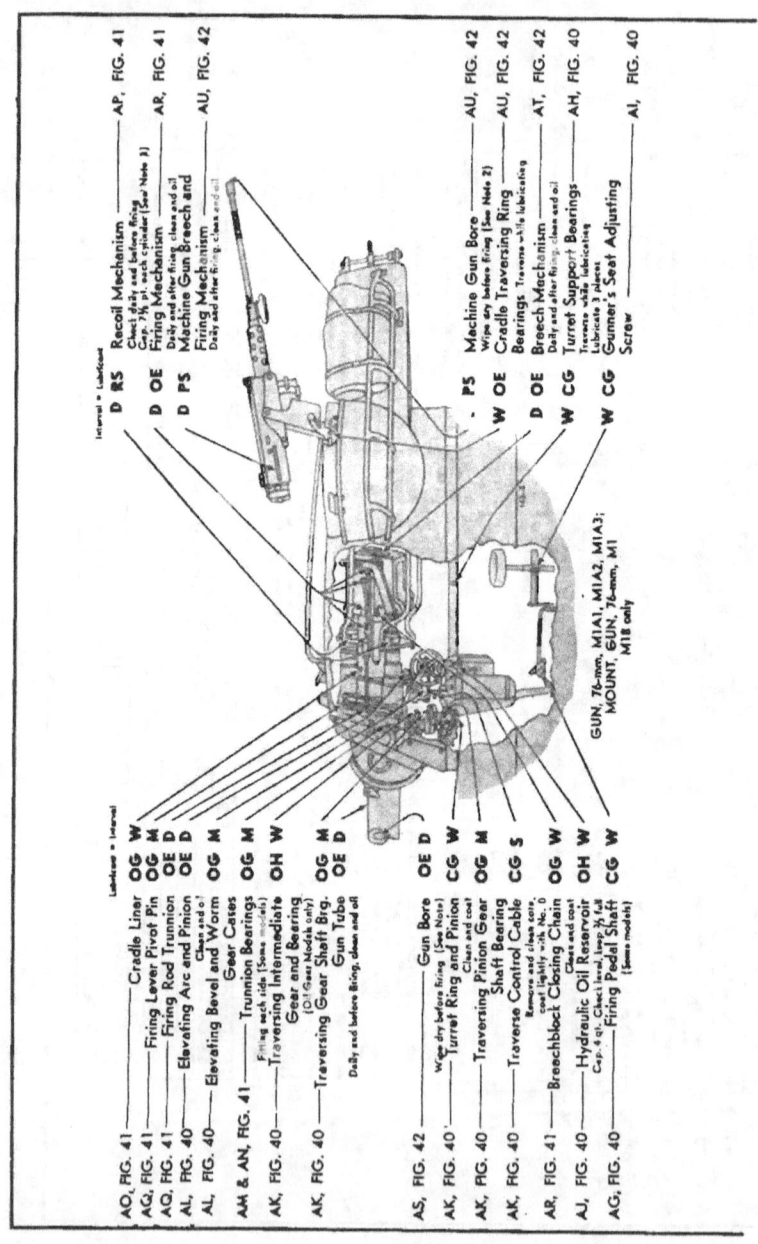

Figure 33— Lubrication

Lubrication

TM 9-755
38

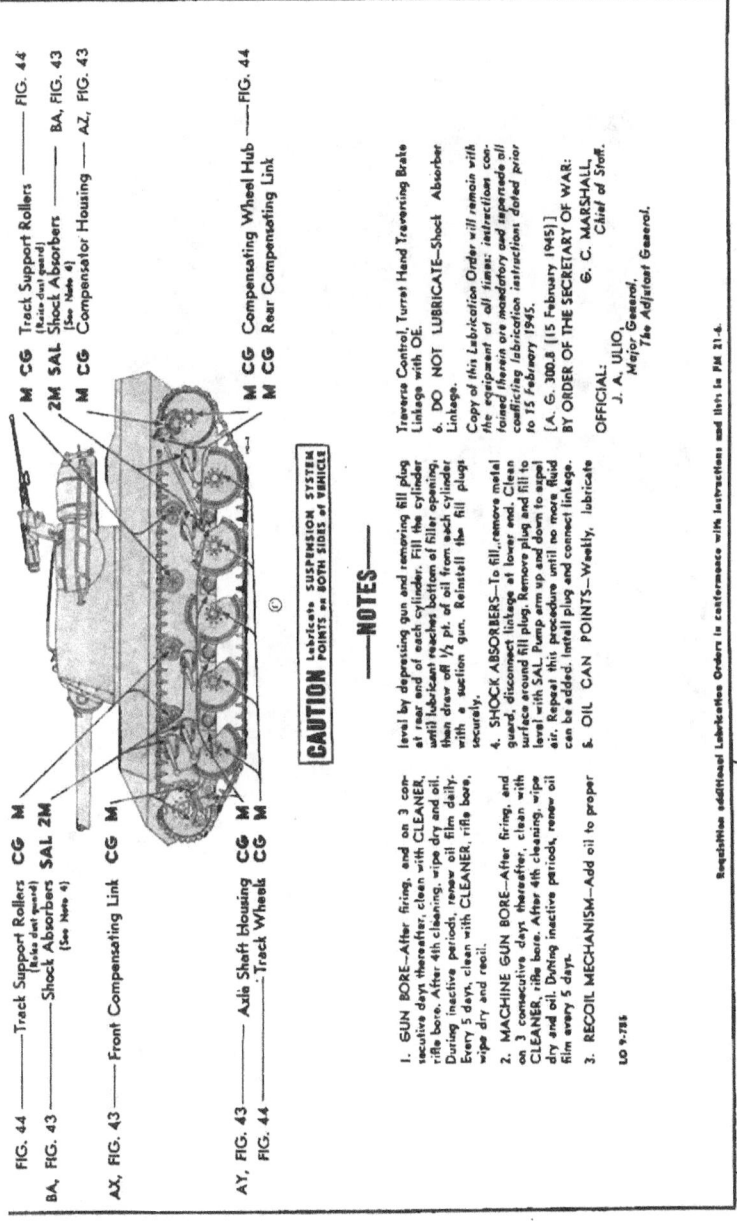

FIG. 44 — Track Support Rollers — FIG. 44
         (Raise dust guard)
BA, FIG. 43 — Shock Absorbers — BA, FIG. 43
              (See Note 4)
AX, FIG. 43 — Front Compensating Link — AZ, FIG. 43
                                         Compensator Housing

M CG   M CG Compensating Wheel Hub — FIG. 44
2M SAL   2M SAL Shock Absorbers
                                             (See Note 4)
                                        M CG Rear Compensating Link

CAUTION Lubricate SUSPENSION SYSTEM POINTS on BOTH SIDES of VEHICLE

— NOTES —

1. GUN BORE—After firing, and on 3 consecutive days thereafter, clean with CLEANER, rifle bore. After 4th cleaning, wipe dry and oil. During inactive periods, renew oil film daily. Every 5 days, clean with CLEANER, rifle bore, wipe dry and recoil.

2. MACHINE GUN BORE—After firing, and on 3 consecutive days thereafter, clean with CLEANER, rifle bore. After 4th cleaning, wipe dry and oil. During inactive periods, renew oil film every 5 days.

3. RECOIL MECHANISM—Add oil to proper level by depressing gun and removing fill plug at rear end of each cylinder. Fill the cylinder until lubricant reaches bottom of filler opening, then draw off 1/2 pt. of oil from each cylinder with a suction gun. Reinstall the fill plugs securely.

4. SHOCK ABSORBERS—To fill, remove metal guard, disconnect linkage at lower end. Clean surface around fill plug. Remove plug and fill to level with SAL. Pump arm up and down to expel air. Repeat this procedure until no more fluid can be added. Install plug and connect linkage.

5. OIL CAN POINTS—Weekly, lubricate Traverse Control, Turret Hand Traversing Brake Linkage with OE.

6. DO NOT LUBRICATE—Shock Absorber Linkage.

Copy of this Lubrication Order will remain with the equipment at all times; instructions contained therein are mandatory and supersede all conflicting lubrication instructions dated prior to 15 February 1945.

[A. G. 300.8 (15 February 1945)]
BY ORDER OF THE SECRETARY OF WAR:
G. C. MARSHALL,
Chief of Staff.

OFFICIAL:
J. A. ULIO,
Major General,
The Adjutant General.

AY, FIG. 43 — Axle Shaft housing — CG M
FIG. 44 — Track Wheels — CG M

LO 9-755

Requisition additional Lubrication Orders in conformance with instructions set forth in FM 21-6.

Order No. 9-755—Back Side

TM 9-755
38

Part Three—Maintenance Instructions

A—BRAKE LEVER AND SHAFT BEARINGS—LEFT

B—BRAKE LEVER AND SHAFT BEARINGS—RIGHT

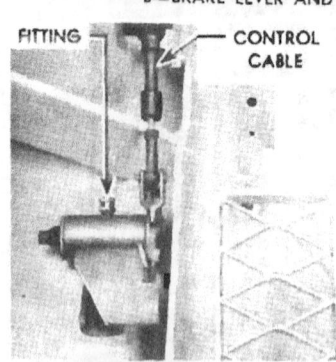
C—BRAKE PEDAL BRACKET AND CABLE

RA PD 344544

Figure 34—Lubrication Points in Driving Compartment

TM 9-755
38

Lubrication

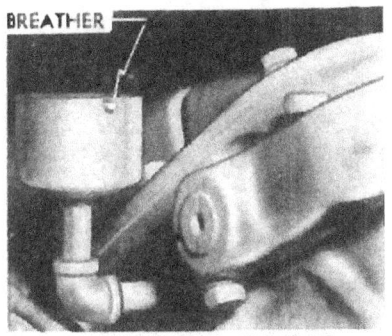
E—DIFFERENTIAL BREATHER

F—DIFFERENTIAL AND TRANSFER CASE OIL SCREEN

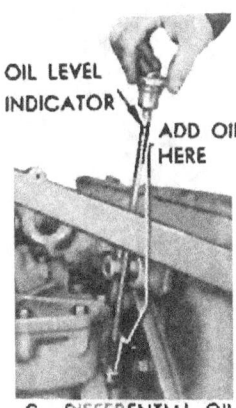
G—DIFFERENTIAL OIL CHECK

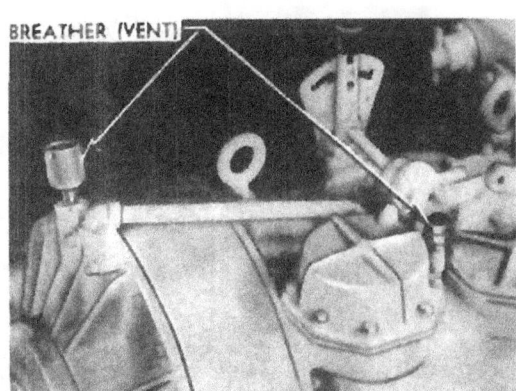
H—TRANSMISSION BREATHERS (VENTS)

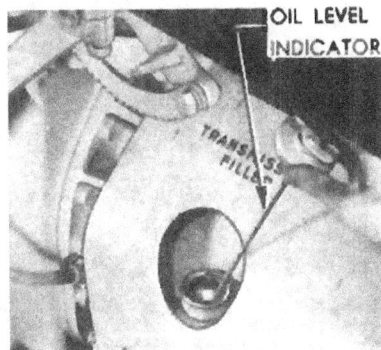
I—TRANSMISSION OIL CHECK

J—TRANSMISSION OIL SCREEN

RA PD 344548

Figure 35—Lubrication Points in Driving Compartment

TM 9-755
38

Part Three—Maintenance Instructions

K—FINAL DRIVE UNIVERSAL JOINT

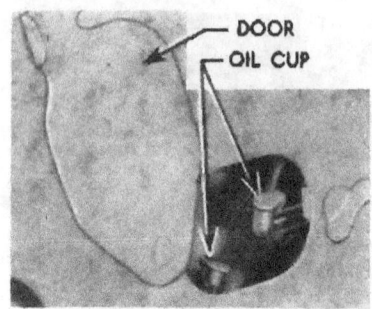
L—ACCELERATOR SHAFT BEARINGS—LEFT END

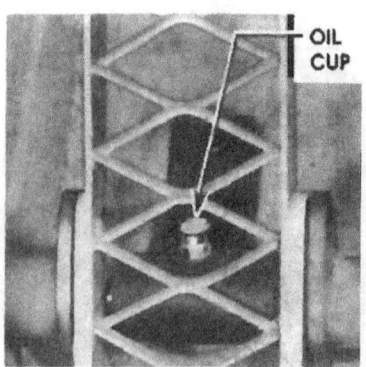
M—ACCELERATOR SHAFT BEARINGS—LEFT PEDAL

N—ACCELERATOR SHAFT BEARINGS—RIGHT PEDAL

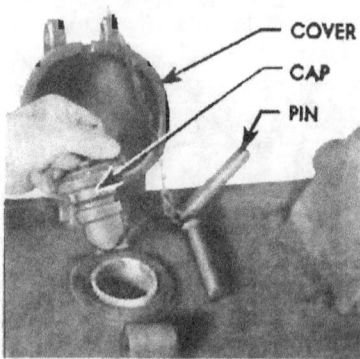
O—AUXILIARY GENERATOR FUEL TANK FILLER

P—AUXILIARY GENERATOR AIR FILTER (CLEANER)

RA PD 344546

Figure 36—Lubrication Points in Driving Compartment

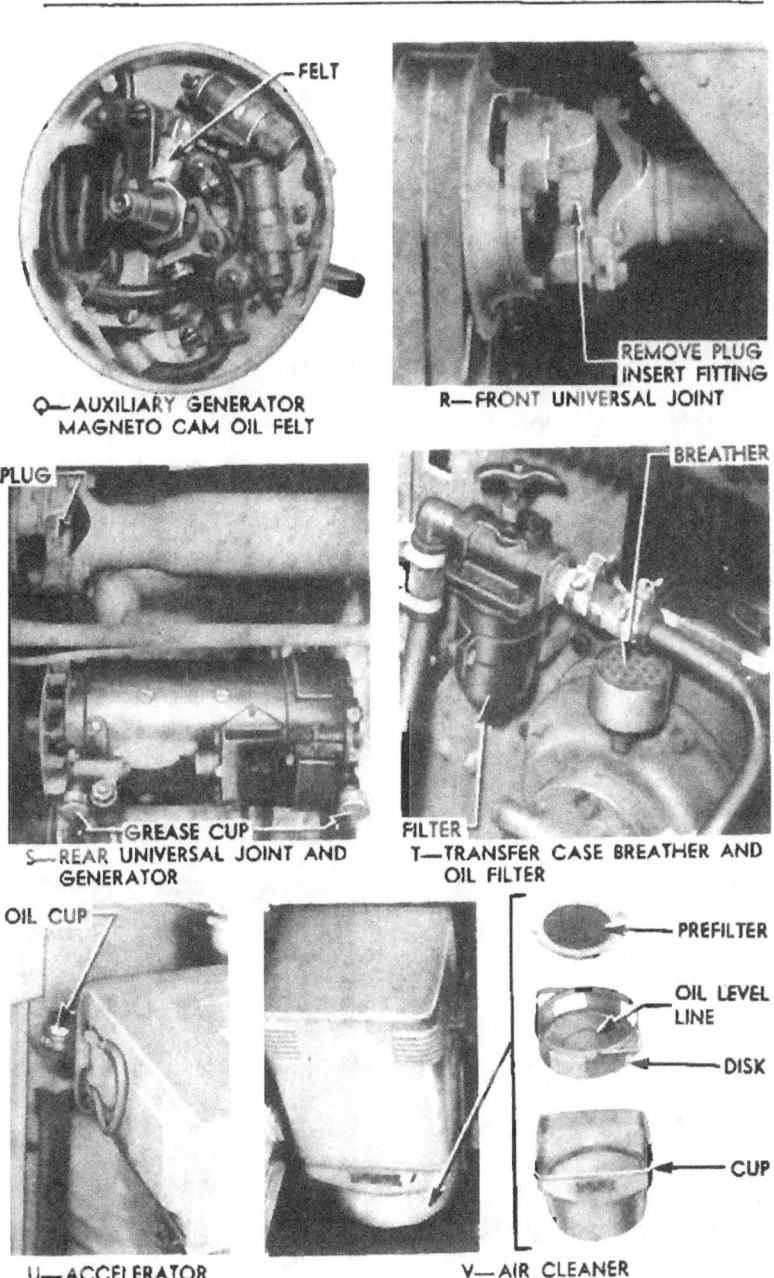

Figure 37—Lubrication Points in Fighting Compartment

TM 9-755
38

Part Three—Maintenance Instructions

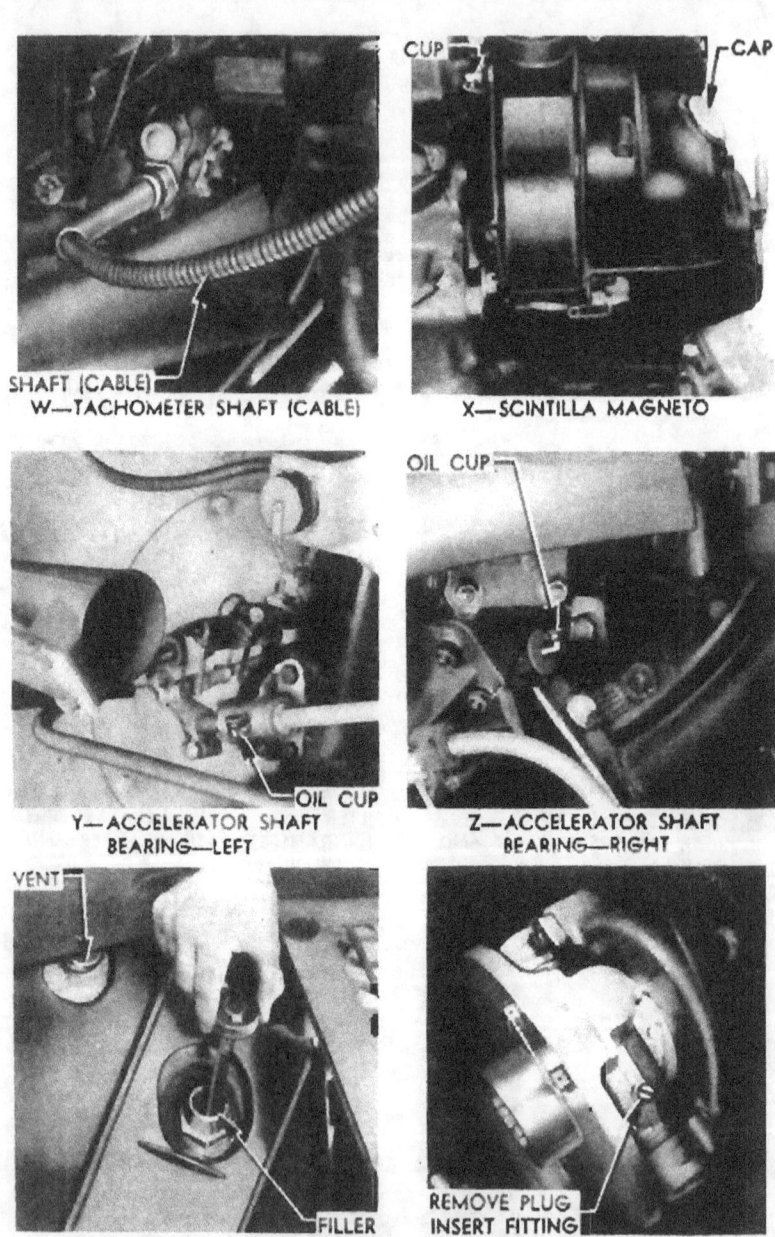

Figure 38—Lubrication Points in Engine Compartment

TM 9-755
38

Lubrication

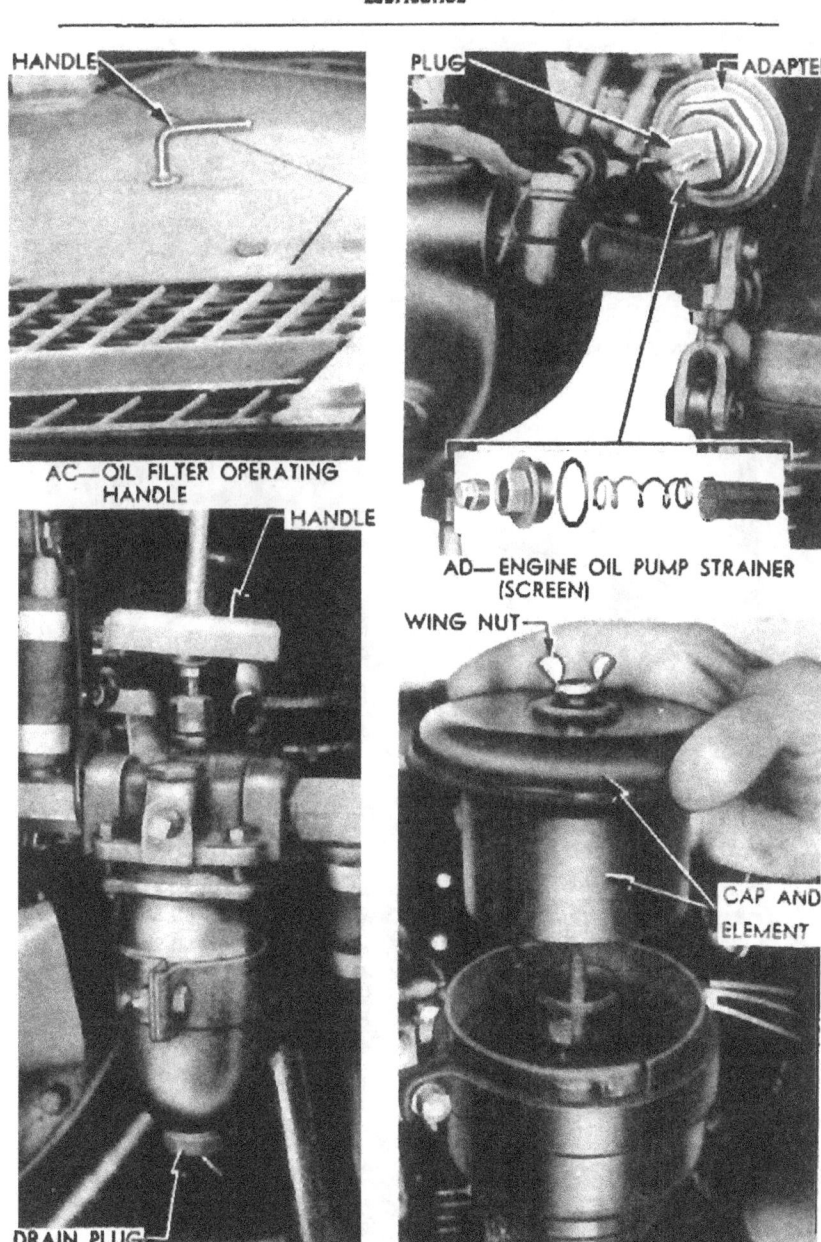

Figure 39—Lubrication Points in Engine Compartment

TM 9-755
38

Part Three—Maintenance Instructions

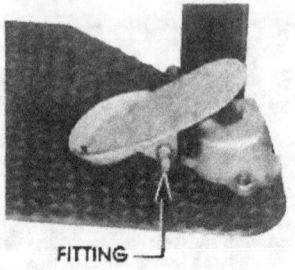

AG—FIRING PEDAL SHAFT

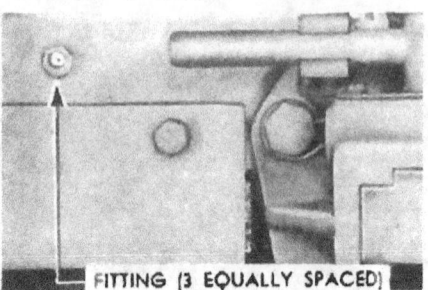

AH—TURRET BEARING LUBRICATION FITTING

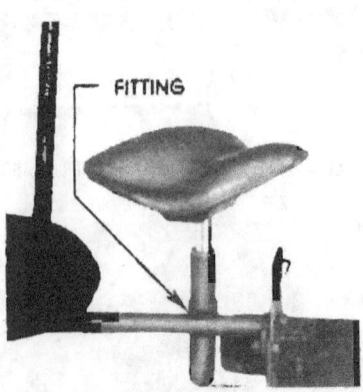

AI—GUNNER'S SEAT ADJUSTING SCREW

AJ—HYDRAULIC OIL RESERVOIR

AK—TURRET TRAVERSING MECHANISM

AL—GUN ELEVATING MECHANISM

RA PD 344580

Figure 40—Lubrication Points in Turret

Lubrication

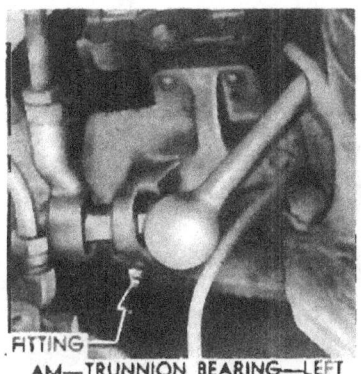

AM—TRUNNION BEARING—LEFT   AN—TRUNNION BEARING—RIGHT

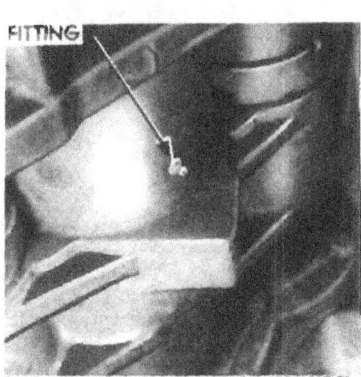

AO—CRADLE LINER (LOWER SIDE)   AP—RECOIL MECHANISM

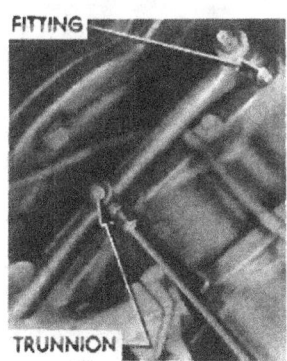
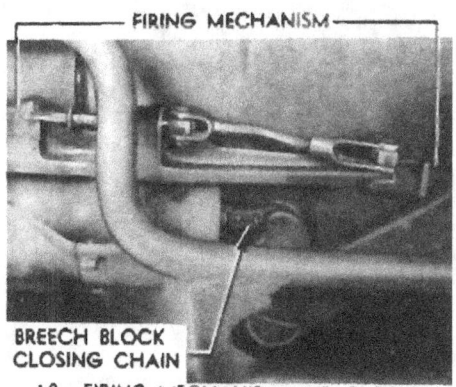

AQ—FIRING LEVER   AR—FIRING MECHANISM AND CLOSING CHAIN

RA PD 244551

Figure 41—Lubrication Points on Gun and Mount

Part Three—Maintenance Instructions

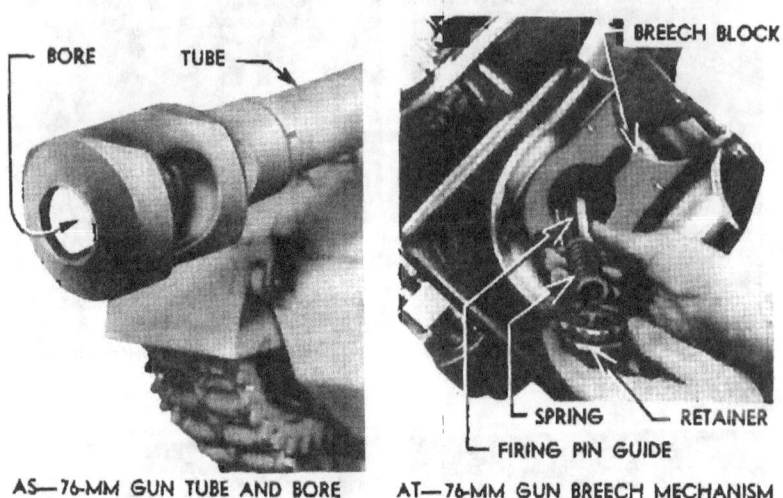

AS—76-MM GUN TUBE AND BORE AT—76-MM GUN BREECH MECHANISM

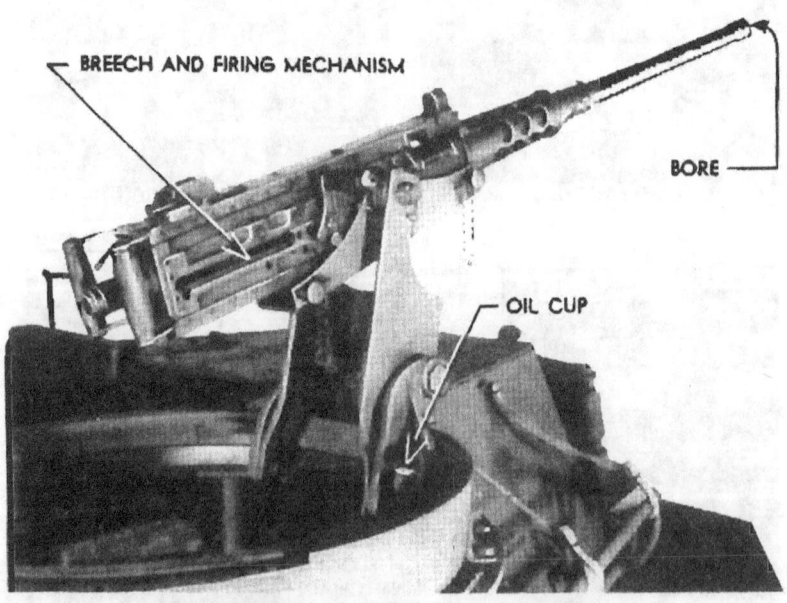

AU—CALIBER .50 MACHINE GUN AND MOUNT

RA PD 344552

Figure 42—Lubrication Points on Guns

Lubrication

AV—FINAL DRIVE FILLER AND DRAIN PLUGS

AW—FINAL DRIVE SHAFT

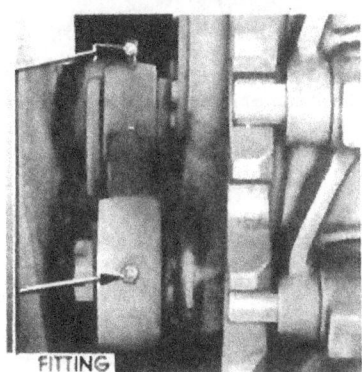
AX—FRONT COMPENSATING LINK

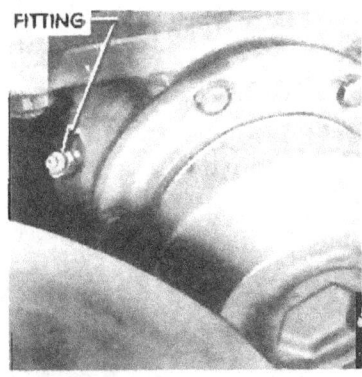
AY—AXLE SHAFT HOUSING

AZ—COMPENSATOR HOUSING

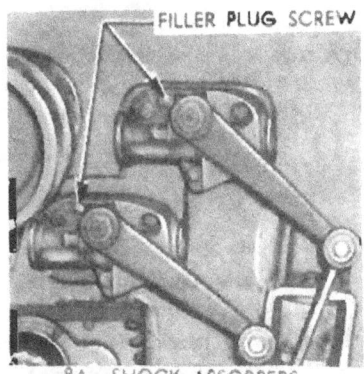
BA—SHOCK ABSORBERS

RA PD 344553

Figure 43—Lubrication Points on Track Suspension System

TM 9-755
38

Part Three—Maintenance Instructions

Figure 44—Lubrication Points on Track and Compensating Wheels, and Support Rollers

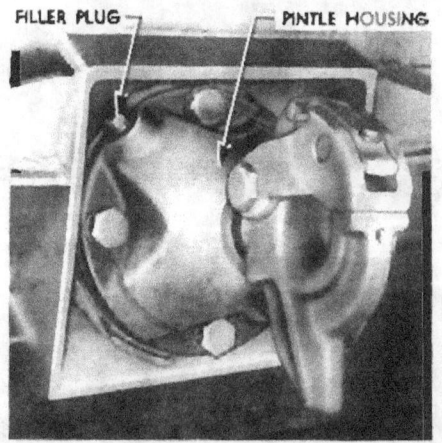

Figure 45—Lubrication Points in Towing Pintle

old lubricant is forced from vent. Exceptions are specified in notes on Lubrication Order.

  c. **Lubricated After Disassembly by Higher Echelon.**

   (1) TURRET TRAVERSING GEARS. Once a year, disassemble

traversing gear unit and clean all parts. Coat bearings lightly with O.D. grease and reassemble in housing. Pack gears with O.D. grease.

(2) ENGINE MAGNETOS (EXCEPT OILERS). At time of magneto removal for inspection or overhaul, magneto rotor bearings will be removed, cleaned and repacked with high temperature special grease. Splined drive coupling will be lubricated by filling with high temperature special grease.

(3) STARTER. Every 6 months, disassemble starter, inspect bearings, replace with new sealed bearings if wear indicates need for replacement.

d. Reports and Records.

(1) Report unsatisfactory performance of materiel to Ordnance Officer responsible for maintenance in accordance with TM 38-250.

(2) A record of lubrication may be maintained in Preventive Maintenance Roster (W.D., A.G.O. Form No. 460).

Section XV

PREVENTIVE MAINTENANCE SERVICES

39. GENERAL INFORMATION.

a. Responsibility and Interval. Preventive maintenance services as prescribed by AR 850-15 are a function of using organization echelons of maintenance, and their performance is the responsibility of commanders of such organizations. These services consist generally of Before, During, At-halt, After-operation, and Weekly Services performed by the crew, and scheduled services to be performed at designated intervals by organizational maintenance personnel.

b. Definition of Terms. General inspection of each item applies also to any supporting member or connection, and is generally a check to see whether the item is in good condition, correctly assembled, secure, or excessively worn.

(1) Inspection for "good condition" is usually an external visual inspection to determine whether the unit is damaged beyond safe or serviceable limits. The term "good condition" is explained further by the following: Not bent or twisted, not chafed or burned, not broken or cracked, not bare or frayed, not dented or collapsed, not torn or cut, not deteriorated.

(2) Inspection of a unit to see that it is "correctly assembled" is usually an external visual inspection to see whether it is in its normal assembled position in vehicle.

(3) Inspection of a unit to determine if it is "secure" is usually an external visual examination; a wrench, hand-feel, or a pry-bar

Preventive Maintenance Services

check for looseness. Such inspection must include any brackets, lock washers, lock nuts, locking wires, or cotter pins used in assembly.

(4) "Excessively worn" will be understood to mean worn beyond serviceable limits, or to a point likely to result in failure if unit is not replaced before the next scheduled inspection.

40. CREW MAINTENANCE (FIRST ECHELON).

a. Purpose. To insure mechanical efficiency it is necessary that the vehicle be systematically inspected at intervals each day it is operated and weekly, so defects may be discovered and corrected before they result in serious damage or failure. Certain scheduled maintenance services will be performed at these designated intervals. Any defects or unsatisfactory operating characteristics beyond scope of first echelon to correct must be reported at the earliest opportunity to designated individual in authority. The services set forth in paragraphs 41, 42, 43 and 44 are those performed by the crew, Before-operation, During-operation, At-halt, and After-operation and Weekly.

b. Use of W. D. Form No. 48. Driver preventive maintenance services are listed on the back of "Driver's Trip Ticket and Preventive Maintenance Service Record" W. D., Form No. 48, to cover vehicles of all types and models. Items peculiar to this vehicle but not listed on W. D., Form No. 48, are covered in manual procedures under items with which they are related. Certain items listed on the form that do not pertain to this vehicle are eliminated from procedures as written into the manual. Every organization must thoroughly train the crew in performing maintenance procedures set forth in this manual, whether they are listed specifically on W. D., Form No. 48 or not. Items listed on W. D., Form No. 48 that apply to this vehicle are expanded in this manual to provide specific procedures for accomplishment of inspections and services. Services are arranged to facilitate inspection and conserve the time of the crew and are not necessarily in the same numerical order as shown on W. D., Form No. 48. Item numbers, however, are identical with those shown on that form.

41. BEFORE-OPERATION SERVICE.

a. Purpose. This inspection schedule is designed primarily as a check to see that vehicle has not been damaged, tampered with, or sabotaged since "After-operation Service was performed. Various combat conditions may have rendered vehicle unsafe for operation and it is the duty of the crew to determine whether vehicle is in condition to carry out any mission to which it is assigned. This operation will not be entirely omitted, even in extreme tactical situations.

b. Procedures. Before-operation Service consists of inspecting items listed below according to procedure described, and correcting or

TM 9-755

Part Three—Maintenance Instructions

reporting any deficiencies. Upon completion of service, results will be reported promptly to designated individual in authority.

(1) ITEM 1, TAMPERING AND DAMAGE. Inspect entire vehicle especially equipment attachments, and armament, for damage that may have occurred from falling debris, shell fire, sabotage or collision since parking. Look through grille into engine compartment for signs of tampering or sabotage such as loosened or damaged accessories, loose fuel or oil pipes or disconnected throttle linkage.

(2) ITEM 2, FIRE EXTINGUISHERS. Examine visible fixed extinguisher lines and horns in engine compartment for security, damage, and correct aiming. Check fixed cylinders and portable fire extinguisher to see they are in good condition and securely mounted. On fixed extinguisher cylinders be sure red safety seal is intact.

(3) ITEM 3, FUEL AND OIL. Check amount of fuel in main and auxiliary (M 18 only) fuel tanks and add fuel as necessary. If auxiliary tank requires filling, follow instructions in lubrication order, paragraph 38 for correct amount and grade of oil to be added to fuel. Check oil level in engine supply tank; add to correct level, if necessary. Any appreciable change in level since After-operation Service should be investigated and reported. Turn oil filter cleaner handle clockwise several complete turns. Check oil levels of transmission, differential and final drive housing, follow instructions in lubrication order, paragraph 38 for correct levels.

(4) ITEM 4, ACCESSORIES AND DRIVES. Inspect carburetor, main generator, starting motor, on M39 Homelite auxiliary unit (M18 only) and blower unit for looseness or damage. Start and test Homelite unit to see that it operates properly. Be sure generator drive belts and oil cooler blower belts are adjusted to have ½-inch finger pressure deflection. NOTE: *Before the following tests are performed, be sure gun traveling lock and turret traversing lock are released.* Test both manual and hydraulic turret traversing mechanisms and gun elevating and firing controls to be sure all mechanisms respond properly. CAUTION: *After completing tests be sure to apply gun traveling lock and turret traversing lock.*

(5) ITEM 6, LEAKS—GENERAL. Check under vehicle, and in engine and fighting compartments for any indications of fuel or oil leaks. Inspect engine accessory mountings, oil filter, visible portions of oil coolers, oil and fuel lines and Oilgear hydraulic unit for indications of leaks. Trace all leaks to source and correct or report them.

(6) ITEM 7, ENGINE WARM-UP. Test for hydrostatic lock by hand cranking; minimum test 50 revolutions with hand crank. Turn on master battery switch and fuel valves and start engine, using all starting precautions outlined in paragraph 17 b and c. Note action of starting mechanism, particularly whether starter has adequate cranking speed and engages and disengages properly without unusual noise. If oil pressure is not indicated immediately, stop engine and correct or report trouble. Set hand throttle so engine will idle at 800 revolutions per minute during warm up (par. 17 b), see that

TM 9-755
41

Preventive Maintenance Services

throttle operates freely and proceed with following Before-operation Services.

(7) ITEM 8, PRIMER. While starting engine, (item 7) observe if primer functions satisfactorily and inspect for loose lines or brackets and traces of leaks.

(8) ITEM 9, INSTRUMENTS.

(a) Engine Oil Pressure Gage. Gage should register a minimum of 50 pounds at idling speed and 50 to 90 pounds at operating speed. If pressure is too low, stop engine and correct or report trouble.

NOTE: *Gage may register as little as 15 pounds at 700 revolutions per minute if oil temperature is above 190°F.*

(b) Ammeter. The ammeter should show high charge until generator has restored to batteries current used in starting the engine; then continue to register slight charge or zero with lights and accessories turned off.

(c) Engine Oil Temperature Gage. Gage should indicate a rise to 100° F before the vehicle is moved and not exceed 190° F under normal operating conditions.

(d) Tachometer. Tachometer should indicate engine revolutions per minute and revolution counter should register accumulating revolutions.

(e) Fuel Gage. Gage should register amount of fuel in each tank when selector switch is used. Normally tanks will be filled before operation and gage should register "F."

(f) Converter Oil Warning Light. Warning (red) light on instrument panel will light when converter is above normal operating temperature. A test switch at left of the warning light is provided to test system. Push in test switch button to be sure warning light is in working order.

(g) Radio, Interphone, and Antenna. Test for proper functioning of radio and interphone. Inspect for security of antenna—look for breaks.

(9) ITEM 10, SIREN AND WINDSHIELD WIPER. If tactical situation permits, test siren for proper operation and tone. When used, inspect driver's hood windshield wiper blade and arm to see they are in good condition and securely attached. Start wiper motor and observe whether blade operates through its full stroke and contacts windshield surface evenly.

(10) ITEM 11, GLASS. Clean all vision device glass and inspect for damage.

(11) ITEM 12, LAMPS (LIGHTS). Clean all lights and examine for looseness and damage. If tactical situation permits, operate all switches and observe if lamps respond satisfactorily.

(12) ITEM 13, WHEEL AND FLANGE NUTS. See that all drive sprocket, compensating wheel, track wheel and support roller assembly and mounting nuts are present and secure.

TM 9-755
41

Part Three—Maintenance Instructions

(13) ITEM 14, TRACKS AND TIRES. Inspect tracks for damage, loose link pin lock key nuts, and see if track tension is satisfactory, paragraph 132 b. Examine track wheel and support roller tires for flat spots or unusual cuts on treads or separation of tires from wheels and rollers. Remove all stones or foreign objects from between tracks, wheels and rollers.

(14) ITEM 15, SPRINGS AND SUSPENSION. Examine torsion bars, suspension arms, volute springs, compensating links and shock absorbers to see they are in good condition and securely mounted.

(15) ITEM 16, STEERING BRAKE LINKAGE. Inspect linkage to see it is in good condition, securely mounted and connected and operates freely. Pull steering brake levers back and see if they meet resistance evenly with approximately one-third ratchet travel in reserve.

(16) ITEM 17, FENDERS AND SAND SHIELDS. Examine fenders and sand shields to see they are in good condition and securely mounted.

(17) ITEM 18, TOWING CONNECTIONS. Inspect pintle hook, and towing shackles or hooks to see they are in satisfactory condition, and tow cable is serviceable and secure.

(18) ITEM 19, HULL AND TARPAULIN. Examine hull for damage, loose attachments and proper operation of driver's entrance and hull escape doors. See that driver's seats are secure and adjustment mechanism operates and latches properly. Inspect tarpaulin and when so equipped, camouflage net, to see they are in good condition and securely attached or stowed.

(19) ITEM 20, DECONTAMINATORS. Inspect decontaminators for closed valve, full charge and secure mounting.

(20) ITEM 21, TOOLS AND EQUIPMENT. Check tools and equipment to see all items are present, in good condition and properly mounted or stowed.

(21) ITEM 22, ENGINE OPERATION. Engine should idle smoothly at 700 revolutions per minute and respond promptly to controls. Accelerate engine several times after it has reached normal operating temperature step (8) (c) above, and note any unusual noise, unsatisfactory operating characteristics or excessive exhaust smoke.

(22) ITEM 23, DRIVER'S PERMIT AND FORM NO. 26. The driver must have his driver's permit on his person, and must be sure that all vehicles and equipment Technical Manuals, Lubrication Orders, Accident-Report Form No. 26 and W. D., A.G.O. Form No. 478, are present, legible and properly stowed. Identification plates attached to ordnance materiel and component assemblies for identification purposes, containing vehicle name, number, manufacturer, instructions and cautions, are made of steel coated with clear lacquer as a rust-preventive measure. When identification plates are found to be in a rusty condition, they should be cleaned thoroughly and heavily coated with additional applications of clear lacquer.

TM 9-755
42

Preventive Maintenance Services

42. DURING-OPERATION SERVICE.

a. **Observations.** While vehicle is in motion, listen for any sounds such as rattles, knocks, squeals, or hums that may indicate trouble. Look for indications of smoke from any part of the vehicle. Be alert for odors indicating overheated components or units (such as generator or brakes), leaks in fuel system or exhaust system, or other trouble. When brakes are used, gear shifted, or vehicle turned, consider this a test and note any unsatisfactory or unusual performance. Watch instruments constantly for unusual behavior indicating possible trouble in systems to which they apply.

b. **Procedures.** During-operation Services consist of observing items listed below according to procedures following each item, and investigating any indications of serious trouble. Note minor deficiencies to be corrected or reported at earliest opportunity, usually next scheduled halt. NOTE: *Before vehicle is moved, be sure gun breach traveling lock and turret lock are applied.*

(1) ITEM 26, STEERING BRAKES. Steering brakes should be in released position when vehicle is moving straight ahead. Note any tendency of vehicle to lead to one side. This usually indicates brake on leading side is tight or track tension is unequal. With vehicle in motion, apply each brake independently and observe if there is normal response without excessive lever travel. Apply both brakes and observe if brakes stop vehicle effectively with approximately one-third lever travel in reserve. With vehicle stopped and levers pulled back for parking, thumb latches should hold levers in applied position.

(2) ITEM 29, TORQMATIC TRANSMISSION. The transmission control lever should move into all positions smoothly, and should operate without unusual noise or vibration, and not slip out of position during operation. Vehicle should start satisfactorily from standing, in any speed range. In low speed range on level ground, vehicle should start to move when engine reaches 800 to 1,000 revolutions per minute.

(3) ITEM 30, TRANSFER. Transfer shifter lever should operate smoothly without looseness or unusual vibration, gears should operate without unusual noise, and not slip out of mesh during operation.

(4) ITEM 31, ENGINE AND CONTROLS. Be on the alert for deficiencies in engine performance such as lack of power, misfiring, unusual noise or stalling, indications of engine oil overheating or unusual exhaust smoke. Accelerate and decelerate engine to see that controls operate freely and are properly adjusted. If radio noise during operation of engine is reported, the driver will cooperate with radio operator in locating interference (par. 65).

(5) ITEM 32, INSTRUMENTS. Observe readings of all pertinent instruments frequently during operation, to see if they indicate proper function of units to which they apply. Speedometer should indicate vehicle speed and odometer register accumulating mileage.

TM 9-755
42-43

Part Three—Maintenance Instructions

(6) ITEM 34, RUNNING GEAR. Listen for any unusual noise from tracks, wheels and rollers, and suspension units.

(7) ITEM 36, (M18, 76MM GUN MOTOR CARRIAGE ONLY) GUNS: MOUNTINGS, ELEVATING, TRAVERSING AND FIRING CONTROLS. NOTE: *Before following tests are performed, be sure gun traveling lock and turret traversing lock are released.* Test both manual and hydraulic turret traversing mechanisms and gun elevating and firing controls to be sure all mechanisms respond properly. CAUTION: *After completing tests be sure to apply gun traveling lock and turret traversing lock.*

43. AT-HALT SERVICE.

a. **Importance.** At-halt Services may be regarded as minimum maintenance procedures, and should be performed under all tactical conditions even though more extensive maintenance services must be slighted or omitted altogether.

b. **Procedures.** At-halt Services consist of investigating any deficiencies noted during operation, inspecting items listed below according to procedures following items, and correcting any deficiencies found. Deficiencies not corrected should be reported to designated individual in authority.

(1) ITEM 38, FUEL AND OIL. Make sure there is adequate fuel and oil to operate vehicle to next scheduled stop; replenish as supply and tactical situation permits.

(2) ITEM 39, TEMPERATURE; HUBS, TRANSFER CASE AND FINAL DRIVE. Place hand cautiously on each track wheel and support roller hub to see if abnormally hot. Check transmission, transfer case and final drives for overheating or excessive oil leaks. Investigate or report any unusual condition noted during operation.

(3) ITEM 40, VENTS. Check vents of transfer case, transmission and final drives to see they are present, secure and whether damaged or clogged.

(4) ITEM 41, PROPELLER SHAFTS. Investigate or report any unusual noise or vibration in propeller shaft assembly noticed during operation.

(5) ITEM 42, SUSPENSION. Look for broken, loose or damaged shock absorber links, suspension arms and volute springs. Remove stones and foreign objects lodged in suspension system or between track wheels.

(6) ITEM 43, STEERING BRAKE LINKAGE. Examine steering brake linkage for damage or looseness, and investigate or report any irregularities noted during operation.

(7) ITEM 44, WHEEL AND FLANGE NUTS. Examine all track wheels, tracks, support rollers, and compensating wheels to see they are secure, not damaged and track tension is satisfactory. Clean out stones and trash from tracks and wheels.

(8) ITEM 46, LEAKS. Look through grille into engine compart-

TM 9-755
43-44

Preventive Maintenance Services

ment, beneath vehicle, and in fighting compartment for indication of leaks. See if oil is leaking from oil tanks, oil coolers, filters, pipes, gear cases or oil gear hydraulic unit.

(9) ITEM 47, ACCESSORIES AND DRIVES. Examine accessories and drives to see they are securely mounted and drive belts are properly adjusted. Examine radio suppression capacitors and bond straps to see if in good condition and securely connected.

(10) ITEM 48, AIR CLEANERS. When operating under extremely dusty or sandy conditions, inspect air cleaners and breather caps at each halt to see if in condition to deliver clean air properly. Service as necessary, according to instructions in lubrication order, paragraph 38.

(11) ITEM 49, FENDERS AND SAND SHIELDS. Inspect fenders and sand shields for damage or looseness.

(12) ITEM 50, TOWING CONNECTIONS. See that towing connections are properly fastened and securely locked. Examine cable for frayed or broken strands. Make sure supports hold cable so as to prevent chafing.

(13) ITEM 51, HULL, TARPAULIN AND/OR CAMOUFLAGE NET. Inspect for damage to hull and attachments. See that entrance and escape doors operate freely and that tarpaulin, when used, and camouflage net are in good condition and securely attached or stowed.

(14) ITEM 52, GLASS. Clean all vision devices and light lenses and inspect for damage and secure mounting.

44. AFTER-OPERATION AND WEEKLY SERVICE.

a. Purpose. After-operation servicing is particularly important because at this time the crew inspects vehicle to detect any deficiencies that have developed, and to correct those they are permitted to handle. They should promptly report results of the inspection to designated individual in authority. If this schedule is performed thoroughly, the vehicle should be ready to roll again on a moment's notice. The Before-operation Service, with a few exceptions, is then necessary only to ascertain if vehicle is in same condition in which it was left upon completion of After-operation Service. After-operation Service should never be entirely omitted, even in extreme tactical situations, but may be reduced to bare fundamental services outlined for At-halt Service, if necessary.

b. Procedures. When performing After-operation Service the crew must remember and consider any irregularities noticed in Before-operation, During-operation, and At-halt Services. After-operation Service consists of inspecting and servicing following items. Those items of After-operation Service marked by an asterisk (*) require additional Weekly Services, procedures for which are indicated in step *(b)* of each applicable item.

(1) ITEM 56, INSTRUMENTS. Before stopping engine, note if instruments indicate proper function of units to which they apply,

TM 9-755
44

Part Three—Maintenance Instructions

and investigate or report any unusual operating conditions of instruments noticed during operation.

(2) ITEM 55, ENGINE OPERATION. Accelerate and decelerate engine, noting any unusual noise or irregular performance. Investigate any deficiencies noted during operation and correct or report them. CAUTION: *Allow engine to run at 800 revolutions per minute 4 to 5 minutes before stopping.* Stop engine by operating fuel shut-off. Open master battery switch and close fuel valves.

(3) ITEM 54, FUEL AND OIL. Refill all fuel and oil tanks. Follow instructions in lubrication order, paragraph 37 when adding oil in filling auxiliary generator fuel tank on M18. CAUTION: *Do not fill fuel and oil tanks to overflowing. Allow room for expansion. See that any fuel or oil used from spare supply cans is replenished.*

(4) ITEM 57, SIREN AND WINDSHIELD WIPERS. If tactical situation permits, test siren for proper operation and tone. If driver's hoods are in use, see that wipers operate properly and inspect blades and arms and glass for good condition and secure mounting.

(5) ITEM 58, GLASS. Clean all vision devices and inspect for damage.

(6) ITEM 59, LAMPS (LIGHTS). Clean all light lenses and examine for broken parts and security of mountings. If tactical situation permits, turn on light switches to see all lamps operate properly.

(7) ITEM 60, FIRE EXTINGUISHERS. Inspect visible lines and horns and fixed and portable cylinders, to see they are in good condition, secure and not leaking. See that horns are not clogged and that fixed horns are aimed properly. If extinguishers have been used or valves have been opened, report them for exchange or refill.

(8) ITEM 61, DECONTAMINATOR. Inspect decontaminator for full charge, damage, and secure mounting. Shake to check contents.

(9) ITEM 62 *BATTERY.

(a) Inspect battery for leaks or damages and see it is securely connected and mounted.

(b) *Weekly.* Clean battery and carrier and inspect for loose or corroded terminals. If terminals are corroded, remove, clean and apply a thin film of grease. Add clean water, if needed, to bring level to ½ inch above plates. NOTE: *Do not add water until battery has cooled off after operation. In freezing temperatures, do not add water until just before vehicle is to be operated.*

(10) ITEM 63, *ACCESSORIES AND BELTS.

(a) Examine generator, starter, oil coolers, auxiliary generator on M18, and blower to see if in good condition, securely mounted and oil cooler lines are not leaking.

(b) *Weekly.* Inspect all accessories to see they are securely mounted and not damaged. Clean all insects and trash from, in, and around accessible oil cooler core air passages.

(11) ITEM 64, *ELECTRICAL WIRING.

(a) Examine all accessible wiring and wiring conduits to see they

Preventive Maintenance Services

are securely supported, connected, and not damaged. Be sure engine wiring harness coupling rings and nuts are secure and bonding connections for radio noise suppression are clean and secure.

(b) Weekly. Check all accessible wiring connections to see they are clean and securely connected. Inspect radio and communication equipment, mountings, connections, and bonding to see if in good condition, clean and secure.

(12) ITEM 65, AIR CLEANERS AND BREATHERS. Inspect to see air cleaners and breathers are in good condition, securely mounted, connected and not leaking. Examine for excessive dirt and proper oil level. Remove air cleaner and crankcase breather reservoir and elements, and service according to instructions in lubrication order, paragraph 38.

(13) ITEM 66, *FUEL FILTER (ON M18 ONLY).

(a) Examine auxiliary generator engine fuel filter to see it is in condition to deliver clean fuel as necessary.

(b) Weekly. Remove sediment bowl and screen and wash in dry-cleaning solvent, reassemble, using new gasket and check for leaks.

(14) ITEM 67, ENGINE CONTROLS. Inspect accelerator, hand throttle and fuel shut-off controls, to see they are securely mounted, not excessively worn, and that they operate freely. Stop engine.

(15) ITEM 68, TRACKS AND TIRES. Examine tracks for damage, excessive looseness, broken or missing link pin lock nuts, and tracks for proper adjustment (par. 132 *b*). Inspect track wheel, support roller, and compensating wheels, to see if in good condition and tires are not separated from wheels. Remove all objects lodged in track and from between tracks and suspension units.

(16) ITEM 69, SUSPENSIONS. Check suspension arms, shock absorbers, compensating arms, and volute springs, for looseness or damage. Check suspension assembly and mounting nuts and screws to see they are present and secure.

(17) ITEM 70, STEERING BRAKE LINKAGE. Inspect levers, latches, linkage and cross shafts for good condition, free operation, adequate lubrication, and secure mounting. Investigate any irregularities noted during operation.

(18) ITEM 71, *PROPELLER SHAFT (WEEKLY ONLY). Examine shaft and universal joints to see if in good condition, secure, and joints are not leaking excessively.

(19) ITEM 72, *VENTS AND BREATHER CAPS.

(a) See that transmission, final drive, transfer case and oil tank vents are not clogged and securely mounted.

(b) Weekly. Remove engine crankcase breather element and transmission, final drive, transfer case and oil tank breathers; wash in dry-cleaning solvent, re-assemble and install securely.

(20) ITEM 73, LEAKS—GENERAL. Examine hull under transmission and transfer case, around pipes, filters and on floor of engine

TM 9-755
44

Part Three—Maintenance Instructions

compartment for indication of fuel or oil leaks. On M18 look at Oilgear hydraulic unit and pipes for indication of oil leaks.

(21) ITEM 76, FENDERS AND SAND SHIELDS. Inspect fenders and sand shields for looseness and damage.

(22) ITEM 77, TOWING CONNECTIONS. Examine pintle hook and towing shackles or hooks to see if in good condition and operate properly. On M39, see all electrical towing connections are in good condition.

(23) ITEM 78, HULL AND TARPAULIN. Inspect entire hull for damage, be sure top and floor escape doors are alined and operate properly. See that all drain plugs and inspection plates are in place and secure. Examine tarpaulin and, when so equipped, camouflage net, to see if in good condition and securely attached or stowed.

(24) ITEM 80, VISION DEVICES. Inspect periscope prisms and windows to see if in good condition, clean, secure in holders, and see that holders are securely mounted. See that lever and locking devices operate freely and are not excessively worn. Check spare prisms and windows also their stowage boxes to see if in good condition, clean and secure. CAUTION: *Prisms should be cleaned only with a soft cloth or brush.*

(25) ITEM 81, TURRET AND GUNS: MOUNTINGS, ELEVATING, TRAVERSING AND FIRING CONTROLS. On M18 gun motor carriage only, be sure gun traveling lock and turret lock are released and proceed as follows: Check all mounted guns to see if secure in their mounts, clean, lightly oiled and, in condition for immediate use. Be sure gun elevating mechanism and firing controls operate properly. Examine exposed wiring to see that it is securely connected, and packing glands, oil lines and drain plugs to see that they are not leaking. Inspect both manual and hydraulic traversing mechanism to see it is in good condition and operates properly. On M39, Armored Utility Tractor only, inspect the cal. .50 gun, seats, grab rails and attachments to see if in good condition and securely mounted.

(26) ITEM 82, *TIGHTEN.

(a) Tighten any unit mounting and assembly nuts or screws where inspection has indicated the necessity.

(b) *Weekly.* Tighten sprocket, compensating and track wheel flange nuts, track link pin lock nuts, universal joint flange, tool and equipment mounting, or any other item where inspection or experience indicates the necessity on a weekly or mileage basis.

(27) ITEM 83, *LUBRICATION.

(a) Oil or grease all points of vehicle where inspection indicates the necessity or indicated on the lubrication Order, paragraph 38, as needing daily lubrication.

(b) *Weekly.* Lubricate all points of vehicle indicated on the lubrication Order, paragraph 38 as necessary on a weekly or mileage basis.

(28) ITEM 84, *CLEAN VEHICLE AND ENGINE.

Preventive Maintenance Services

(a) Remove all refuse from interior of vehicle. Wipe up oil or fuel drippings from driving and engine compartments and on M18, turret compartment. See that engine compartment grilles are clear of obstructions.

(b) Weekly. Wash exterior of vehicle and remove all dirt and mud. If washing is impractical, wipe as clean as possible and watch for bright spots on exterior finish that would cause glare.

(29) ITEM 85, *TOOLS AND EQUIPMENT.

(a) Check tools and equipment stowage lists, Section III, to be sure all items are present, in serviceable condition and properly stowed or mounted.

(b) Weekly. Clean tools and equipment; mount or stow securely in proper location on, or in, vehicle.

45. ORGANIZATIONAL MAINTENANCE (SECOND ECHELON).

a. **Frequency.** The frequency of preventive maintenance services outlined herein is considered a minimum requirement for normal operation of vehicles. Under unusual operating conditions such as extreme temperatures, severe dust, sandy or extremely wet terrain, it may be necessary to perform certain maintenance services more frequently.

b. **First Echelon Participation.** The crews should accompany their vehicles and assist mechanics while periodic second echelon preventive maintenance services are performed. Ordinarily the vehicle should be presented for a scheduled preventive maintenance service in a reasonably clean condition; that is, it should be dry, and not caked with mud or grease to such an extent that inspection and servicing will be seriously hampered. However, vehicle should not be washed or wiped thoroughly clean, because certain types of defects, such as cracks, leaks, and loose or shifted parts or assemblies, are more evident if surfaces are slightly soiled or dusty.

c. **Sources of Additional Information.** If instructions other than those contained in general procedures in subparagraph d, or the specific procedures in subparagraph i, which follow, are required for proper performance of a preventive maintenance service or for correction of a deficiency, they may be secured from other sections of this manual or from designated individual in authority.

d. **General Procedures.** These general procedures are basic instructions which are to be followed when performing services on items listed in specific procedures. NOTE: *Second echelon personnel must be thoroughly trained in these procedures so that they will apply them automatically.*

(1) When new or overhauled subassemblies are installed to correct deficiencies, care must be taken to see if clean, correctly installed, and properly lubricated and adjusted.

(2) When installing new lubricant retainer seals, a coating of the lubricant should be wiped over sealing surface of lip of seal. When

TM 9-755
45

Part Three—Maintenance Instructions

new seal is a leather seal, it should be soaked in SAE 10 engine oil at least 30 minutes. The oil should be warm, if practical. Then, leather lip should be worked carefully by hand before installing seal. Lip must not be scratched or marred.

e. Definition of Terms. Refer to paragraph 39 h.

f. Special Services. These are indicated by repeating item numbers in columns which show interval at which services are to be performed and show that the parts or assemblies are to receive certain mandatory services. For example, an item number in one or both columns opposite a TIGHTEN procedure, means actual tightening of object must be performed. The special services include:

(1) ADJUST. Make all necessary adjustments in accordance with pertinent section of this manual, special bulletins, or other current directives.

(2) CLEAN. Clean units of vehicle with dry-cleaning solvent to remove excess lubricant, dirt, and other foreign material. After parts are cleaned, rinse in clean solvent and dry thoroughly. Take care to keep parts clean until reassembled and be certain to keep cleaning solvent away from rubber or other material which it will damage. Clean protective grease coating from new parts since this material is usually not a good lubricant.

(3) SPECIAL LUBRICATION. This applies both to lubrication operations that do not appear on vehicle lubrication order and to items that do appear on order but should be performed in connection with maintenance operations if parts have to be disassembled for inspection or service.

(4) SERVE. This usually consists of performing special operations, such as replenishing battery water, draining and refilling units with oil, and changing or cleaning the oil filter, air cleaner, or cartridges.

(5) TIGHTEN. All tightening operations should be performed with sufficient wrench torque (force on wrench handle) to tighten unit according to good mechanical practice. Use a torque-indicating wrench where specified. Do not overtighten, as this may strip threads or cause distortion. Tightening will always be understood to include correct installation of lock washers, lock nuts, lock wire, or cotter pins provided to secure tightening.

g. Special Conditions. When conditions make it difficult to perform all preventive maintenance procedures at one time, they can sometimes be handled in sections, planning to complete all operations within the week if possible. All available time at halts and in bivouac areas must be utilized, if necessary, to assure maintenance operations are completed. When time is limited by tactical situation, items with Special Services in columns, should be given first consideration.

h. Work Sheet. The numbers of the preventive maintenance procedures that follow are identical with those outlined on W.D., A.G.O. Form No. 462, which is the "Preventive Maintenance Service Work Sheet for Full Track and Tank-like Wheeled Vehicles." Certain items on

Preventive Maintenance Services

work sheet that do not apply to this vehicle are not included in procedures in this manual. In general, the numerical sequence of items on work sheet is followed in manual procedures, but in some instances there is deviation for conservation of mechanic's time and effort.

i. **Specific Procedures.** Procedures for performing each item in semi-monthly and monthly maintenance procedures, whichever shall occur first, are described in following chart. Each page of chart has two columns at its left edge corresponding to monthly and the semi-monthly maintenance respectively. Very often a particular procedure does not apply to both scheduled maintenances. In order to determine which procedure to follow, look down column corresponding to maintenance due, and wherever an item number appears, perform operations indicated opposite the number. NOTE: *Those procedures preceded by an asterisk (*) require additional services at each third monthly operation.*

MAINTENANCE		
Monthly	Semi-Monthly	
		ROAD TEST
		NOTE: *When the tactical situation does not permit a full road test, perform those items which require little or no movement of vehicle. When a road test is possible, it should be for preferably 4 and not over 6 miles.*
1	1	Before-operation Service. Perform Before-operation Service as outlined in paragraph 41.
2	2	Instruments and Gages. Check as follows:
		A. ENGINE OIL PRESSURE. Watch oil pressure gage to see it registers sufficient pressure for safe operation of engine (50 pounds minimum at 800 revolutions per minute). Continue to observe oil pressure throughout road test at various speed ranges. CAUTION: *If oil pressure is excessively low at any time, stop engine immediately.*
		B. AMMETER. With a fully charged battery, reading should show charge for only a short time after transfer case is engaged, then slightly above zero with all lights and electrical accessories off. If battery charge is low, ammeter will indicate charge for a longer period of time.
		C. SPEEDOMETER AND ODOMETER. Inspect speedometer for proper miles per hour reading, excessive fluctuation of hand, or unusual noises. See that odometer registers accumulating mileage correctly.
		D. TACHOMETER AND REVOLUTION COUNTER. Inspect for proper reading without excessive hand fluctuation or unusual noise. See that it registers accumulating revolutions properly.
		E. ENGINE OIL TEMPERATURE. Inspect gage to see if it operates properly and if engine oil temperature is

TM 9-755
45

Part Three—Maintenance Instructions

MAINTENANCE	
Monthly	Semi-Monthly
3	3
5	5
7	7
9	9

normal throughout road test. CAUTION: *If oil temperature becomes excessive (more than 190°F) stop vehicle to allow engine to cool and determine cause of heating. Normal temperature is 150 to 190 degrees F.*

F. CONVERTER OIL TEMPERATURE. Observe if converter oil temperature indicator shows red at any time during road test. If so, stop vehicle and determine cause.

G. FUEL GAGE. Operate selector switch to both R and L hand, observe if fuel gage indicates approximate amount of fuel in each tank.

Windshield Wiper and Siren. When in use, check security of mounting. Inspect wiper blades and connections for proper operation. Test siren.

Brakes (Levers, Braking Effect, Steering Action). With vehicle stopped, pull back on steering brake levers. If brakes are properly adjusted, levers should meet resistance with approximately one-third ratchet travel in reserve. If either lever travels beyond this position, turn corresponding brake adjuster to right (clockwise) to tighten.

Accelerate vehicle to a moderate speed in low gear, release accelerator, apply both steering brakes, and observe whether they stop vehicle properly. Apply steering brakes independently and see that they steer vehicle properly.

NOTE: *If vehicle tends to lead in either direction it usually indicates either tight brake adjustment, which will cause drag and excessive wear of brake lining, or unequal track tension, caused by improper adjustment or worn or sprung sprocket.*

Torqmatic Transmission and Transfer Case (Control Lever Action, Vibration, Noise). Shift through each speed range of transmission. Observe if control lever operates properly and if there are unusual vibrations in any speed range that might indicate loose mountings or improper operation.

Engine and Mountings (Idle, Acceleration, Power, Noise, Vibrations and Oil Consumption). Idle. With vehicle stopped, observe if engine runs smoothly at normal idling speed (700 rpm). Throughout road test, observe if there is any tendency of the engine to stall when accelerator is released and hand throttle closed.

ACCELERATION, POWER, VIBRATION AND NOISE. Test engine for normal acceleration, and pulling power in

Preventive Maintenance Services

MAINTENANCE		
Monthly	Semi-Monthly	
		each speed range. While testing in high range, accelerate from low speed with wide-open throttle up to 40 miles per hour and listen for unusual engine noise, "ping" or vibration that might indicate loose, damaged, excessively worn, or inadequately lubricated engine parts or accessories.
		NOTE: *It is not necessary to drive vehicle any great distance at wide open throttle.* During road test, look for excessive smoke from exhaust, or from engine compartment. An abnormal blue smoke at engine exhaust usually indicates excessive oil consumption. At completion of road test, oil supply should be checked to see if engine has been consuming an excessive amount of oil.
14	14	Noise and Vibration (Engine, Mountings, Accessories, Clutch, and Exhaust). While accelerating and decelerating engine, listen for unusual noise in engine or accessories. Notice if there is excessive vibration that may indicate loose engine mountings or accessories.
		CAUTION: *Before stopping engine, allow it to idle at 700 revolutions per minute for 5 minutes to reduce temperature of heads.*
10	10	Unusual Noise (Propeller Shaft, Universal Joints, Differential and Final Drives, Sprockets, Wheels and Tracks). During road test, listen for any unusual noise indicating damaged, defective, or loose parts, or inadequate lubrication.
15	15	Track Tension. Check track tension by placing a ¼-inch spacer on No. 1 and No. 3 track support rollers between tire and track. With spacer in place, No. 2 track support roller should turn freely if track is properly adjusted (par. 132 b).
11	11	Temperatures (Final Drives, Hubs of Sprockets, Track Support Rollers and Track Wheels). After operating, check by hand-feel for any abnormal temperature in above items. NOTE: *If proper location is selected for this check, time will be saved in performing item 12.*
12	12	Gun Elevating and Traversing Mechanism. On M18 only, place vehicle in a position where it is tilted (sidewise) about 10 degrees. Traverse turret through full 360 degree range by both hand and power controls, check for indications of looseness or binding. With gun pointed forward, elevate it through its entire range with hand controls, check for binding, excessive lash, or erratic action.

TM 9-755
45

Part Three—Maintenance Instructions

MAINTENANCE		
Monthly	Semi-Monthly	MAINTENANCE OPERATIONS
13	13	**Leaks.** Open engine compartment door, check within compartment and underneath vehicle for indications of oil or fuel leaks, also check hydraulic traversing unit in M18 only for oil leaks. NOTE: *Do not tighten flexible fittings on fuel and oil lines unless actual leak is noted.*
16	16	**Fuel Tank Pump Test.** Tanks must be at least one-quarter full when testing pumps. Test left tank pump as follows: With engine stopped, turn on fuel pumps switch, open left fuel shut-off valve and note reading on oil dilution pressure gage on bulkhead (fig. 30). Gage should read 5 to 7 pounds. Test right tank pump as follows: Close right fuel shut-off valve, disconnect right fuel pipe at feed pipe check valve and attach fuel pump tester to pipe. Turn on tank pump switch, open right fuel shut-off valve and note reading on tester. Pressure should be same as for left tank described above. If tester is not available, test pump with hose (par. 51 e (2)).
17	17	**Crankcase; Leaks.** Stop engine and open master battery switch. Inspect crankcase, accessory mounting, and oil tank for leaks and see if oil is at correct level.
17	17	**SERVE.** Immediately after stopping the engine and completing the above inspection, drain engine oil tank and refill with specified oil according to instructions in lubrication notes, paragraph 38.
		NOTE: *Service oil filter, item 54, before starting engine.*
18	18	**Siren, Paint and Markings, and Shackles.** Inspect siren for condition and security of mounting. Examine towing shackles or hooks for excessive wear. Look for rust spots or polished surfaces on hull that may cause reflections, and all markings to see they are legible.
19	19	**Bottom of Hull (Drain Plugs and Escape Door).** Inspect drain plugs for condition and tightness. See that escape door latch operates properly and is adequately lubricated, and all drain plugs are tight. Apply a few drops of oil to escape door latches.
20	20	**Differential and Final Drives.** Inspect housings for condition and leakage; check lubricant level. See that all assembly and mounting bolts are secure.
		NOTE: *If organizational records indicate change of lubricant is due, drain and refill with specified oil at this time, according to lubrication order, paragraph 37.*

Preventive Maintenance Services

MAINTENANCE		
Monthly	Semi-Monthly	
20		TIGHTEN all external assembly and mounting bolts securely.
21	21	Track, Links and Lock Pins. Inspect to see they are in good condition, correctly assembled, and secure. Pay particular attention to loose or excessively worn links and pins and to alinement of lock key nuts (fig. 163). Tighten and aline all track link pin lock key nuts securely with tension wrench set to 120 foot-pounds. Nuts must be set with two flat sides parallel to track to prevent nuts striking.
22	22	Compensating Idler Wheels. Inspect for good condition, correct assembly and security of mounting, also for excessive wear. Note whether idler bearing seals are leaking excessively.
		NOTE: *Since no vents are used, lubrication will force grease out through seals. This does not necessarily indicate seals are defective.*
22		TIGHTEN all assembly and mounting bolts securely.
23	23	Suspension Arms, Shock Absorber Links, Axle Housings. Examine suspension arms and shafts for excessive grease leaks at axle housing, see that housing cover bolts are secure. Pry up on each track wheel hub to see that torsion bar is not broken. Disconnect shock absorber link and work up and down to see that absorber is operating properly.
23		TIGHTEN all assembly and mounting bolts securely.
24	24	Wheels, Track Support Rollers and Tires. Inspect these items for good condition, correct assembly, and secure mounting. Pay particular attention to see that tire rubber has not separated from rim, and tires are not cut, torn or excessively worn. Inspect for excessive lubricant leaks from bearings.
		NOTE: *Lubrication will force some lubricant through seal. This does not indicate defective seal.*
24		Jack up track wheels as described in paragraph 136 b, and examine bearings for looseness and end-play. Spin wheels and listen for any unusual noise. Tighten all assembly and mounting bolts securely.
25	25	Sprockets (Hubs, Teeth and Nuts). Inspect for good condition and correct assembly and security of mounting bolts. Inspect sprocket teeth for excessive wear and if shaft flange gaskets or oil seals are leaking lubricant ex-

TM 9-755
45

Part Three—Maintenance Instructions

MAINTENANCE	
Monthly	Semi-Monthly
26	26
27	27
28	28
30	
30	
31	
32	
34	34

cessively. If sprocket teeth are excessively worn, sprocket should be replaced or reversed (par. 127).

TIGHTEN assembly and mounting bolts securely.

Track Tension. Check track tension by placing a ¾-inch spacer on No. 1 and No. 3 track support rollers between tire and track. With spacers in place, No. 2 track support roller should turn freely. If necessary, adjust by tightening track adjuster (turn right to tighten, paragraph 132 b). Tighten locking nuts securely.

Top Armor, Turret (M18 Only), Paint and Markings, Caps, Grilles, Doors, Covers and Latches, and Antenna Mast. Inspect to see that these items are in good condition and secure; and door hinges and latches operate properly, are not excessively worn, and are adequately lubricated. Examine paint for rust spots, or polished surfaces that may cause reflections. Check all vehicle markings to see they are legible.

Filler Caps and Gaskets (Fuel and Oil). Inspect to see they are in good condition, secure, and not leaking. Be sure gaskets are in place and serviceable.

Engine Removal (When Required). SERVE. (par. 75). Remove engine on monthly maintenance service, only if inspections made in items 9, 13, 14 and a check of records on oil consumption indicate definite need.

CLEAN. Exterior of engine and dry thoroughly, taking care to keep cleaning solvent away from electrical wiring, terminal boxes and equipment.

Valve Mechanism (Clearances, Lubrication, Cover Gaskets, Rocker Boxes and Push-Rod Housings).

NOTE: *Perform item 31 only when engine is removed from vehicle* (fig. 65). Adjust valve clearances to 0.006 in., also inspect valve tappets, rocker arms, shafts and valve springs to see they are in good condition, correctly assembled and secure, and oil is going to rocker arms and shafts properly. Inspect rocker arm and shafts for excessive wear, or if rocker arm rollers have flat spots. Inspect rocker box covers for condition, for serviceable gaskets and for condition of push rod housings.

Spark Plugs. SERVE. Replace spark plugs with new or properly reconditioned plugs. Space points to 0.018 to 0.020 in.

Generator and Starting Motor. Inspect for good condition and security of mounting; if wiring connections are secure; if generator is correctly alined with drive belt,

Preventive Maintenance Services

MAINTENANCE		
Monthly	Semi-Monthly	
34		and pulley on drive shaft. Be sure that generator armature brush radio suppression capacitors are in good condition, and securely connected.
		*Remove commutator inspection cover and examine commutator for good condition; see that brushes are free in brush holders, clean, not excessively worn; that brush connections are secure, that wires are not broken or chafing. Clean commutator end of generator and starting motor by blowing out with compressed air. Tighten starter and generator mounting bolts securely.
37	37	Magnetos. Inspect condition, security of mounting, and for evidence of oil leaks at mounting pad gaskets. Remove breaker point inspection covers to see points are not pitted, are clean, well alined with mating surfaces, and engaging squarely. Be sure radio suppression shielding ring nuts are securely connected.
37		Adjust magneto breaker point caps to 0.012 in. (par. 88 d). Breaker points requiring adjustment must first be removed and smoothed with proper hone.
38	38	Ignition Wiring and Harness. Inspect for good condition, cleanliness, correct assembly, tight connections, security of mountings and for chafing against other engine parts. Clean all exposed ignition wiring with a dry cloth. NOTE: *Do not disturb connections unless they are actually loose.* Check all radio suppressions shielding ring nuts and shielding support internal-external toothed washers to see they are securely connected and mounted.
39	39	Booster Coil. Examine booster coil and radio suppression shielding conduits to see they are in good contion, clean, securely mounted and connected.
40	40	Engine (Oil Pumps, Sump, Oil Screens and Lines, Accessory Case, Crankcase, Fuel Screens and Lines, Control Linkage). Inspect to see these items are in good condition and secure; oil is not leaking from oil pumps, sump, lines, accessory case, or crankcase. Inspect fuel lines to engine for leaks. NOTE: *Do not tighten Sealflex fittings on fuel and oil lines unless leak is detected.* Remove oil pump screen, clean thoroughly in solvent, dry and reinstall.
40		SERVE. Drain off old oil from the engine sump.
40		TIGHTEN accessible assembly, mounting bolts and screws securely.
42	42	Breather Caps. Remove cleaner element from engine breather (fig. 57), clean element and body in solvent

Part Three—Maintenance Instructions

MAINTENANCE		
Monthly	Semi-Monthly	
		and dry. Inspect for good condition. Dip element in engine oil and reinstall securely. Remove breather units from final drive, transmission and transfer case; wash in dry-cleaning solvent and reinstall.
43	43	**Air Cleaners.** Inspect air cleaner parts (fig. 37) to see if in good condition. Clean reservoirs and elements (both) in solvent and drain. Fill reservoirs to correct level with clean oil, lubrication order (par. 38). Reassemble cleaners, making certain all gaskets are in good condition and in place, giving special attention when mounting, to see that cleaners are pressed firmly in place against air duct seals and securely fastened.
44	44	**Carburetor, Throttle, Linkage, Governor, and Primer.** Inspect for good condition, correct assembly, and security of mounting. See that carburetor does not leak; throttle control linkage is not excessively worn, is properly adjusted to fully open throttle, operates freely and that governor is properly sealed. Also see that lines of priming system are in good condition, secure, and not leaking. Remove screen from carburetor fuel inlet; clean in solvent, dry and reinstall (par. 93).
45	45	**Manifolds (Intake and Exhaust).** Inspect to see manifolds and gaskets are in good condition, correctly assembled, and secure. Tighten intake pipes at their flanges, clamps, and gland-packing nuts. Check for indications of leaks by looking for excessive carbon streaks.
46	46	**Cylinders.** Inspect to see they are in good condition and secure and if there are indications of oil leakage or blow-by around studs or gaskets. Inspect cylinders to see if the cooling fins are clogged. CAUTION: *Cylinder hold-down nuts should not be tightened unless there is a definite indication of looseness or leaks. If tightening is necessary, use a torque wrench and tighten to 350 to 375 inch-pounds tension.*
46		CLEAN excess deposits of dirt or grease from between and around cylinder cooling fins.
47	47	**Engine (Cowling, Air Deflectors, Flywheel, Fan, Steady Bar and Support Beam).** Inspect to see they are in good condition, correctly assembled, and securely mounted. *Be sure cowling inspection covers are all in place.*
47		TIGHTEN all accessible mounting and assembly bolts or screws securely.
50	50	**Accessory Drives (Belts, Pulleys, Shafts, and Couplings).** Inspect to see they are in good condition, correctly assembled, and secure; generator and blower drive

Preventive Maintenance Services

MAINTENANCE		
Monthly	Semi-Monthly	
		belts and pulleys are alined, not excessively worn; and if belts are frayed, oil-soaked, or bottoming in pulleys. Adjust all drive belts to ½-inch deflection when depressed with finger tips. Lock adjustment devices securely.
51	51	**Engine Compartment, Bulkhead, and Control Linkage.** Check to see that engine compartment, including bulkhead, is in good condition and clean, and control linkage in engine compartment in good condition and securely connected and mounted.
51		CLEAN. Clean engine compartment thoroughly.
53	53	**Fuel (Tanks, Pipes and Pumps).** Inspect to see they are in good condition, correctly assembled, securely mounted and not leaking. Examine right and left fuel tank to hull radio suppression bond straps to see they are in good condition and securely connected.
53		TIGHTEN all fuel tank mountings and brackets securely. NOTE: *Do not tighten Sealflex pipe fittings unless actual leak is detected.*
53		SERVE. With tanks nearly empty, drain water and sediment from each fuel tank by removing drain plugs and allowing fuel to drain until it runs clean. Tighten plugs securely and be sure they do not leak.
52	52	**Engine Oil (Tank, Cooler, Pipes and Fittings).** Inspect to see they are in good condition, correctly assembled, securely mounted, and not leaking. Check oil level with lever indicator and inspect sample of oil on indicator for grit, water, or dilution. See that filler cap and gasket are in good condition and sealed properly. CAUTION: *Do not tighten Sealflex oil or fuel pipe fittings unless actual leak is detected.*
52		SERVE. Drain engine oil tank, and clean tank (par. 84 b). When thoroughly drained, reinstall plug and fill tank to proper level with specified engine oil. See lubrication notes, paragraph 38.
54	54	**Engine Oil Filter.** Inspect oil filter to see it is in good condition, secure, and not leaking. Turn handle clockwise several complete turns, remove drain plug, and drain off contents.
54	54	CLEAN AND SERVE. Remove oil filter cover and element and, without disassembling, clean in solvent. Dry and inspect for damage. If element is serviceable, reinstall; if not, replace entire filter assembly.

TM 9-755
45

Part Three—Maintenance Instructions

MAINTENANCE		
Monthly	Semi-Monthly	
56	56	**Oil Coolers, Filter, Screens, and Pipes (On Transfer Unit, Transmission, Differential).** Examine oil coolers, including cores and connecting pipes to see they are in good condition, secure and do not leak. Clean all insects and trash from core air passage. Drain differential oil strainer (par. 124 a), F, figure 35. Turn transfer case oil filter handle several complete turns clockwise. (T, fig. 37).
56		SERVE. Remove differential oil strainer screen and transmission oil screen and transfer filter element, clean in dry-cleaning solvent. See lubrication order, (par. 38). CAUTION: *Do not tighten Sealflex oil pipe fittings unless actual leak is detected.*
57	57	**Exhaust Pipes and Mufflers.** Inspect to see they are in good condition, securely assembled and mounted.
57		TIGHTEN all mounting bolts and connections securely.
58	58	**Engine Mountings.** Examine mountings and radio suppression bond straps from engine to mounting ring to see they are in good condition and secure.
58		TIGHTEN all mounting bolts securely.
60	60	**Fire Extinguisher System (Cylinders, Valves, Pipes, Horns, and Mountings).** Remove fixed cylinders and weigh to determine charge. If a cylinder weighs less than 9 pounds plus weight of cylinder stamped on head, it should be replaced. Inspect valves and general condition, blow out with compressed air. Reinstall cylinders and tighten securely. Inspect control cables and handles to see they are in good condition and not corroded or frozen. Inspect lines and horns for condition and security of mounting. Disconnect control cables and lubricate as needed to assure free operation. Be sure horns are aimed correctly. CAUTION: *Before working on fixed fire extinguisher system, safety lock pins must be installed in valve heads on tanks. Pins should be removed when work is completed to put system in operating condition.* Tighten all assembly and mounting bolts and screws.
61		**Engine (Install Mountings, Lines and Fittings, Wiring, Control Linkage, and Oil Supply).** SERVE. If engine was removed for repair or replacements, reinstall at this time (see par. 76). Tighten mountings and radio suppression bond straps securely and properly connect all fuel and oil pipes, wiring, and control

Preventive Maintenance Services

MAINTENANCE		
Semi-Monthly	Monthly	
63	63	linkage which were disconnected when engine was removed. **Batteries (Cables, Hold-Downs, Carrier, Gravity and Voltage, Switch and Fuel Valves).** Remove battery and clean and dry thoroughly. Inspect battery hold-down clamps for good condition. Clean battery carrier and paint if corroded, clean battery terminals, scrape until bright, and grease lightly. Make a high rate discharge test of battery to see if all cells are in satisfactory condition. A true test cannot be made if gravity reading is below 1.225. If cells vary more than 30 per cent, report condition. Bring electrolyte up to level of ½ inch above plates by adding distilled or clean water. Reinstall battery and connect cables securely.
64	64	**Accelerator (Linkage).** Examine accelerator and connecting linkage to see it is in good condition, opens throttle fully, is securely connected and operates freely.
65	65	**Starter, Primer, and Instruments.** Observe all starting precautions outlined in paragraph 17 b and c. Start engine, observing primer operation; note whether action of starting motor is satisfactory, particularly if starter drive engages and operates properly without unusual noise and has adequate cranking speed, and if engine starts readily. As engine starts, see that all instruments operate properly, and particularly oil pressure and ammeter indications are satisfactory. NOTE: *Ammeter will not show charge, unless transfer case gears are engaged.*
66	66	**Leaks (Engine Oil and Fuel).** Inspect within engine compartment to see if oil is leaking from engine, oil filter, or lines; and if there are leaks from fuel system.
67	67	**Magneto Timing.** Check magneto timing. Set timing to 25 degrees before top dead center (par. 87).
68	68	**Regulator Units (Connections, Voltage, Current, and Cut-out).** Inspect for good condition and see if all connections and mountings are secure. Examine main and auxiliary regulator radio suppression capacitors, to see they are in good condition and securely connected.
	68	**Test.** With unit at normal operating temperature, connect low voltage circuit tester to regulator and see if voltage regulator, current regulator and cut-out properly control generator output. Follow instructions which accompany test instrument.
69	69	**Engine Idle.** Observe if engine idles smoothly at normal idle speed. Adjust speed to 700 revolutions per minute by means of throttle stop screw. Close hand

TM 9-755
45

Part Three—Maintenance Instructions

MAINTENANCE	
Monthly	Semi-Monthly

		throttle all the way. Move mixture adjusting lever in direction which "leans" mixture until engine idle becomes rough due to misfiring. Then slowly move lever in opposite direction to enrich mixture until "roughness" disappears and engine idles smoothly. Do not turn farther than necessary for smooth idle. If this adjustment increases or decreases engine idle speed, reset idle speed to 700 revolutions per minute. NOTE: *Idle adjustment should be made only after engine has reached normal operating temperature.*
71	71	**Fighting Compartment (Seats, Safety Straps, Crash Pads, Stowage Boxes, Ammunition Boxes, Clips and Racks).** Inspect to see these items are in good condition, securely assembled and mounted; fighting compartment is clean, paint in satisfactory condition; adjusting mechanisms of seats operate properly and are adequately lubricated. Pay particular attention whether dividers and shell pads are all present and properly installed in ammunition boxes and racks, and clips have sufficient tension to hold shells securely.
72	72	**Turret Lock, M18 Only.** Examine to see lock is in good condition and secure; turret can be traversed easily when lock is released; and turret can be locked and released properly.
73	73	**Periscopes.** Examine periscope prisms and windows to see if in good condition, clean, secure in holders, and holders are securely mounted; their traversing, elevating, and locking devices operate freely and are not excessively worn. See if periscope window cleaner blades are in good condition and secure. Examine spare prisms and windows and their stowage boxes to see if in good condition, clean and secure. CAUTION: *Prisms should be cleaned only with a soft cloth or brush.*
75	75	**Brakes (Steering Levers, Latches, Linkage, and Shafts).** Inspect steering brake levers, linkage, and shafts to see if in good condition, securely connected, mounted, and not excessively worn. Test latches for proper operation. Apply steering brake levers and observe whether they both begin to meet resistance just before reaching a vertical position and have one-third ratchet travel in reserve when applied.
75	75	TIGHTEN all assembly and mounting bolts securely.
77	77	**Differential, Universal Joints and Breathers.** Examine accessible part of the differential case in driver's compartment to see if in good condition; that all mounting and assembly bolts or cap screws are secure; and there

TM 9-755
Preventive Maintenance Services

MAINTENANCE		
Monthly	Semi-Monthly	
77		are no oil leaks. Inspect breathers to see if in good condition, secure, and not clogged.
77		*Remove sheet metal covers and inspect differential to final drive universal joints to see if in good condition and securely connected.
77		SPECIAL LUBRICATION (UNIVERSAL JOINTS). Remove screw plugs and insert fitting. Fill joint until grease just starts to come out at any one of the seals. Reinstall covers. Remove, clean, and reinstall differential breathers.
77		TIGHTEN all external assembly and mounting bolts and screws.
78	78	Transmission (Vents and Seals). Inspect transmission to see if in good condition, outside parts securely assembled and mounted; and there are no oil leaks. Remove, clean, and reinstall breathers.
78		TIGHTEN all external assembly and mounting bolts and screws securely.
79	79	Transfer Unit (Seals and Breather). Inspect transfer unit to see if in good condition, securely assembled and mounted; note whether oil is leaking from case, pipes or seals. Remove and inspect breather, clean and reinstall.
79		TIGHTEN all external assembly and mounting bolts and screws securely.
80	80	Transmission and Transfer Case Controls and Linkage. Inspect to see transmission control lever and transfer case lever operate properly, are in good condition, correctly assembled, securely connected, and not excessively worn.
81	81	Propeller Shaft (Joints, Alinement, and Flanges). Inspect propeller shaft to see if in good condition, correctly and securely assembled and mounted; universal joints are properly alined, and not excessively worn.
81		TIGHTEN universal joint assembly and companion-flange bolts securely.
82	82	Hand Crank and Cover. Inspect to see if in good condition, secure, and not excessively worn; and crank hole cover or guard is in good condition and secure.
84	84	Compass (Fluid and Lamp). Examine compass to see if in good condition and secure; look for low level or indications of bubbles in the fluid bowl. Fill fluid bowl with ethyl alcohol, if needed. See that compass lamp and switch operate properly.

TM 9-755
45

Part Three—Maintenance Instructions

MAINTENANCE		
Monthly	Semi-Monthly	
85	85	Lamps and Switches (Head, Tail, Blackout, Internal). Test to see that switches and lamps operate properly. Inspect to see if all lamps are in good condition and secure, and check for broken lenses and discolored reflectors.
85		ADJUST headlight beams (par. 151 e).
86	86	Wiring (Junction and Terminal Blocks and Boxes). Inspect to see all exposed electrical wiring and conduits, terminal blocks and boxes are in good condition, well supported and securely connected. Examine all radio suppression bond straps and shielding ring nuts to see if securely connected.
87		Collector (Slip) Ring (Brushes, Heads, Cylinder, and Cover, on M18 Only). With master battery switches off, remove collector ring cover (fig. 246), and examine to see if items are all in good condition and clean. Make certain that brushes contact cylinder properly under normal spring tension; and leads are securely connected and not chafing. Examine all radio suppression units to see if in good condition and securely connected. Reinstall cover securely.
88	88	Radio Bonding (Suppressors, Filters, Condensers and Shielding). See that all units not covered in the foregoing specific procedures are in good condition and securely mounted and connected. Be sure all additional noise suppression bond straps and internal-external-tooth washers listed in paragraph 176 g and h, are inspected for looseness or damage, and see that contact surfaces are clean. If objectionable radio noise from vehicle has been reported, make test in accordance with paragraph 65. If cleaning and tightening of mountings and connections and replacement of defective radio noise suppression units does not eliminate trouble, radio operator will report the condition to the designated individual in authority.
		AUXILIARY GENERATOR
89	89	Engine (Crankcase, Fan and Housing, Cylinder Shield, Mountings, and Exhaust Pipe). Inspect to see if items are in good condition and secure.
90	90	Spark Plug. SERVE. Replace the spark plug with a new or reconditioned one using a new gasket and make sure gap is set to 0.025 inch. Clean adapter baffle. If corroded, replace baffle.

Preventive Maintenance Services

MAINTENANCE	
Monthly	Semi-Monthly
91	91

Magneto (Points, Wiring, and Shield). Inspect to see if items are in good condition, correctly assembled, and securely mounted; interior of magneto and rotor arm are in good condition and clean; and breaker points are clean and not uneven or pitted (see figure 192). Adjust breaker point gap with points fully open to 0.020 inch. Apply a few drops of oil to magneto cam wick.

92	92

Carburetor and Air Cleaner. Inspect to see if in good condition, securely mounted, and not leaking. Close fuel supply valve, remove air cleaner element and screen in carburetor fuel-inlet connection, clean in solvent, and dry thoroughly. Dip air cleaner element in engine oil, drain and reassemble. Reinstall carburetor inlet screen.

93	93

Fuel (Filter, Line, Tank and Cap). Examine to see if in good condition, secure, and not leaking. Clean fuel strainer sediment bowl and screen and reinstall, using new gasket if necessary. Open fuel supply valves.

94	94

Generator (Commutator, Brushes, Control Box, Wiring). Remove brush head cover plate and examine commutator to see if in good condition, clean, and not excessively worn. See that brushes are clean, free in their holders, properly spring loaded, and not excessively worn. Inspect control box and buttons, ammeter, and wiring to see if in good condition, correctly assembled and connected and secure. Examine generator to hull radio suppression bond straps to see if in good condition and securely connected.

94	

*CLEAN. At each third monthly service, clean commutator by placing a strip of flint paper, 2/0 over a block of wood of correct size and with engine running slowly, press flint paper against the commutator until it is clean (fig. 194). Blow out dust with compressed air.

95	95

Operation (Engine, Generator, Ammeter and Leaks). Start engine, observing if it starts easily and runs at normal speed, and listen for any unusual noise. See if generator output is satisfactory. Examine fuel and lubrication systems for leaks with engine running.

ARMAMENT

126	126

Guns, Mounts, Traversing and Elevating Mechanism, and Firing Controls (M18 Only). Inspect to see these items are in good condition, clean, well lubricated, correctly and securely assembled and not excessively worn. Check to see manual and electric firing controls are in

TM 9-755
45

Part Three—Maintenance Instructions

MAINTENANCE	
Monthly	Semi-Monthly

good condition and secure, paying particular attention to wiring, switches, and connections. Operate each firing control, both manual and electric, to see that they function properly.

Operate gun hand elevating controls through entire range to see that they function properly. Traverse turret by hand to see if there is any binding and that turret can be turned through its entire 360-degree range. See if brake is effective when brake release lever is released. Inspect electric hydraulic traversing system, including motor, pump, reservoir, wiring and operating controls to see if in good condition, correctly assembled, secure, operate properly, and are not excessively worn. Be sure radio suppression capacitors on motor brush holders and on motor switch are in good condition and securely connected. Examine hydraulic system for leaks and for proper level of oil. Add oil, hydraulic, if required, according to lubrication order (par. 38). Make an operating check of traversing system by closing motor switch, placing shifting lever in power operating position and turning pump control handle to right, left, and neutral positions to see if traversing mechanism responds properly and any overrun is properly controlled. Tighten all assembly and mounting bolts and screws securely. Be sure ground straps at both ends of drag link (fig. 227 and 228) are securely connected.

127	127	**Recoil Control.** Inspect cylinders for leaks or damage and check level of recoil oil. Add oil, recoil, if required, according to lubrication order (par. 38). NOTE: *Recoil operating checks must be made under firing conditions and in accordance with instructions, paragraph 210.*
128	128	**A. A. Guns, Mounts, Traversing and Elevating Mechanism.** On M18 and M39, inspect to see that they are in good condition, clean, secure, and adequately lubricated; and if mechanism operates freely.
128		TIGHTEN all assembly and mounting bolts securely.
129	129	**Spare Gun Barrels and Parts.** See if they are present, in good condition, and properly stowed.

TOOLS AND EQUIPMENT

| 130 | 130 | **Tools (Vehicle, Kit and Pioneer).** Check standard vehicle tools against stowage lists (par. 6) to see all vehicle and pioneer tools are present, in good condition and properly stowed or mounted. Any tools mounted on outside of vehicle, having bright or polished surfaces should be painted or otherwise treated to prevent glare or reflections. |

TM 9-755
45

Preventive Maintenance Services

Maintenance		
Monthly	Semi-Monthly	
131	131	**Equipment.** Check special equipment items against vehicle stowage list (par. 7) to see if all are present, in serviceable condition and properly stowed or mounted.
132	132	**Spare Track Links.** Inspect to see if all are present, in good condition and properly stowed or mounted.
134	134	**Decontaminators.** Examine to see they are in good condition, secure and fully charged. Make latter check by shaking cylinder.
		NOTE: *The solution must be renewed every 3 months, as it deteriorates.*
135	135	**Fire Extinguisher (portable).** Inspect to see it is fully charged, in good condition and securely mounted. Weigh cylinder to determine if it is fully charged. If it weighs less than 3½ pounds, cylinder should be replaced with fully charged one.
136	136	**Publications and Form No. 26.** Check to see if vehicle manuals, Lubrication Order, and Accident-Report Form No. 26, and MWO and Major Unit Assembly Replacement Record W.D., A.G.O. Form 478 are present, legible, and properly stowed.
137	137	**Vehicle Lubrication.** If due, lubricate in accordance with lubrication order (par. 38), and current lubrication directives, using only clean lubricant and omitting items that have had special lubrication during this service. Replace damaged or missing fittings, vents, flexible lines, or plugs.
138	138	**Modifications (Modification Work Orders Completed).** Inspect vehicle and organizational records to determine that all Modification Work Orders have been properly completed, and entered on W.D., A.G.O. Form 478. Enter any replacement of Major Unit Assembly made at time of this service.
139	139	**Final Road Test.** Make a final road test, rechecking items 2 to 15 inclusive. Confine this road test to the minimum distance necessary to make satisfactory observations. While testing vehicle, operate it in a normal manner. NOTE: *Correct or report any deficiencies found during final road test.*

TM 9-755
46-47

Part Three—Maintenance Instructions

Section XVI

TROUBLE SHOOTING

46. GENERAL.

a. This section contains trouble shooting information and tests to help determine the causes of some of the troubles that may develop in vehicles used under average climatic conditions (above 32° F.). Each symptom of trouble given under individual unit of system is followed by a list of possible causes of the trouble. The tests necessary to determine which one of possible causes is responsible for the trouble are explained after each possible cause.

47. ENGINE.

a. *Reference.* Engine troubles most commonly encountered are described in this paragraph. The subjects covered are listed below for quick reference:

Engine will not turn when hand cranked	Subpar. b.
Engine will turn when hand cranked but not with starter	Subpar. c.
Engine turns but will not start	Subpar. d.
Engine does not develop full power	Subpar. e.
Engine misfires	Subpar. f.
Excessive oil consumption	Subpar. g.
Engine will not stop	Subpar. h.
Engine overheating	Subpar. i.
Engine rich on idle	Subpar. j.
Engine noisy	Subpar. k.

b. **Engine Will Not Turn When Hand Cranked.**

(1) HYDROSTATIC LOCK OR SEIZURE. If engine cannot be turned by hand crank, remove rear spark plugs from two lower cylinders and attempt to turn engine with hand crank to expel excess liquid from combustion chambers and relieve hydrostatic lock. Conditions which cause hydrostatic lock are described in paragraph 19. For hydrostatic lock due to oil, but not caused by improper stopping of engine, clean oil tank outlet check valve (par. 84 d and e). If liquid drained from cylinders is gasoline, replace carburetor (par. 93 i and j).

(2) INTERNAL DAMAGE. If engine cannot be turned by hand crank, with spark plugs removed, seizure due to internal damage is indicated; notify higher authority.

c. **Engine Will Turn When Hand Cranked But Not With Starter.**

(1) NO CURRENT IN BATTERY CIRCUIT. Check battery circuit (par. 58 b).

(2) STARTER SYSTEM INOPERATIVE. Check starter system (par. 49 b).

Trouble Shooting

d. **Engine Turns But Will Not Start.**

(1) INOPERATIVE BOOSTER COIL. The booster coil will cause engine to fire intermittently while being cranked with magnetos turned off, or inoperative. If engine does not fire intermittently, press booster coil circuit breaker button to make sure circuit breaker is set; then listen over engine compartment while another man holds booster switch in "ON" position. Coil will give buzzing sound if operating; if no buzz can be heard, check booster coil (par. 50 b).

(2) INOPERATIVE MAGNETOS. If engine fires on booster coil but not on magnetos, disconnect cable from a spark plug and turn magneto switch to "BOTH" position. Crank engine with starter while holding spark plug cable contact ¼ inch from cylinder or other grounded metal part. If spark does not jump the ¼-inch gap, check magnetos (par. 50 c).

(3) INOPERATIVE ENGINE PRIMER PUMP. Open left tank fuel shut-off valve and turn on fuel tank pumps. Disconnect one primer tube at engine intake pipe and operate primer pump to see if fuel squirts out of tube. If no fuel squirts out of tube, loosen inlet pipe coupling nut at primer pump to see if fuel is reaching pump. If fuel is reaching primer pump, replace pump (par. 97). If no fuel is reaching pump, check left fuel tank pump (par. 51 e (1)).

(4) INOPERATIVE ENGINE FUEL PUMP. Loosen hose clamp and disconnect fuel pump to carburetor coupling hose. Open fuel shut-off valves, turn on fuel tank pumps, turn magneto and booster coil switches to "OFF" positions, and crank engine with starter. If fuel does not flow freely from engine fuel pump, check fuel system (par. 51).

e. **Engine Does Not Develop Full Power.**

(1) USE OF IMPROPER TYPE OF FUEL. Use only U.S. Army Specification 2-103A above freezing; use only U.S. Army Specification 2-103C below freezing.

(2) IMPROPERLY ADJUSTED PEDAL TO THROTTLE LINKAGE. Check linkage, adjust if necessary (par. 94).

(3) OIL TEMPERATURE TOO HIGH. Incorrect grade of oil used. Oil cooler core clogged with dirt; clean core (par. 81 or 82).

(4) WEAK MAGNETO IGNITION. Check magnetos (par. 50 c).

(5) INCORRECT IGNITION TIMING. Check timing and reset if necessary (par. 87). If engine vibrates, pay particular attention to synchronization of magnetos.

(6) WRONG MAGNETOS INSTALLED. Check magnetos and replace if wrong type (par. 88 a, b, e and f).

(7) PREIGNITION. If proper octane fuel is being used and ignition system is functioning satisfactorily, spark plug gaskets may be too thin or spark plugs may be wrong type (use only Champion 63S); otherwise, internal engine troubles would be indicated. Notify higher authority.

TM 9-755

Part Three—Maintenance Instructions

(8) IMPROPER VALVE ADJUSTMENT. Check clearance and adjust if necessary (par. 73).

(9) IMPROPER CARBURETOR ADJUSTMENT. Check carburetor adjustment (par. 93 d).

(10) ACCELERATING PUMP STEM STICKING. If engine lacks power on acceleration, accompanied by spitting from leanness, free up pump stem (par. 93 f).

(11) CLOGGED GAS STRAINER IN CARBURETOR. If engine is overheating or indicates shortage of fuel, clean gas strainer in carburetor (par. 93 e).

(12) WRONG CARBURETOR INSTALLED. Check carburetor and replace if wrong type (par. 93 b, c, i and j).

(13) INCORRECT GOVERNOR SETTING. Place transmission in neutral, with transfer case clutch engaged, and open throttle to wide open position. If engine does not reach speed of 2,400 revolutions per minute for R975-C1 engine, or 2,500 revolutions per minute for R975-C4 engine, disconnect rod at governor valve lever and tie lever in wide open position. Accelerate engine gradually to wide open throttle position (but not over 2,500 revolutions per minute); if speed of 2,400 revolutions per minute is then reached on R975-C1 engine, or 2,500 revolutions per minute on R975-C4 engine, governor operation is at fault. Adjust governor (par. 95 d) or replace (par. 95 b and c).

(14) LOW ENGINE COMPRESSION OR IMPROPER VALVE TIMING. If engine does not develop full power with fuel reaching combustion chambers, adequate ignition and sufficient oil of proper grade in engine oiling system, low compression, or improper valve timing would be indicated. Notify higher authority.

f. **Engine Misfires.**

(1) FAULTY IGNITION SYSTEM. Test magnetos (par. 50 c), test spark plugs and ignition wiring (par. 50 d).

(2) INCORRECT CARBURETOR ADJUSTMENT. Adjust carburetor (par. 93 d).

(3) RESTRICTED FUEL FLOW. See paragraph 51.

(4) WATER IN FUEL. Remove two drain plugs at the bottom of the carburetor and inspect for water (par. 93 e).

(5) LOW ENGINE COMPRESSION. See subparagraph e (14) above.

(6) INTAKE PIPE AIR LEAKS. Test for air leaks by squirting small amount of fuel around packing nut while engine is idling. If this causes engine to accelerate or operate erratically, packing is leaking air. Tighten packing nut with wrench (41-W-1537) and repeat test. If packing still leaks, replace packing (par. 70 b).

g. **Excessive Oil Consumption.** See paragraph 48 d.

h. **Engine Will Not Stop.** If engine is idling above 700 revolutions per minute, engine may not stop when fuel idle cut-off switch is pressed.

(1) HAND THROTTLE CONTROL NOT FULLY RELEASED. Release throttle control (par. 14 e).
(2) THROTTLE STOP SCREW IMPROPERLY ADJUSTED. Adjust to 700 revolutions per minute (par. 93 d).
(3) THROTTLE LINKAGE STICKING. Check linkage and correct sticking (par. 94).
(4) FAULTY FUEL CUT-OFF. Inspect idle fuel cut-off switch and wiring (par. 51 h). Replace degasser assembly if faulty (par. 93 h).

i. Engine Overheating.
(1) USE OF IMPROPER TYPE OF FUEL. Use only U.S. Army Specification 2-103A above freezing; use only U.S. Army Specification 2-103C below freezing.
(2) CYLINDER COOLING FINS CLOGGED WITH DIRT. Remove engine and clean out dirt.
(3) AIR INLET OR OUTLET GRILLE OBSTRUCTED. Remove covers or items stowed on air inlet or outlet grille.
(4) OIL TEMPERATURE TOO HIGH. See paragraph 48 c.
(5) LATE IGNITION TIMING. Check timing and reset if necessary (par. 87).
(6) WRONG MAGNETOS INSTALLED. Check magnetos and replace if wrong type (par. 88 a, b, c and f).
(7) WRONG CARBURETOR INSTALLED. Check carburetor and replace if wrong type (par. 93 b, c, i and j).
(8) FUEL SYSTEM CLOGGED WITH GUM. Notify higher authority.

j. Engine Rich on Idle.
(1) INCORRECT CARBURETOR ADJUSTMENT. Adjust carburetor (par. 93 d).
(2) LEAKING PRIMER PUMP. Disconnect primer tube at intake pipe, plug opening and run engine; if fuel comes out of primer tube, replace primer pump (par. 97 b and c).
(3) IMPROPERLY ADJUSTED DEGASSER. Adjust degasser (par. 93 g).

k. Engine Noisy.
(1) LOW OIL PRESSURE. Improper grade of oil or improperly adjusted relief valve (par. 78 b).
(2) ENGINE OVERHEATING. See subparagraph i above.
(3) INTERNAL DAMAGE. Notify higher authority.

48. ENGINE OILING SYSTEM.

a. Reference. Oiling system troubles most commonly encountered are described in this paragraph. The subjects covered are listed below for quick reference.

Low or no oil pressure	Subpar. b.
Oil temperature too high	Subpar. c.
Excessive oil consumption	Subpar. d.
Oil pipe coupling hoses blow off	Subpar. e.
Oil pressure or temperature gage unit inoperative	Subpar. f.

b. **Low or No Oil Pressure.**

(1) LACK OF OIL. Fill oil tank.

(2) EXTERNAL LEAKS AT OIL PIPES OR CONNECTIONS. Tighten or replace connections. Replace damaged pipes.

(3) PRESSURE RELIEF VALVE INCORRECTLY ADJUSTED OR LEAKING. Adjust relief valve (par. 78 b). If valve cannot be adjusted to give proper pressure, remove relief valve parts and inspect for sticking or damaged parts. Clean parts, replace damaged parts, install and adjust.

(4) CLOGGED OIL PUMP SUCTION STRAINER. Remove and clean (par. 79 b).

(5) PRESSURE GAGE INOPERATIVE. See paragraph 62 f.

(6) OIL PUMP INOPERATIVE. If oil pressure is not obtained after completing steps (1) through (5) above, replace the oil pressure and oil scavenge pump assembly (par. 78 c and d).

c. **Oil Temperature Too High.**

(1) LOW OIL SUPPLY. Fill oil tank.

(2) OIL FILTER OBSTRUCTING OIL FLOW. Clean filter (par. 80).

(3) OIL COOLER CORE EXTERNALLY CLOGGED WITH DIRT. Clean cooler core (par. 81 b or 82 b).

(4) TEMPERATURE GAGE INOPERATIVE. See paragraph 62 g.

(5) ENGINE OVERHEATING. See paragraph 47 i.

(6) INOPERATIVE COOLER BY-PASS VALVE. If by-pass valve remains open when oil is hot, oil will not circulate through core and be cooled. If cooler core is noticeably cooler than the outlet fitting, remove by-pass valve and test in oil at 160° F to 165° F. Valve should open at this temperature and close when allowed to cool off.

(7) OIL COOLER CORE INTERNALLY OBSTRUCTED. Loosen oil pipe connection at inlet to oil tank. If normal oil flow is not evident with engine running, loosen oil pipe connection at oil filter outlet. If normal flow of oil is found at this point, oil cooler core is clogged internally. Replace cooler core (par. 81 or 82).

d. **Excessive Oil Consumption.**

(1) DILUTED OIL. When engine oil is diluted for sub-zero operation, oil consumption will be greater than normal (par. 27 c (4)).

(2) OIL VISCOSITY TOO LOW. Drain oil tank and fill with specified grade of oil.

(3) EXTERNAL OIL LEAKS. Inspect for oil leaks at all oil pipe connections, oil sump drain plugs, rocker box hose connections, push rod housing adapters, oil tank drain plug and outlet check valve. Tighten parts where leaks are found; replace coupling hoses where necessary.

(4) WORN INTERNAL ENGINE PARTS. If large volume of blue smoke comes from muffler tail pipes, worn piston rings or other internal parts are indicated; notify higher authority.

Trouble Shooting

e. Oil Pipe Coupling Hoses Blow Off.

(1) LOOSE HOSE CLAMPS OR DETERIORATED HOSES. Tighten hose clamps; replace deteriorated hose.

(2) STRAP BRACKETS LOOSE, OR NOT INSTALLED. Strap brackets are located at oil pump outlet and oil filter inlet hose connections, and at inlet connections to oil cooler. Tighten brackets to oil pipes; install brackets if missing.

(3) OIL TOO HEAVY. Drain oil tank and fill with oil of specified grade. In sub-zero weather dilute oil as specified in paragraph 27 c.

f. Oil Pressure or Temperature Gage Inoperative. See paragraph 62 f or g.

49. STARTER SYSTEM.

a. Reference. Starter troubles most commonly encountered are described in this paragraph. The subjects covered are listed below for quick reference.

   Starter will not operate .... Subpar. b
   Starter operates but will not crank engine .... Subpar. c
   Slow cranking speed .... Subpar. d

b. Starter Will Not Operate.

(1) MASTER SWITCH NOT CLOSED. Close master switch (fig. 15).

(2) TRANSMISSION SHIFT LEVER NOT IN NEUTRAL. Place in neutral position.

(3) STARTER CIRCUIT BREAKER OPEN. Press button to close circuit breaker (A, fig. 16).

(4) NO CURRENT IN BATTERY CIRCUIT. Turn on inside lights and close starter switch. If lights go out when switch is closed, check battery circuit (par. 58 b).

(5) STARTER NEUTRAL SAFETY SWITCH DEFECTIVE OR INCORRECTLY TIMED. Check switch and time or replace as necessary (par. 149).

(6) NO ENERGIZING CURRENT TO STARTER RELAY. After checking steps (1) through (5) above, remove starter relay junction box cover and tighten relay mounting screws and terminal nuts. If screws or nuts are corroded, remove and clean to insure metal-to-metal contact. If starter is still inoperative disconnect red wire from lower terminal (1, fig. 200), connect one lead of test light (fig. 47) to wire, other lead to ground. If test light burns when starter switch is held at "ON" position, current is being delivered to starter relay. If light does not burn, remove sponson side opening cover and disconnect conduit from main trunk receptacle on instrument panel (5, fig. 46). Using test light, test relay wiring for continuity between relay junction box and socket (J) in conduit connector plug at instrument panel. If wiring is satisfactory, test starter switch (par. 61 j).

(7) STARTER RELAY INOPERATIVE. If energizing current is being delivered to starter relay (step (6) above), connect the positive (+)

TM 9-755
49-50

Part Three—Maintenance Instructions

lead of a voltmeter to relay terminal post to which wire from starter junction box is attached (2, fig. 200); connect voltmeter negative (—) lead to ground on hull. Hold starter switch at "ON" position and note voltmeter reading. If no reading is obtained, check wiring and connections between relay junction box, starter junction, and master switch box. If reading is obtained, connect voltmeter positive (+) lead to opposite terminal at top of relay, hold starter switch at "ON" position and note voltmeter reading. If voltmeter readings show a difference in voltage between the two top terminals, relay contacts are burned; replace starter relay (par. 148).

(8) STARTER INOPERATIVE. If preceding test (steps (1) through (7) above) fails to show cause of trouble, check wiring and connections between starter relay junction box and starter. If satisfactory, starter is inoperative and must be replaced (par. 147).

c. Starter Operates But Will Not Crank Engine.

(1) HYDROSTATIC LOCK. See paragraph 47 b (1).

(2) WEAK CURRENT IN BATTERY CIRCUIT. Check battery circuit (par. 58 b).

(3) INTERNAL ENGINE SEIZURE. Notify higher authority.

d. Slow Cranking Speed.

(1) ENGINE OIL TOO HEAVY. Drain oil tank and fill with specified grade. In sub-zero weather dilute engine oil (par. 27 c).

(2) HIGH ELECTRICAL RESISTANCE. Weak current in battery circuit (par. 58 b), wrong size starter wires (fig. 49 or 52), faulty starter relay or starter (subpar. b (7) or (8) above).

(3) STARTER WORN OUT. Starter will be excessively noisy; replace (par. 147).

50. IGNITION SYSTEM.

a. Reference. Ignition troubles most commonly encountered are described in this paragraph. Subjects covered are listed below for quick reference.

Booster coil does not function	Subpar. b
Weak or inoperative magnetos	Subpar. c
Magneto, booster, or starter switch faulty	Subpar. d
Spark plugs faulty	Subpar. e

b. Booster Coil Does Not Function. Listen over engine compartment while another man holds booster switch in "ON" position. Coil will give a buzzing sound if operating; if no sound can be heard, test for following causes.

(1) CIRCUIT BREAKER OPEN. Press button to set circuit breaker (A, fig. 16). If button snaps out, wiring or booster coil is shorted. If test in step (2) below shows that wiring is satisfactory, shorted coil is indicated.

(2) NO CURRENT TO BOOSTER COIL. Remove air outlet grille (par. 183 a) and open hull roof door. Remove primary terminal

Trouble Shooting

cover (fig. 92), disconnect primary wire from coil terminal, connect one lead of test light (fig. 47) to wire and other lead to ground on hull. If light burns when booster switch is held in "ON" position, current is being delivered to coil and fault is in coil or its ground connection (step (3) below). If light does not burn, the wiring or switch is faulty. Disconnect magneto wire conduit at rear junction box (fig. 235), connect one test light lead to box receptacle pin "C" and other lead to ground on hull. If light now burns when booster switch is held at "ON," the wire in magneto conduit is defective. If light still does not burn, remove sponson side opening cover, disconnect conduit from switch conduit connector (fig. 204). Using test light, test wire (C) for continuity between rear junction box and front end of conduit. If wire is satisfactory, test booster switch (par. 62 j).

(3) BOOSTER COIL NOT GROUNDED, OR DEFECTIVE. If current is being delivered to booster coil (step (2) above), leave one test light lead connected to primary wire and connect other lead to engine mounting ring. If light does not burn when booster switch is held at "ON" position, inspect engine ground strap (fig. 225). If light burns, connect ground lead to coil housing, making sure to secure a good metal-to-metal contact. If light does not burn when booster switch is held at "ON" position, remove coil, clean mounting surfaces, install coil and tighten attaching screws securely. If light burns, however, coil is defective; replace coil (par. 90).

c. **Weak or Inoperative Magnetos.** Remove air inlet grille (par. 183 d), open roof rear door, remove cowl cover over No. 1 spark plug, and disconnect cables from front and rear spark plugs in No. 1 cylinder. Turn magneto switch to (L) position and hold rear spark plug cable contact ¼ inch from cylinder or other grounded part while cranking engine with starter. If a spark does not jump the ¼-inch gap, the left magneto or wiring is faulty. Turn magneto switch to "R" position and hold the front spark plug cable contact ¼ inch from cylinder while cranking engine with starter. If a spark does not jump the ¼-inch gap, the right magneto or wiring is faulty. Check faulty magneto for the following causes.

(1) FAULTY WIRING. Disconnect magneto wire conduit at rear junction box (fig. 235) and repeat test on faulty magneto; if spark now jumps the ¼-inch gap, the wiring between rear junction box and instrument panel is at fault. If spark does not jump the gap, remove radio shield, disconnect ground wire from (P) terminal on terminal block (par. 88 e (1) and (2)), install distributor blocks and repeat test on faulty magneto. If spark now jumps the ¼-inch gap, the ground wire in magneto conduit is at fault; if spark does not jump gap, the magneto is at fault (see step (2) below). If test indicates that wiring between rear junction box and instrument panel is at fault, remove sponson side opening cover and disconnect conduit from switch conduit connector (fig. 205). Using test light (fig. 47) test magneto wire for continuity between rear junction box and front end of conduit, connecting to wire (A) for left magneto and wire (B) for right magneto. If wire is satisfactory, test magneto switch (par. 62 j).

Part Three—Maintenance Instructions

(2) FAULTY MAGNETO. If test in step (1) above indicates faulty magneto, inspect and adjust breaker points (par. 88 d), inspect distributor blocks and connections of spark plug cables for corrosion, then repeat test of magneto. If spark does not jump the ¼-inch gap, the magneto is faulty; replace magneto (par. 88 e and f).

d. Spark Plugs Faulty. Start and warm up engine (par. 17 b and c). With transmission in neutral, set hand throttle control for engine speed of 1,800 revolutions per minute and run engine until exhaust manifold is hot. Place a few drops of oil on each exhaust pipe close to cylinder and note whether oil fails to burn off; this will indicate any cylinders that are not firing. Check spark plugs and cable connections at cylinder that is not firing.

(1) SPARK PLUG FOULED. Replace spark plug (par. 89).

(2) WRONG TYPE PLUG. Wrong type plug may fire but give improper ignition. Install Champion, type 63S plug.

(3) SPARK PLUG CABLE FAULTY. Notify higher authority.

51. FUEL SYSTEM.

a. Reference. The troubles most commonly encountered in the fuel system are described in this paragraph. The subjects covered are listed below for quick reference.

Fuel does not reach carburetor	Subpar. b.
Fuel does not reach cylinders	Subpar. c.
Shortage of fuel at high speed	Subpar. d.
Fuel tank pumps inoperative	Subpar. e.
Engine fuel pump inoperative	Subpar. f.
Degasser improperly adjusted	Subpar. g.
Idle fuel cut-off inoperative	Subpar. h.
Fuel gage inoperative	Subpar. i.

b. Fuel Does Not Reach Carburetor.

(1) FUEL SHUT-OFF VALVES CLOSED. Open fuel shut-off valves (par. 14 a).

(2) NO FUEL IN TANKS. Check by gage on instrument panel; fill tanks if empty.

(3) CLOGGED FUEL TANK VENTS. Remove fuel tank caps; if fuel then reaches carburetor, vents in caps are clogged. Clean caps in dry-cleaning solvent or replace caps if unable to clear the vents.

(4) FUEL TANK PUMPS INOPERATIVE. See subparagraph e below.

(5) ENGINE FUEL PUMP INOPERATIVE. See subparagraph f below.

c. Fuel Does Not Reach Cylinders.

(1) PRIMER PUMP INOPERATIVE. An inoperative engine primer pump is evident only when hard starting is experienced. Open left fuel shut-off valve and turn on fuel tank pumps. Disconnect a primer tube from one of the intake pipes and operate primer pump to see whether fuel squirts out of tube. If no fuel squirts out of tube, dis-

Trouble Shooting

connect inlet pipe at primer pump to see whether fuel is reaching pump. If fuel reaches primer pump, replace pump (par. 97). If fuel does not reach primer pump, check supply pipe and left fuel tank pump (subpar. e below).

(2) THROTTLE VALVE NOT OPENING PROPERLY. Check throttle valve operation and, if necessary, adjust linkage (par. 94).

(3) CARBURETOR IDLE ADJUSTMENT TOO LEAN. Evident only at engine speeds below approximately 1,200 revolutions per minute. Adjust carburetor (par. 93 d).

(4) CARBURETOR GAS STRAINER CLOGGED. Clean or replace strainer (par. 93 e).

(5) CARBURETOR JETS CLOGGED. Evident only at engine speeds above approximately 1,200 revolutions per minute. Replace carburetor (par. 93 i and j).

d. **Shortage of Fuel at High Speeds.** If engine lacks power, overheats, and detonates at high speed, shortage of fuel (extreme lean mixture) is indicated.

(1) CARBURETOR GAS STRAINER RESTRICTED. Clean or replace strainer (par. 93 e).

(2) INSUFFICIENT FUEL PUMP PRESSURE. Test fuel pumps (subpar. e and f below).

(3) CARBURETOR JETS RESTRICTED. Replace carburetor (par. 93 i and j).

(4) WRONG CARBURETOR INSTALLED. See paragraph 93 b and c; replace carburetor if wrong type (par. 93 i and j).

e. **Fuel Tank Pumps Inoperative.** Since the engine fuel pump will supply sufficient fuel for normal operation when fuel is cold, inoperative fuel tank pumps become evident only when fuel is hot; engine then will overheat and give other indications of lack of fuel. If inoperative fuel tank pumps are indicated, operate vehicle with first one fuel shut-off valve closed, then the other, to determine which pump is inoperative. Test inoperative fuel tank pump and circuit as described in following steps. NOTE: *Tank must be at least one quarter full when testing pump.*

(1) TESTING LEFT FUEL TANK PUMP. Press circuit breaker button (Y, fig. 16) to make certain that circuit breaker is set. With engine stopped, turn on fuel pumps switch, open left fuel shut-off valve and note reading on oil dilution pressure gage on bulkhead (fig. 30); gage should read 5 to 7 pounds. Turn switch on and off several times to make sure that pump will start at different positions of armature. If no pressure is indicated, test pump wiring circuit (step (3) below); if low pressure is indicated, remove fuel tank pump and screen for inspection and cleaning or replacement (par. 98).

(2) TESTING RIGHT FUEL TANK PUMP. Press circuit breaker button (Y, fig. 16) to make certain that circuit breaker is set. With engine stopped, close right fuel shut-off valve, disconnect fuel pipe at feed pipe check valve, and connect a fuel pump pressure gage to fuel pipe. Turn on fuel pumps switch, open right fuel shut-off valve and

note reading on gage. Gage readings and action to be taken are the same as for left pump (step (1) above). If fuel pump pressure gage is not available, attach hose to fuel pipe of sufficient length to reach fuel tank filler neck; with pump operating fuel should flow freely from hose.

(3) TESTING FUEL TANK PUMP WIRING CIRCUIT. If fuel tank pump does not show pressure (step (1) or (2) above), turn pumps switch off, and close fuel shut-off valves. Open hull rear door (par. 181 a) and disconnect wire from fuel pump cable terminal (par. 98 b (3), (4) and (6)). Connect one lead of test light (fig. 47) to pump wire and other lead to ground on hull. Turn fuel pumps switch on; if test light burns the wiring circuit is satisfactory; replace fuel pump (par. 98 h). If test light does not burn, disconnect main trunkline conduit from instrument panel and test wiring circuit "H" for left pump, or circuit "P" for right pump, as described in paragraph 64. If wiring is satisfactory, test circuit breaker, wiring, and fuel pumps switch in instrument panel (par. 61 d, e, and k) and replace parts found to be faulty.

f. **Engine Fuel Pump Inoperative.** Test the engine fuel pump after making certain that fuel tank pump screens are clean (subpar. e above). Close fuel shut-off valves and turn fuel tank pumps off. Remove square head plug from carburetor body at gas strainer (P, fig. 97) and attach a fuel pump pressure gage to carburetor body. Open fuel shut-off valves and start engine. With engine running at 1,800 revolutions per minute, engine fuel pump pressure should be 4½ to 7 pounds with fuel at temperature of 100° F or less; pressure will be lower at higher fuel temperatures. With engine running at idling speed (700 revolutions per minute), fuel pump pressure should be not less than 3½ pounds. Replace engine fuel pump (par. 96) if pressures are below these specifications.

g. **Degasser Improperly Adjusted.** If engine backfires, surges, or rolls on deceleration, adjust degasser (par. 93 g).

h. **Idle Fuel Cut-off Inoperative.** If engine will not stop when carburetor idle fuel cut-off switch is pressed (H, fig. 16), make certain that hand throttle control is fully released and engine is idling at 700 revolutions per minute; also press cut-off circuit breaker button (A, fig. 16) to make certain that circuit breaker is set. If throttle and circuit breaker are properly set, turn off fuel pumps switch and close fuel shut-off valves; engine will stop when carburetor runs out of fuel. Make the following tests to locate faulty parts.

(1) TESTING FUEL CUT-OFF IN DEGASSER. Remove terminal cap from degasser (fig. 104), attach one lead of test light (fig. 47) to terminal post and other lead to ground on engine. If test light does not burn when fuel cut-off switch is pressed, test fuel cut-off circuit (step (2) below). If test light burns when fuel cut-off switch is pressed, current is reaching degasser. Remove degasser (par. 93 h) and cut-off needle in carburetor. Check needle and its seat in carburetor for dirt which might hold needle open. Place needle in degasser

and hold it with finger, ground degasser to engine with jumper wire (fig. 47) and have another man press cut-off switch. If needle is pushed outward when switch is pressed, the cut-off solenoid in degasser is operative and dirt on needle seat was cause of trouble. If needle does not move when switch is pressed, replace degasser assembly.

(2) TESTING FUEL CUT-OFF WIRING CIRCUIT. Disconnnect main trunkline conduit from instrument panel and test wiring circuit (L) as described in paragraph 64. If wiring is satisfactory, test circuit breaker, wiring, and fuel cut-off switch in instrument panel (par. 62 d, e and k) and replace parts found to be faulty (circuit breaker, par. 161 b; switch, par. 160 f).

i. Fuel Gage Inoperative. See paragraph 62 h.

52. REAR TRANSFER CASE.

a. Reference. The most commonly encountered transfer case troubles described in this paragraph are listed below for quick reference.

Transfer case overheats	Subpar. b.
Oil flows out of breather	Subpar. c.
Oil leaks at oil seals	Subpar. d.
Oil leaks at gaskets	Subpar. e.
Clutch jumps out of engagement	Subpar. f.
Abnormal gear noise	Subpar. g.

b. Transfer Case Overheats.

(1) LACK OF OIL CAUSED BY AIR LEAK. An air leak at any connection in pipes between differential and differential oil pump will cause pump to lose its prime and fail to pump oil. Examine all connections and correct any leaks found.

(2) LACK OF OIL CAUSED BY DIFFERENTIAL OIL PUMP LOSING ITS PRIME. This condition may occur after oil pipes have been disconnected or pump has been removed. Prime pump (par. 119 e).

(3) LACK OF OIL CAUSED BY PLUGGED PASSAGE IN PUMP SHAFT OR TRANSFER CASE INPUT SHAFT. If oil passages controlled by check rod become plugged, oil will not enter transfer case. Remove oil pump and check rod (par. 119 c and d) and clean passages.

(4) MISALINEMENT OF ENGINE ON MOUNTINGS. Misalinement of engine on mountings, causing abnormal pressure on transfer case input shaft, will cause binding of gears and bearings. Check and correct engine alinement (par. 76 e).

c. Oil Flows Out of Breather. If engine is run for extended period with transfer case clutch disengaged, transfer case may fill and overflow because transfer case pump is out of operation. Otherwise, condition is caused by failure of pump or restriction in suction or return pipes.

(1) SUCTION OR RETURN PIPES RESTRICTED BY DENTS. Inspect pipes and replace if damaged.

(2) SUCTION OR RETURN PIPES CLOGGED. Disconnect at both ends and blow out.

(3) TRANSFER CASE OIL PUMP HAS LOST ITS PRIME. Disconnect oil pipe at pump, inject oil into pump, and connect pipe.

(4) TRANSFER CASE OIL PUMP WORN OUT. Replace pump (par. 120 c).

d. Oil Leaks at Oil Seals.

(1) TRANSFER CASE TOO FULL OF OIL. Same causes as described in subparagraph c, above.

(2) OIL SEALS WORN OUT. Notify higher authority.

e. Oil Leaks at Gaskets.

(1) LOOSE ATTACHING SCREWS. Tighten screws.

(2) DEFECTIVE GASKET. Replace, if oil pump gaskets (par. 119 or 120). If other gaskets, notify higher authority.

f. Clutch Jumps Out of Engagement.

(1) SHIFTER ROD POPPET BALL SPRING WEAK OR BROKEN. Notify higher authority.

(2) WORN CLUTCH GEAR TEETH. Notify higher authority.

g. Abnormal Gear Noise.

(1) LACK OF OIL. Refer to subparagraph b, above.

(2) MISALINEMENT OF ENGINE ON MOUNTINGS. Check and correct alinement (par. 76 e).

(3) WORN OR BROKEN GEARS OR BEARINGS. Notify higher authority.

53. UNIVERSAL JOINTS AND PROPELLER SHAFT.

a. Abnormal Backlash.

(1) LOOSE UNIVERSAL JOINT ATTACHING SCREWS. Tighten screws (par. 106 and 107).

(2) WORN TRUNNION BEARINGS OR SPIDER. Replace worn parts (par. 106 and 107).

b. Abnormal Vibration in Propeller Shaft.

(1) LOOSE UNIVERSAL JOINT ATTACHING SCREWS. Tighten screws (par. 106 and 107).

(2) WORN UNIVERSAL JOINT, TRUNNION BEARINGS OR SPIDER. Replace worn parts (par. 106 and 107).

(3) BENT PROPELLER SHAFT. Replace shaft (par. 106 and 107).

54. TORQMATIC TRANSMISSION.

a. Reference. The most commonly encountered transmission troubles described in this paragraph are listed below for quick reference.

Operating road test	Subpar. b.
Slipping or sluggish drive when shift is made	Subpar. c.
Power loss or slipping in transmission	Subpar. d.

Trouble Shooting

Drives in some ranges but not in others	Subpar. e.
Failure to drive	Subpar. f.
Vehicle moves forward with shift lever in neutral	Subpar. g.
Steaming, smoking, oil foaming at breather	Subpar. h.
Transmission overheats	Subpar. i.
Abnormal noise	Subpar. j.
Oil leaking into differential	Subpar. k.

   b. **Operating Road Test.** Considerable time will be saved, and more accurate results obtained in diagnosing trouble in a torqmatic transmission by first making a thorough operating road test of the vehicle. Before transmission test is made, be sure that engine is properly tuned and throttle linkage correctly adjusted so that engine will operate at near governed speed (2,400 rpm for C1 engine; 2,500 rpm for C4 engine) to apply full power to torque converter. Vehicle should be operated in reverse and all forward speed ranges under various conditions of load so that performance of the transmission in each speed range can be noted. Manual shift lever must work freely and have a positive detent in every speed range position. Operator must always move lever completely from one detent point to another; holding lever in between shift points will damage the shift mechanism. Any unusual condition such as power loss, slippage, noise, over-heating or oil foaming should be noted, together with particular speed ranges in which the condition occurs. Reference to the following subparagraphs c through k will assist in determining cause of trouble and remedy required.

   c. **Slipping or Sluggish Drive When Shift is Made.** In a normal operating transmission all gear changes will be very positive when the manual shift lever is moved from one position to another. Soft shifts and sluggish drive through transmission immediately after lever is shifted, followed by normal drive, usually indicates that bands are starting to slip before taking hold.

   (1) BOTH BANDS STARTING TO SLIP. All shifts will be soft and sluggish. Adjust bands (par. 110).

   (2) SECOND SPEED BAND STARTING TO SLIP. First to second and third to second shifts will be soft and sluggish. Adjust front band (par. 110).

   (3) THIRD AND REVERSE BAND STARTING TO SLIP. The second to third shift will be soft and sluggish. Adjust rear band (par. 110).

   (4) LOW CONTROL OIL PRESSURE. All shifts will be soft and sluggish. Low pressure is caused by internal leaks in hydraulic control system. Notify higher authority.

   d. **Power Loss or Slipping in Transmission.**

   (1) NO OIL IN TORQUE CONVERTER. Refer to subparagraph f (1) below.

   (2) LOW OIL LEVEL. Check and fill transmission.

   (3) WATER IN TRANSMISSION. If drive through transmission is satisfactory when it is cold, but power loss occurs when it is hot, and

steam flows out of breather; there is water in the transmission. Drain and refill.

(4) BOTH BANDS SLIPPING. If drive through transmission is positive in first range but power loss occurs in other ranges; both bands are slipping. Adjust bands (par. 110); notify higher authority if bands cannot be adjusted to hold.

(5) SECOND SPEED BAND SLIPPING. If drive through transmission is positive in all other ranges but power loss occurs in second; the second speed band (front) is slipping. Adjust band (par. 110); notify higher authority if band cannot be adjusted to hold.

(6) THIRD AND REVERSE BAND SLIPPING. If drive through transmission is positive in first and second ranges but power loss occurs in third and reverse; the third and reverse band (rear) is slipping. Adjust band (par. 110); notify higher authority if band cannot be adjusted to hold.

(7) CLUTCHES SLIPPING. If drive through transmission is positive in all forward ranges but not in reverse; outer clutch is slipping. If drive through transmission is positive in reverse but power loss occurs in all forward ranges; inner clutch is slipping. In either case notify higher authority.

e. Drives in Some Ranges But Not in Others.

(1) BOTH BANDS FAIL TO APPLY. If vehicle drives in first range but not in any other; both bands are loose or worn out. Adjust bands (par. 110); notify higher authority if bands cannot be adjusted to hold.

(2) SECOND SPEED BAND FAILS TO APPLY. If vehicle drives in all ranges but second; the second speed band (front) is loose or worn out. Adjust band (par. 110); notify higher authority if band cannot be adjusted to hold.

(3) THIRD AND REVERSE BAND FAILS TO APPLY. If vehicle drives in first and second range but not in third and reverse; the third and reverse band (rear) is loose or worn out. Adjust band (par. 110); notify higher authority if band cannot be adjusted to hold.

(4) CLUTCHES FAIL TO APPLY. If vehicle drives in all forward ranges but not in reverse; outer clutch fails to apply. If vehicle drives in reverse but not in any forward range; the inner clutch fails to apply. In either case notify higher authority.

f. Failure to Drive.

(1) NO OIL IN CONVERTER. Oil slowly drains out of converter into transmission case when transmission is not operating. If vehicle has been parked for a long period of time, engine and transmission must be operated until transmission pump fills converter before attempting to move vehicle, otherwise failure to drive or excessive power loss will occur until converter fills.

(2) LOW OIL LEVEL. Check and fill transmission.

Trouble Shooting

g. **Vehicle Moves Forward with Shift Lever in Neutral.**
(1) OUTER CLUTCH NOT RELEASING. Notify higher authority.
(2) PLANET GEARS SEIZED OR LOCKED. Notify higher authority.

h. **Steaming, Smoking, Oil Foaming at Breather.**

(1) WATER IN TRANSMISSION. Water in transmission oil will cause steaming at breather usually accompanied by loss of power through transmission when it is hot. Drain and refill.

(2) TRANSMISSION OVERHEATING. An overheating transmission will cause smoking at breather. Refer to subparagraph i below.

(3) DRAGGING SECOND SPEED BAND. If transmission smokes excessively when operated in third gear and reverse but does not smoke when operated in second range; the second speed band is dragging. Check band adjustment (par. 110). If condition continues with band properly adjusted, notify higher authority.

(4) TRANSMISSION OIL PUMPS SUCKING AIR. Any condition which allows the oil pumps to suck air will cause oil to foam and flow out of breather; notify higher authority.

i. **Transmission Overheats.** If transmission overheats, oil temperature warning light on instrument panel will burn red. Steaming, smoking, or oil foaming at breather does not definitely indicate overheating (subpar. h above).

(1) BLOWER DRIVE BELTS LOOSE OR BROKEN. Adjust tension (par. 122 a) or install new belts (par. 122 b and c).

(2) OIL COOLER CORE AIR PASSAGES RESTRICTED WITH DIRT. Remove cooler inlet duct service cover (par. 121 a) and blow dirt out of air passages with air stream. A soft brush and dry-cleaning solvent may be used to loosen dirt.

(3) OIL COOLER CORE OIL PASSAGES RESTRICTED WITH DIRT. Replace core (par. 121 e and f) for service by higher authority.

(4) BINDING OR BROKEN INTERNAL PARTS. This condition will usually be accompanied by abnormal noise (subpar. j below). Notify higher authority.

j. **Abnormal Noise.**

(1) WORN OR BROKEN TRANSFER CASE GEARS OR BEARINGS. Any unusual noise originating within transmission transfer case may be determined by running at idling speed, with rear transfer case clutch engaged and transmission control lever in neutral. Under these conditions both rear transfer case and transmission transfer case will be in operation. Since these units are separated from each other, although connected by the propeller shaft, it should be possible to determine which unit is causing unusual noise by use of a stethoscope or sounding rod. As a further test to eliminate the possibility of the noise originating in the engine, disengage rear transfer case clutch, and run engine idling at the same speed as before. Both transfer units will then be out of operation and if noise continues it is caused by the

TM 9-755
54-55

Part Three—Maintenance Instructions

engine. If noise is found to be in transmission transfer case notify higher authority.

(2) PLANET GEARS WORN OR BROKEN. If engine is accelerated beyond 800 revolutions per minute with rear transfer case clutch engaged and transmission control lever in neutral, power flow through the torque converter will cause the planet gears to rotate without transmitting power. If an unusual noise develops under this condition, test planet gear for unusual noise under load by driving the vehicle on a road that is smooth and level as possible to eliminate differential gear noise. Shift from one speed range to another and carefully note any pronounced change in noise level. If tests indicate planet gear noise, notify higher authority.

k. Oil Leaking Into Differential.

(1) LEAKING OUTPUT SHAFT OIL SEALS. Evidenced by oil level dropping in transmission and raising in differential. Notify higher authority.

55. CONTROLLED DIFFERENTIAL.

a. Reference. The most commonly encountered differential troubles described in this paragraph are listed below for quick reference.

Steering brakes do not hold	Subpar. b.
Parking brakes do not hold	Subpar. c.
Brakes do not steer vehicle properly	Subpar. d.
Differential overheating	Subpar. e.
No drive through differential	Subpar. f.
Abnormal noise	Subpar. g.

b. Steering Brakes Do Not Hold.

(1) LOOSE BRAKE SHOE ADJUSTMENT. Adjust brakes (par. 114).

(2) BRAKE CONTROLS IMPROPERLY ADJUSTED. Check linkage and adjust (par. 116 c).

(3) BRAKE SHOE LININGS WORN OUT. Replace brake shoes (par. 115 c) if linings are worn so that proper adjustment (par. 114 b) is no longer possible.

(4) BRAKE SHOE LININGS GLAZED. If brakes and linkage are properly adjusted (par. 114) but brakes do not hold, or hold only when abnormal pressure is applied to hand levers; brake shoe linings are probably glazed. Replace brake shoes (par. 115 c); buffing or grinding linings is not effective.

(5) BRAKE CONTROL LEVERS IMPROPERLY INSTALLED. Install levers on brake shaft so that index marks on levers and shafts are alined.

(6) BRAKE SHAFTS REVERSED DURING ASSEMBLY. If right and left brake shafts are installed in wrong sides of carrier; low braking efficiency will result. Notify higher authority.

c. Parking Brakes Do Not Hold.

(1) BRAKES DO NOT HOLD. Refer to subparagraph b. above.

(2) LOCKING PAWL IMPROPERLY ADJUSTED. Adjust (par. 116 b).

(3) LOCKING PAWL OR QUADRANT WORN. Replace worn parts.

d. Brakes Do Not Steer Vehicle Properly.
(1) BRAKES DO NOT HOLD. Refer to subparagraph b, above.
(2) DIFFERENTIAL GEARS DAMAGED OR IMPROPERLY ASSEMBLED. Notify higher authority.

e. Differential Overheating.
(1) LOW OIL LEVEL. Check and fill differential.
(2) OIL COOLER CORE AIR PASSAGES RESTRICTED WITH DIRT. Remove cooler inlet duct service cover (par. 119 c) and blow dirt out of air passages with air stream. A soft brush and dry-cleaning solvent may be used to loosen dirt.
(3) OIL COOLER CORE OIL PASSAGES RESTRICTED WITH DIRT. Replace core (par. 121) for service by higher authority.
(4) BLOWER BELTS LOOSE OR BROKEN. Adjust tension (par. 122 a) or install new belts (par. 122 b and c).
(5) THERMOSTAT BYPASS—VALVE INOPERATIVE. Replace core (par. 121).
(6) OIL NOT CIRCULATING THROUGH COOLING SYSTEM. Clean oil strainer (par. 124 a), clean filter (par. 124 b), check pump inlet line for air leaks, prime pump (par. 119 e), replace pump if worn (par. 119 c and d).

f. No Drive Through Differential.
(1) BROKEN DRIVE SHAFT. Notify higher authority.
(2) DIFFERENTIAL GEARS SEIZED OR BROKEN. Evident when one steering brake is applied. Notify higher authority.

g. Abnormal Noise. Some gear noise will be evident whenever differential is in operation. If noise becomes abnormal check the following possible causes:
(1) LOW OIL LEVEL. Check and fill.
(2) OIL NOT BEING PUMPED TO RING AND PINION GEARS. Clean oil strainer (par. 124 a), clean filter (par. 124 b), check pump inlet line for air leaks or prime pump (par. 119 e), replace pump if worn (par. 119 c and d).
(3) WORN OR BROKEN INTERNAL PARTS. Notify higher authority.

56. FINAL DRIVE AND UNIVERSAL JOINTS.

a. Abnormal Gear Noise. Some gear noise is always present during operation of final drives. Abnormal gear noise should be investigated as soon as possible, to avoid damage to mechanism.
(1) NO LUBRICANT IN FINAL DRIVE. Lubricate.
(2) WORN OR BROKEN GEARS OR BEARINGS. Replace final drive assembly (par. 129).

b. Abnormal Heat. Same causes as for gear noise (subpar. a above).

TM 9-755
56-57

Part Three—Maintenance Instructions

c. **Abnormal Backlash.**

(1) LOOSE UNIVERSAL JOINT ATTACHING SCREWS. Remove shields, remove lock wires, tighten screws. Install lock wires and shields.

(2) WORN UNIVERSAL JOINT TRUNNION BEARINGS OR SPIDER. Replace worn parts (par. 128).

(3) WORN FINAL DRIVE PINION SHAFT OR INTERNAL PARTS. Replace pinion shaft or final drive assembly (par. 129).

d. **No Drive to Track Through Final Drive.**

(1) BROKEN PINION SHAFT. Replace shaft (par. 129).

(2) BROKEN INTERNAL PARTS. Replace final drive assembly (par. 129).

e. **Oil Leaks from Final Drive.**

(1) LOOSE FILLER OR DRAIN PLUG. Tighten plug (AV, fig. 43).

(2) LOOSE CARRIER COVER OR BEARING RETAINER ATTACHING BOLTS. Tighten.

(3) WORN OIL SEALS OR BROKEN GASKETS. Replace final drive assembly (par. 129).

57. TRACKS AND SUSPENSION.

a. **Reference.** The most commonly encountered troubles, described in this paragraph, are listed below for quick reference.

Vehicle leads to one side	Subpar. b.
Abnormal wear of track link guide lugs	Subpar. c.
Rapid wear of track wheel tires	Subpar. d.
Thrown track	Subpar. e.
Inoperative track support roller	Subpar. f.
Vehicle sags to one side	Subpar. g.
Inoperative shock absorber	Subpar. h.

b. **Vehicle Leads to One Side.**

(1) UNEQUAL TRACK TENSION. Unequal track tension will cause vehicle to lead to the side having the tighter tension. Adjust both tracks to equal tension (par. 132 h).

(2) WORN OR DISTORTED FINAL DRIVE SPROCKETS. Replace (par. 127).

(3) WORN TRACK LINK PINS IN ONE TRACK. If track link pins in one track are worn much more than pins in other track, vehicle will lead to the side on which pins are worn least. If leading is excessive, replace worn pins or install new track (par. 132).

(4) NEW TRACK ON ONE SIDE, WORN TRACK ON OTHER SIDE. If new track is installed on one side and other track is well broken in, vehicle will lead to side of new track. Replace worn track (par. 132) if leading is excessive.

(5) TRACK SUSPENSION MISALINED. If track suspension is misalined due to sprung hull plates, or suspension components; vehicle will lead to one side. Notify higher authority.

c. Abnormal Wear of Track Link Guide Lugs.

(1) EXTENDED OPERATION WHERE VEHICLE IS TILTED SIDEWAYS. If vehicle is operated for extended periods where it is tilted sideways, rapid wear of all guide lugs will result.

(2) BENT TRACK WHEEL, COMPENSATING WHEEL, OR FINAL DRIVE SPROCKET. Replace track wheel (par. 136), compensating wheel (par. 135), or final drive sprocket (127) as required.

(3) BENT SUPPORT ARM. Replace support arm and axle shaft housing (par. 138).

(4) TRACK SUSPENSION MISALINED. If track suspension is misalined due to sprung hull plates or suspension components, rapid wear of guide lugs will result. Notify higher authority.

d. Rapid Wear of Track Wheel Tires. Tire wear is normally more rapid on No. 5 track wheels than on other track wheels.

(1) STONES OR OTHER FOREIGN MATERIAL IMBEDDED BETWEEN WHEEL DISKS. This condition will interfere with track guide lugs and cause track wheel to scuff on track, causing wear of tire. Remove all foreign material.

(2) BENT TRACK WHEEL. Replace wheel (par. 136).

(3) BENT SUPPORT ARM. Replace support arm and axle shaft housing (par. 138).

(4) DAMAGED SURFACE ON TRACK LINKS. Replace damaged links (par. 132).

e. Thrown Track.

(1) SKIDDING AND ROUGH HANDLING ON TURNS. Review driving instructions and precautions (par. 18).

(2) ROCKS WEDGED BETWEEN TRACK WHEEL DISKS. Clean rocks out.

(3) EXCESSIVELY LOOSE OR WORN TRACK. Adjust track tension (par. 132 b) or replace track (par. 132 d and f).

(4) FRONT TORSION BAR BROKEN. Replace bar (par. 137).

f. Inoperative Track Support Roller.

(1) FOREIGN MATERIAL BETWEEN ROLLER AND TRACK. Clean foreign material out.

(2) DRY BEARINGS. Lubricate bearings (par. 38).

(3) SEIZED BEARINGS. Replace bearings (par. 134).

g. Vehicle Sags to One Side.

(1) BROKEN TORSION BAR. Replace bar (par. 137).

(2) WRONG INSTALLATION OF TORSION BARS. If torsion bars are installed on wrong side of vehicle, or support arms are not set to specifications when torsion bars are installed, vehicle will sag. Install torsion bars correctly (par. 137).

h. Inoperative Shock Absorber. If a shock absorber does not become warm during operation; it is an indication that it is not functioning properly.

(1) FLUID LOW. Fill with fluid (par. 139 c).
(2) FLUID LEAKING AT SEALS. Replace absorber (par. 139 c).
(3) SCORED OR DAMAGED INTERNAL PARTS. Replace absorber (par. 139 c).

58. BATTERIES.

a. Reference. The troubles most commonly encountered with the batteries and battery circuit are described in this paragraph. The subjects covered are listed below for quick reference.

Low current in battery circuit .... Subpar. b.
No current in battery circuit .... Subpar. c.
Batteries do not stay charged .... Subpar. d.

b. Low Current in Battery Circuit.

(1) BATTERY CIRCUIT TERMINALS LOOSE OR CORRODED. Clean and tighten terminals on batteries (par. 142 h). Clean and tighten ground cable connection to hull. Clean and tighten connections in battery junction box and master switch box.

(2) BATTERY FLUID LOW. Check fluid level and add water (par. 142 c).

(3) BATTERIES PARTIALLY DISCHARGED. Check specific gravity (par. 142 b) and recharge batteries if necessary.

(4) BATTERY CABLES WORN OR CORRODED. Inspect cables and replace if worn or corroded so as to reduce capacity.

(5) MASTER SWITCH CONTACTS BURNED OR CORRODED. Test switches (par. 63 b) and replace if necessary.

c. No Current in Battery Circuit.

(1) BATTERY CIRCUIT TERMINALS LOOSE OR CORRODED. See subparagraph b (1) above.

(2) BATTERIES DISCHARGED. Check specific gravity (par. 142 b); recharge or replace batteries if necessary.

(3) BATTERY CABLES CHAFED THROUGH OR BROKEN. Replace unserviceable cables (par. 142 f).

(4) CUT-OUT RELAY STICKING CLOSED. If relay is sticking closed, ammeter will show heavy discharge when generator is not running; replace regulator (par. 145 b).

(5) BATTERY CABLES GROUNDED. Install two fully charged batteries and connect ground and battery-to-battery cables only (par. 142 g and h). Test for ground in 12-volt and 24-volt circuits as follows.

(a) *Test for Ground in 12-Volt Circuit.* Turn 12-volt master switch and all other 12-volt circuit switches off. Touch the 12-volt positive cable to positive (+) terminal of the grounded battery; if a flash is produced, the 12-volt circuit is grounded between battery and master switch. Test this circuit as described in paragraph 64.

(b) *Test for Ground in 24-Volt Circuit.* Turn 24-volt master switch and all other 24-volt circuit switches off. Touch the 24-volt positive cable to battery positive (+) terminal; if a flash is produced, the 24-volt circuit is grounded between battery and master switch. Test this circuit as described in paragraph 64.

d. **Batteries Do Not Stay Charged.**

(1) EXCESSIVE USE OF ELECTRICAL EQUIPMENT. Use auxiliary generator to keep batteries charged when current requirements exceed the capacity of engine generator.

(2) LOW GENERATOR CHARGING RATE. See paragraph 59 b.

(3) HIGH RESISTANCE IN BATTERY CIRCUIT. Clean and tighten all connections between master switch and batteries, battery terminals, and battery ground cable connection to hull.

(4) CUT-OUT RELAY STICKING CLOSED. If relay is sticking closed, ammeter will show heavy discharge when generator is not running; replace regulator (par. 145).

(5) DEFECTIVE BATTERIES. Test with high rate discharge tester and replace if necessary (par. 142 g and h).

59. ENGINE GENERATOR.

a. **Reference.** The troubles most commonly encountered with engine generator and regulator are described in this paragraph. Subjects covered are listed below for quick reference.

Low charging rate	Subpar. b.
Unsteady charging rate	Subpar. c.
Generator does not charge	Subpar. d.
Improperly adjusted regulator	Subpar. e.
Ammeter shows discharge	Subpar. f.

b. **Low Charging Rate.** If batteries are fully charged, generator charging rate is reduced by the regulator to a point where the ammeter indicates practically nothing; therefore, the ammeter reading should not be taken as an indication of unsatisfactory generator condition. Run engine at 1,450 revolutions per minute for 25 to 30 minutes to bring engine generator and regulator to operating temperature, then reduce engine to idling speed. Remove cap from master switch box receptacle, connect positive (+) lead of a test voltmeter to positive (+) terminal in receptacle, and connect negative (—) lead of voltmeter to ground on the hull. Slowly increase engine speed to 1,450 revolutions per minute, and turn 24-volt master switch off; the voltmeter then should read 28 to 30.4 volts. If voltmeter reads above 30.4 volts, the regulator is improperly adjusted; replace regulator (par. 145). If voltmeter reads below 28 volts, inspect for the following causes:

(1) GENERATOR DRIVE BELTS SLIPPING. Adjust belts (par. 143 a).

(2) LOOSE OR CORRODED TERMINALS. Inspect, clean, and tighten all wiring connections in terminal box on generator, at regulator in main filter box, in starter junction box, and in master switch box. Check generator ground wire connections.

TM 9-755
59

Part Three—Maintenance Instructions

(3) GENERATOR COMMUTATOR DIRTY, OR BRUSHES STICKING. Inspect and clean commutator and brushes (par. 143 d).

(4) REGULATOR IMPROPERLY ADJUSTED. Repeat the open circuit voltage test described above after completing steps (1) through (3); if voltmeter does not read 28 to 30.4 volts, replace regulator (par. 145).

(5) DEFECTIVE GENERATOR. Repeat open circuit voltage test after replacing regulator in step (4) above; if voltmeter does not read 28 to 30.4 volts, replace generator (par. 143 e and f).

c. **Unsteady Charging Rate.**

(1) GENERATOR DRIVE BELTS SLIPPING. Adjust belts (par. 143 a).

(2) LOOSE CONNECTIONS IN GENERATOR CIRCUIT. See subparagraph b (2) above. Also inspect ammeter wiring connections.

(3) DIRTY COMMUTATOR, STICKING BRUSHES, HIGH COMMUTATOR MICA. Inspect and clean commutator and brushes (par. 143 d). If mica is high between commutator segments, replace generator (par. 143 e and f).

(4) LOOSE OR FAULTY REGULATOR. Inspect regulator and ground strap and tighten, if loose. If this does not correct unsteady charging rate, replace regulator (par. 145).

(5) DEFECTIVE GENERATOR. If preceding steps (1) through (4) do not correct trouble, replace generator (par. 143 e and f).

d. **Generator Does Not Charge.**

(1) TRANSFER CASE CLUTCH NOT ENGAGED. Engage clutch.

(2) BROKEN DRIVE BELTS. Replace belts (par. 143 e and f).

(3) OPEN CIRCUIT. Test wiring between generator and regulator, between regulator and starter junction box, and between junction box and master switch box (par. 64). Inspect for loose or broken generator ground wire.

(4) DEFECTIVE REGULATOR. Replace regulator (par. 145).

(5) DEFECTIVE GENERATOR. If preceding steps (1) through (4) do not correct trouble, replace generator (par. 143 e and f).

e. **Improperly Adjusted Regulator.** An improperly adjusted regulator is indicated if voltmeter does not read 28 to 30.4 volts on open circuit test (subpar. b above). Sticking cut-out relay points are indicated if ammeter shows heavy discharge when generator is not operating (subpar. f below). The cut-in point of the cut-out relay may be tested in the following manner: Warm up generator and regulator, and connect test voltmeter as described in subparagraph b above. Connect one lead of test light (fig. 47) to positive (+) terminal of master switch box receptacle and other lead to ground on hull. Slowly increase engine speed until test light just starts to burn; voltmeter reading should be 25.6 volts to 26.5 volts. If regulator assembly appears to be improperly adjusted in any respect; replace the assembly (par. 145).

Trouble Shooting

f. **Ammeter Shows Discharge.** In M39 vehicles equipped with winterization heater (par. 26), the ammeter will show discharge when heater unit is operating. Under all other conditions the ammeter should not show discharge in either the M18 or M39 vehicle. If ammeter shows discharge with engine generator or auxiliary generator operating, check for reversed ammeter wires at shunt in starter junction box, terminal block in master switch box, or at ammeter in instrument panel. If ammeter shows discharge when neither generator is operating, check for short circuits between generator and shunt in starter junction box; if none are found, cut-out relay points are sticking closed. Disconnect wires from engine generator regulator and then the auxiliary generator regulator; replace regulator which is causing ammeter to show discharge (par. 145).

60. AUXILIARY GENERATOR.

a. **Reference.** Auxiliary generator troubles most commonly encountered are described in this paragraph. Subjects covered are listed below for quick reference.

Engine fails to start	Subpar. b.
Engine starts, then stops	Subpar. c.
Engine fails to run at full speed	Subpar. d.
Engine runs irregularly or misses	Subpar. e.
Engine overheats	Subpar. f.
Low charging rate	Subpar. g.
Generator does not charge	Subpar. h.

b. **Engine Fails to Start.**

(1) DEFECTIVE STARTING SWITCH IN CONTROL BOX. Inspect switch and replace if necessary.

(2) SPARK PLUG FOULED BY CARBON OR LEAD DEPOSIT. Remove and clean spark plugs (par. 144 a).

(3) SPARK PLUG POINT GAP INCORRECT. Adjust to 0.025-inch gap (par. 144 a).

(4) SPARK PLUG POINTS BADLY WORN OR PORCELAIN CRACKED. Replace spark plug (par. 144 a).

(5) WRONG TYPE SPARK PLUG. Use Champion HO-14S spark plugs.

(6) SPARK PLUG ADAPTER HOLES PLUGGED. Remove spark plugs and clean adapter holes (par. 144 a).

(7) MAGNETO CONTACT POINTS OUT OF ADJUSTMENT. Adjust to 0.020-inch gap (par. 144 b).

c. **Engine Starts, Then Stops.**

(1) NO FUEL IN TANK. Fill tank.

(2) FUEL SHUT-OFF COCKS CLOSED OR CLOGGED. Open or clean shut-off cocks.

(3) SCREEN IN FUEL STRAINER CLOGGED. Remove bowl and clean screen (par. 144 d).

(4) WATER OR DIRT IN FUEL. Drain and clean tank.

TM 9-755
60

Part Three—Maintenance Instructions

---

(5) FUEL LINE CLOGGED. Clean out fuel lines.

d. **Engine Fails to Run at Full Speed.**

(1) IMPROPER CARBURETOR ADJUSTMENT. Adjust carburetor (par. 144 c).

(2) SCREEN IN FUEL STRAINER PARTIALLY CLOGGED. Remove bowl and clean screen (par. 144 d).

(3) WATER IN FLOAT NEEDLE CHAMBER. Drain float needle chamber.

(4) CARBURETOR NOZZLE CLOGGED. Remove and clean nozzle.

(5) FLOAT STUCK. Remove bowl cover and clean.

(6) FLOAT NEEDLE WORN. Replace float needle.

e. **Engine Runs Irregularly or Misses.**

(1) MAGNETO CONTACT POINTS OUT OF ADJUSTMENT. Adjust to 0.020-inch gap (par. 144 b).

(2) MAGNETO CONTACT POINTS PITTED OR WORN. Dress or replace (par. 144 b).

(3) LOOSE CONNECTIONS IN MAGNETO. Tighten connections (par. 144 b).

(4) BROKEN SPARK PLUG CABLE. Replace spark plug cable.

(5) MAGNET WEAK, COIL OR CAPACITORS WEAK. Notify higher authority.

f. **Engine Overheats.**

(1) SPARK PLUG ADAPTER CLOGGED. Clean or replace (par. 144 a).

(2) MUFFLER OR EXHAUST PIPES CLOGGED. Replace muffler or exhaust pipe.

(3) CYLINDER PORTS CLOGGED WITH CARBON. Notify higher authority.

g. **Low Charging Rate.** If batteries are fully charged, generator charging rate is reduced by the regulator to a point where the instrument panel ammeter and control box ammeter will indicate practically nothing; therefore, ammeter reading should not be taken as an indication of unsatisfactory generator condition. Turn 24-volt master switch on, start the auxiliary generator, and run it for 25 to 30 minutes to bring regulator to operating temperature. Note control box ammeter reading, then turn on all vehicle lights and turret traversing electric motor and note control box ammeter reading. Charging rate should increase in proportion to amount of current used as electrical units are turned on. If batteries are low, the ammeter reading will be higher than if batteries are near full charge. If charging rate is low, remove cap from master switch box receptacle. Connect the positive (+) lead of a test voltmeter to the positive (+) terminal in receptacle and connect negative lead of voltmeter to ground on hull. Turn 24-volt master switch off; the voltmeter then should read 28 to 30.4 volts. If voltmeter reads above 30.4 volts, the regulator is improperly adjusted; replace regulator (par. 145). If voltmeter reads below 28 volts, inspect for the following causes.

(1) ENGINE NOT UP TO FULL SPEED. See subparagraph d above.

(2) DIRTY COMMUTATOR. Clean commutator (par. 144 f).

(3) BRUSHES NOT PROPERLY SEATED, OR WORN OUT. Seat brushes (par. 144 f); replace if worn out.

(4) LOOSE CONNECTIONS IN CONTROL BOX. Tighten connections.

(5) LOOSE OR CORRODED TERMINALS. Inspect, clean, and tighten all wiring connections in auxiliary generator junction box, at regulator in main filter box, in starter junction box, and in master switch box. Check generator ground strap connections.

(6) REGULATOR IMPROPERLY ADJUSTED. Repeat the open circuit voltage test described above after completing steps (1) through (5); if voltmeter does not read 28 to 30.4 volts, replace the regulator (par. 145).

(7) DEFECTIVE GENERATOR. Repeat the open circuit voltage test after replacing the regulator in step (6) above; if voltmeter does not read 28 to 30.4 volts, replace the generator (par. 144 h and i).

h. Generator Does Not Charge.

(1) CONTROL BOX CIRCUIT BREAKER OPEN. Press reset button. If button snaps out, a ground exists in charging circuits; test circuit (par. 64).

(2) DIRTY COMMUTATOR. Clean commutator (par. 144 f).

(3) BRUSHES NOT PROPERLY SEATED, STUCK IN HOLDERS, OR WORN OUT. Free up and seat brushes (par. 144 f); replace if worn out.

(4) LOOSE OR DIRTY CONNECTIONS, OR BROKEN WIRES. Check wiring and connections (par. 64). Clean and tighten loose connections. Repair or replace broken wires.

(5) DEFECTIVE REGULATOR. Replace regulator (par. 145).

(6) DEFECTIVE GENERATOR. Replace generator (par. 144 h and i).

61. LIGHTING SYSTEM.

a. Reference. Those troubles most commonly encountered in the lighting system are described in this paragraph. Subjects covered are listed below for quick reference.

All vehicle lights inoperative	Subpar. b.
All lights burn dimly	Subpar. c.
All outside lights inoperative	Subpar. d.
Headlights, blackout lights, or blackout marker lights inoperative	Subpar. e.
Taillights or stop lights inoperative	Subpar. f.
Instrument panel or compass light inoperative	Subpar. g.
Hull or turret dome lights inoperative	Subpar. h.
Trailer blackout taillight or stop light inoperative —M39 vehicle	Subpar. i.

Part Three—Maintenance Instructions

b. **All Vehicle Lights Inoperative.** If neither outside or inside lights burn when light switches are turned on, check to make sure 24-volt master switch is turned on (fig. 15) and press both lights circuit breaker buttons to set circuit breakers (M, N, fig. 16). If lights still fail to burn, turn fuel gage switch on; if fuel gage registers, the trouble is in instrument panel wiring (par. 62 c), but if fuel gage does not register check following possible causes.

  (1) NO CURRENT IN BATTERY CIRCUIT. See paragraph 58 c.
  (2) DEFECTIVE MASTER SWITCH. Bridge across 24-volt switch terminals with jumper wire; if lights burn, replace master switch (par. 167 a).
  (3) LOOSE OR CORRODED CONNECTIONS OR OPEN CIRCUIT. Disconnect master switch box conduit from instrument panel (fig. 205) and test circuits (A) and (D) as described in paragraph 64. If wiring circuit is satisfactory test circuit breakers, wiring, and switches in instrument panel (par. 62).

c. **All Lights Burn Dimly.**

  (1) LOOSE OR CORRODED BATTERY CABLE TERMINALS. Inspect, clean and tighten battery cable terminals (par. 142 f). Inspect, clean and tighten connections in battery junction box and master switch box.
  (2) BATTERIES DISCHARGED. Test battery specific gravity (par. 142 h); recharge or replace batteries (par. 142 d, g and h).
  (3) DEFECTIVE MASTER SWITCH. Bridge across 24-volt switch terminals with jumper wire; if lights burn properly, replace master switch (par. 167 a).
  (4) LOOSE OR CORRODED CONNECTIONS IN WIRING CIRCUIT TO INSTRUMENT PANEL. See subparagraph b (3) above.

d. **All Outside Lights Inoperative.** If no outside lights burn at any position of lights switch, press outside lights circuit breaker button to make sure breaker is set. Turn on instrument panel and dome lights; if these lights do not burn refer to subparagraph b above. If inside lights burn, trouble is most probably in instrument panel wiring, outside lights circuit breaker, or lights switch; test as described in paragraph 62.

e. **Headlights, Blackout Lights, or Blackout Marker Lights Inoperative.** If neither headlights nor taillights burn, refer to subparagraph d above. If taillights burn, substitute headlight or blackout light assemblies from another vehicle; if lights then burn, replace lamp-units in inoperative light assemblies (par. 151). If lights do not burn with substitute units, remove cover from headlight junction box. Attach one lead of test light (fig. 47) to terminal "B" for headlight test, terminal "D" for blackout light test, or terminal "E" for blackout marker test (fig. 233), and connect other test lead to ground on hull. If test light burns with switch at "HD-LT" position for headlight test, "BO-DR" position for blackout light test, or "BO-MK" position for blackout marker test, the circuit to headlight junction box is satisfactory. Using test light and 24-volt battery, test wiring connections between junction box and light sockets if current is reaching junction box, or between junction box and instrument panel if current

TM 9-755
61

Trouble Shooting

is not reaching junction box. If the later circuit is satisfactory, test outside lights circuit breaker, wiring, and switch in instrument panel (par. 62).

f. **Taillights or Stop Lights Inoperative.** If neither headlights nor taillights burn, refer to subparagraph b above. If stop lights do not burn, test stop light switches and check adjustment of operating cams (par. 154 c (2)). Replace inoperative taillight or stop light lamp-units (par. 152 a). If replacement lamp-units do not correct trouble, test wiring from lights to instrument panel as described in paragraph 64. If wiring circuit is satisfactory, test outside lights circuit breaker, wiring, and switch in instrument panel (par. 62).

g. **Instrument Panel or Compass Light Inoperative.** Instrument panel and compass lights are connected through panel light switch and inside lights circuit breaker (P, N, fig. 16). If any one light burns the trouble is due to burned out lamp or open circuit between lamp and switch; replace lamp; if this does not correct trouble test wiring circuit (par. 64 for compass wiring; par. 62 for instrument panel wiring). If hull dome lights burn, but panel and compass lights do not burn, test panel light switch (par. 62 k). If hull dome lights, panel and compass lights are inoperative, test inside lights circuit breaker (par. 62 d).

h. **Hull or Turret Dome Lights Inoperative.** Hull dome lights are connected to inside lights circuit breaker in instrument panel (N, fig. 16). If neither hull dome lights nor panel lights burn, test circuit breaker (par. 62 d). If panel lights burn, replace lamp in inoperative dome light. If this does not correct trouble, test for open circuit between instrument panel and dome light (par. 64). The turret dome lights are connected to the dome lamp circuit breaker in turret wiring switch box (fig. 243). If one dome light burns, replace lamp in inoperative light. If neither light burns, press circuit breaker button; if this does not correct trouble turn on traverse motor master switch to determine whether current is reaching switch box. Replace circuit breaker, or check for open circuit to switch box as required.

i. **Trailer Blackout Taillight or Stop Light Inoperative, M39 Vehicle.** Lighting system in Vehicle M39 is designed to operate only one blackout taillight and one blackout stop light in a trailer—either 6-volt or 12-volt lamps. If trailer is equipped with more lights, the additional lights must be disconnected. If equipped with proper number of lights and correct lamps, and lights are inoperative, check all vehicle outside lights. If any vehicle outside lights are also inoperative, make corrections specified in pertinent subparagraphs above. If all vehicle lights are satisfactory, check for following causes relative to trailer lights.

(1) TRAILER CABLE CONNECTOR PLUG NOT PROPERLY INSTALLED IN RECEPTACLE. Inspect receptacle and connector plug for dirty or corroded contacts; clean as required. Insert connector plug fully into receptacle.

TM 9-755
61-62

Part Three—Maintenance Instructions

---

(2) LIGHTS SWITCH NOT AT PROPER POSITION. Trailer stop and taillights operate only when light switch is at "BO-MK" or "BO-DR" position. Turn switch to proper position.

(3) NO CURRENT TO TRAILER ELECTRIC RECEPTACLE. Install a 6-volt or 12-volt 3-candlepower lamp in test light. Connect one lead of test light to ground terminal in receptacle (lower left terminal, viewed from rear). Connect other lead to (TL) terminal (upper left, viewed from rear) for taillight test, or to (SL) terminal (lower right, viewed from rear) for stop light test. If test light burns when lights switch is at (BO-MK) or (BO-DR) position, vehicle wiring is satisfactory and trouble is in trailer cable or trailer wiring. If light does not burn, remove test light lead from ground terminal and connect it to ground on hull. If light burns, electrical receptacle is not properly grounded by ground wire (fig. 226); inspect and tighten ground wire connections. If test light still does not burn, test wiring circuits (step (4) below).

(4) OPEN CIRCUIT BETWEEN INSTRUMENT PANEL AND RECEPTACLE. Apply vehicle brakes if trailer stop light circuit is being tested. Remove covers from tail and stop light blackout resistor above instrument panel. Connect one lead of 24-volt test light to ground on hull; connect other lead to terminal on rear side of resistor to which black-natural-tracer wire is connected for test of taillight circuit, or to which red wire is connected for test of stop light circuit. If test light burns with lights switch at (BO-MK) or (BO-DR) position, circuit from instrument panel is satisfactory. If light does not burn, disconnect trailer stop and taillight conduit from instrument panel and test wires (B) and (C) in conduit and instrument panel (fig. 208). If circuit from instrument panel to resistor is satisfactory, install a 6-volt or 12-volt lamp in test light and move test light lead to terminal on front side of resistor to which the same code color wire is attached. (Black-natural tracer for taillight; red for stop light.) If test light burns, resistor is satisfactory; otherwise, replace the resistor. If resistor and circuit to resistor are found to be satisfactory, test wire of same code color from resistor to receptacle as described in paragraph 64. If wiring circuit is satisfactory but no current comes through the receptacle (step (3) above), replace receptacle (par. 174).

62. INSTRUMENTS AND SIGNAL SENDING UNITS.

a. Reference. Troubles most commonly encountered with parts in the instrument panel, and signal sending units which control certain instruments, are described in this paragraph. The subjects covered are listed below for quick reference.

General	Subpar. b.
Circuit breaker open or defective	Subpar. c.
Wires and connections faulty	Subpar. d.
Ammeter inoperative	Subpar. e.
Engine oil pressure gage inoperative	Subpar. f.
Engine oil temperature gage inoperative	Subpar. g.

Trouble Shooting

Fuel gage or switch inoperative .... Subpar. h.
Headlight, blackout light, and taillight
   switch inoperative .... Subpar. i.
Magneto, booster and starter switch
   inoperative .... Subpar. j.
Miscellaneous switches inoperative .... Subpar. k.
Converter oil temperature warning light
   inoperative .... Subpar. l.
Speedometer or tachometer inoperative .... Subpar. m.

   b. General. When tests of an electrical unit and its wiring connections, as described in pertinent paragraphs in this section, indicate that the fault is in the instrument panel, associated instruments and wiring within panel assembly must be tested for operation and continuity as described in the following subparagraphs. When the proper amphenol receptacle is accessible with sponson side opening cover removed, circuits within the panel may be tested for continuity by connecting a test light to the proper pin in receptacle (fig. 46 and 207 or 208). If receptacle is not accessible through opening in sponson, remove instrument panel face plate (par. 156). When wires affecting a given circuit are found to be loose or damaged, it is advisable to examine all other wires in the panel assembly to guard against future trouble with other electrical units.

   c. Circuit Breaker Open or Defective. Test any circuit breaker without removal from instrument panel face plate by connecting test light (fig. 47.) and battery in series with both terminals of circuit breaker. Press button which should remain set, test light will burn if circuit breaker is closed and satisfactory for use. If reset button will not stay in when pressed, test light does not burn, or circuit breaker shows evidence of having been overheated or damaged, replace the part (par. 161).

   d. Wires and Connections Faulty. Before any instrument is tested or replaced, inspect and test all wires within instrument panel which are connected to it or are a part of circuit. All wires affecting a given instrument are shown on diagram in panel case (fig. 207 or 208) and may be identified by code color and letters on receptacle pins where indicated. Inspect wires for loose or corroded connections, broken insulation or other damage. Clean and tighten loose connections and repair or replace damaged wires. Using test light (fig. 47) connected to 24-volt battery, test wires for continuity by connecting test leads to both ends of wire. Replace wire if open circuited.

   e. Ammeter Inoperative. If batteries are near full charge ammeter hand will show very slight movement with either generator running. Start engine and set speed at 1,450 revolutions per minute. If ammeter does not show charge, turn on lights and other electrical units to increase generator output; ammeter should show increased charge under these conditions. If vehicle is equipped with auxiliary generator, run auxiliary generator with engine generator stopped. Readings on ammeters in instrument panel and generator control box will be approximately the same if both instruments are in good condition. If tests indicate that ammeter is inoperative, test ammeter

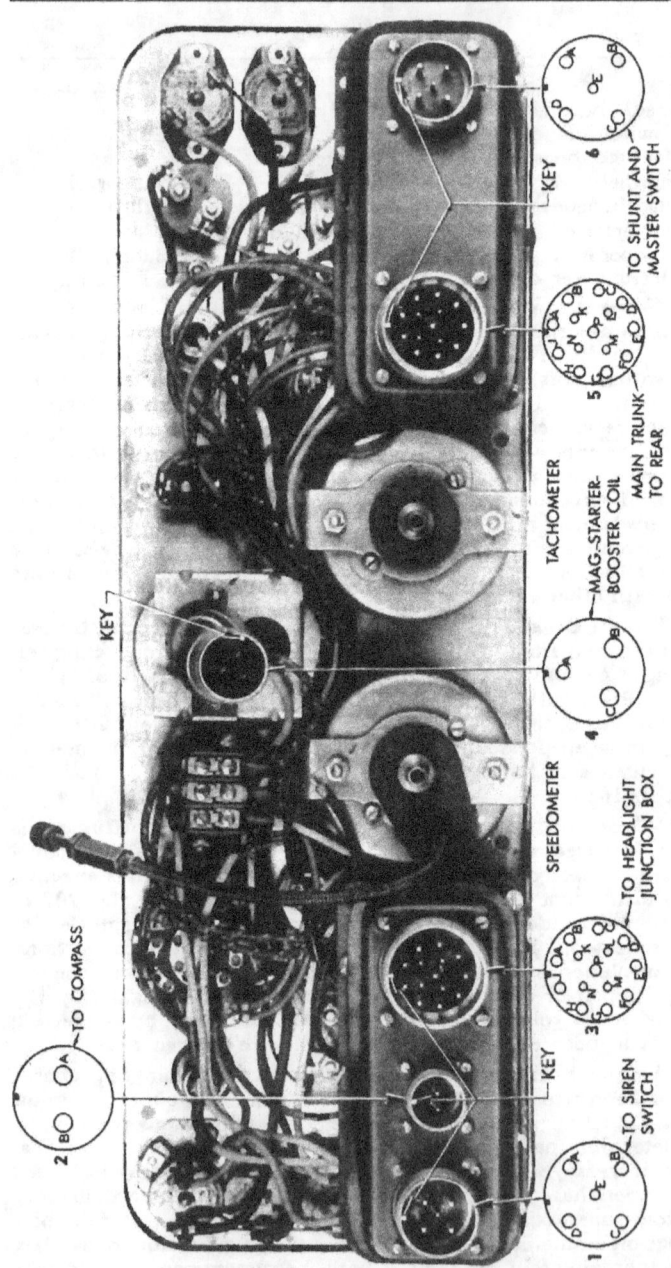

Figure 46—Instrument Panel Wire and Multiple Pin Locations—M18

wires from shunt in starter junction box to terminal block in master switch box, from terminal block to shunt and master switch receptacle on instrument panel, and from receptacle pins (B) and (C) to ammeter (fig. 207 or 208), using test light as described in paragraph 64 and subparagraph e below. If all wires are satisfactory, replace ammeter (par. 157).

f. **Engine Oil Pressure Gage Inoperative.** Press "GAGES" circuit breaker button on instrument panel to make certain that circuit breaker is set. If circuit breaker is set, unscrew captive screws to disconnect conduit housing from gage unit and detach wire from unit (fig. 249). If oil pressure gage hand goes to high pressure end of scale when the wire is disconnected, and fluctuates when the wire is intermittently grounded to engine, gage and wiring are satisfactory and gage unit is faulty; replace gage unit (par. 170 c). If gage unit is not faulty, disconnect main trunkline conduit from instrument panel and test wiring circuit (K) as described in paragraph 64. If wiring is satisfactory, test circuit breaker and instrument panel wiring to gage (subpar. c and b above); replace gage if circuit breaker and wiring are satisfactory (par. 158).

g. **Engine Oil Temperature Gage Inoperative.** Press "GAGES" circuit breaker button on instrument panel to make certain that circuit breaker is set. If circuit breaker is set, unscrew captive screws to disconnect conduit housing from gage unit and detach wire from unit (fig. 248). If gage hand moves to low temperature end of scale when wire is disconnected, and moves to high temperature end of scale when wire is grounded to hull, gage and wiring are satisfactory and gage unit is faulty; replace gage unit (par. 170 b). If gage unit is not faulty, disconnect main trunkline conduit from instrument panel and test wiring circuit (N) as described in paragraph 64. If wiring is satisfactory, test circuit breaker and instrument panel wiring to gage (subpar. c and d above); replace gage if circuit breaker and wiring are satisfactory (par. 158).

h. **Fuel Gage or Switch Inoperative.** Press "GAGES" circuit breaker button on instrument panel to make certain that circuit breaker is set. If fuel gage indicates empty when fuel gage switch is at "LEFT" or "RIGHT" position and fuel tank is known to be full, a short or ground exists in the corresponding fuel gage tank unit or fuel gage circuit. If fuel gage indicates full when switch is on either tank position and fuel tank is known to be empty, an open circuit exists in fuel gage tank unit or fuel gage circuit. For either condition, check fuel gage tank unit (step (1) below). If fuel gage indicates full with switch at "OFF" position, the fault is in fuel gage or switch in instrument panel; check switch and gage (step (3) below).

(1) TESTING FUEL GAGE TANK UNIT. Disconnect wire from fuel gage tank unit which appears to be faulty (par. 170 d) and turn switch on. If fuel gage indicates full when wire is disconnected, and indicates empty when wire is grounded to hull, wiring and instruments in panel are satisfactory and tank unit is faulty; replace tank unit (par. 170 d). If fuel gage does not indicate as specified. wiring or instruments are at fault; test wiring (step (2) below).

TM 9-755

Part Three—Maintenance Instructions

(2) TESTING TANK UNIT WIRING FROM INSTRUMENT PANEL. Disconnect main trunk-line conduit from instrument panel and test wiring circuit (F) for left tank, or circuit (G) for right tank, as described in paragraph 64. If wiring is satisfactory, test fuel gage circuit in instrument panel (step (3) below).

(3) TESTING FUEL GAGE CIRCUIT IN INSTRUMENT PANEL. Test circuit breaker and instrument panel wiring to fuel gage and switch (subpar. c and d above). Test fuel gage switch by making connections specified in following table; if test light does not act as shown under, Test Light Indication, the switch is faulty and must be replaced (par. 160 e). If circuit breaker, wires, and switch test satisfactory, replace fuel gage (par. 158).

Test Light Connections to Switch Terminals	Switch Lever Position	Test Light Indication
No. 5 and No. 1-2-3-4 (joined by bus bar)	OFF	OFF
	LEFT	ON
	RIGHT	ON
No. 6 and No. 7-8 (joined by bus bar)	OFF	OFF
	LEFT	ON
	RIGHT	OFF
No. 6 and No. 9-10 (joined by bus bar)	OFF	OFF
	LEFT	OFF
	RIGHT	ON

l. Headlight, Blackout Light, and Taillight Switch Inoperative. Test outside lights circuit breaker (subpar. d above) and test wiring and lights switch as specified in the following steps.

(1) TESTING WIRING IN PANEL. Inspect and test light wiring as described in subparagraph e above, connecting test light leads as shown in following table (see fig. 207 or 208).

Switch Terminal	Headlight Junction Receptacle Pin	Color of Wire	Lights Affected
HT	B	Orange-green tracer	Both headlights
BOD	D	Black-Nat. tracer	B. O. headlight
BHT	E	Orange	B. O. markers
SS	F	Grey	Stop lights
SW	G	Green	Stop lights
	Main Trunk Receptacle Pin		
HT	B	Orange-green tracer	Service tail and stop lights
S	C	Tan	Service tail and stop lights
BS	D	Red	B. O. stop light
BHT	E	Black-Nat. tracer	Left B. O. taillight
BHT	N	Orange	Right B. O. taillight
BAT	Lights Circuit Breaker	Yellow	All lights
	Trailer Receptacle Pin—M39 Only		
TT	B	Nat.-black and red tracer	Trailer B. O. taillight
SS	C	Grey	Trailer stop light

Trouble Shooting

(2) TESTING LIGHTS SWITCH. Connect a jumper wire between switch terminals (SS) and (SW) and connect test light with 24-volt battery as specified in following table. When placing switch lever in positions specified, work lever several times to test whether switch is making a positive contact. If test light does not burn when connected under all conditions indicated in table, replace switch assembly (par. 160 b).

Switch Terminals	Switch Lever Position	Vehicle Lights Affected
BAT and BS	BO-DR	B. O. stop light
BAT and HT	HD-LT	Both headlights and service taillight
	BO-DR	Trailer B. O. Taillight (M39)
BAT and TT	BO-DR	B. O. markers, B. O. taillights
	BO-MK	B. O. markers, B. O. taillights, B. O. stop lights
BAT and BHT	BO-DR	B. O. headlight (left)
BAT and BOD		
BAT and S	STOP-LT	Service stop light (left)
	HD-LT	Service stop light (left)
BAT and SS	BO-MK	Trailer B. O. stop light (M39)

j. **Magneto, Booster, and Starter Switch Inoperative.** Test booster coil and starter control circuit breaker (subpar. c above), test wiring and switch assembly as described in the following steps.

(1) TESTING WIRING IN PANEL. Inspect and test wiring in instrument panel as described in subparagraph d above, connecting test light leads at the following points (fig. 207 or 208).

(a) Bus bar at lights circuit breakers and bus bar at booster coil circuit breaker, to test black wire.

(b) Booster coil circuit breaker and two-wire terminal on idle cut-off switch, to test grey wire.

(c) Terminal block post nearest starter switch and pin "H" in headlight junction receptacle, to test red-natural-tracer wire.

(d) Pin "J" in headlight junction receptacle and middle post on terminal block, to test red-natural-tracer wire.

(e) Middle post on terminal block and pin "J" in main trunk receptacle, to test red-natural-tracer wire.

(2) TESTING SWITCH AND CABLE ASSEMBLIES (fig. 207). Connect test light with 24-volt battery as specified in following table. When placing switch levers in positions indicated, work lever several times to test whether switch is making a positive contact. If test light does not act as shown under Test Light Indication, proceed to step (3) below.

Test Light Connections	Switch and Position	Test Light Indication
2-wire terminal on idle cut-off switch and terminal "C" in switch cable connector	Booster switch—"ON"	ON

TM 9-755
62

Part Three—Maintenance Instructions

Test Light Connections	Switch and Position	Test Light Indication
2-wire terminal on idle cut-off switch and post nearest starter switch on terminal block	Starter switch—"ON"	ON
Ground on panel face plate and terminal "A" in switch cable connector	Magneto switch—"OFF" "L" "R" "BOTH"	ON OFF ON OFF
Ground on panel face plate and terminal "B" in switch cable connector	Magneto switch—"OFF" "L" "R" "BOTH"	ON ON OFF OFF

(3) TESTING SWITCH CABLE AND CONNECTIONS. If tests in step (2) above indicate faulty switch or cable, remove switch assembly, remove switch cover, and inspect wires for loose or corroded connections to switch terminals. Test each wire with test light. Replace cable assembly or other wires if found faulty. If wires are satisfactory, replace switch assembly.

k. **Miscellaneous Switches Inoperative.** Switches referred to in this subparagraph are those not covered separately in subparagraphs h, i and j above. When a switch appears to be inoperative, test circuit breakers and wires connected to it (subpar. c and d above). Connect test light with 24-volt battery to both terminals of switch and work switch a number of times to test whether it is opening and closing properly. The panel lights switch is a rheostat type. It is off, when knob is turned clockwise as far as it will go; on, when knob is turned counterclockwise; light should burn brighter as knob is turned farther counterclockwise. Replace any switch found to be faulty (par. 160).

l. **Converter Oil Temperature Warning Light Inoperative.**

(1) LIGHT BURNS WHEN TRANSMISSION IS COLD. Disconnect conduit and wire from oil temperature sending unit on transmission transfer case to cooler elbow. If warning light goes out when wire is disconnected from unit, replace the unit (par. 170 a). If warning light does not go out, disconnect headlight junction conduit from instrument panel and test circuit "K" as described in paragraph 64. If wire circuit is satisfactory, test inside lights circuit breaker, panel wiring, and test switch (subpar. c, d and k above).

(2) LIGHT DOES NOT BURN WHEN TRANSMISSION IS HOT. The warning light should burn when transmission oil reaches 280° F to 290° F. If light does not burn when oil is known to be at these temperatures or above, check the following causes:

(a) *Circuit Breaker Open.* Press inside lights circuit breaker button.

(b) *Lamp Burned Out.* Test lamp by pressing test switch and replace lamp if necessary.

Trouble Shooting

(c) *Temperature Sending Unit or Wiring Defective.* Test sending unit and wiring (step (1) above).

m. Speedometer or Tachometer Inoperative.
(1) INSTRUMENT HAND FLUCTUATES. Shaft improperly connected (fig. 72).
(2) BROKEN SHAFT. Replace shaft and connect properly (fig. 72).
(3) FAULTY INSTRUMENT. Replace (par. 159).
(4) DRIVE GEAR BROKEN. Replace broken parts.

63. ELECTRICAL EQUIPMENT.

a. General. Troubles commonly encountered on electrical equipment, not covered in other paragraphs of this section, are as follows:

Master switch inoperative .... Subpar. b.
Siren inoperative .... Subpar. c.
Trailer electric brake inoperative—M39 vehicle Subpar. d.

b. Master Switch Inoperative. If vehicle lights are dim or inoperative, and starter fails to crank engine or has low cranking speed, the 24-volt master switch should be tested as a possible source of trouble. If lights are dim or inoperative, remove switch box cover and bridge across master switch terminals with a jumper wire; if this corrects the light condition the switch is faulty. A further test may be made with a test voltmeter. Connect negative (—) lead of test voltmeter to ground on the hull, and connect positive (+) lead to the lower terminal of 24-volt master switch. Turn switch on and crank engine with starter while noting voltmeter reading. Move positive (+) lead to upper terminal of switch and note voltmeter reading while cranking engine with starter. If voltmeter reading is higher at lower terminal than at upper terminal, there is high resistance in the switch; replace switch (par. 167 a).

c. Siren Inoperative. If siren fails to operate when siren switch is pressed, push siren circuit breaker button (L, fig. 16) to make certain that breaker is set. Ground the siren to hull with a jumper wire; if siren then operates, inspect and tighten siren attaching screws to secure a positive ground connection. If siren does not operate when grounded, remove headlight junction box cover, connect one lead of test light to junction box terminal post (C) (fig. 233), connect other lead to ground on hull. If test light burns when siren switch is pressed, check siren wire for broken or dirty terminal; if terminal and wire are satisfactory replace siren (par. 171). If test light does not burn, remove sponson side opening cover and disconnect siren switch box conduit from instrument panel (fig. 205). Connect jumper wire to pins (A) and (C) in receptacle on instrument panel; if siren operates, trouble is in wires to switch or switch is faulty. If siren does not operate with jumper wire between receptacle pins (A) and (C), disconnect headlight junction box conduit from instrument panel and test wire circuit (C) to junction box with test light. If wire is satisfactory, test siren circuit breaker and siren circuit wiring in instrument panel (par. 62 d and e).

TM 9-755
63-64

Part Three—Maintenance Instructions

d. **Trailer Electric Brake Inoperative, M39 Vehicle.**

(1) TRAILER CABLE CONNECTOR PLUG NOT PROPERLY INSTALLED IN RECEPTACLE. Inspect receptacle and connector plug for dirty or corroded contacts; clean as required. Insert connector plug fully into receptacle.

(2) NO CURRENT TO TRAILER ELECTRIC RECEPTACLE. Connect one lead of test light to ground terminal in receptacle, lower left terminal, viewed from rear. Connect other lead to (BK) terminal, lower right, viewed from rear. Have another man operate brake pedal through entire range and then, with pedal depressed, operate knob on controller through entire range. If test light does not burn when pedal is depressed, remove test light lead from ground terminal in receptacle, attach it to ground on hull and repeat test. If light burns with test light grounded to hull, inspect and tighten receptacle ground wire connections (fig. 226). If test light does not burn in either test, test wiring circuits (step (3) below).

(3) OPEN CIRCUIT BETWEEN MASTER SWITCH BOX AND RECEPTACLE. Using test light (fig. 47), progressively test wiring circuit between master switch box and receptacle in the following sequence.

(a) Connect one lead of test light to ground on hull; connect other lead to terminal marked "MSTR-SW" in brake resistor box (fig. 252). If test light does not burn, replace red-natural-tracer wire to master switch box.

(b) Connect one lead of test light to ground on hull and other lead to terminal marked "CTR" in resistor box. If test light does not burn, replace resistor (par. 173).

(c) Move test light lead from "CTR" to terminal marked "BRK-CTR" and depress brake pedal. If test light burns, test natural colored wire between resistor box and trailer electric receptacle; if test light does not burn, test brake controller (step (d) below).

(d) Remove sponson side opening cover and remove terminal cover from brake controller. Ground one lead of test light to hull and connect other lead to terminal marked "BAT" on controller. If light does not burn, replace wire between resistor box and controller terminal "BAT." If light burns, however, move test lead from "BAT" to terminal marked "BK." If test light does not burn when brake pedal is depressed, replace brake controller (par. 172); if light burns, however, replace wire between "BK" terminal and resistor box.

64. VEHICLE WIRING, JUNCTION BOXES, AND GROUND STRAPS.

a. **General.** When a test of wiring leading to an electrical unit is specified in other paragraphs of this section, all wires between the source of current and unit, as well as all intervening connections and junction or terminal boxes are included. Where ground straps or wires are used to assure positive grounding of the unit, these must be examined also to make certain that a complete electrical circuit exists. Location, identification, and replacement of wires, cables and conduits are covered in paragraph 164. Location and replacement of

Trouble Shooting

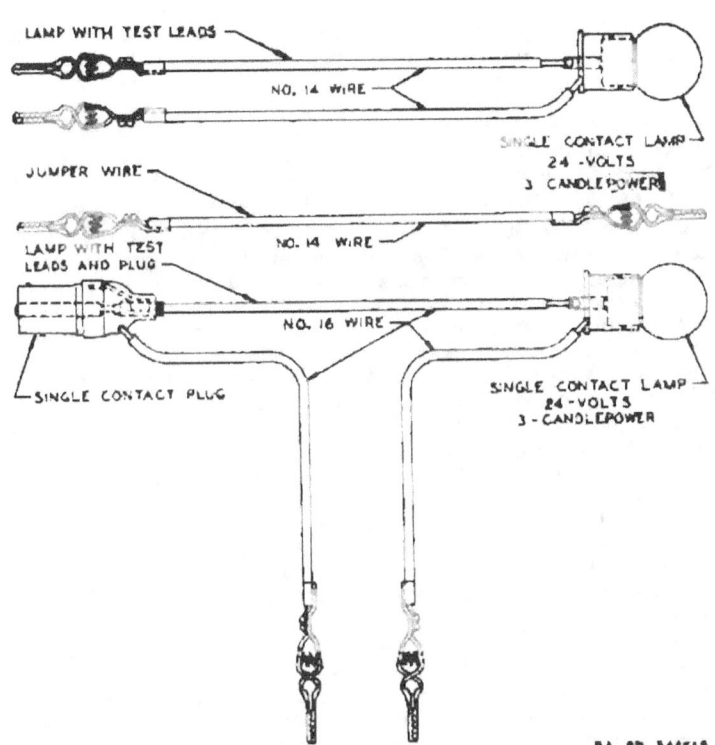

Figure 47—Test Light and Jumper Wire

junction and terminals boxes are covered in paragraph 166. Location and replacement of ground straps and wires are covered in paragraph 165.

b. Identification of Individual Wires. Where a number of wires are run through one conduit, the conduits are usually closed at the ends by a multiple pin connector for easy attachment to equipment. Each wire in the conduit is attached to an individual connector pin or socket which is identified by a letter on the connector; both ends of the wire have the same identification letter. Where multiple pin connectors are not used, the individual wires may be identified by their color. Any wire may be identified at connecting points for test purposes by referring to the vehicle wiring diagrams, figures 48, 49 and 50 for M18 vehicle, or figures 51, 52 and 53 for M39 vehicle.

c. Test Equipment. Figure 47 illustrates two simple test lights and a jumper wire, with which all wiring tests can be made. One test light shown consists of a 24-volt lamp in a socket provided with two test leads. This light may be connected in series with wires or equipment, or may be used with the jumper wire and a 24-volt battery where separate current is required for a test. The other test light shown includes a single-contact plug which may be installed in any outlet socket on vehicle to use current from vehicle batteries for test purposes. Either test light may be used for a test, depending upon the source of battery current. The jumper wire may be used for grounding wires or equipment for test purposes.

d. *Test Procedure.* Turn both master switches off. Disconnect the wire, or conduit containing the wire, at the electrical unit and at point where the circuit connects to battery current, in order to test all wires and connections in circuit at the same time. When current for circuit being tested comes through instrument panel, remove sponson side opening cover and disconnect affected conduit from receptacle on back of instrument panel (fig. 46 and 205). Using a test light (subpar. c above), test wiring circuit as follows:

(1) TEST FOR GROUNDED CIRCUIT. Connect one lead of test light to battery end of wire circuit and connect other test lead to positive terminal of a 24-volt battery whose negative terminal is grounded to hull. Disconnect wire from electrical unit and place it so that its terminal does not touch any metal part. Move or shake wire on conduit where accessible while observing test light. If test light does not burn, the circuit is not grounded; if test light burns or flashes, a ground exists and must be located and corrected (step (3) below).

(2) TEST FOR OPEN CIRCUIT. Using jumper wire, ground the wire where disconnected from electrical unit. With test light connected as in step (1) above, move or shake wire or conduit where accessible while observing test light. If test light burns steadily, circuit is closed and is satisfactory for use; if test light does not burn steadily, circuit is open and must be corrected (step (3) below).

(3) LOCATING AND CORRECTING FAULT IN CIRCUIT. If the preceding tests indicate a fault in wiring circuit, disconnect wires or conduits at intermediate junction or terminal boxes so that each individual wire and connection can be inspected and tested with test light until fault is found. Grounds are usually caused by crushed conduits or damaged insulation. Open circuits are caused by broken wires or connections, or by loose or corroded connections.

65. RADIO INTERFERENCE.

a. *General.* The radio set is designed to automatically suppress ordinary interference experienced in radio reception. Make sure that set is properly tuned, adjusted and installed so that interference normally suppressed will not be detected. With vehicle stopped, determine source of interference by making tests of vehicle electrical systems as outlined in the following subparagraphs. When source of the interference has been determined, and corrections have been made, also check electrical equipment to prevent further development of interference. Inspect all conduits for damage, loose or

Trouble Shooting

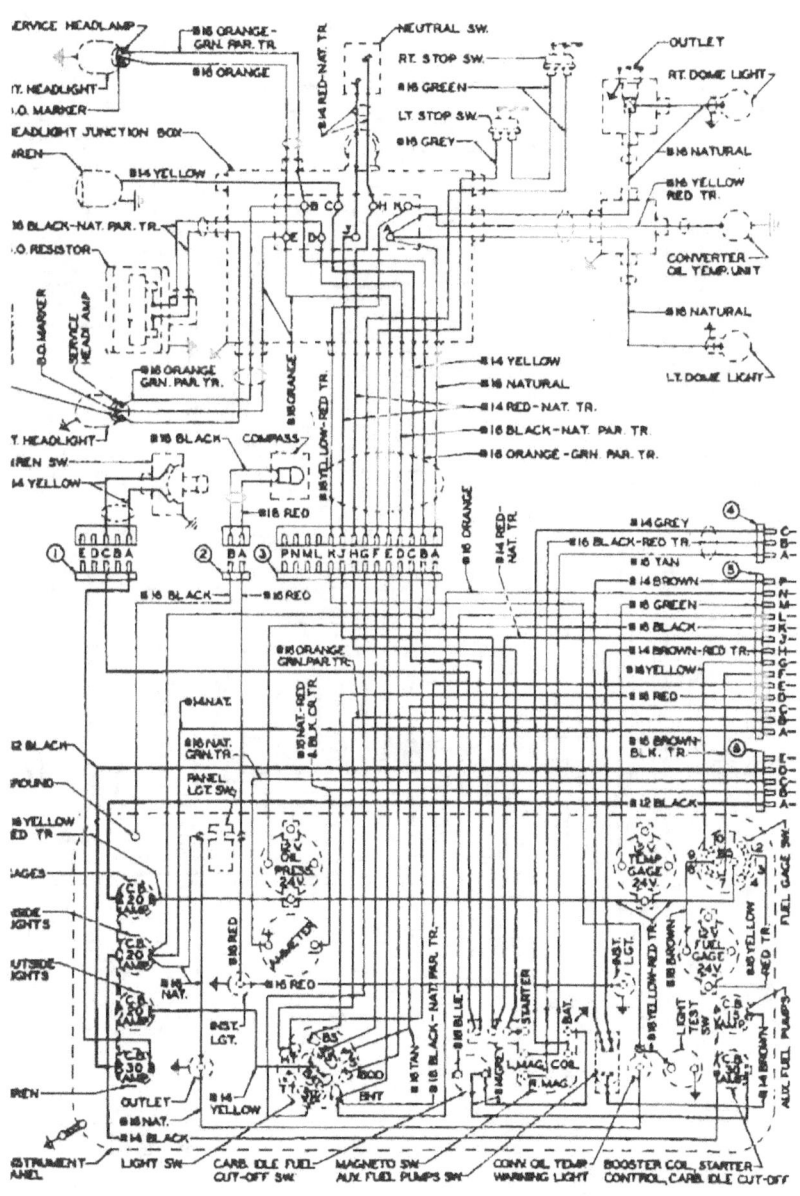

Figure 48—Wiring Diagram—Front Section—M18

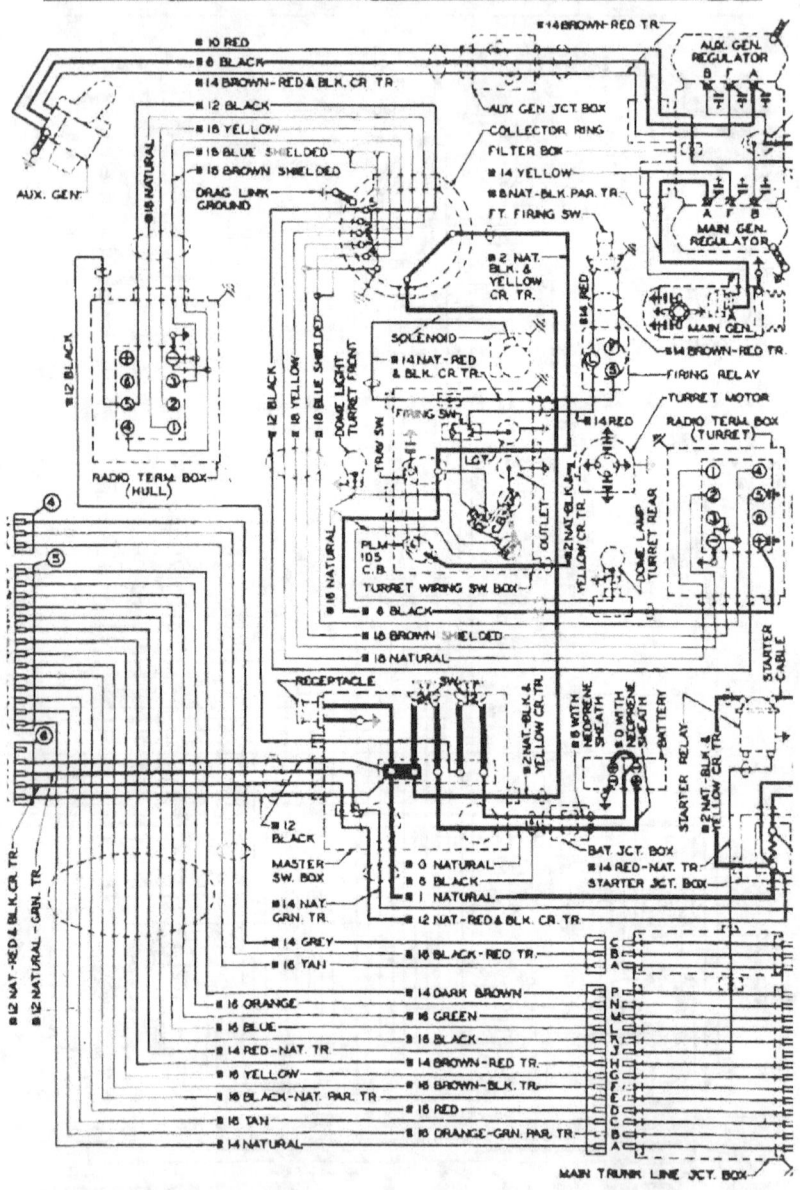

Figure 49—Wiring Diagram—Center Section—M18

Trouble Shooting

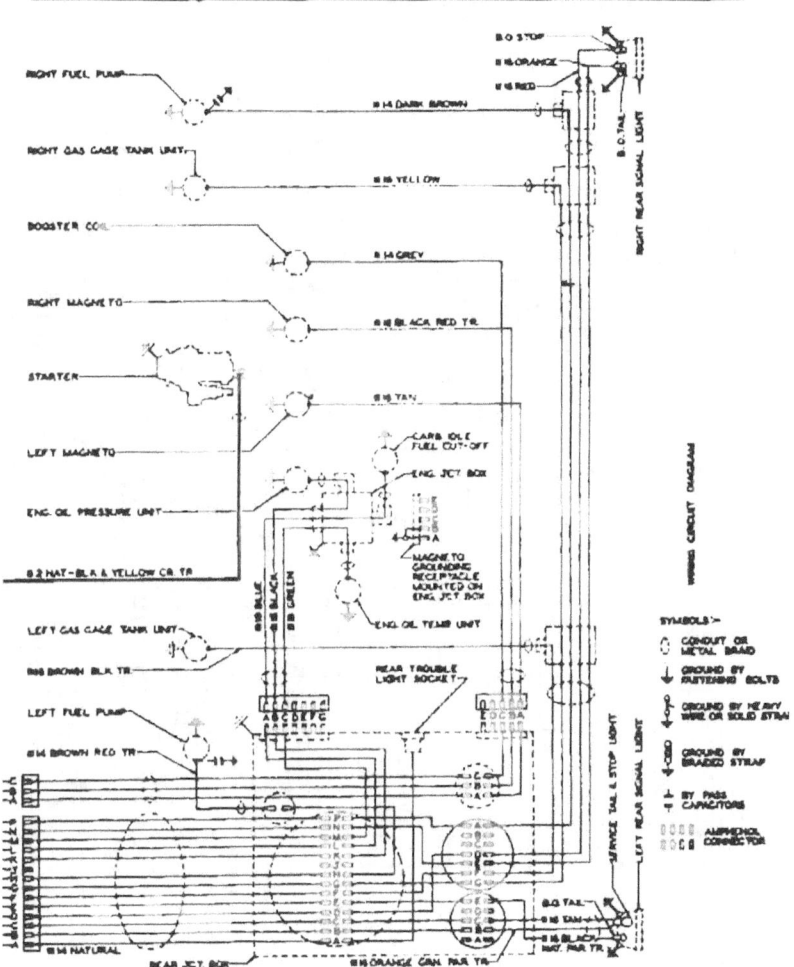

Figure 50—Wiring Diagram—Rear Section—M18

TM 9-755

Part Three—Maintenance Instructions

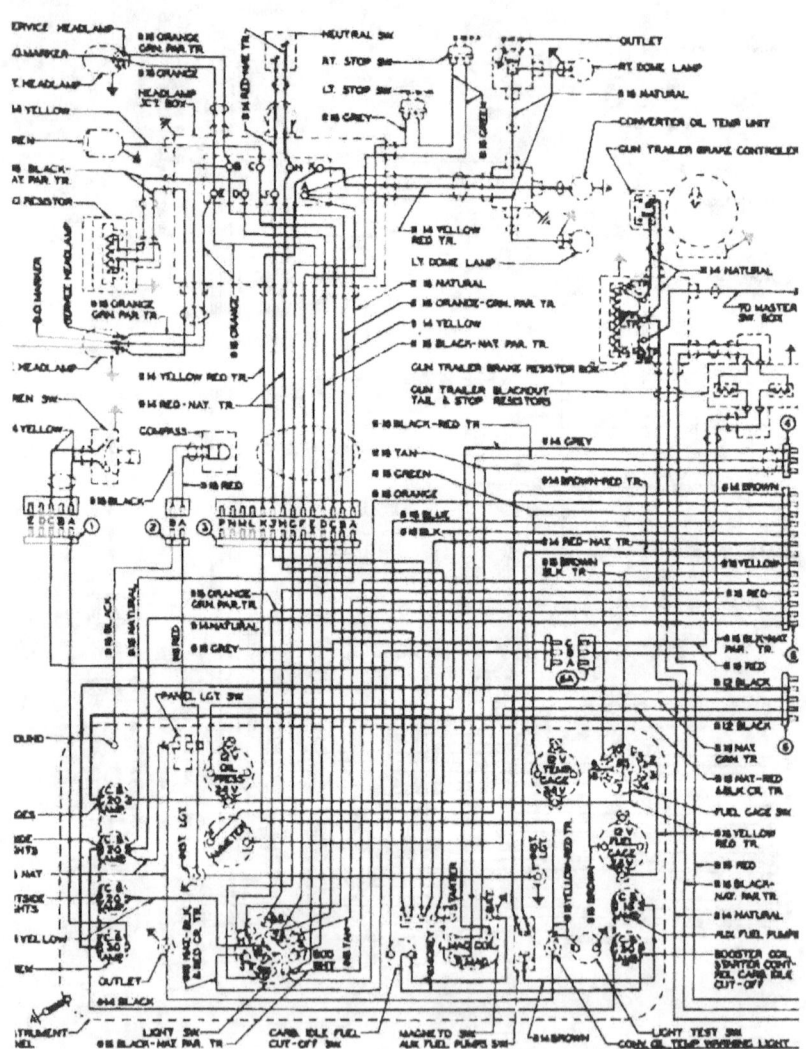

Figure 51—Wiring Diagram—Front Section—M39

Trouble Shooting

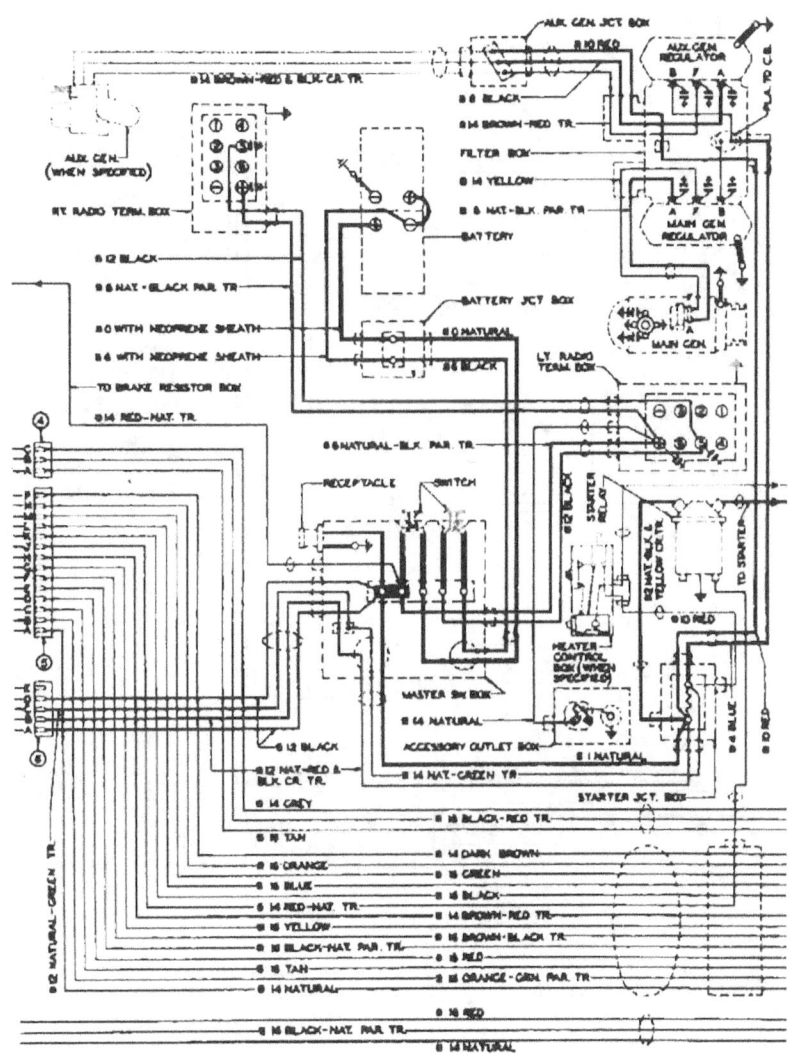

Figure 52—Wiring Diagram—Center Section—M39

TM 9-755

Part Three—Maintenance Instructions

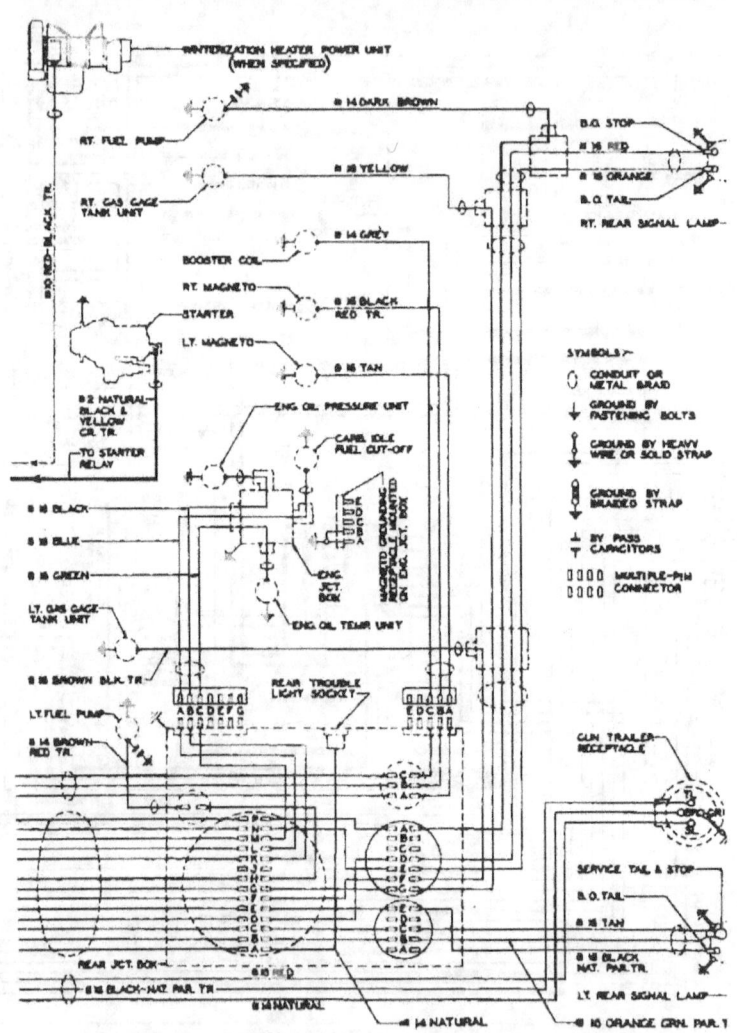

Figure 53—Wiring Diagram—Rear Section—M39

Trouble Shooting

corroded bond connections or ground straps. Check all wiring connections in the terminal boxes for loose terminals, corroded or loose connections. Make sure terminal box covers are securely fastened and make good contact with the boxes. Faulty connections, conduits, or bonds in any part of the electrical system may transmit interference to radio from any source in the vehicle. Test the batteries to be sure that they are properly charged, connected and not damaged. Faulty batteries can cause improper operation of radio so that suppression features of the tubes and circuits will not be effective.

b. **Radio Interference with Engine Running (Vehicle Stopped).**

(1) IGNITION SYSTEM. If a rhythmic "popping" noise is heard in the radio receiver with changes in frequency as the speed of the engine is increased or decreased, the trouble is probably caused by ignition system: Burned magneto points, faulty condenser, loose contacts, broken wiring or improperly gapped spark plugs. Run the engine on first one magneto and then on the other to determine which unit is at fault. Inspect all shielding, bonding and wiring for damage or looseness. Replace damaged conduits, tighten loose connections and correct any faulty bonding as necessary. If interference is still present, test all spark plugs in the faulty circuit for faulty contacts and improperly spaced points. Inspect high tension contacts in the magnetos. Make necessary repairs or replacement to correct the trouble.

(2) ENGINE GENERATOR. If a "whining" or "squealing" sound is detected in radio receiver while engine generator is running, and sound varies in pitch with changes in generator speed, the trouble is probably in engine generator. Stop engine, disengage transfer case clutch (par. 102) and start engine again. If interference is eliminated with generator not running, inspect generator ground wire to see that it is properly connected to clean surfaces with external-toothed lock washers. If satisfactory, remove cover from generator terminals and inspect connections. If connections are tight and there is no obvious damage to conduit containing generator leads, replace capacitors (par. 177 b (1)).

(3) ENGINE GENERATOR REGULATOR. If an intermittent "clicking" sound is heard in receiver while generator is running, and disappears when generator is not running, the trouble may be found in engine generator regulator. Speed up engine to approximately 1,000 revolutions per minute and note intensity of one of the dome lights. If the noise can be heard in receiver just as the light starts to increase in brilliance, the trouble is in regulator. Replace regulator condensers (par. 177 b (2)). If replacement of condensers does not correct trouble, replace regulator (par. 145).

c. **Radio Interference With Turret Operating, M18 Vehicle.**

(1) INTERRUPTION OF DYNAMOTOR OPERATION. If dynamotor speed varies or it operates intermittently, effecting performance of transmitter, interphone amplifier or receiver, inspect all connections from set to radio terminal box in turret. If all connections are satis-

TM 9-755
65-66

Part Three—Maintenance Instructions

factory, inspect wiring and connections from master switch box to turret slip ring box and from slip ring box to turret wiring switch box and turret radio terminal box. Tighten all connections in circuits. If trouble is not corrected by inspection and corrections of circuits, notify higher authority.

(2) NOISE IN RADIO WITH TURRET TRAVERSING ELECTRIC MOTOR RUNNING. If radio interference is experienced when electric motor is started or while it is running, inspect all wiring and connections between turret slip ring box and turret wiring switch box, and tighten. Make sure that turret ground straps on drag link are properly connected. If no trouble is found in circuit, replace condenser in turret wiring switch box (par. 168). If noise is not eliminated by replacement of condenser, replace electric motor (par. 198).

d. Radio Interference with Auxiliary Generator Only in Operation.

(1) LOOSE SPARK PLUG SHIELDING CONDUIT. Tighten conduit.

(2) LOOSE CONNECTIONS IN CONTROL BOX. Tighten connections.

(3) EXCESSIVELY DIRTY COMMUTATOR. Clean commutator (par. 144 f).

(4) DEFECTIVE CAPACITOR AT STARTER SWITCH. Replace capacitor.

(5) DEFECTIVE CONDENSERS AT REGULATOR. Replace condensers.

(6) DEFECTIVE ENGINE GROUND STRAP. Clean and tighten or replace.

(7) DEFECTIVE GROUND STRAPS AT REGULATOR. Clean and tighten, or replace.

(8) DEFECTIVE HIGH TENSION SUPPRESSOR. Replace suppressor.

e. Radio Interference Only When Vehicle Is in Motion. If no interference exists when individual circuits are in operation with vehicle not in motion, as described in subparagraphs above, and interference develops when vehicle is set in motion, it is caused either by loose bonds or conduit coupling nuts or by track static. Check and tighten all coupling nuts and bond attaching screws as described in subparagraph a above. If interference is not eliminated, notify higher authority.

66. TURRET TRAVERSING MECHANISM.

a. Reference. Turret traversing troubles most commonly encountered are described in this paragraph and are listed below for quick references:

Turret does not turn freely by hand traversing mechanism	Subpar. b.
Electric motor fails to operate	Subpar. c.
Electric motor runs, but turret cannot be turned in either direction	Subpar. d.

172

Trouble Shooting

Turret traverse speed is low in both directions	Subpar. e.
Turret traverse speed is low in one direction only	Subpar. f.
Turret can be traversed in one direction only	Subpar. g.
Turret creeps in one direction while vehicle is in horizontal position	Subpar. h.
Turret drifts excessively when vehicle is not in horizontal position	Subpar. i.
Unsteady or sluggish turret operation	Subpar. j.
Abnormal noise in hydraulic pump during operation	Subpar. k.
Oil leaking from hydraulic motor or pump	Subpar. l.
Abnormal noise in hydraulic motor or adapter	Subpar. m.

b. **Turret Does Not Turn Freely by Hand Traversing Mechanism.** Before investigating hydraulic traversing mechanism for malfunctions, place shifting lever (fig. 20) in down position, turn turret lock handle to "FREE" position, and traverse turret manually to make certain that it will rotate freely and smoothly in both directions. Failure to traverse freely and smoothly may be caused by one of the following conditions:

(1) MAIN DRIVE SHAFT PINION IMPEDED BY FOREIGN MATTER ON RACE RING GEAR. Inspect and clean.

(2) TURRET RACE BALL BEARINGS DRY. Lubricate bearings (par. 38).

(3) BRAKE DRAGGING. Adjust brake (par. 195 e).

(4) TURRET LOCK PAWL CONTACTING RACE RING GEAR. Adjust clearance (par. 194).

(5) FOREIGN MATTER OR SCORED PARTS IN HAND TRAVERSING MECHANISM. Replace mechanism (par. 195 b and c).

(6) TURRET RACE BALL BEARINGS OR RINGS DAMAGED. Notify higher authority.

c. **Electric Motor Fails to Operate.**

(1) MASTER SWITCH IN SWITCH BOX IS TURNED OFF. Close master switch (par. 15 a).

(2) TRAVERSE MOTOR MASTER SWITCH TURNED OFF. Turn switch to "ON" position.

(3) MOTOR SWITCH CIRCUIT BREAKER IS OPEN. Reset circuit breaker by pressing button.

(4) BATTERIES DISCHARGED. Test batteries (par. 142 b). If discharged, charge or replace batteries (par. 142 g and h).

(5) BATTERY OR OTHER WIRES DAMAGED OR MAKING POOR CONNECTIONS (fig. 49). Inspect wires and connections at following points: Battery to ground; battery to battery junction box; junction box to master switch box; master switch box to slip ring box; terminals and brushes in slip ring box; slip ring box through drag link to master switch in turret wiring switch box; master switch to circuit breaker bus bar; circuit breaker to electric motor. Clean and tighten loose or corroded connections; replace damaged wires.

(6) ELECTRIC MOTOR BURNED OUT. Replace motor (par. 198).

TM 9-755
66

Part Three—Maintenance Instructions

d. Electric Motor Runs, But Turret Cannot Be Turned In Either Direction.

(1) TURRET LOCK IN "LOCK" POSITION. Turn lock handle to "FREE" position (fig. 19).

(2) SHIFTING LEVER DOWN IN MANUAL POSITION. Apply turret lock, push shifting lever up (fig. 20), then disengage turret lock.

(3) LOW OIL LEVEL IN RESERVOIR. Check for loose connections on oil tubes and tighten securely. Fill reservoir until level is two-thirds of way up on inspection window.

(4) BATTERIES LOW. Recharge or replace batteries (par. 142).

(5) LOOSE OR DIRTY ELECTRICAL CONNECTIONS. Refer to subparagraph c (5) above.

(6) GEAR PUMP RELIEF VALVE PLUNGER STICKING IN OPEN POSITION. Remove and clean plunger (par. 199 c).

(7) GEAR PUMP RELIEF VALVE PLUNGER SPRING BROKEN OR FATIGUED. Replace spring (par. 199 c).

(8) CONTROL BOX VALVE PLUNGER STUCK. Clean and free up plunger (par. 197 c).

(9) HIGH PRESSURE RELIEF VALVE PLUNGER HELD OPEN BY FOREIGN MATTER. Remove and clean plunger (par. 199 b).

(10) HIGH PRESSURE RELIEF VALVE PLUNGER SPRING BROKEN OR FATIGUED. Replace spring (par. 199 b).

(11) HYDRAULIC PUMP SHAFT COUPLING BROKEN. Replace coupling (par. 197 e and f).

(12) PUMP INTERNAL CONTROL PARTS DAMAGED. Replace pump (par. 197 e and f).

(13) HYDRAULIC MOTOR INTERNAL PARTS DAMAGED. Replace hydraulic motor (par. 196).

(14) FOREIGN MATTER OR BROKEN PARTS IN HYDRAULIC MOTOR ADAPTER. Replace adapter (par. 196 b and c).

e. Turret Traverse Speed is Low In Both Directions.

(1) BATTERIES LOW. Recharge or replace batteries (par. 142).

(2) LOOSE OR DIRTY ELECTRICAL CONNECTIONS. Refer to subparagraph c (5) above.

(3) LOW OIL LEVEL IN RESERVOIR, ALLOWING AIR TO ENTER SYSTEM. Check for loose connections on oil tubes and tighten securely. Fill reservoir until level is two-thirds of way up on inspection window.

(4) CHECK VALVES LEAKING. Remove and clean valves (par. 197 d).

(5) HIGH PRESSURE RELIEF VALVE LEAKING. Remove and clean valve (par. 199 b).

(6) GEAR PUMP RELIEF VALVE LEAKING OR PLUNGER STUCK IN OPEN POSITION. Remove and clean valve (par. 199 c).

(7) GEAR PUMP RELIEF VALVE PLUNGER SPRING BROKEN OR FATIGUED. Replace spring (par. 199 c).

Trouble Shooting

(8) TURRET DOES NOT TURN FREELY. Refer to subparagraph b above.

(9) FOREIGN MATTER OR SCORED PARTS IN HYDRAULIC MOTOR ADAPTER. Replace adapter (par. 196 b and e).

f. Turret Traverse Speed is Low in One Direction Only.

(1) EITHER CHECK VALVE IS LEAKING. Remove and clean valve (par. 197 d).

(2) GEAR PUMP RELIEF VALVE PLUNGER LEAKING. Remove and clean valve (par. 199 e).

(3) GEAR PUMP RELIEF VALVE PLUNGER SPRING BROKEN OR FATIGUED. Replace spring (par. 199 e).

(4) PUMP CONTROL HANDLE IMPROPERLY ADJUSTED. Adjust control handle (par. 197 h).

(5) PUMP INTERNAL CONTROL PARTS IMPROPERLY ADJUSTED. Replace pump (par. 197 e and f).

g. Turret Can Be Traversed in One Direction Only.

(1) EITHER CHECK VALVE STUCK IN OPEN POSITION. Remove and clean valve (par. 197 d).

(2) FOREIGN MATTER LODGED IN TRAVERSING MECHANISM. Replace traversing mechanism (par. 195 b and e).

h. Turret Creeps in One Direction While Vehicle is in Horizontal Position.

(1) PUMP CONTROL HANDLE NOT ADJUSTED TO NEUTRAL POSITION. Adjust control handle (par. 197 b).

i. Turret Drifts Excessively When Vehicle is Not in a Horizontal Position.

NOTE: *When hydraulic traversing mechanism is operating with control handle in neutral position and vehicle is tilted, the unbalanced load of the 76-mm gun will cause turret to drift slowly until gun reaches the lowest position. Under these conditions do not turn off the traverse motor master switch until turret lock is placed in "LOCK" position; otherwise, the turret will drift rapidly until 76-mm gun reaches the lowest position, and injury to personnel may result. Whenever power traversing is not necessary, turn turret lock handle to "LOCK" position.*

(1) PUMP CONTROL HANDLE NOT ADJUSTED TO NEUTRAL POSITION. Adjust control handle (par. 197 b).

(2) PUMP SUCKING AIR, DUE TO LOW OIL LEVEL IN RESERVOIR. Fill reservoir until level is two-thirds of way up on inspection window.

(3) PUMP SUCKING AIR, DUE TO LOOSE GEAR PUMP OIL TUBE CONNECTIONS. Tighten connections (No. 3 and 13, fig. 274).

(4) PUMP SUCKING AIR, DUE TO CRACKED GEAR PUMP HOUSING. Replace pump (par. 197 e and f).

TM 9-755

Part Three—Maintenance Instructions

(5) CHECK VALVES LEAKING. Remove and clean valve (par. 197 d).

(6) EXCESSIVE INTERNAL SLIP IN PUMP, DUE TO WORN PARTS. Replace pump (par. 197 e and f).

(7) EXCESSIVE INTERNAL SLIP IN HYDRAULIC MOTOR, DUE TO WORN PARTS. Replace motor (par. 198 b and c).

j. **Unsteady or Sluggish Turret Operation.** When vehicle is level, turret traversing should be steady; however, when vehicle is tilted, traversing will vary in speed as 76-mm gun changes position.

(1) BATTERIES LOW. Recharge or replace batteries (par. 142).

(2) LOOSE OR DIRTY ELECTRICAL CONNECTIONS. Refer to subparagraph c (5) above.

(3) BINDING BETWEEN MAIN DRIVE SHAFT PINION AND RACE RING GEAR. Adjust for proper lash (par. 193 d).

(4) CHECK VALVES LEAKING. Remove and clean valve (par. 197 d).

(5) STICKING RELIEF VALVE PLUNGERS. Remove and clean plungers (par. 199 b and c).

(6) LOW OIL LEVEL IN RESERVOIR. Fill reservoir until level is two-thirds of way up on inspection window.

(7) NO LUBRICANT IN GEAR MECHANISM OR ADAPTER. Lubricate gear mechanism (par. 38). Fill adapter with ⅕ pint of hydraulic oil.

k. **Abnormal Noise in Hydraulic Pump During Operation.**

(1) LOW OIL LEVEL IN RESERVOIR. Fill reservoir until level is two-thirds of way up on inspection window.

(2) LOOSE GEAR PUMP SUCTION TUBE, CONNECTIONS. Tighten connections (No. 3 and 13, fig. 274).

(3) WORN INTERNAL PARTS. Replace pump (par. 197 e and f).

l. **Oil Leaking from Hydraulic Motor or Pump.**

(1) LOOSE OIL TUBE CONNECTIONS. Tighten connections.

(2) DEFECTIVE GASKETS. Replace motor (par. 196 b and c) or pump (par. 197 e and f).

(3) SHAFT OIL SEAL WORN OUT. Replace motor (par. 196 b and c) or pump (par. 197 e and f).

m. **Abnormal Noise in Hydraulic Motor or Adapter.**

(1) LOW OIL LEVEL IN RESERVOIR. Fill reservoir until level is two-thirds of way up on inspection window.

(2) LACK OF LUBRICANT IN ADAPTER. Fill adapter with ⅕ pint of hydraulic oil.

(3) WORN OR BROKEN INTERNAL PARTS. Replace hydraulic motor and adapter (par. 196 b and c).

Section XVII

ENGINE DESCRIPTION AND MAINTENANCE

67. DESCRIPTION AND DATA.

a. Description (figs. 54 and 55). The engine assembly, generally referred to as the engine in this manual, consists of an engine mounted upon a support with all accessories to form a unit power plant ready to install in the vehicle and connect to the power train, external tanks, conduits, mufflers, etc. In addition to the engine and support, the assembly includes the following accessories: starter; magnetos and booster coil; connecting wires, conduits and junction box; primer distributor and pipes; fuel pump, carburetor, governor and throttle box; exhaust manifold; fan, universal joint and flange; oil filter and oil cooler; fuel and oil pipes.

(1) ENGINE TYPE, AND MODELS USED. The engine is a single-row, 9-cylinder, static radial, air-cooled, 4-cycle Continental engine. M18 vehicles having serial numbers 1 through 1350 use Model R975-C1 engines. M18 vehicles having serial numbers from 1351 up, and all M39 vehicles, use Model R975-C4 engines. Since the C1 and C4 engines are not interchangeable because of differences in oil coolers and connecting oil and fuel pipes, replacement engines must be of the same model as originally installed in vehicle.

(2) DIRECTIONAL DESIGNATIONS, ROTATION, AND FIRING ORDER. Throughout this manual, universal joint and fan end of the engine is designated as "front", and magneto and carburetor end is designated as "rear". The "right" and "left" sides of the engine are as viewed from the rear. Viewed from the rear, the crankshaft turns in a clockwise direction and cylinders are numbered in a clockwise direction starting with top cylinder as No. 1. Following this designation, firing order of engine is 1,3,5,7,9,2,4,6,8.

(3) ENGINE MOUNTINGS AND SUPPORT. A support tube securely anchored to engine front crankcase (fig. 56) provides a means of mounting front end of engine on support through rubber mountings and brackets (figs. 54 and 62). A mounting ring bolted to engine rear crankcase provides a means of mounting rear end of engine on support through rubber mountings and cages (fig. 60). Rubber mountings provide flexibility to absorb vibration and shock. Support is a welded structure which provides a means of anchoring engine to hull and holding it in alinement with transfer case. Welded structure includes two air inlet tubes which make connections to carburetor and to carburetor air duct and air cleaners mounted on vehicle bulkhead. Oil cooler of the C1 engine is mounted on front end of support. Four ball-bearing rollers installed on lower corners of support provide for rolling engine assembly into or out of hull upon mounting rails on hull floor and hull rear door.

(4) CONNECTION TO TRANSFER CASE. A double universal joint connects flywheel on front end of engine crankshaft to a splined flange

TM 9-755
67

Part Three—Maintenance Instructions

Figure 54—C1 Engine Assembly—Front View

which engages the splined transfer case input shaft. The universal joint permits movement of engine upon its rubber mountings, but it is not intended to compensate for misalinement of engine crankshaft and transfer case input shaft.

  b. Data.

  (1) GENERAL.

Make .... Continental
Model,
  M18 vehicle, serial numbers 1 through 1350 .. .. R 975-C1
  M18 vehicle, serial numbers 1351 up, all M39 vehicles   R 975-C4
Number of cylinders 9
Bore and stroke .... 5.00 in. x 5.50 in.
Piston displacement .... 973 cu in.
Compression ratio .... 5.7 to 1

178

Engine Description and Maintenance

Figure 55—C4 Engine Assembly—Front View

Rated brake horsepower,
  Model R 975-C1 .... 350 @ 2,400 rpm
  Model R 975-C4 .... 400 @ 2,400 rpm
Governed speed (with load) .... 2,400 rpm
Crankshaft rotation (viewed from rear end) .... Clockwise
Overall diameter of engine .... 45 in.
Shipping weight of engine assembly (complete)
  Model R 975-C1 .... 1400 lb
  Model R 975-C4 .... 1505 lb

    (2) VALVE CLEARANCE.
Adjust valves—engine cold .... 0.006 in.

    (3) IGNITION.
Magneto, make and type .... Scintilla, VAG9DFA
Booster Coil, make and model .... Delco-Remy, 1115482
Spark plugs, make and type .... Champion, 63S

    (4) FUEL SYSTEM.
Carburetor, make and model .... Stromberg, NAR9G
Fuel specification .... U.S. Army 2-103A
Engine fuel pump, make and model .... A.C., BF

TM 9-755
67-68

Part Three—Maintenance Instructions

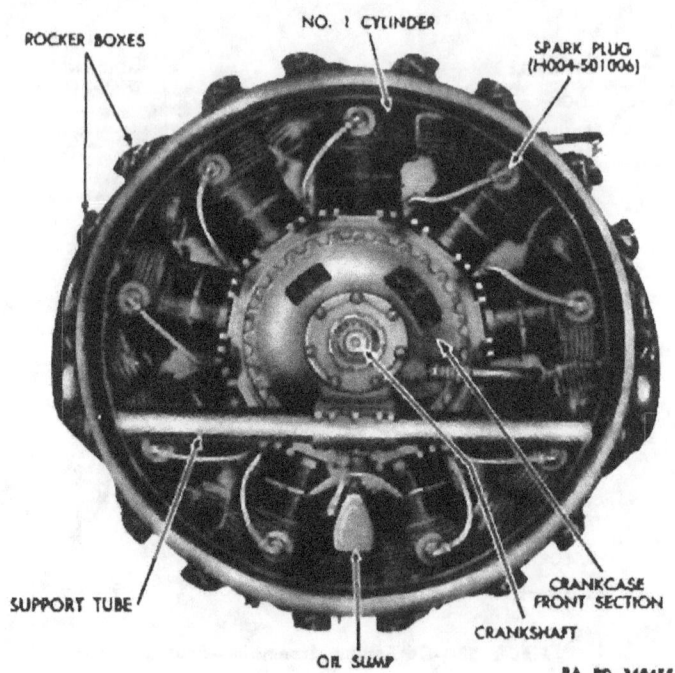

Figure 56—C1 Engine With Fan and Flywheel Removed—Front View

Fuel tank pumps, make and model .... Carter, P571S
Governor, make and model .... Pierce, MA1522

(5) LUBRICATION.

Type .... Dry-sump, pressure feed
Oil pressure at operating speed (1800 to 2,400 rpm) .... 50 to 90 psi
Engine oil temperature at normal operating speed   150° F to 190° F

68. CRANKCASE BREATHER.

a. **Description.** A crankcase breather system is provided for engine ventilation. A breather assembly is mounted above the rear crankcase by means of a clamp and support attached to the rocker boxes of No. 1 cylinder. The breather is connected to openings in main and rear crankcase sections by an elbow, a tube, and connecting hoses. The breather contains a filter element which may be removed for cleaning. Breather assembly (G104-1526041) on the C1 engine is not interchangeable with breather assembly (2220-203415) on the C4 engine.

TM 9-755
68-69

Engine Description and Maintenance

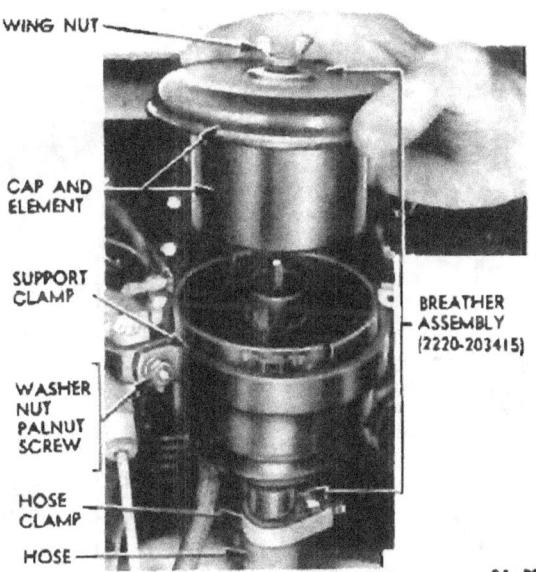

Figure 57—Removing Crankcase Breather Cap and Element

b. **Cleaning Breather Element (fig. 57).** Unscrew captive wing nut and pull upward on breather cap to remove cap and element. Wash cap and element in dry-cleaning solvent by sloshing unit up and down in solvent, then allow to drain and dry. Install element and cap and tighten wing nut.

c. **Breather Removal.** Loosen upper clamp on hose at lower end of breather. Remove two palnuts, hexagon nuts, plain washers, and cap screws which attach clamp to breather support, and pull breather out of hose.

d. **Breather Installation.** Push end of breather tube far enough into hose to insure a full bearing of hose clamp around end of tube. Anchor breather to support by installing clamp with two cap screws (¼ in. — 28 x 25/32 in.), plain washers, hexagon nuts, and palnuts (fig. 57). Tighten hose clamp.

69. EXHAUST MANIFOLD.

a. **Description (fig. 58).** The exhaust manifold is divided into right and left sections which are composed of exhaust pipes joined together by clamps to allow for expansion and contraction. Each exhaust pipe has a welded flange by which it is attached to the

TM 9-755
69
Part Three—Maintenance Instructions

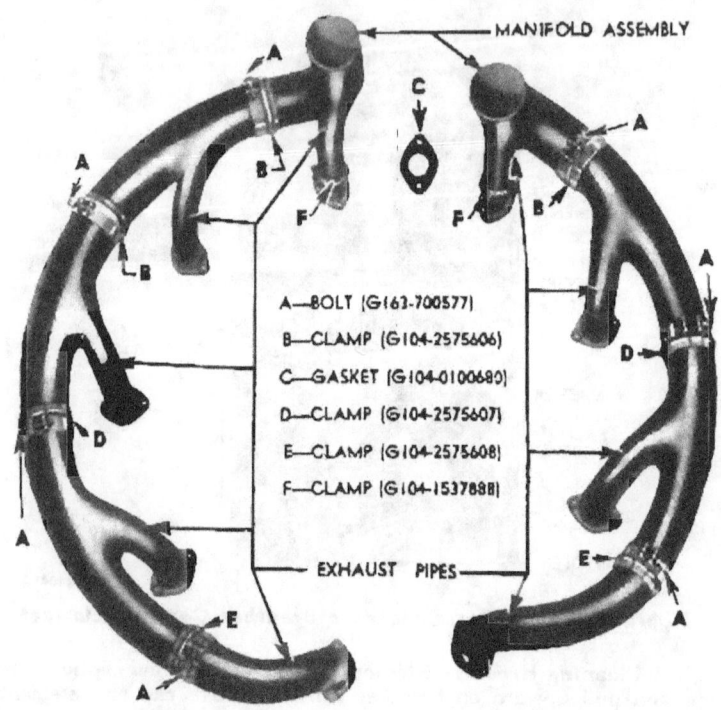

Figure 58—Exhaust Manifold Assembly

exhaust elbow on the engine cylinder. A copper-asbestos gasket is used between flange and elbow.

b. **Removal.** Move engine out upon hull rear door (par. 75). If heater air tube is installed on engine, remove two $3/8$-inch cap screws, nuts and lock washers which attach air tube upper bracket to engine mounting ring, remove one $3/8$-inch cap screw, nut, lock washer, and plain washer which attach air tube clamp to bracket on accessory drive case, and remove air tube with bracket and clamp attached. On some engines, exhaust pipe flanges are attached to exhaust elbows with safety nuts on studs; on others, flanges are attached with slotted nuts secured by locked wires. Remove all nuts and remove complete manifold section from engine, then remove gaskets.

c. **Installation.** Before installation of manifold sections, inspect joint surfaces of exhaust elbows and exhaust pipe flanges to make sure they are clean and flat, then place a new gasket over studs on

Engine Description and Maintenance

each exhaust elbow. Slightly loosen exhaust pipe clamp bolts and install each manifold section on engine. Attach exhaust pipe flanges with safety nuts (5/16 in.—24) if studs are not drilled for lock wire; use slotted nuts secured by lock wire if studs are drilled. Tighten exhaust pipe clamp bolts firmly, then loosen them 1½ to 2 turns to allow a slight creep at each joint. Secure each bolt with a lock wire. If heater air tube was removed from engine, install it by attaching lower clamp to bracket on accessory drive case with one cap screw (⅜ in.—24 x 1 in.), plain washer, lock washer, and nut, and attaching upper bracket to engine mounting ring with two cap screws (⅜ in.—24 x 1 in.), lock washers and nuts. Move engine into place and complete installation (par. 76).

70. INTAKE PIPES.

a. **Description.** An individual seamless steel pipe conducts the fuel-air mixture to each cylinder from distribution chamber in engine crankcase. Each pipe is joined to distribution chamber by means of a rubber packing and packing nut, and is joined to cylinder intake port by a flange and vellumoid gasket attached by three cap screws. The joints at both ends of each intake pipe must be airtight, and the pipes must be free of cracks or dents.

b. **Packing Replacement.** Move engine out upon hull rear door (par. 75). Replace each packing, as required, in the following manner. Unscrew packing nut using packing nut wrench (41-W-1537) and slide nut up on pipe (fig. 59). Remove old packing from port around end of pipe and clean all particles of old packing out of port. With a razor blade or sharp knife, cut the new packing on a 45-degree diagonal so that the feather edge will point in the direction that packing nut will turn when it is being tightened against packing. Spread packing and place it around intake pipe, then carefully push it down into place with packing nut. Start packing nut with fingers to insure proper engagement of threads, then tighten nut using packing nut wrench (41-W-1537) until threads on nut are about flush with crankcase. CAUTION: *Excessive tightening will distort packing and cause air leaks.* Move engine into place and complete the installation (par. 76). Before closing hull rear door, test the intake pipe joints for air leaks (par. 47 f (6)).

c. **Intake Pipe Removal.** Move engine out upon hull rear door (par. 75). Remove each intake pipe, as required, in the following manner. Remove two nuts and lock washers which attach priming tube support clips to clamp on intake pipe. Loosen clamp around push rod housing upper hose and slide clamp down on housing so that push rod housing can be pushed slightly to one side. Remove the three 5/16-inch cap screws which attach intake pipe flange to cylinder and remove intake pipe and flange from engine. Loosen screw and remove priming tube clamp from intake pipe. Remove old packing from distribution chamber port and clean out all particles of packing that may adhere to the metal.

TM 9-755
70-71

Part Three—Maintenance Instructions

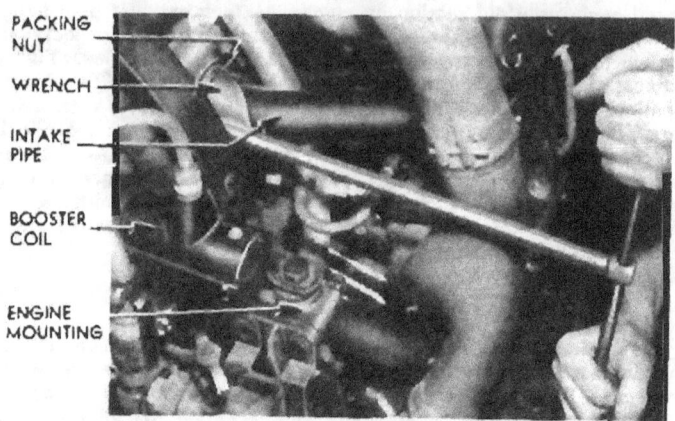

Figure 59—Removing Intake Pipe Packing Nut With Wrench 41-W-1537

d. **Intake Pipe Installation.** Install each intake pipe, as required, in the following manner. Place flange over intake pipe, install priming tube clamp below the flange, then place packing nut and new packing on pipe. Insert end of pipe into distribution chamber port, place a new vellumoid gasket between flange and cylinder and attach flange to cylinder with three cap screws (5/16 in.—24) and lock washers, leaving screws slightly loose. Push packing down into distribution chamber port with packing nut and start nut with fingers to insure proper engagement of threads. Tighten flange cap screws to 10-15 foot-pounds tension, and tighten packing nut using packing nut wrench (41-W-1537) (fig. 59) until threads on nut are about flush with crankcase. CAUTION: *Excessive tightening will distort packing and cause air leak.* Attach priming tube clip to clamp on intake pipe with two nuts and lock washers. Move engine into place and complete installation (par. 76). Before closing hull rear door, test intake pipe joints for air leaks (par. 47 f) (6).

71. ENGINE MOUNTINGS.

a. **Description.** Each engine mounting consists of a heavy steel inner sleeve, a rubber bushing, and an outer sleeve, all vulcanized together to form one unit. The inner sleeve is longer than the rubber bushing and outer sleeve so that it may be clamped tight by the mounting bolt while leaving bushing free to flex under load. The outer sleeve is a press fit in the bracket or cage in which it is installed. Front mountings are pressed into mounting brackets which are inserted into ends of support tube, and inner sleeves are anchored

Engine Description and Maintenance

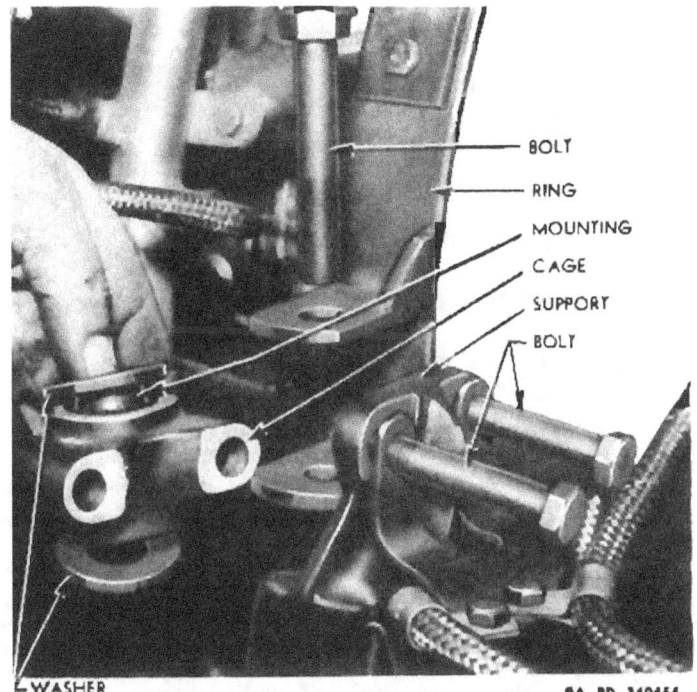

Figure 60—Rear Mounting, Cage, and Washers

to engine support by bolts and flat washers (fig. 62). Rear mountings are pressed into cages which are bolted to engine support, and inner sleeves are anchored to mounting ring by bolts (fig. 60). By this arrangement, the weight of the engine is supported by rubber portions of the mounts. Large rubber washers are placed around the extended ends of the inner sleeves to act as snubbers between cages and mounting brackets and the parts to which they are anchored.

b. **Rear Mounting Removal** (fig. 60). Move engine out upon hull rear door (par. 75). Attach engine lifting sling (41-S-3831-835) (fig. 68) and hoist, and lift engine just enough to relieve load on mountings. Remove either mounting as required in the following manner. Remove the 5/8-inch bolt and safety nut which anchors the mount and cage to mounting ring. Remove two 1/2-inch bolts and safety nuts which attach cage to engine support and remove cage and mounting washers. Press or punch mounting out of cage, applying pressure against outer steel sleeve.

c. **Rear Mounting Installation** (fig. 60). Rest replacer guide

TM 9-755
71

Part Three—Maintenance Instructions

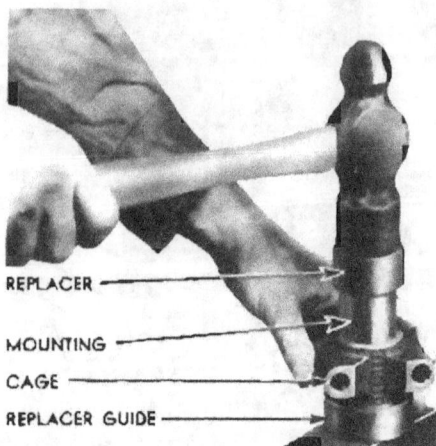

Figure 61—Installing Rear Mounting in Cage With Replacer 41-R-2397-155

on a firm support and place cage over the guide. Place new mounting and replacer (41-R-2397-155) over opening in cage and drive mounting into cage (fig. 61). Place a rubber mounting washer on each side of cage with tapered holes in washers fitting tapered ends of mounting inner sleeve, slide cage and washers between brackets on mounting ring and insert bolt (5/8 in.—18 x 3½ in.). Attach cage to engine support with two bolts (½ in.—20 x 3½ in.) and safety nuts tightened to 70-80 foot-pounds tension. Install safety nut on 5/8-inch bolt and tighten to 75-85 foot-pounds tension. Lower engine and remove hoist and lifting sling. Move engine into place and complete installation (par. 76).

d. **Front Mounting Removal** (fig. 62). Move engine out upon hull rear door (par. 75). Attach engine lifting sling (41-S-3831-835) (fig. 68) and hoist, and lift engine just enough to relieve load on mountings. Remove either mounting, as required, in the following manner. Remove 7/8-inch bolt which is anchored to engine support by a safety nut at lower end, then remove flat washer and rubber washer from top of mounting bracket and rubber washer under bracket. Tap mounting bracket out of end of support tube. Press or punch the mounting out of the bracket, applying pressure against the outer steel sleeve.

e. **Front Mounting Installation** (fig. 62). Install mounting in mounting bracket by the use of mounting replacer (41-R-2397-150) in the same manner as described for installing rear mounting in cage (subpar. c above). Install mounting bracket in end of support tube

TM 9-755
71-72

Engine Description and Maintenance

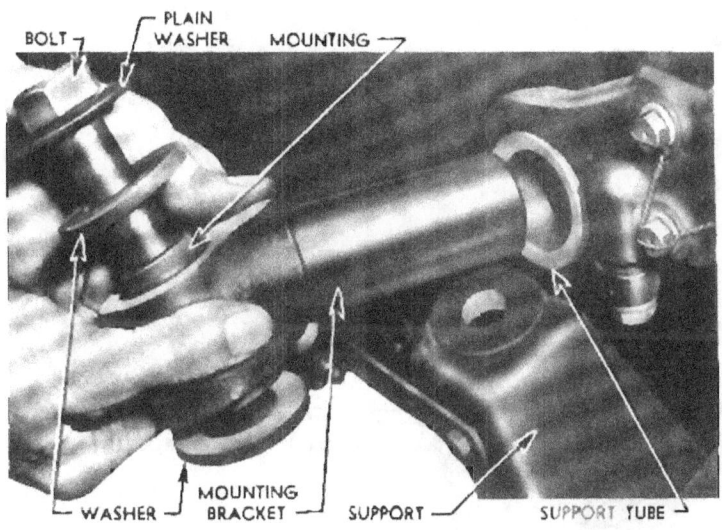

Figure 62—Front Mounting, Mounting Bracket, and Washer

and place a rubber mounting washer on upper and lower sides of bracket, with tapered holes in washers fitting tapered ends of mounting inner sleeve. Place large flat washer over the bolt (⅞ in.—14 x 4½ in.), insert bolt through mounting and hole in engine support and secure it with a safety nut tightened to 90-100 foot-pounds tension. Lower the engine and remove hoist and lifting sling. Move engine into place and complete installation (par. 76).

72. VALVE ROCKER ASSEMBLIES.

a. Description. Each intake and exhaust valve is actuated by a rocker assembly located in a rocker box which is an integral part of the cylinder. The rocker assembly consists of a machined steel forging which is forked at one end to contain a roller, bored laterally near the center to contain a double roller bearing, and machined at the other end to contain an adjusting screw secured by a lock nut. The rocker assembly is placed in rocker box so that roller contacts valve stem, and bearing inner race is clamped to rocker box by a special bolt which passes through inner race and both sides of rocker box. The rocker is actuated by a push rod which bears against adjusting screw. The adjusting screw provides a means of adjusting valve clearance (par. 73).

b. Removal. The following procedure covers removal of any one intake or exhaust rocker assembly.

TM 9-755
72

Part Three—Maintenance Instructions

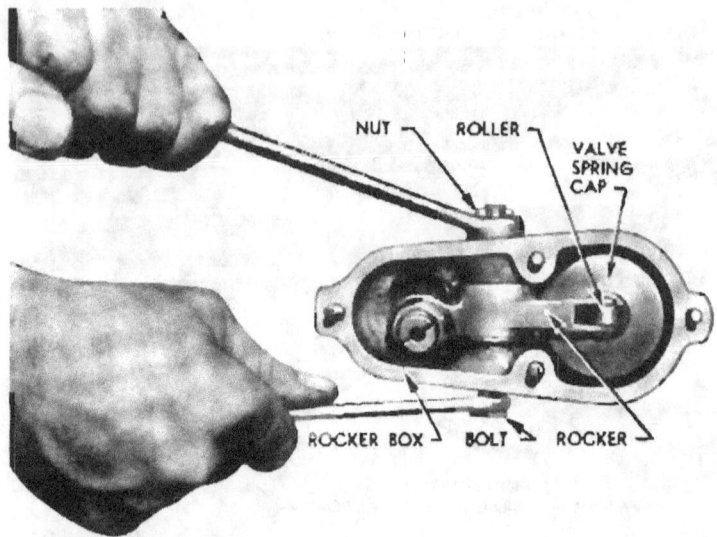

Figure 63—Removing Rocker Hub Bolt

(1) Move engine out upon hull rear door (par. 75).

(2) If rocker assembly is to be removed from No. 5 or No. 6 cylinder, lower the hull door plate if it is hinged (par. 181 c). If door plate is not hinged, raise engine by means of engine lifting sling (41-S-3831-835) and hoist (fig. 68) and place blocks under engine support for safety.

(3) On C4 engine only, if rocker assembly is to be removed from No. 5 or No. 6 cylinder it is necessary to remove the oil sump. Loosen clamps and slide connecting hoses up on sump scavenger tee and sump adapter. Remove $\frac{5}{8}$-inch safety nuts from two studs which attach front end of sump to rocker box drain fitting. Remove two $\frac{1}{4}$-inch bolts, plain washers, spacers, and safety nuts which attach rear end of sump to rocker boxes, pull sump to rear and remove it from engine.

(4) Remove four $\frac{1}{4}$-inch stud nuts, lock washers, and flat washers and remove cover and gasket from rocker box. Where two covers are connected by a drain hose, remove both covers at the same time; do not disconnect hose unless it needs to be replaced.

(5) Crank engine until rocker to be removed is free from push rod pressure. Pull cotter pin, hold hub bolt head with wrench and remove nut and washer from bolt (fig. 63). Slide bolt and washer from rocker box and lift out rocker assembly.

(6) When removing rocker assembly from a lower cylinder, also remove push rod, tappet ball socket, and socket spring so that these parts will not drop out and be damaged or lost.

c. **Installation.** The following procedure covers installation of any one intake or exhaust rocker assembly. Intake and exhaust rocker assemblies are not interchangeable.

(1) Install socket spring, tappet ball socket, and push rod in push rod housing in the order named, if these parts were removed.

(2) Place 1/8-inch washer on hub bolt with concave side toward bolt head. Place rocker assembly in box with push rod engaged in adjusting screw and insert hub bolt through rocker box and rocker bearing from side nearest centerline of cylinder. Place 3/16-inch washer over bolt with concave side outward and install slotted nut. Hold hub bolt head with wrench and tighten nut to 20-25 foot-pounds tension (fig. 63). On C4 engine only, install cotter pin (1/8 in. x 3/4 in.).

(3) Adjust valve clearance to 0.006 inch (par. 73).

(4) Place a new gasket and the cover over studs on rocker box, install flat washers, lock washers, and nuts (1/4 in. —28) on studs and tighten nuts evenly. Where rocker box covers are joined by a drain hose install joined covers together; do not disconnect hose. Tighten hose clamps.

(5) On C4 engine only, install oil sump if this was removed. Place a new gasket over studs on front end of sump and push studs through holes in rocker box drain fitting. Attach rear end of sump to rocker box with two bolts (1/4 in. —28 x 1 21/32 in.) having plain washers under heads, spacers between rocker box and sump bracket, and safety nuts. Install safety nuts on studs at front end of sump. Slide connecting hoses to full bearing on sump fittings and tighten hose clamps.

(6) Lower engine to hull door, if raised, or lift hinged door plate and attach it to rails with two bolts.

(7) Move engine into place and complete installation (par. 76).

73. VALVE CLEARANCE ADJUSTMENT.

a. **Remove Engine, Spark Plugs, and Rocker Box Covers.** Move engine out upon hull rear door (par. 75). Lower hull rear door plate if it is hinged type (par. 181 c). Remove rear spark plugs (par. 89). Remove four 1/4-inch stud nuts lock washers, and flat washers which attach each rocker box cover and remove all covers and gaskets. CAUTION: *Thoroughly clean all dirt from covers and joints before removing covers from cylinders, to avoid getting dirt into engine.* Where two covers are connected by a drain hose, remove both covers at the same time; do not disconnect hose unless it needs to be replaced.

b. **Set Valves for Clearance Adjustment.** Each valve must be fully closed and valve tappet roller must be at lowest point on cam when valve clearance is checked or adjusted. This condition is ob-

TM 9-755

Part Three—Maintenance Instructions

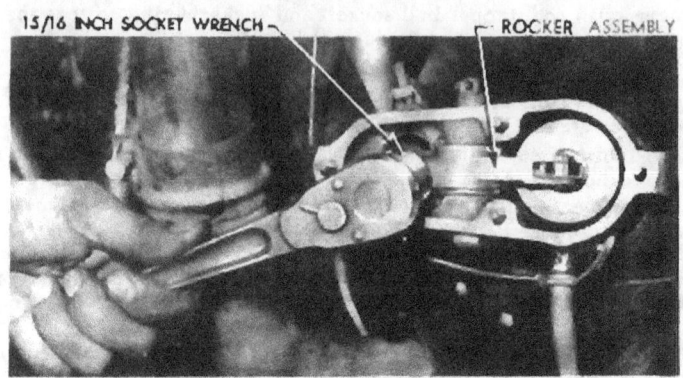

Figure 64—Loosening Rocker Adjusting Screw Lock Nut

tained for both valves in a given cylinder by cranking engine until piston nears top of its compression stroke, as indicated by air rushing out of spark plug port. Check or adjust both intake and exhaust valve clearances on one cylinder (subpar. c below) before proceeding to next cylinder. Start with No. 1 cylinder, then crank engine as required to properly set valves for adjustment on succeeding cylinders, following the engine firing order of 1-3-5-7-9-2-4-6-8. Note that the odd numbered cylinders are taken in order, then the even numbered cylinders starting with No. 2.

c. **Check and Adjust Valve Clearance.** The rocker roller is held in contact with valve stem by a spring in tappet which pushes outward on push rod. Press on adjusting screw end of rocker to compress this spring, and insert in 0.006-inch feeler gage (41-G-412-77) between valve stem and rocker roller. The gage should slide between the parts smoothly, with just enough drag to cause rocker roller to turn. If clearance is not correct, loosen adjusting screw lock nut with a $1\frac{5}{16}$-inch socket wrench (fig. 64) and turn adjusting screw as required to secure proper clearance, using screw driver (41-S-1725) (fig. 65). While holding adjusting screw stationary with screw driver, tighten lock nut with an end wrench until screw is securely locked, then tighten lock nut to 65-75 foot-pounds tension with socket wrench. Recheck clearance after tightening lock nut to make certain that adjustment did not change. Readjust if clearance is not correct.

d. **Install Rocker Box Covers.**

(1) Original intake and exhaust rocker box covers on the C1 engine are not interchangeable; however, replacement cover (G104-7002367) may be installed on either intake or exhaust rocker box. C1 engine rocker box covers are not interchangeable with C4 engine

Engine Description and Maintenance

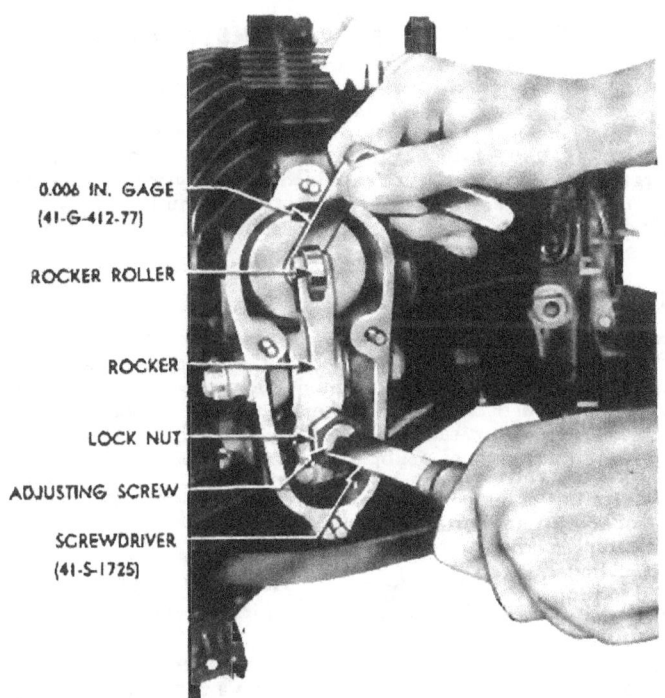

Figure 65—Adjusting Valve Clearance

covers. Four different rocker box covers are used on the C4 engine, as follows:

Stock Number	Name	Cylinders
G163-7006983	Intake valve rocker box cover (plain type)	1, 2, 3, 6, 9
G163-7006984	Exhaust valve rocker box cover (plain type)	1, 2, 5, 8, 9
G163-7006985	Intake valve rocker box cover (drain type)	4, 5, 7, 8
G163-7006986	Exhaust valve rocker box cover (drain type)	3, 4, 6, 7

(2) Place a new gasket and the cover over studs on rocker boxes, and secure covers with flat washers, lock washers, and nuts (¼ in. —28) on studs. Tighten nuts evenly. Where rocker box covers are joined by drain hoses, install joined covers together; do not disconnect hoses. Tighten hose clamps.

c. **Install Spark Plugs and Engine.** Install spark plugs (par. 89). Lift hinged type door plate and attach it to rails with two bolts. Move engine into place and complete installation (par. 76).

TM 9-755
74-75

Part Three—Maintenance Instructions

Section XVIII
ENGINE REMOVAL AND INSTALLATION

74. COORDINATION WITH HIGHER AUTHORITY.

a. Replacement of this major assembly with a new or rebuilt unit is normally a third echelon operation, but may be performed in an emergency by second echelon, provided authority for performing this replacement is obtained from appropriate commander. Tools needed for operation which are not carried in second echelon may be obtained from a higher echelon of maintenance.

75. REMOVAL.

a. **Need for Removal.** In addition to removal for replacement, engine must be removed when a complete cleaning or inspection is necessary, or for performance of maintenance operations described in this manual which cannot be accomplished with engine in place in vehicle. For these operations it is usually sufficient to move engine out upon the hull rear door (subpar. b through i below).

b. **Turn Off Master Switch and Close Fuel Shut-off Valves.** Turn off 24-volt switch in master switch box (par. 15 a). Close all fuel shut-off valves (par. 14 a).

c. **Open Hull Doors, Remove Grilles and Plate.**

(1) Open hull rear door and support it on turnbuckle hooks (par. 181 a).

(2) On M18 vehicle, turn turret so that extension is clear of air inlet grille and engage turret lock. Remove air inlet and outlet grilles (par. 183 d and e).

(3) Remove three cap screws, open hull rear roof door and tie it so it cannot fall.

(4) Remove nuts and lock washers from four ⅜-inch cap screws which attach two muffler clamps to mounting brackets on hull upper plate. Cut lock wires, remove six ½-inch attaching bolts and lift rear hull plate away from hull.

d. **Remove Mufflers and Connector** (fig. 116). Apply a liberal quantity of penetrating oil to muffler joints and clamps. Remove nuts and lock washers from four ⅜-inch cap screws which attach muffler clamps to mounting brackets on hull. Loosen nuts on saddle clamp U-bolts, separate mufflers from connector and remove mufflers from hull. Loosen nuts on saddle clamp U-bolts and remove connector from exhaust manifold outlet pipes.

e. **Remove Fire Extinguisher Horns and Supports** (J, K, fig. 66 and B, C, fig. 67). Unscrew fire extinguisher pipe coupling nuts from elbows on the lower rear fire extinguisher horns. Remove two 5/16-inch cap screws and external-tooth lock washers which attach

Engine Removal and Installation

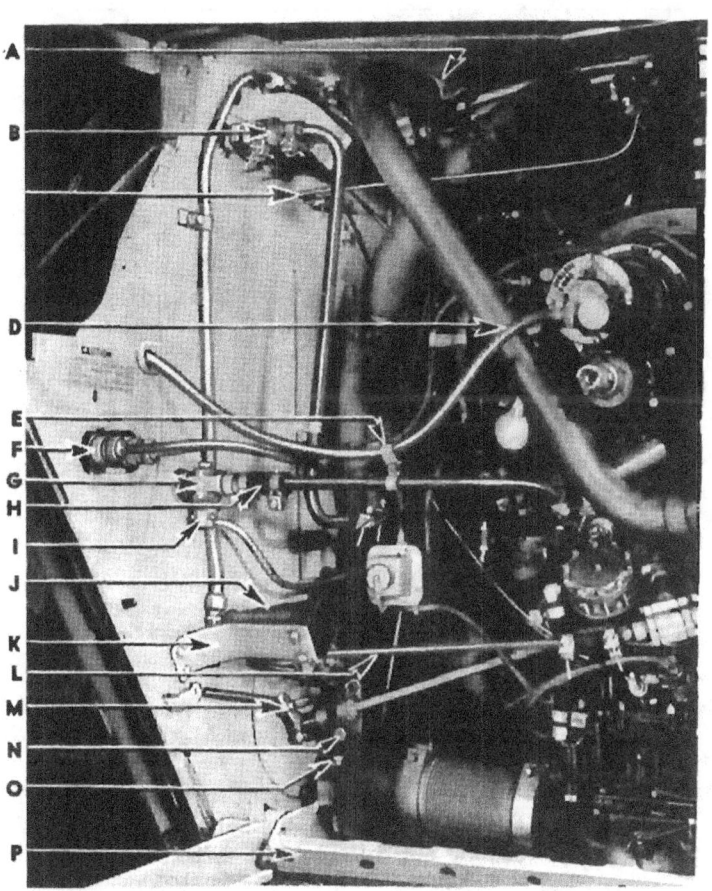

A—BRACKET
B—OUTLET OIL PIPE
C—PRIMER PIPE
D—STARTER WIRE
E—CLIP
F—WIRING CONNECTIONS
G—OIL PIPE TEE
H—INLET OIL PIPE
I—TEMPERATURE GAGE UNIT
J—HORN
K—HORN SUPPORT
L—FUEL PIPE
M—THROTTLE ROD
N—SUPPORT ALINEMENT BOLT
O—SUPPORT REAR BOLT
P—FLOOR BRACE

RA PD 340469

Figure 66—Items To Disconnect on Left Side When Removing Engine

TM 9-755

Part Three—Maintenance Instructions

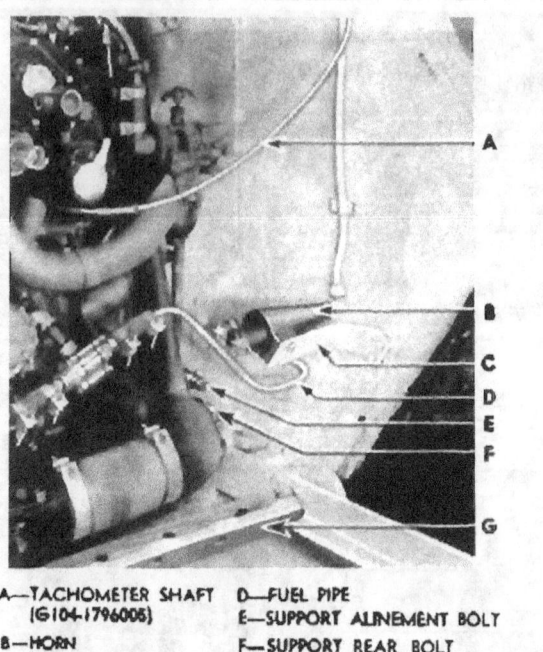

A—TACHOMETER SHAFT (G104-1796005)
B—HORN
C—HORN SUPPORT
D—FUEL PIPE
E—SUPPORT ALINEMENT BOLT
F—SUPPORT REAR BOLT
G—FLOOR BRACE

Figure 67—Items To Disconnect on Right Side When Removing Engine

each horn support to hull, and remove horns with supports attached.

f. **Disconnect Tachometer Shaft and Engine Wiring.**

(1) Unscrew coupling nut to disconnect tachometer shaft (A, fig. 67) and tie shaft up where it will not be damaged.

(2) Unscrew coupling nuts and disconnect two conduits from rear junction box (F, fig. 66). Unscrew cover from receptacle on engine junction box and connect magneto conduit securely to receptacle (fig. 89). If junction box is not equipped with grounding receptacle, install ground plug on conduit connector plug.

(3) Unscrew captive screws which attach conduit housing to temperature gage unit on oil tank outlet tee and pull wire out of gage unit (I, fig. 66).

(4) Cut lock wire, unscrew cap from terminal shield on starter and unscrew conduit coupling nut from shield (fig. 198). Remove ⅜-inch brass nut, internal-tooth lock washer, plain washer and starter wire terminal from terminal stud in shield. Remove bolt which attaches conduit support clip to the lower oil pipe (E, fig. 66).

Engine Removal and Installation

Figure 68—Lifting Engine Assembly With Sling 41-S-3831-835

g. **Disconnect Oil and Fuel Pipes and Throttle Rod.**

(1) Disconnect oil pipes from fittings at top and bottom of oil tank. These pipes are connected either by coupling hoses or by Sealflex compression nuts (B, H, fig. 66).

(2) Remove tee from check valve at lower end of oil tank (G, fig. 66). NOTE: *If oil leaks out of check valve, move plunger in valve slightly with a blunt tool to make it seat properly* (fig. 85).

(3) Disconnect right and left fuel pipes at feed pipe check valve and at fittings on tanks (L, fig. 66 and D, fig. 67). These pipes are connected either by coupling hoses or by Sealflex compression nuts.

(4) Loosen clamp and disconnect primer pipe at coupling hose on hull wall (C, fig. 66).

(5) Remove throttle return spring. Remove cotter pin and clevis pin to disconnect throttle rod from throttle cross shaft lever (M, fig. 66).

TM 9-755
75-76

Part Three—Maintenance Instructions

Figure 69—Auxiliary Rollers 41-R-2743 Installed

h. **Remove Rear Roof Support and Floor-to-Door Brace.**

(1) Remove six ½-inch bolts which attach ends of rear roof support and remove support.

(2) Remove eight ½-inch bolts which anchor the floor-to-door brace to hull floor and remove the brace.

i. **Move Engine Out Upon Hull Rear Door.**

(1) If a heater tube is installed on engine, loosen the two bolts at top of support bracket (A, fig. 66).

(2) Cut lock wires and remove two bolts which anchor rear corners of engine support to skid pads on mounting rails.

(3) Unscrew the engine support alinement bolts which extend rearward along each side of engine support.

(4) Roll engine out upon hull rear door.

j. **Remove Engine From Hull Rear Door.** Attach engine lifting sling (41-S-3831-835) using care to engage rear hook securely to mounting ring, attach hoist to sling and lift engine from hull rear door (fig. 68). Install engine mounting auxiliary rollers (41-R-2743) over rollers on engine support and tighten clamp screws securely (fig. 69). These auxiliary rollers provide additional clearance under the carburetor air scoop when engine assembly is placed upon a floor.

76. INSTALLATION.

a. **Clean Engine Oil Tank.** If a new or rebuilt engine is being installed, drain engine oil tank. Flush out tank with dry-cleaning solvent, agitating the solvent with a clean stick to loosen sediment, then drain and dry out tank with air to remove all sediment and solvent. Fill tank with new oil (par. 38 d).

Engine Removal and Installation

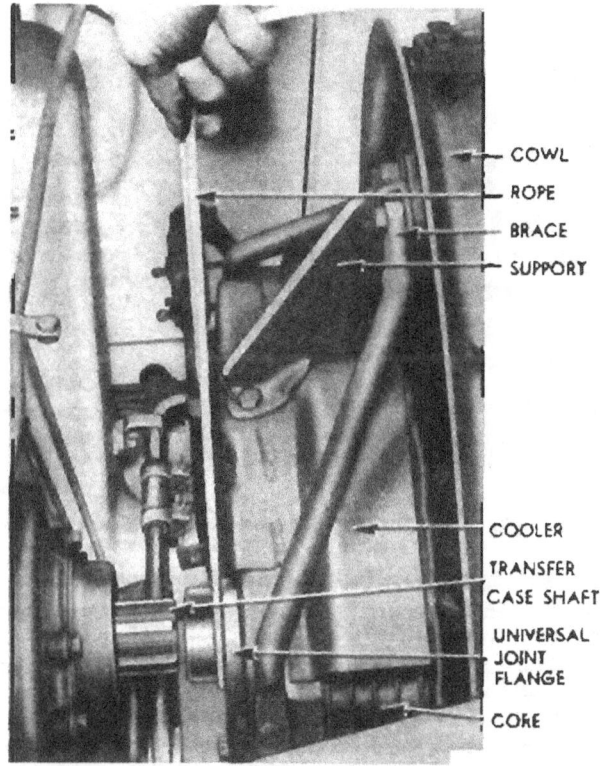

Figure 70—Engaging Universal Joint Flange With Transfer Case Input Shaft

b. **Place Engine on Hull Rear Door.** Attach engine lifting sling (41-S-3831-835) using care to engage rear hook securely to mounting ring, attach hoist to sling and lift engine into position over hull rear door (fig. 68). Remove engine mounting auxiliary rollers (41-R-2743) if these are installed on support rollers (fig. 69). Lower engine until support rollers rest on mounting rails on door, and remove hoist and lifting sling.

c. **Install Air Duct Seals** (fig. 100). Apply a light coat of general purpose grease on air inlet tubes in engine support, and in the flared openings of the carburetor air ducts mounted on rear side of bulkhead. Install new air duct seals on forward ends of air inlet tubes in engine support.

197

TM 9-755
76

Part Three—Maintenance Instructions

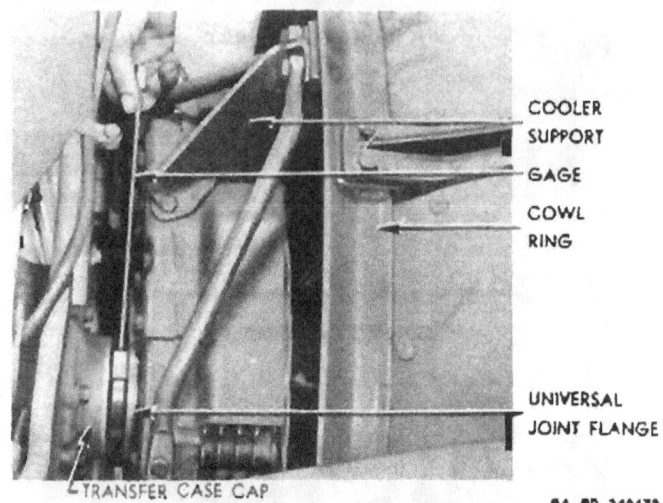

Figure 71—Checking Engine Alinement With Gage 41-G-13-300

d. **Move Engine Into Hull and Install Support Attaching Bolts.**

(1) Roll engine forward until universal joint flange is close to the transfer case input shaft.

(2) Place a sling around universal joint flange (fig. 70) and hold flange in alinement with input shaft, then slowly move engine forward while engaging splines of flange with splines of input shaft.

(3) When engine support engages alinement brackets on hull, start the alinement bolts into nuts in bracket, but do not tighten. Install two bolts (⅝ in.—18 x 1¾ in.) through rear corners of engine support, then tighten front and rear bolts securely (N, fig. 66).

e. **Check Engine Alinement.**

(1) Check the clearance between front surface of the universal joint flange and rear surface of transfer case input shaft rear bearing cap, using alinement gage (41-G-13-300) as shown in figure 71. The clearance must not be more than ⅜ inch or less than 5/16 inch.

(2) If clearance is not within these limits, remove attaching bolts, move engine out upon rear door, and remove skid pads from mounting rails. If clearance was more than ⅜ inch, add one or more shims (1/32 in. thick) under skid pads; if clearance was less than 5/16 inch, remove one or more shims and install skid pads.

(3) Roll engine into hull and install support attaching bolts (subpar. d above). Check alinement and repeat shimming operation if necessary. NOTE: *If alinement cannot be secured by shims, hull floor or mounting rails are bent; notify higher authority.*

Engine Removal and Installation

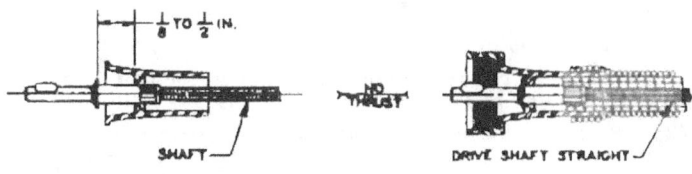

Figure 72—Tachometer Shaft and Casing Adjustment

(4) After alinement is secured and support attaching bolts are securely tightened, secure rear bolts with lock wire.

f. **Install Floor-to-Door Brace and Rear Roof Support.**

(1) Install floor-to-door brace (P, fig. 66) and anchor it to hull floor with eight bolts (½ in.—20 x 1¾ in.) and lock washers tightened to 50-60 foot-pounds tension.

(2) Install rear roof support and anchor it to roof plate and brackets on hull with six bolts (½ in.—20 x 1½ in.) tightened to 50-60 foot-pounds tension.

g. **Connect Throttle Rod and Check Throttle Opening.**

(1) Connect throttle rod to throttle cross shaft lever by installing clevis pin through rod yoke and inner hole in lever. Attach throttle return spring to outer hole in lever and to bracket on hull (M, fig. 66). Check accelerator pedal to throttle linkage for proper adjustment (par. 94).

h. **Connect Oil and Fuel Pipes.** In early production M18 vehicles, oil and fuel pipes are connected with Sealflex fittings which must be connected and tightened as described in paragraph 125. Later production vehicles use coupling hoses and hose clamps to connect the pipes.

(1) Connect primer pipe at coupling hose on hull wall and tighten clamp (C, fig. 66).

(2) Connect right and left fuel pipes at feed pipe check valve and at fittings on fuel tanks and tighten clamps (L, fig. 66 and D, fig. 67).

(3) Install tee in check valve at lower end of oil tank and tighten tee with temperature gage unit pointing down (G, fig. 66).

(4) Connect oil pipes to fittings at top and bottom of oil tank and tighten clamps (B, H, fig. 66).

i. **Connect Tachometer Shaft** (A, fig. 67). Grasp shaft casing and turn shaft by hand to adjust its length so that distance between flange on shaft and seat in casing is 1/8 inch to 1/2 inch as shown in figure 72. Hold parts in this position while connecting shaft to engine, then tighten coupling nut firmly.

j. **Connect Engine Wiring.**

(1) Insert end of wire into oil temperature gage unit, attach conduit housing to unit and tighten two screws (fig. 248).

(2) Place starter wire terminal on terminal stud through side opening in terminal shield and secure it with a zinc-plated plain washer, a cadmium-plated external-tooth lock washer, and a brass nut (3/8 in. — 16). Screw conduit coupling nut firmly on terminal shield. Install cap in shield and secure it with a lock wire (fig. 198). Attach starter wire clip to clip on oil pipe (E, fig. 66).

(3) Disconnect magneto wiring conduit from grounding receptacle on engine junction box, or remove ground plug. Connect this conduit to rear receptacle in rear junction box and connect other engine wiring conduit to front receptacle of junction box (F, fig. 66).

k. **Install Fire Extinguisher Nozzles and Supports** (J, K, fig. 66 and B, C, fig. 67). Start fire extinguisher pipe coupling nuts on elbows of fire extinguisher nozzles, then attach each nozzle support to hull with two cap screws (3/16 in. — 24 x 5/8 in.) and external-tooth lock washers. Tighten attaching screws and coupling nuts securely.

l. **Install Mufflers, Connector, and Rear Hull Plate.**

(1) Place saddle clamps over front ends of connector and place ends of connector over outlet pipes of exhaust manifold; do not tighten clamps.

(2) Place saddle clamps over rear ends of connector and place mufflers in position with ends engaging ends of connector.

(3) Lift hull rear plate into position and attach it to hull with six bolts (1/2 in. — 20 x 1 3/4 in.) tightened to 50-60 foot-pounds tension. Install lock wire through bolt heads.

(4) Install the four clamps which anchor the mufflers to mounting brackets on hull plate and hull and tighten clamp bolts securely.

(5) Turn clamps at mufflers so that saddles are on top and tighten U-bolts. Turn clamps at exhaust manifold so that saddles are

on top, position connector on manifold pipes so that flexible sections are in a neutral position and tighten U-bolts.

m. Test Engine and Controls. Start and warm up engine as described in paragraph 17 b and c. In addition to tests prescribed in that paragraph, also check all pipes and connections for fuel or oil leaks. After engine is warmed up and checked, stop the engine and perform the after-operation services as described in paragraph 17 i and j.

n. Install Grilles and Close Doors.

(1) Close hull rear roof door and anchor it with three cap screws (½ in. — 20 x 1½ in.) tightened to 50-60 foot-pounds tension.

(2) Install air inlet and outlet grilles (par. 183 d and e). On M18 vehicle, turn turret so that 76-mm gun points straight forward and engage turret lock.

(3) Close hull rear door (par. 181 b).

o. Record of Replacement. If engine was replaced, record the replacement on W.D., A.G.O. 478 M.W.O. and Major Unit Assembly Record.

Section XIX

ENGINE OILING SYSTEM

77. DESCRIPTION AND DATA.

a. Description (figs. 73 and 74). Engine is oiled by a dry-sump, pressure-feed system which includes an external oil tank, an oil pressure and scavenge pump, oil sumps, filter, oil cooler, and connecting oil pipes.

(1) The oil tank is a component part of left fuel tank and is connected to oil pump on the engine by a pipe. A spring-loaded check valve is attached at the tank outlet to prevent oil draining into oil pump and flooding engine crankcase while engine is not running. When engine is running, oil pump suction opens the check valve so that oil is drawn from tank into engine.

(2) The oil pressure and scavenge pump combines two pumps in one unit. The pressure pump draws oil from tank and forces it under pressure to the various parts of the engine. The pressure is regulated by an adjustable relief valve in pump. After oil has circulated through engine and has drained into oil sump it is withdrawn by scavenge pump which has greater capacity than pressure pump; therefore, sump remains relatively empty. In the C1 engine a separate scavenge pump is used in addition to scavenge pump combined with pressure pump. The C1 engine has one oil sump and C4 engine has two oil sumps, all located at lower side of engine.

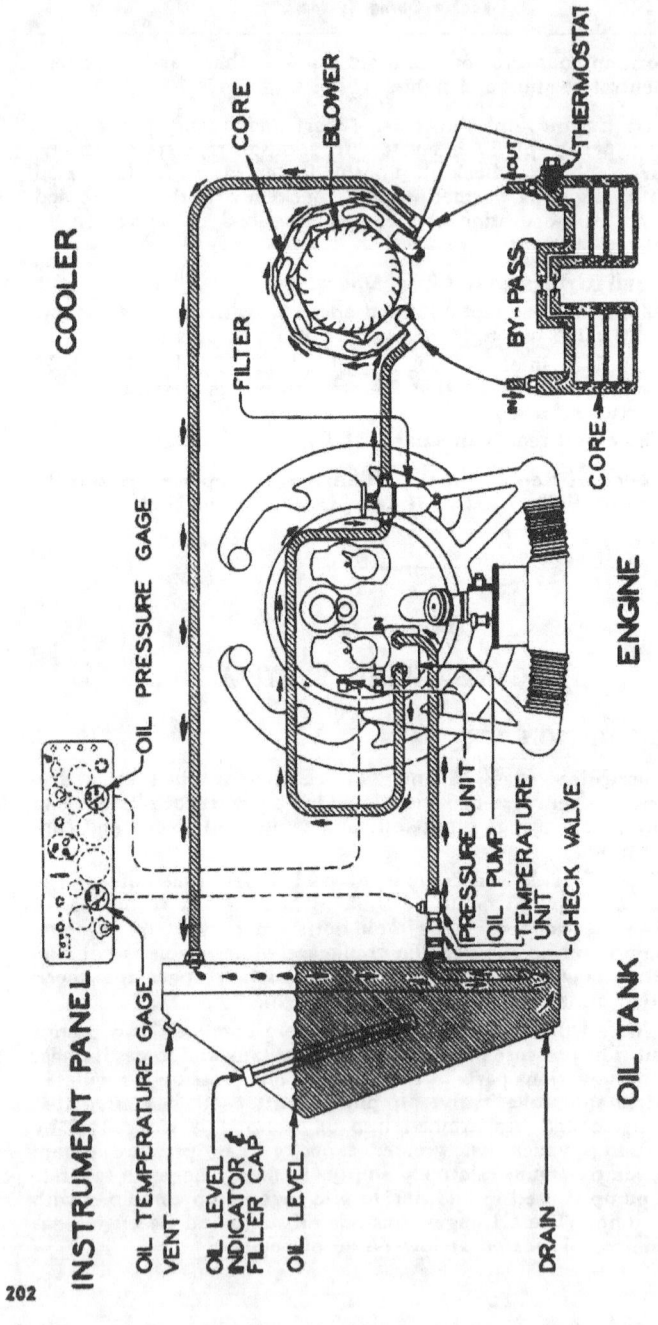

Figure 73—C4 Engine Oiling System—Schematic View

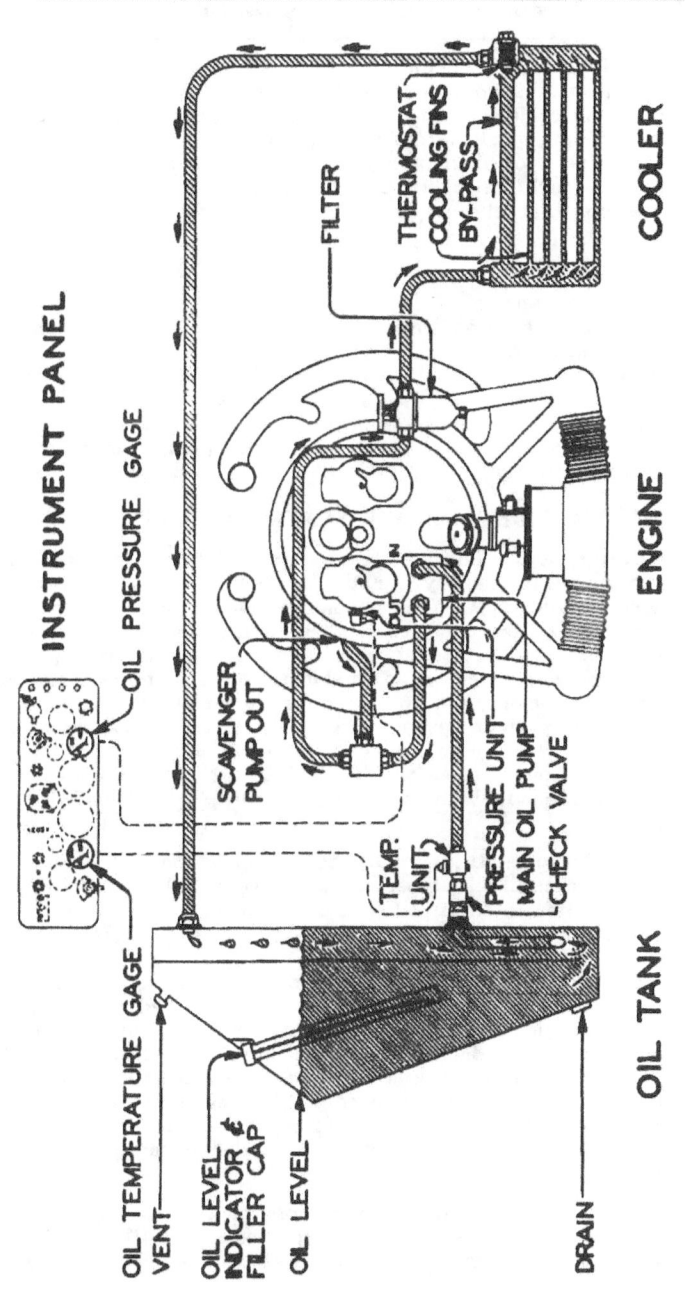

Figure 74—C1 Engine Oiling System—Schematic View

TM 9-755
77-78

Part Three—Maintenance Instructions

(3) Oil removed from sump is forced by scavenge pump through an external oil filter and an oil cooler, from which it returns to oil tank. The oil cooler contains a thermostatically-controlled by-pass valve to facilitate quick warming up of engine oil. When the oil is cold, the valve is open to permit oil to return to oil tank without circulation through the cooling passages in core. As oil heats up the valve closes so that oil must circulate through cores before returning to oil tank.

b. Data.

Engine oil tank capacity . . . . 44 qt
Oil inlet and outlet connections . . . ¼ in. std. pipe
Location of oil tank
   drain plug . . . Behind inspection plug in rear end of hull
Filter type . . . Purolator
Oil cooler type . . . Air-cooled core
Location of oil strainers . . . In oil pump and oil sumps
Oil pressure regulation . . . Adjustable relief valve

78. OIL PRESSURE AND OIL SCAVENGE PUMP.

a. **Description.** The oil pressure and oil scavenge pump assembly combines two gear type pumps in one unit so that both pumps are driven by same shaft. The scavenge section of unit has a capacity approximately 25 percent greater than pressure section, to assure removal of oil from the engine sump as fast as it accumulates. The pump assembly includes an adjustable relief valve to regulate oil pressure within engine, and an oil strainer to remove foreign matter before oil enters pump gears. The pump assembly also includes gears which operate governor which is mounted on rear end of pump body. The oil pump assembly (G104-2994152) on the C1 engine is not interchangeable with oil pump assembly (G163-7005781) on C4 engine.

b. **Oil Pressure Adjustment.** Start engine (par. 17 b and c) and warm it up until oil temperature is between 160° F and 180° F before attempting to adjust oil pressure. Cut lock wire and remove oil pressure relief valve adjusting screw cap (L, fig. 75). Hold relief valve body (E) with wrench (41-W-636-620) and loosen lock nut (K). Set engine speed at 1,800 revolutions per minute. Adjust oil pressure to 65 pounds per square inch by turning adjusting screw (I) in a clockwise direction to increase pressure or in a counterclockwise direction to decrease pressure. When correct pressure has been obtained, hold adjusting screw stationary and tighten lock nut (K) securely. Increase engine speed to 2,400 revolutions per minute; oil pressure should be 75 to 80 pounds per square inch at this speed. Install relief valve cap (L) with new gasket, tighten cap securely and install lock wire.

c. **Oil Pump Removal.** Loosen hose clamp (or compression nut) on oil pipe connections to oil pump, loosen clamp bolt on strap bracket on outlet pipe, and disconnect pipes from pump. Remove clevis

Engine Oiling System

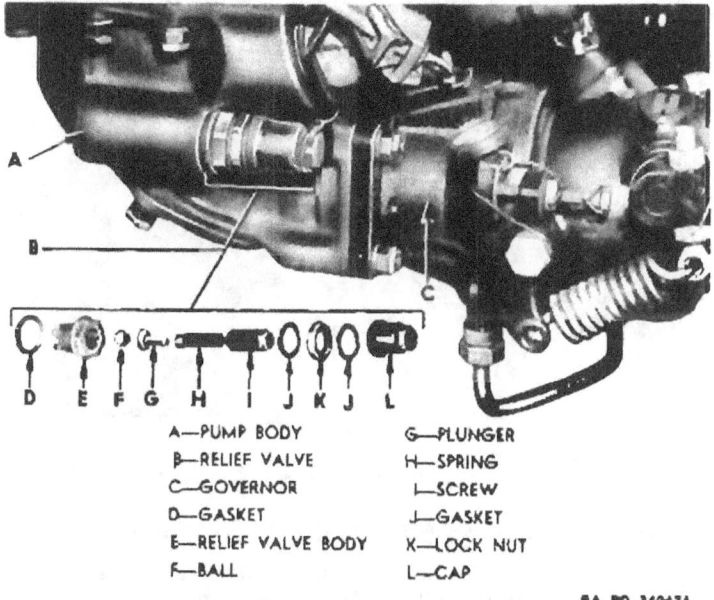

A—PUMP BODY
B—RELIEF VALVE
C—GOVERNOR
D—GASKET
E—RELIEF VALVE BODY
F—BALL
G—PLUNGER
H—SPRING
I—SCREW
J—GASKET
K—LOCK NUT
L—CAP

Figure 75—Oil Pressure Relief Valve

pin to disconnect governor linkage at governor box lever (fig. 106) and disconnect oil drain tube from governor. Remove four ¼-inch nuts palnuts and plain washers from studs that hold governor to oil pump and remove governor. Remove nuts and palnuts from eight studs (¼ x 28) and plain washers that attach the oil pump to accessory case, using special wrench (41-W-1577-500) as shown in figure 76. Support pump with hand while removing the last nut and remove pump. CAUTION: *Care must be taken to keep foreign matter out of crankcase after pump is removed.*

   d. **Installation of Oil Pump.** Clean surface of crankcase pad and pump body. Install a new gasket over studs and install oil pump. Install eight plain washers, hexagon nuts (¼ x 28), and palnuts on studs and tighten securely with special wrench (41-W-1577-500) as shown in figure 76. Install governor as described in paragraph 95 d. Install oil pipe nipples in pump body and fill inlet side of pump with engine oil, connect oil pipes with strap bracket attached to outlet pipe and tighten hose clamps or compression nuts. Start engine and check oil pressure; adjust relief valve if required (subpar. b above).

TM 9-755
79

Part Three—Maintenance Instructions

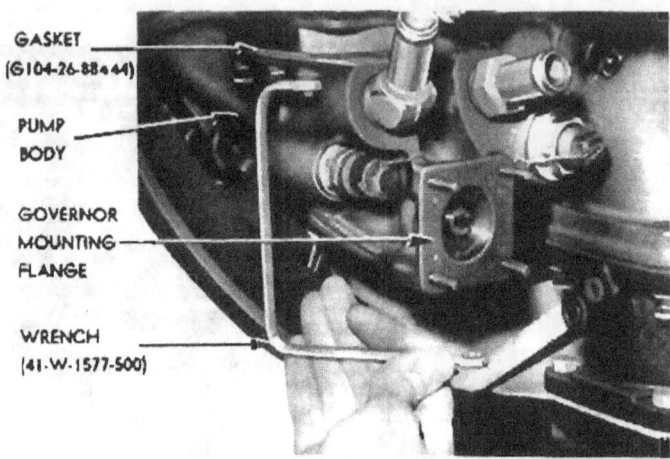

Figure 76—Removing Oil Pressure and Oil Scavenger Pump

79. OIL STRAINERS AND OIL SUMP DRAIN PLUG.

a. Description. A strainer is located in the inlet side of oil pump body to remove foreign matter before oil enters pump. Strainers are located in oil sumps to strain oil before it enters scavenger pump. A drain plug is located in front end of oil sump on C1 engine.

b. Cleaning Oil Pump Oil Suction Strainer (fig. 77). Cut lock wire and unscrew adapter from pump body, then remove gasket, spring and strainer. Wash parts in dry-cleaning solvent and dry with air stream. Insert strainer and spring into pump body and install adapter and gasket. Tighten adapter securely and install lock wire.

c. Cleaning C1 Engine Oil Sump Strainer (B, fig. 78). The oil sump strainer is located in rear end of oil sump. Remove carburetor (par. 93 i). Cut lock wire and remove two nuts which attach lower exhaust pipe flange to exhaust elbow; cut lock wire and loosen exhaust pipe clamp bolt and remove exhaust pipe from engine. Cut lock wire, unscrew plug from sump and remove spring and strainer. Wash parts in dry-cleaning solvent and dry with air stream. Wipe all metal particles from magnetic plug. Place strainer and spring in sump, install plug with new gasket and tighten securely. Install lock wire through plug. Install exhaust pipe with nuts and lock wire. Tighten exhaust pipe clamp then loosen bolt 1½ to 2 turns and secure with lock wire. Install carburetor (par. 93 j).

d. Removal and Installation of C1 Engine Oil Sump Drain Plug (A, fig. 78). The drain plug in front end of the oil sump can

Engine Oiling System

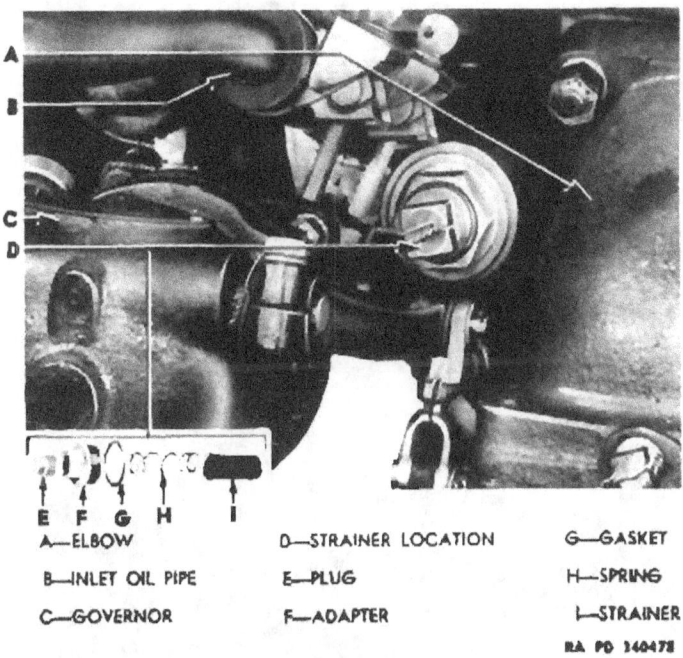

A—ELBOW D—STRAINER LOCATION G—GASKET
B—INLET OIL PIPE E—PLUG H—SPRING
C—GOVERNOR F—ADAPTER I—STRAINER

RA PD 340478

Figure 77—Engine Oil Pump Oil Suction Strainer

only be removed when engine is out of hull. Remove lower right cowl inspection cover. Cut lock wire and remove drain plug. After draining is complete, install drain plug with a new gasket, tighten and secure with a lock wire.

c. Cleaning C4 Engine Oil Sump Strainers (fig. 79). The strainer in each oil sump can be removed only when engine is removed from hull. Cut lock wires and unscrew strainer assemblies from both sumps. Wash parts in dry-cleaning solvent and dry with air stream. Place a new gasket over each strainer, screw strainer plug securely into sump and install lock wire.

80. OIL FILTER.

a. Description (fig. 80). Engine oil is cleaned by an oil filter located on right rear side of engine. It is connected in oil return pipe between scavenge pump and oil cooler. The filtering element is a cylinder made of a spirally wound crimped metallic ribbon. The spaces between the windings are small enough to prevent passage of particles larger than 0.005 inch but will permit oil to pass through. Particles separated from oil remain on the outside of element, from

TM 9-755
80

Part Three—Maintenance Instructions

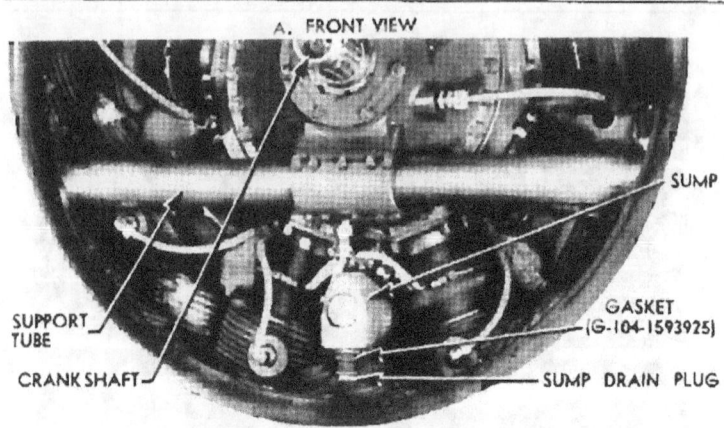

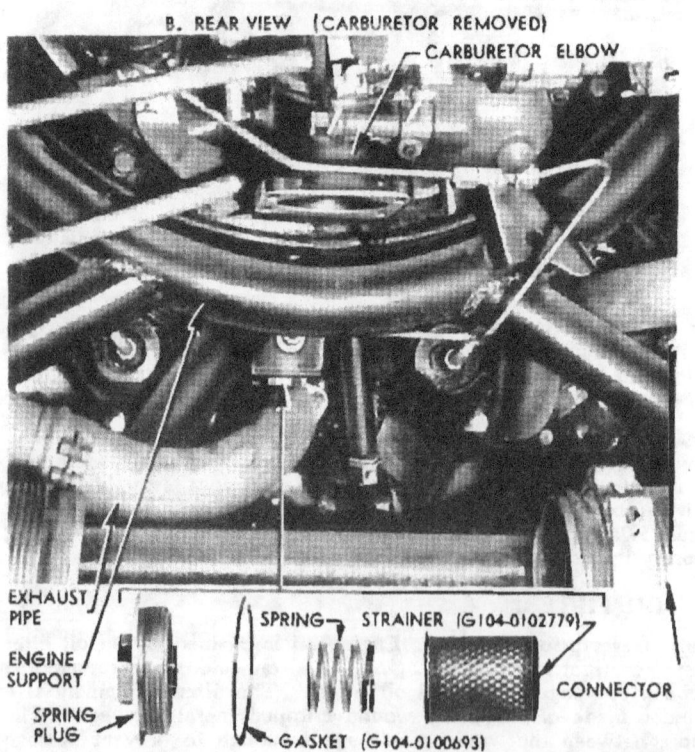

Figure 78—C1 Engine Oil Sump Strainer and Drain Plug

TM 9-755
80

Engine Oiling System

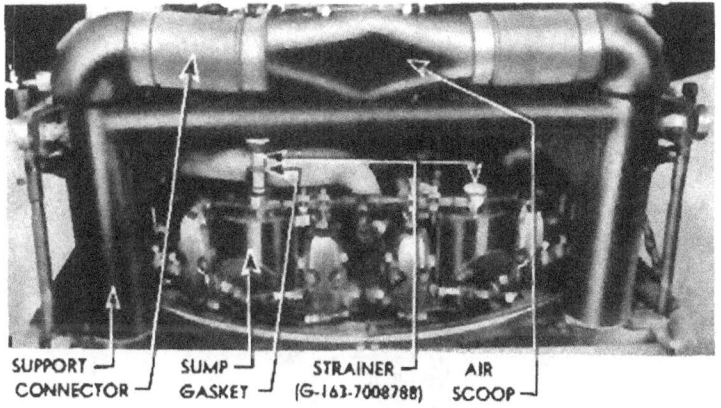

Figure 79—C4 Engine Oil Sump Strainer—Bottom View

which they are cleaned by a knife-blade scraper when handle on top of filter is rotated. An operating handle which extends through rear roof door is provided for turning filter handle from outside engine compartment.

b. **Cleaning and Draining.** Rotate filter handle one complete turn clockwise to clean dirt from filter element. Remove the drain plug at bottom of filter case to drain out sludge and sediment. After draining, install drain plug in case with new gasket.

c. **Removal, Disassembly, and Cleaning (fig. 80).** Loosen the hose clamps (or compression nuts) and disconnect the two oil pipes. Loosen ⅜-inch cap screw and lock nut in clamp which anchors filter case to mounting bracket and remove filter. Remove four stud nuts lock washers and remove case from filter element. Care should be taken to protect filter element from damage while handling. Wash filter element in dry-cleaning solvent, using a nonmetallic brush to remove dirt particles from the outside of element. Force air stream through the outlet fitting after removing element from solvent. Clean cavity in element cover and clean inside of case.

d. **Assembly and Installation.** Place a new gasket in position, install filter element and knife assembly in filter case. Install four nuts (⁵⁄₁₆ in.—24) and lock washers on studs and tighten evenly. Place filter in clamp on mounting bracket, locating it so that opening marked "IN" is toward left side of vehicle. Tighten cap screw (⅜ in. —24) and nut in bracket clamp. Connect two oil pipes with strap bracket attached to inlet pipe and tighten hose clamps (or compression nuts). If there is evidence of leakage at packing around stem of handle, tighten packing gland.

TM 9-755
81

Part Three—Maintenance Instructions

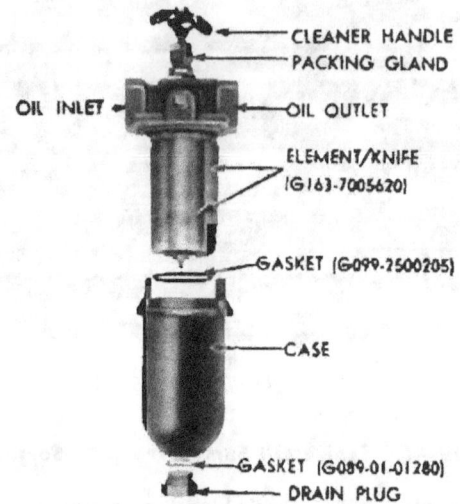

Figure 80—Oil Filter Disassembled

81. C1 ENGINE OIL COOLER.

a. **Description.** The oil cooler on the C1 engine assembly consists of one cooler core which is mounted on front end of engine support (fig. 54) and connected to oil filter and oil tank by pipes having flexible connections at joints. Cooling air is drawn through cooler core air passages by engine fan.

b. **Cleaning Cooler.** When cooler air passages become restricted with dirt, direct an air stream through air passages from rear side, or in opposite direction to normal air flow. Brush cooler with a nonmetallic brush and dry-cleaning solvent to soften dirt, if necessary, and blow out with air stream. If air passages cannot be properly cleaned by this method, remove cooler so that it will be more accessible for cleaning (subpar. c below).

c. **Removal.** Move engine out upon hull rear door (par. 75). Loosen hose clamps (or compression nuts) and disconnect oil pipes at both ends of cooler. Remove eight 1/4-inch cap screws and safety nut while attaching cooler to supports, and remove cooler.

d. **Installation.** Place cooler in position with outlets up and attach it to supports with eight cap screws (1/4 in.—28 x 3/4 in.) and safety nuts. Connect pipes and tighten hose clamps. If pipes are fitted with Sealflex compression nuts, connect and tighten as described

Engine Oiling System

Figure 81—Removing Oil Cooler From C4 Engine

in paragraph 125. Move engine into place and complete installation (par. 76).

82. C4 ENGINE OIL COOLER.

a. Description (fig. 55). Oil cooler on the C4 engine assembly consists of six cooler cores connected together and supported upon mounting plates to form a circular unit (fig. 82). This unit is attached to cowl ring by three bracket supports so that cooler cores surround oil cooler blower, which forces cooling air outward through the air passages in cores. The cooler is connected to oil filter and to oil tank by pipes having flexible coupling hoses at joints.

b. Cleaning Cooler Cores. When core air passages become restricted with dirt, direct an air stream inward through passages, or in opposite direction to normal air flow. Brush cores with a non-

TM 9-755

Part Three—Maintenance Instructions

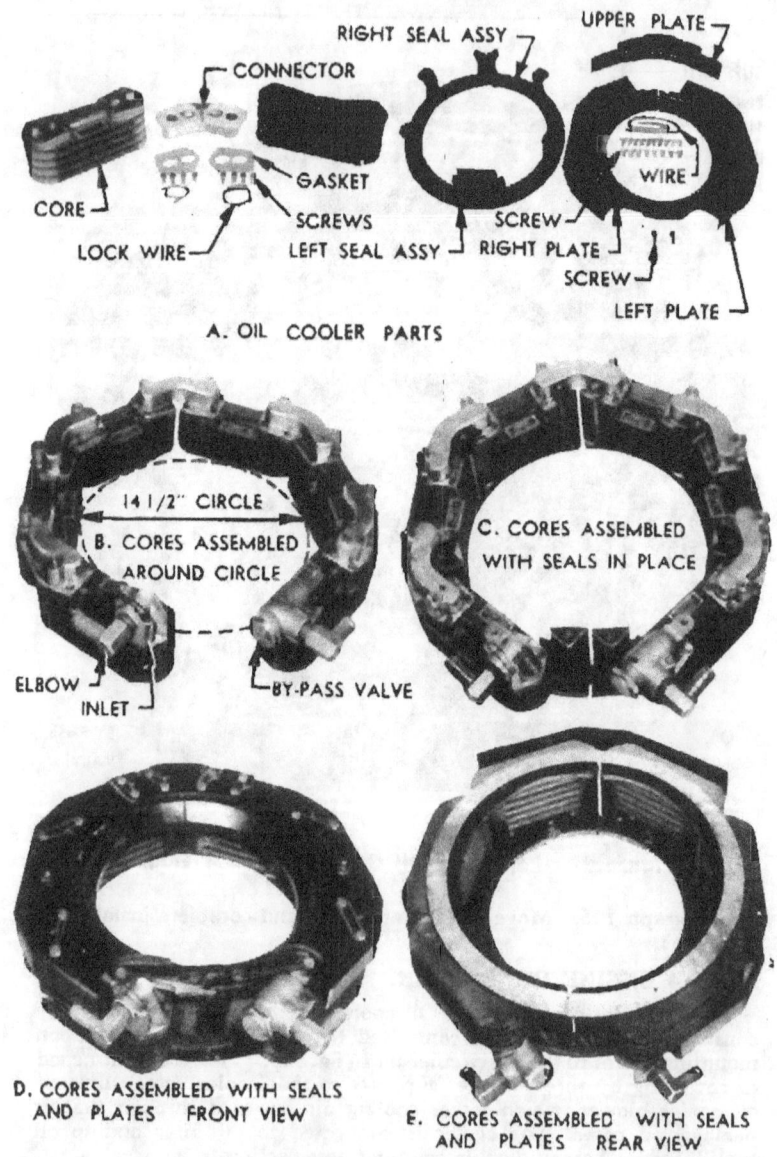

Figure 82—Assembly of C4 Engine Oil Cooler

metallic brush and dry-cleaning solvent to soften dirt, if necessary, and blow out with air stream. If air passages cannot be properly cleaned by this method, remove the cooler so that it will be more accessible for cleaning (subpar. c or g below).

c. **Removal and Installation Procedures.** The oil cooler can be removed and installed more easily if the engine is out of hull; however, if engine is not being removed for inspection or other maintenance operations it is not necessary to remove it in order to remove and install the cooler. Removal and installation procedures to meet either position of engine are given in the following subparagraphs.

d. **Removal, With Engine Out of Hull** (fig. 81). Remove cap screw and nut which attach strap bracket to clamp on inlet pipe, loosen hose clamps and disconnect pipes at upper coupling hoses. Cut lock wires, remove six ⅜-inch cap screws which attach cooler supports and braces to cowl ring while supporting cooler to prevent it falling upon blower, then move cooler straight outward away from engine.

e. **Disassembly.**

(1) Remove ⅜-inch cap screws and lock washers which attach pipe clips to mounting plates. Loosen bolt which clamps strap bracket to inlet pipe, loosen hose clamps and remove both oil pipes and coupling hoses.

(2) Cut lock wires and remove four ⅜-inch cap screws which attach two braces to mounting plates. Cut lock wires and remove six ⅜-inch cap screws which attach three mounting bracket supports to mounting plates (fig. 81).

(3) Cut lock wires and remove 20 ⅜-inch cap screws which attach mounting plates to core connections, core inlet, by-pass valve, and seals. Remove plates and seals (fig. 82).

(4) Cut lock wires and remove 16 ¼-inch cap screws which attach core inlet, core connectors, and by-pass valve to the cores; separate these parts and remove all gaskets.

f. **Assembly.**

(1) Soak the vellumoid gaskets in oil for 5 to 7 minutes. Clean joint surfaces of cores, connectors, by-pass valve and core inlet.

(2) Draw a circle 14½ inches in diameter upon a smooth level surface. Place the six cores upon surface with the openings upward and inner edges touching circle (B, fig. 82).

(3) Place gaskets and core connectors in position to connect adjoining cores and attach each connector to each core with four cap screws (¼ in.—28 x $2^{1}\!/_{32}$ in.) left slightly loose. Check position of cores around circle, then tighten all cap screws evenly to 10-12 foot-pounds tension. Install lock wires through adjacent pairs of screws.

(4) Install core inlet and gasket with two cap screws (¼ in.—28 x $2^{1}\!/_{32}$ in.) and two cap screws (¼ in.—28 x $2^{8}\!/_{32}$ in.). Install by-pass valve and gasket with three cap screws (¼ in.—28 x $2^{1}\!/_{32}$ in.) and

one cap screw ($\frac{1}{4}$ in.—28 x $2\frac{5}{32}$ in.). Tighten all screws evenly to 10-12 foot-pounds tension. Install lock wires through adjacent pairs of screws (B, fig. 82).

(5) Place right and left seal assemblies in position under core assembly (C, fig. 82). Place right, left, and upper mounting plates over core assembly. Start 20 cap screws ($\frac{3}{8}$ in.—16 x $1\frac{1}{8}$ in.) through mounting plates into seals, core connectors, core inlet, and by-pass valve, then tighten all screws evenly to 28-30 foot-pounds tension. Install lock wires through adjacent pairs of screws (D, fig. 82).

(6) Attach each of three mounting bracket supports to a mounting plate with two cap screws ($\frac{3}{8}$ in.—16 x $1\frac{1}{16}$ in.). Attach a brace to each side mounting plate with two cap screws ($\frac{3}{8}$ in.—16 x $1\frac{1}{8}$ in.). Tighten all cap screws to 28-30 foot-pounds tension and install lock wires (fig. 81).

(7) Connect pipe and coupling hose to elbow on by-pass valve and tighten hose clamps. Place large end of strap bracket over elbow on core inlet, install coupling hose on elbow, insert end of pipe through end of strap bracket into coupling hose, then tighten hose clamps and clamp bolt in bracket. Attach both pipes to mounting plates with clips secured by cap screws ($\frac{3}{8}$ in.—16 x $\frac{5}{8}$ in.) and lock washers.

g. **Installation, with Engine out of Hull** (fig. 81). Lift cooler into position, using care not to damage cores against blower, and support cooler while installing a cap screw ($\frac{3}{8}$ in.—16 x $1\frac{1}{16}$ in.) through lower hole of each lower support into cowl ring. Install four cap screws ($\frac{3}{8}$ in.—16 x $1\frac{1}{8}$ in.) through braces and supports into cowl ring and tighten all cap screws. Rotate crankshaft and check for clearance between blower and cooler, shifting cooler, if necessary, to obtain clearance all around. Clearances between cooler cores and rear flange of engine fan, and between fan and mounting plates, must be $\frac{1}{8}$ inch to $\frac{1}{4}$ inch. Tighten all support cap screws to 28-30 foot-pounds tension and install lock wires. Connect oil pipes at upper coupling hoses and tighten hose clamps. Attach strap bracket to clamp on inlet pipe with cap screw, lock washer and nut.

h. **Removal, With Engine in Hull.** The cooler assembly may be divided at top center and one half removed without disturbing the other half, or both halves may be removed as required, by the following procedure:

(1) Turn magneto switch to "OFF" position.

(2) On M18 vehicle, turn turret so that extension is clear of inlet grille and engage turret lock. Remove air inlet grille (par. 183 d). On M39 vehicle, remove grille guard which is attached to sponson extensions by four $\frac{5}{16}$-inch cap screws, plain washers, and lock washers.

(3) Remove cap screw and nut which attaches strap bracket to clamp on inlet pipe, loosen hose clamps and disconnect oil pipes at upper coupling hoses.

(4) Cut lock wires, remove two $\frac{3}{8}$-inch cap screws which attach

TM 9-755
82

Engine Oiling System

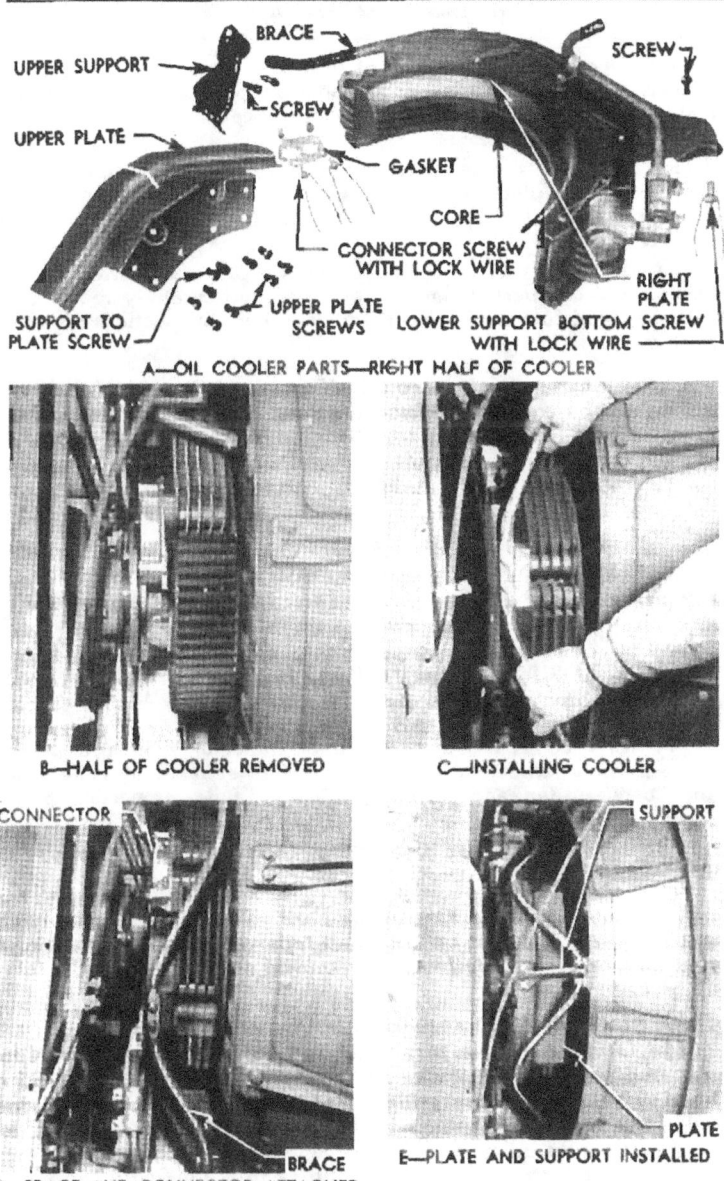

Figure 83—Installation of Half Section of C4 Engine Oil Cooler—Engine in Hull

upper support and upper ends of braces to cowl ring, remove two ⅜-inch cap screws which attach upper support to upper mounting plate and remove the support. Loosen the cap screws which anchor braces to mounting plates.

(5) Cut lock wire, remove eight cap screws (⅜ in.) which anchor upper mounting plate and remove plate. Cut lock wire and remove four cap screws (¼ in.) which attach the right end of top connector to core.

(6) Cut lock wire and remove the two cap screws which attach the brace and lower support to cowl ring and carefully remove the half section of cooler. Repeat on other side if both sections are to be removed.

i. **Installation With Engine In Hull (fig. 83).** If both half sections of the oil cooler were removed, install each section in the same manner, using the following procedure for installation of one section.

(1) Insert lock wires through heads of all attaching screws and bend wires into a "U" (A, fig. 83). The wires will make it easier to hold and start the screws, and make final locking easier.

(2) Carefully move cooler section into place and support it with a punch through upper hole in brace while installing a cap screw (⅜ in.—16 x 1$^{1}/_{16}$ in.) through lower hole of lower support into cowl ring. Install a cap screw (⅜ in.—16 x 1⅛ in.) through brace and lower support into cowl ring. Do not tighten either cap screw.

(3) Place a new gasket between core and connector and install four attaching cap screws (¼ in.—28 x $^{21}/_{32}$ in.). Do not tighten cap screws.

(4) Shift cooler section to obtain clearance of ⅛ inch to ¼ inch between cooler cores and rear flange of blower fan, and same clearance between fan and mounting plate. While supporting cooler section in this position, tighten the ⅜-inch support attaching screws to 28-30 foot-pounds tension, and tighten the ¼-inch connector attaching screws to 10-12 foot-pounds tension.

(5) After attaching screws are tightened, turn crankshaft and check to make sure that blower does not strike cords; a slight noise will be made if blower contacts cores. Twist lock wires of adjacent cap screws together.

(6) Install upper mounting plate with eight cap screws (⅜ in.—16 x ⅝ in.). Attach upper support to the upper mounting plate with two cap screws (⅜ in.—16 x 1$^{1}/_{16}$ in.). Attach support and upper ends of braces to cowl ring with two cap screws (⅜ in.—16 x 1⅛ in.). Tighten all cap screws, including two at middle of brace, to 28-30 foot-pounds tension and install lock wires through adjacent pairs of screws.

(7) Connect oil pipes at upper coupling hoses and tighten hose clamps. Attach strap bracket to clamp on inlet pipe with cap screw, lock washer and nut.

(8) Start and warm up engine (par. 17 b and c). Check all cooler connections and cores for oil leaks.

Engine Oiling System

Figure 84—Removing Oil Cooler Blower From C4 Engine

(9) Install inlet air grille (par. 183 d). On M39 vehicle, install grille guard and attach it to sponson extensions with four cap screws ($\frac{5}{16}$ in.—24 x $\frac{5}{8}$ in.), plain washers and lock washers.

83. OIL COOLER BLOWER—C4 ENGINE.

a. **Description.** The oil cooler blower on C4 engine is a sirocco type fan which forces cooling air radially outward through air passages of cooler cores which surround blower. The blower is mounted upon engine fan plate and rotates at crankshaft speed.

b. **Removal** (fig. 84). Move engine out upon hull rear door (par. 75). Remove oil cooler (par. 82 d). Remove cowl inspection cover and remove twelve ½-inch bolts, plain washers, and safety nuts which attach blower, fan plate and fan to flywheel. Remove blower, being careful not to let plate and fan fall off engine.

c. **Installation.** Place blower in position against fan plate, install 12 bolts (½ in.—20 x $1\frac{23}{32}$ in.), plain washers and safety nuts and tighten to 80-100 foot-pounds tension. NOTE: *Some engines use longer bolts, which require spacers. If spacers are used, install them under heads of all bolts.* Install oil cooler (par. 82 g). Move engine into place and complete installation (par. 76).

84. OIL TANK.

a. **Description.** Oil tank is formed in rear end of left fuel and oil tank assembly. A filler with oil level indicator rod attached, and a

TM 9-755
84

Part Three—Maintenance Instructions

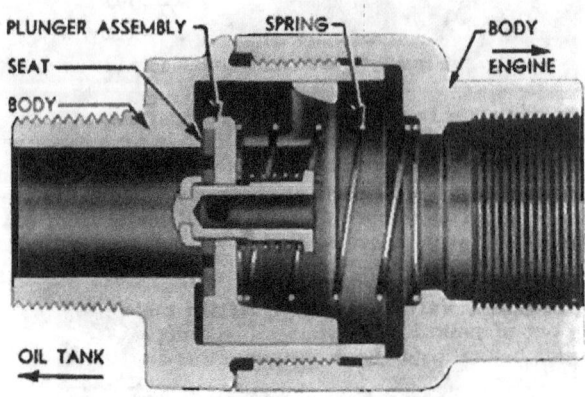

Figure 85—Cross Section of Engine Oil Tank Outlet Check Valve

vent to relieve pressure in tank as oil heats up, are located in upper rear side of tank (AA, fig. 38). They are reached by opening filler screen door in hull roof (C, fig. 258). A drain plug in lower rear corner of tank is reached by removing a plug in hull rear plate. Oil is drawn from near bottom of tank by an internal outlet pipe welded to outlet port located slightly below middle of tank. Oil is returned to tank through a port at top, where it flows into a vertical internal passage having an outlet near bottom of tank.

b. **Cleaning Tank.** Remove plug in hull and drain plug in tank while oil is hot, in order to remove as much sediment as possible with oil. Install drain plug and fill tank approximately ½ full (5 gal.) with S.A.E. 10 oil. Agitate oil with a clean stick or rod to loosen sediment, then thoroughly drain oil from tank. Securely install plugs in tank and hull, and fill tank to proper level with specified engine oil (par. 38 d (6)).

c. **Tank Removal and Installation.** The removal and installation procedure is same as for left fuel tank, paragraph 99.

d. **Outlet Check Valve Disassembly and Cleaning** (fig. 85). Outlet check valve is located in a recess in the oil tank so that it is not possible to remove valve by applying a wrench to male body which is screwed into tank outlet port. It is usually necessary, therefore, to disassemble valve while it is attached to tank, in order to clean plunger seat or to replace plunger or spring with parts from a stock valve assembly.

(1) Drain oil tank. Open hull rear door (par. 181 a).

(2) Disconnect oil pipe from outlet tee. Unscrew captive screws to disconnect conduit housing from oil temperature gage unit and pull wire out of unit (fig. 249). Remove tee from check valve.

TM 9-755
84-85

Engine Oiling System

(3) Remove rear junction box inspection cover which is attached with five 5/16-inch cap screws and two safety nuts. Remove three 5/16-inch cap screws and special washers which attach rear edge of fuel tank insulator, and pull insulator out far enough to provide access to check valve.

(4) Unscrew outer body of valve from body attached to oil tank, and remove spring and plunger assembly. If body unscrews from oil tank, clamp this part in vise and disassemble valve.

(5) Thoroughly clean all parts of check valve with dry-cleaning solvent, being careful to remove all particles of metal from plunger seat. If plunger seat is damaged so that it will not seal properly, replace plunger assembly.

e. **Outlet Check Valve Assembly.** If the male body of check valve came out of tank during disassembly (subpar. d (3) above), assemble valve before installing it on tank; otherwise, assemble valve on tank, as follows:

(1) Place plunger in male body on tank, with rubber seat inward toward tank (fig. 85). Place spring in recess in female body and hold it in place with finger inserted through outlet port. Slide female body through hole in fuel tank insulator and felt washer, and screw it securely upon body on tank. Check to make sure spring is properly seated against plunger.

(2) Attach rear edge of fuel tank insulator with three cap screws (5/16 in.—24 x 3/4 in.) and special washers. Install junction box inspection cover with five cap screws (5/16 in.—24 x 3/4 in.) and special washers, with safety nuts on two upper screws.

(3) Coat threads of tee with joint and thread compound, type (A), and install it on check valve with temperature gage unit pointing down. Connect oil pipe to tee and tighten hose clamps (or compression nuts). Insert wire in gage unit, connect conduit housing to gage and tighten the housing captive screws (fig. 249).

(4) Fill oil tank. Start engine and check for oil leaks. Close hull rear door (par. 181 b).

85. ENGINE OIL PIPES AND FITTINGS.

a. **Description.** Oil pipes are made of steel tubing formed to shape. All oil pipes on the C4 engines are joined together and to various fittings by coupling hoses secured with hose clamps. Strap brackets which anchor two pipes together are used in connection with coupling hoses to prevent oil pressure blowing hose off at both ends of pump-to-filter pipe and at both ends of cooler inlet-to-cooler pipe. Oil pipes on the C1 engines may have either coupling hoses or Sealflex type fittings. The Sealflex fittings are of same type as described in paragraph 125, and are to be connected and tightened in same manner.

b. **Inspection and Replacement of Oil Pipes.** If inspection of oil pipes reveals that they are contacting and chafing against other parts, they should be repositioned to provide proper clearance or should be insulated to prevent damage. All supporting clips and brackets must be in place and securely mounted. Crushed or restricted oil pipes should be replaced. When oil pipes are installed make certain that they are clean internally, and avoid bending to facilitate installation. Operate engine and test for oil leaks at all connections before closing hull rear door.

Section XX

IGNITION SYSTEM

86. DESCRIPTION, CIRCUITS, AND DATA.

a. Description. The ignition system, which furnishes dual ignition to each cylinder, consists of the following components.

(1) MAGNETOS. Two Scintilla magnetos, type VAG9DFA, are mounted on the engine rear crankcase. They are driven in a counter-clockwise direction at 1 1/8-crankshaft speed by engine accessory drive gears. The right-hand magneto fires front spark plugs; left-hand magneto fires rear spark plugs. An automatic advance mechanism is incorporated in magneto; there is no manual means of advancing spark range.

(2) SPARK PLUGS AND IGNITION CABLE ASSEMBLY. Two Champion shielded spark plugs, type 63S, are installed in each cylinder; one in front side and one in rear. The spark plugs are connected to distributor blocks in magnetos by rubber insulated ignition cables which are inclosed in a metal harness to protect wiring and to effect radio interference suppression.

(3) BOOSTER COIL. A Delco-Remy booster coil, Model 1115482, is used to provide additional ignition spark when starting engine. It is a vibrator-type high tension induction coil in a metal housing which serves as a shield as well as a means of attaching unit to engine mounting ring.

(4) SWITCHES AND WIRING. The magneto and booster coil switches on instrument panel are described in paragraph 16 e. They are connected to magnetos and to booster coil by wires inclosed in conduits.

(5) BOOSTER COIL CIRCUIT BREAKER. The booster coil circuit breaker in instrument panel is described in paragraph 16 b.

b. Ignition Circuit (fig. 86).

(1) MAGNETOS. When magneto switch is in "BOTH" position and engine is running, or being cranked, both magnetos generate high tension current which is conducted to spark plugs by distributor cylinder and distributor blocks in magnetos and ignition cables. A separate wire is connected between ground terminal on each magneto and magneto switch on instrument panel. When switch is in the "OFF"

Ignition System

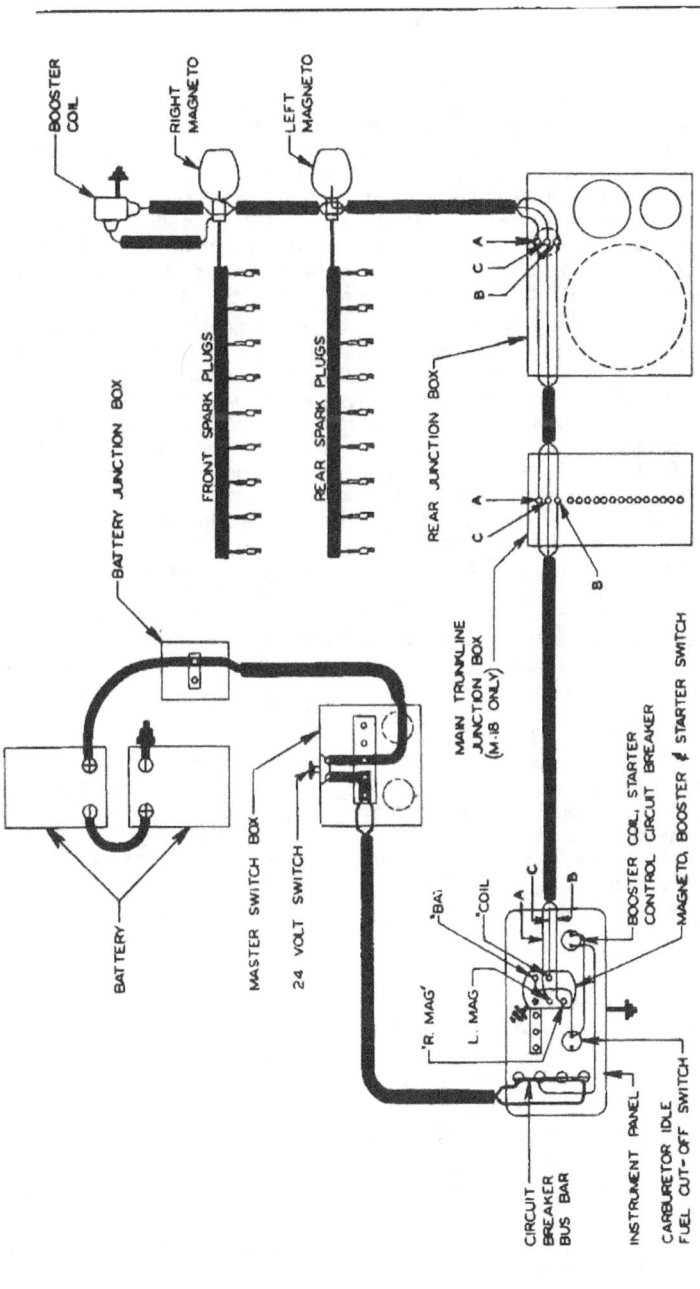

Figure 86—Ignition Circuit

position, these wires provide a direct path to ground for primary current in each magneto, which prevents generation and delivery of high tension current to spark plugs. When switch is at "R" with engine running, left magneto is grounded, but right magneto delivers current to front spark plugs. When switch is at "L" with engine running, right magneto is grounded but left magneto delivers current to rear spark plugs. Grounding wires "A" left magneto, "B" right magneto (fig. 86), are inclosed in same conduit and make connections at engine junction box, rear junction box, main trunkline junction box (M-18 only) and to terminals marked "R MAG" and "L MAG" on switch assembly.

(2) BOOSTER COIL. Primary winding of booster coil receives 24 volt current from battery through a single-wire ground-return circuit. Wires in a conduit connect the 24-volt battery circuit in master switch box to a common circuit breaker bus bar in instrument panel. Wires in instrument panel connect bus bar to booster coil circuit breaker, and connect circuit breaker to "BAT" terminal on booster switch. The "COIL" terminal on booster switch is connected to primary terminal of booster coil by wiring (C, fig. 86) which is inclosed in same conduits and connected at same junction boxes as magneto grounding wires (subpar. b (1), above). When booster switch is in "ON" position, high tension current induced in the booster coil is conducted by a wire to the "H" terminal (fig. 88) in right magneto, and through magneto distributor cylinder and blocks to front spark plugs. Booster segment is located so that it trails magneto segment on distributor cylinder to give a retarded spark with booster coil current. Booster coil is grounded to engine.

c. Data.

Ignition firing order .... Clockwise cyl. 1-3-5-7-9-2-4-6-8
Magneto rotation .... Counterclockwise
Maximum engine rpm drop with one magneto
 (engine loaded) .... 100 rpm from 2,400 rpm
Both magnetos timed at .... 25 deg B.T.C.
Breaker point gap .... 0.012 in.
Spark plug gap .... 0.018 to 0.020 in.

87. IGNITION TIMING.

a. General. Both Scintilla magnetos on the R975-C1 and R975-C4 engines are timed at 25 degrees before top center in full advance position. The procedure described in this paragraph may be used to check and correct ignition timing with engine either in or out of hull.

b. Placing Engine in Timing Position. CAUTION: *If engine is in hull, make certain that magneto switch is in "OFF" position. If engine is out of hull, make certain that magnetos are grounded (par. 88 c).*

(1) If engine is in hull, open hull rear door (par. 181 a) and remove air inlet grille (par. 183 d).

Ignition System

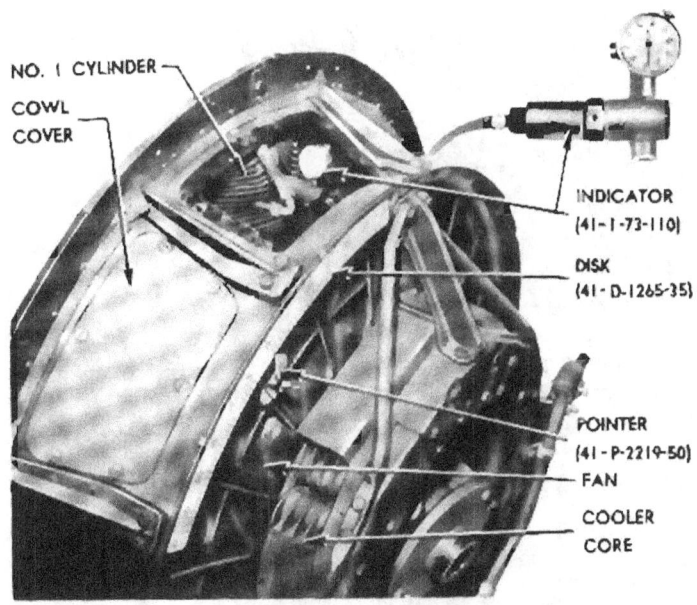

Figure 87—Engine Timing Disk, Pointer, and Top Dead Center Indicator Installed

(2) Remove cowl cover and remove front spark plug from No. 1 cylinder. Turn engine by fan until No. 1 piston is near top of compression stroke, which will be indicated by air rushing out of spark plug port.

(3) Install top dead center indicator (41-I-73-100) in No. 1 front spark plug port, with arm pointing down and dial indicator up (fig. 87).

(4) Rotate crank shaft very slowly clockwise until piston strikes indicator arm, causing dial indicator hand to move. Continue rotating crankshaft very slowly until indicator hand just stops moving.

(5) Install a timing disk (41-D-1265-35) on cowl ring and pointer (41-P-2219-50) on fan blade so that tip of pointer is at "0" mark on timing disk.

(6) Turn crank shaft one-eighth turn counterclockwise by fan, then turn it clockwise until pointer is at "25" mark on timing disk. This is timing position of engine.

c. **Checking and Setting Magnetos for Correct Timing** (fig. 88).

(1) Remove two cap screws and plain washers which attach

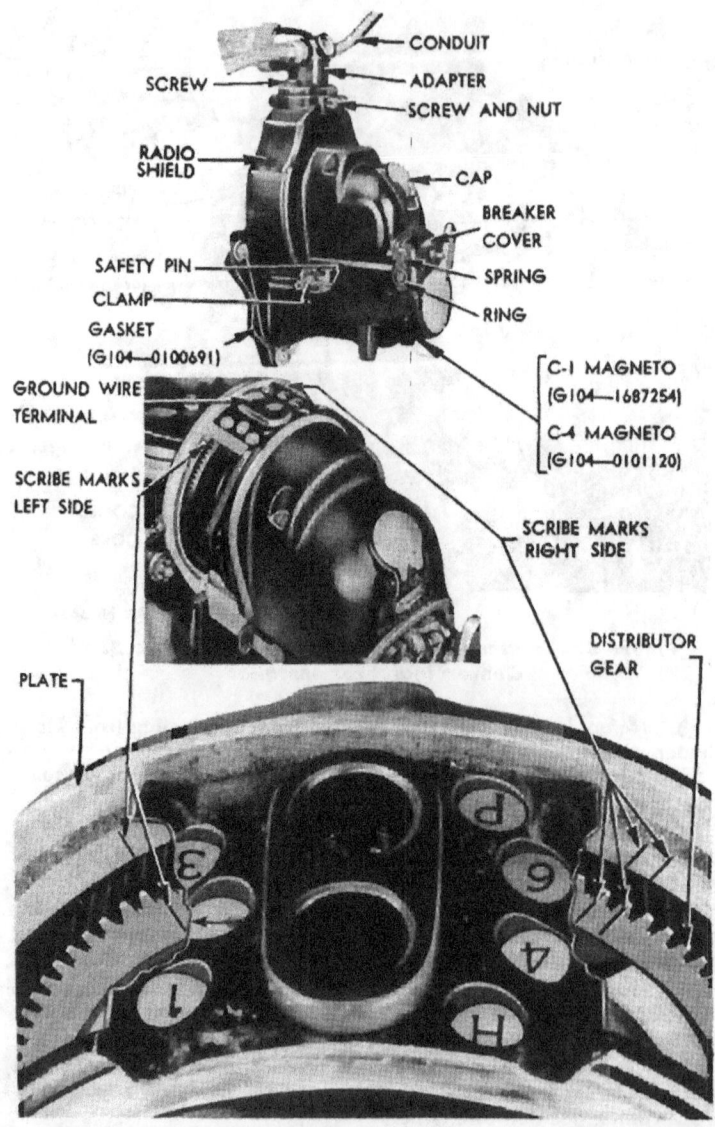

Figure 88—Magneto Assembly and Timing Scribe Marks

adapter to radio shield on each magneto and remove two cap screws, plain washers, and nuts which hold halves of radio shield together at upper end.

(2) Remove safety pins, swing clamps down, and remove half of radio shield which is on outer side of each magneto. Lift distributor blocks out of magnetos so that timing scribe marks can be seen.

(3) The right magneto is correctly timed if the two scribe marks on the distributor gear are in *exact* alinement with the two scribe marks on the magneto front end plate (fig. 88).

(4) The left magneto is correctly timed if single scribe mark on distributor gear is in *exact* alinement with single scribe mark on magneto front end plate (fig. 88).

(5) If either magneto is not correctly timed, remove lock wire and loosen the nuts on magneto mounting studs. Move magneto through range provided in slotted holes in mounting flange until scribe marks on distributor gear and on front end plate are in *exact* alinement; then tighten mounting stud nuts and install lock wire. NOTE: *Both magnetos must be set exactly alike for proper synchronization.*

(6) Mark exact position of pointer on fan blade, then remove pointer. Turn engine nearly two complete revolutions, then install pointer in position it previously occupied on fan. Turn engine clockwise only, until pointer is again at the "25" mark on timing disk. Magneto timing scribe marks must be in *exact* alinement, as previously set.

d. **Removing Tools and Installing Parts.**

(1) Remove timing pointer, disk, and top dead center indicator.

(2) Install spark plug and cowl cover.

(3) Install half of radio shield on each magneto and attach it to other half with two cap screws, plain washers, and nuts. Swing clamps up into place and install safety pins. Attach adapter to each radio shield with two cap screws and plain washers.

(4) If engine is in hull, install air inlet grille (par. 183 d) and close hull rear door (par. 181 b).

88. MAGNETOS.

a. **Non-interchangeability of Magnetos.** Do not interchange magnetos between R975-C1 and R975-C4 engines. Both magnetos are timed at 25 degrees before top center in full advance position; however, advance curve on the C4 engine magneto goes up much more rapidly than on the C1. Use of a C4 engine magneto on a C1 engine will result in overheating, detonation, and probable damage to engine at slower operating speeds. Likewise, use of a C1 engine magneto on a C4 engine will greatly reduce power available at operating speeds slower than normal. The more rapid spark advance in C4 engine is made allowable only because of additional cooling area provided on C4 engine cylinders.

TM 9-755
88

Part Three—Maintenance Instructions

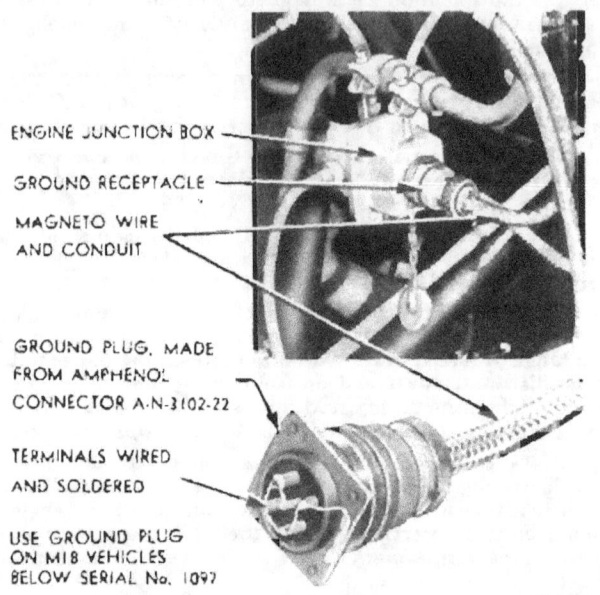

Figure 89—Magneto Ground Receptacle and Ground Plug

b. **Identification.** Magnetos used on R975-C4 engines are being identified by painting a large white "C4" on side of coil cover. This paint may wear off in use so actually only positive identification is to be found on specification plate.

(1) R975-C1 ENGINE MAGNETOS. Scintilla magnetos (G104-1687254) for R975-C1 engine have one of the following designations stamped on specification plate, and all parts are interchangeable between these two types:

    Type: VAG9DFA, Manufacturer's Drawing: 2-1071-4
    Type: VAG9DFA, Manufacturer's Drawing: 2-1071-6

(2) R975-C4 ENGINE MAGNETOS. Scintilla magnetos (G104-0101120) for R975-C4 engine have one of the following designations stamped on specification plate, and all parts are interchangeable between these three types:

    Type: VAG9DFA, Manufacturer's Drawing: 2-1091-2
    Type: VAG9DFA, Manufacturer's Drawing: 2-1071-3
    Type: VAG9DFA, Manufacturer's Drawing: 2-1071-7

c. **Magneto Ground Receptacle and Ground Plug** (fig. 89). When engine is out of hull, or magneto wires are disconnected from rear junction box for any reason, the magnetos must be grounded for

TM 9-755
88

Ignition System

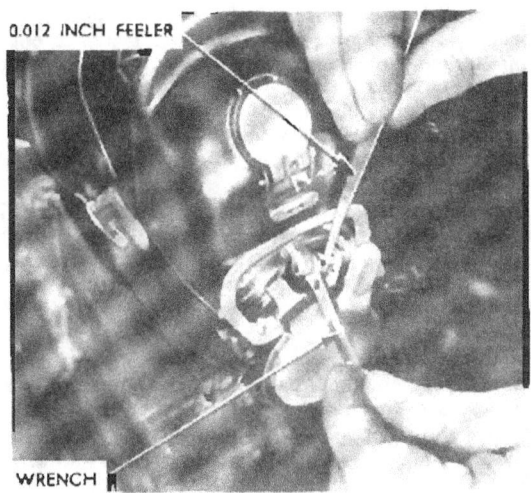

Figure 90—Adjusting Breaker Contact Points

safety reasons. If the magnetos are not grounded, and a small amount of fuel remains in carburetor or cylinders, engine may start if crankshaft is turned, with possible injury to personnel or damage to engine. M18 vehicles beginning with serial number 1097 and all M39 vehicles, have a ground receptacle installed on engine junction box cover, to which magneto wire and conduit must be connected immediately after it is disconnected from rear junction box. If vehicle is not equipped with ground receptacle, install a ground plug on magneto wire and conduit as shown in figure 89. CAUTION: *Make certain that conduit coupling nut is securely tightened on ground receptacle or ground plug.* When receptacle is not in use keep it covered with cap which screws on receptacle and is attached to junction box by a chain.

d. **Breaker Contact Point Inspection and Adjustment.** Remove rings, unhook springs, and remove breaker cover from magneto (fig. 88). Turn engine with hand crank until magneto contact points are held wide open by a cam lobe. Carefully inspect contact points and if pitted or burned notify higher authority; do not attempt to adjust pitted or burned points. Check gap between points with a 0.012-inch feeler gage. If gap is more or less than 0.012-inch, loosen the lock nut on the long contact screw with a contact point wrench and adjust the screw until there is 0.012-inch clearance between contact points; then tighten the lock nut (fig. 90). Install breaker cover, snap springs into place and install rings.

227

TM 9-755
88

Part Three—Maintenance Instructions

e. **Magneto Removal.** Remove and install one magneto at a time, if possible, so that other magneto can be used to locate timing position of engine. When magnetos are once timed to engine, timing cannot change unless a mechanical failure occurs. The procedure in this subparagraph covers removal of either right or left magneto, with engine either in or out of hull. CAUTION: *If engine is out of hull make certain that magnetos are grounded to prevent starting of engine* (subpar. c above). *If engine is in hull, make certain that magneto switch is at "OFF" position.*

(1) Remove two cap screws and plain washers which attach adapter to radio shield, and remove two cap screws, plain washers, and nuts which hold halves of radio shield together at upper end. Remove safety pins from radio shield clamps, swing clamps down and remove shield from magneto.

(2) Loosen conduit coupling nuts on adapter, and disconnect booster wire (right magneto only) and ground wire from terminal block by lifting the rubber lock and unscrewing terminal screws. Lift distributor blocks out and disconnect ignition cables from blocks.

(3) Remove the outside half of radio shield and lift out the distributor blocks from other magneto which will remain on engine (step (1) above).

(4) Turn engine by hand crank until scribe marks on distributor gear and front end plate are in *exact* alinement on magneto that is to remain on engine (fig. 88). This places engine in timing position for later installation of magneto.

(5) Cut lock wire and remove slotted nuts from mounting studs of magneto that is to be removed, and remove magneto and gasket from engine.

f. **Magneto Installation.** The procedure in this subparagraph covers installation of either right or left magneto, with engine either in or out of hull. CAUTION: *Make certain that magneto switch is in "OFF" position, or that magnetos are grounded* (subpar. c above).

(1) Check magneto that is on engine to make certain that scribe marks are in alinement, as set in step (4) of subparagraph e above. If both magnetos were removed, or it appears advisable to check timing of both magnetos, place engine in timing position as described in paragraph 87.

(2) Check breaker contact point adjustment of magneto to be installed (subpar. d above).

(3) Remove adapter and radio shield from replacement magneto and remove the distributor blocks.

(4) Place a new gasket (G104-0100691) on magneto mounting flange, with a small amount of heavy grease to hold it in place. Lightly lubricate the splined end of magneto rotor shaft with high temperature grease.

(5) Turn magneto rotor shaft until the scribe marks on distributor gear and front end plate are in alinement (fig. 88), hold gear in this position and install magneto over mounting studs on engine.

Install plain washers and slotted nuts, leaving nuts just loose enough so magneto can move on crankcase.

(6) Move the magneto through range provided in slotted holes in mounting flange until scribe marks on distributor gear and on front end plate are in *exact* alinement, then tighten mounting stud nuts and install lock wire. NOTE: *If alinement of scribe marks cannot be secured within the range allowed by slotted holes, remove magneto, turn distributor gear one complete revolution to change position of spline, install magneto again and aline scribe mark.*

(7) Turn engine two complete revolutions until scribe marks on one magneto are in *exact* alinement, then check to make certain that scribe marks on other magneto are also in *exact* alinement.

(8) Loosen set screws in distributor blocks, push each ignition cable into terminal having same number as number on cable, and tighten all set screws. The numbers on distributor blocks indicate firing sequence of magneto and not firing sequence of engine. The ignition cables attached to distributor blocks connect to cylinders in the following order:

Distributor Block Terminal Number	Cylinder Number
1	1
2	3
3	5
4	7
5	9
6	2
7	4
8	6
9	8

(9) Place distributor blocks in magneto, with block having five terminals on left side. Slide rubber lock over booster wire (right magneto only) and ground wire, connect ground wire to terminal (P, fig. 88) and connect booster wire to terminal (H) in terminal block at top of magneto. Tighten terminal screws and push rubber lock down over screws.

(10) Install both halves of radio shield on magneto and attach them together with two cap screws, plain washers, and nuts. Swing clamps up into position and install safety pins. Attach adapter to radio shield with two cap screws and plain washers, and tighten conduit coupling nuts.

89. SPARK PLUGS.

a. Removal. To remove rear spark plugs, open hull rear door (par. 181 a). To remove front spark plugs, move engine out upon hull rear door (par. 75) and remove cowl covers. To reach lower front plug, lower hull door plate if hinged (par. 181 c), or raise engine with sling (fig. 68). Unscrew coupling nut and disconnect elbows from spark plugs. Pull out on ignition cables until contacts are free from plugs. Remove spark plugs, using a deep 1-inch socket wrench and an extension handle. Insert extension handle between fan blades when removing front plugs.

TM 9-755
89-90

Part Three—Maintenance Instructions

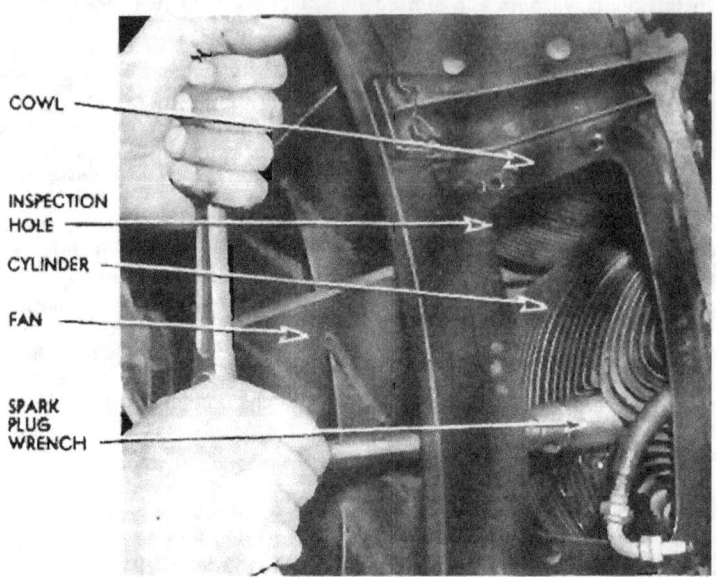

Figure 91—Installing Front Spark Plug

b. **Installation.** Check gaps in all spark plugs (H004-501006) and set to 0.018-inch to 0.020-inch if necessary. NOTE: *If spark plugs do not have the gaskets assembled on them, install gasket (G104-0501275) on plugs for C1 engine, and install gasket (G163-700-5769) on plugs for C4 engine.* Coat threads of plugs with antiseize compound, install plugs in cylinders and tighten to 37 to 40 foot-pounds tension, using a deep 1-inch socket wrench and extension handle (fig. 91). Insert contacts into plugs using care to avoid chafing contacts, connect elbows to spark plugs and tighten coupling nuts. Install cowl covers and install engine if removed (par. 76). Close hull rear door (par. 181 b).

90. BOOSTER COIL.

a. **Removal** (fig. 92). Open hull rear door (par. 181 a). Disconnect conduit coupling nut and pull high tension wire out of coil. Remove terminal cover, loosen screw which anchors primary wire to terminal, disconnect conduit coupling nut and remove primary wire. Remove booster coil which is attached to engine mounting ring with two screws, plain washers, and external-tooth lock washers.

b. **Installation** (fig. 92). Attach booster coil (163-0139400) to engine mounting ring with two round-head machine screws (10 —

TM 9-755
90-91

Fuel and Air Intake and Exhaust Systems

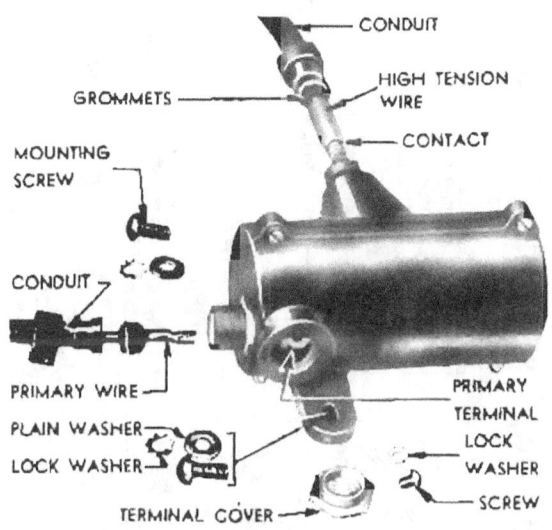

Figure 92—Booster Coil and Connections

32 x ¼ in.), plain washers, and external-tooth lock washers. NOTE: Make certain that lock washers cut through paint to make a positive ground connection. Connect primary wire to terminal and tighten screw. Connect conduit to coil and tighten coupling nut. Install primary terminal cover. Insert contact on high tension wire into terminal in coil, connect conduit and tighten coupling nut. Close hull rear door (par 181 h).

Section XXI
FUEL AND AIR INTAKE AND EXHAUST SYSTEMS

91. DESCRIPTION AND DATA.

a. General Description. The fuel and air intake systems include all parts up to distribution chamber in engine crankcase, from which fuel-air mixture is conducted to cylinders by intake pipes. Since these two systems merge into one at the carburetor, description will be simplified by division into parts supplying fuel to carburetor, parts supplying air to carburetor, and parts controlling fuel-air mixture. Exhaust system includes all parts leading outward from exhaust manifold.

b. Description of Fuel System (fig. 93). Fuel system consists of the following components:

TM 9-755
91

Part Three—Maintenance Instructions

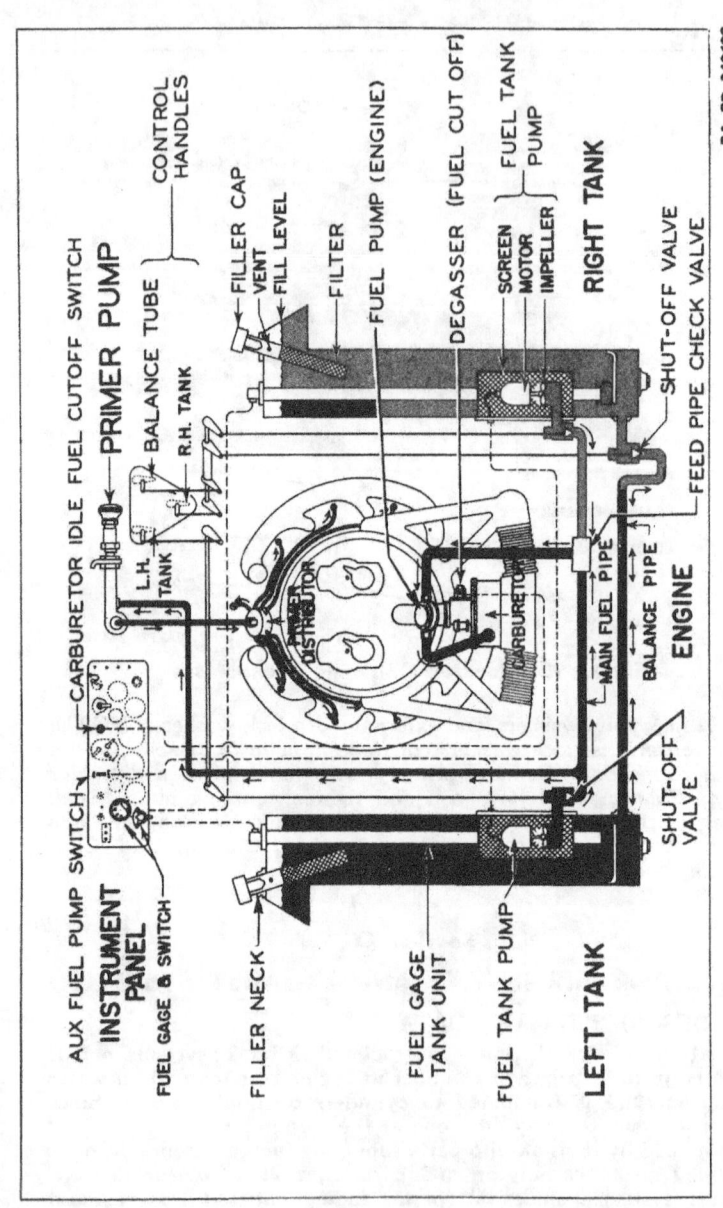

Figure 93—Fuel System—Schematic View

Fuel and Air Intake and Exhaust Systems

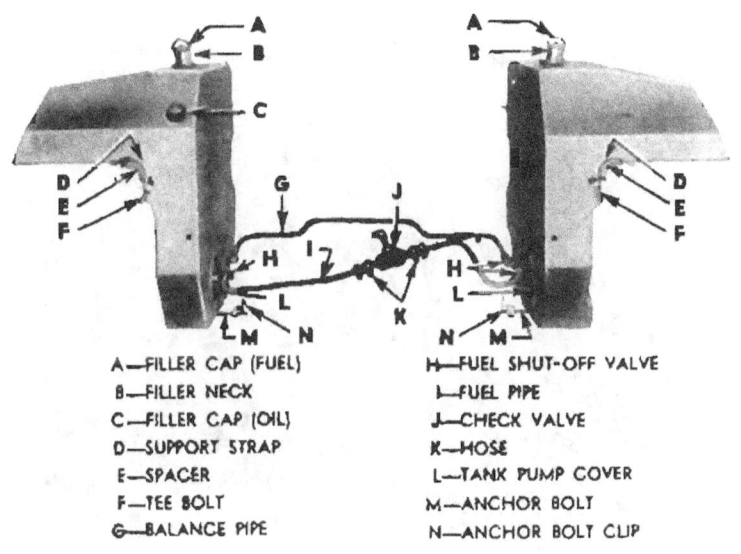

A—FILLER CAP (FUEL)
B—FILLER NECK
C—FILLER CAP (OIL)
D—SUPPORT STRAP
E—SPACER
F—TEE BOLT
G—BALANCE PIPE
H—FUEL SHUT-OFF VALVE
I—FUEL PIPE
J—CHECK VALVE
K—HOSE
L—TANK PUMP COVER
M—ANCHOR BOLT
N—ANCHOR BOLT CLIP

RA PD 340434

Figure 94—Fuel Tanks, Pipes, and Valves

(1) FUEL TANKS (fig. 94). A fuel tank is mounted in hull extending into sponson, on each side of engine compartment. Right tank has a capacity of 90 gallons; left tank, in which engine oil tank is formed, has a capacity of 75 gallons. Both tanks are covered, and shielded from heat of engine, by insulators composed of steel plates covered with insulating material. A vented expansion chamber and a breather in the filler cap provides for expansion of fuel due to heat. Each tank is filled separately through a filler neck provided with a removable filler cap located under a hinged cover on hull roof. Fuel entering the filler neck passes through a filter before passing into tank. A fuel gage tank unit located in each tank and connected to fuel gage on instrument panel provides a means of indicating fuel level (par. 163 h).

(2) FUEL TANK PUMPS (fig. 112). An electric pusher-type fuel pump is located near the bottom in each fuel tank. The pump is fully submerged in fuel and is surrounded by a large screen. These pumps supply a large volume of fuel almost instantly at a steady non-pulsating pressure, when starting the engine in either sub-zero or extremely high temperatures. Since pumps push fuel from the tanks, fuel supply from these pumps to carburetor is kept under pressure which raises boiling point and reduces possibility of vaporization and vapor lock. Location of pumps within fuel tanks permits pumps to liberate large volumes of vapor into vented tanks.

TM 9-755
91

Part Three—Maintenance Instructions

Figure 95—Primer Pipe Distributor on Engine

(3) FUEL PIPES AND VALVES (fig. 94). The fuel tanks are connected together by a balance pipe to permit maintenance of approximately equal fuel level in both tanks, and also to permit use of fuel in both tanks in case one fuel tank pump becomes inoperative. The fuel tank pumps are connected by pipes to a feed pipe check valve through which fuel from both tanks enters the engine fuel pump (step (4) below). This check valve prevents fuel from being pumped from one tank into the other in case one fuel tank pump develops less pressure than the other, or becomes inoperative. The balance pipe and two fuel pipes are provided with shut-off valves which are controlled by handles on bulkhead (fig. 9). Use of valves is explained in paragraph 14 a.

(4) ENGINE FUEL PUMP (fig. 108). A diaphragm type mechanical fuel pump is mounted on engine from which it is actuated by a camshaft. Fuel from both fuel tank pumps enters engine fuel pump through feed pipe check valve, and passes from engine fuel pump to carburetor through a coupling hose. If fuel tank pumps become inoperative, or are not turned on, engine fuel pump will supply sufficient fuel for normal operation when the fuel is cold; however, when the fuel is hot, vapor lock will occur and fuel-air mixture will become too lean for safe operation of engine. The lean mixture will cause excessive cylinder head temperature which will damage engine if vehicle is operated any distance without fuel tank pumps in operation.

Fuel and Air Intake and Exhaust Systems

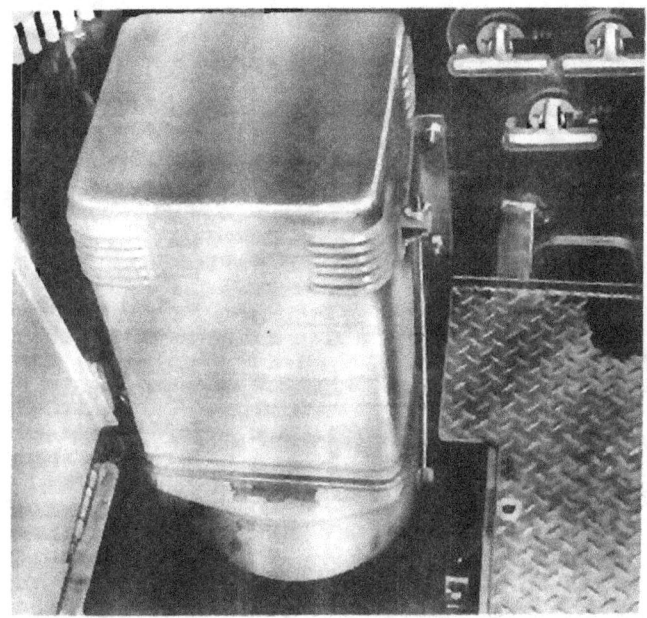

Figure 96—Air Cleaner Installed—Right Side—M18

(5) PRIMING SYSTEM. A priming system is provided to inject a spray of fuel into intake pipes to aid in starting engine. The hand operated primer pump (fig. 10) in front of driver's seat draws fuel from left fuel tank and forces it through a pipe to primer pipe distributor (fig. 95) on engine from which it is distributed through tubes to upper intake pipes.

c. Description of Air Intake System. Two heavy-duty oil-bath air cleaners are used to remove dust from air before it enters carburetor. Cleaners are mounted on front face of bulkhead on right and left sides of fighting compartment (M18), or crew compartment (M39) (fig. 96). The air outlets of cleaners connect to air ducts mounted on rear face of bulkhead, which connect to air inlet tubes welded into engine support assembly. The rear ends of air inlet tubes are connected to scoop on bottom of carburetor through flexible air inlet joints. All joints in air inlet system are closed by rubber seals to prevent entrance of dust and water.

d. Fuel-Air Control System. The fuel and air are mixed in a combustible ratio by carburetor, and volume of fuel-air mixture supplied to engine is controlled by carburetor throttle valve and its operating linkage. A governor connected to a governor valve box

TM 9-755

Part Three—Maintenance Instructions

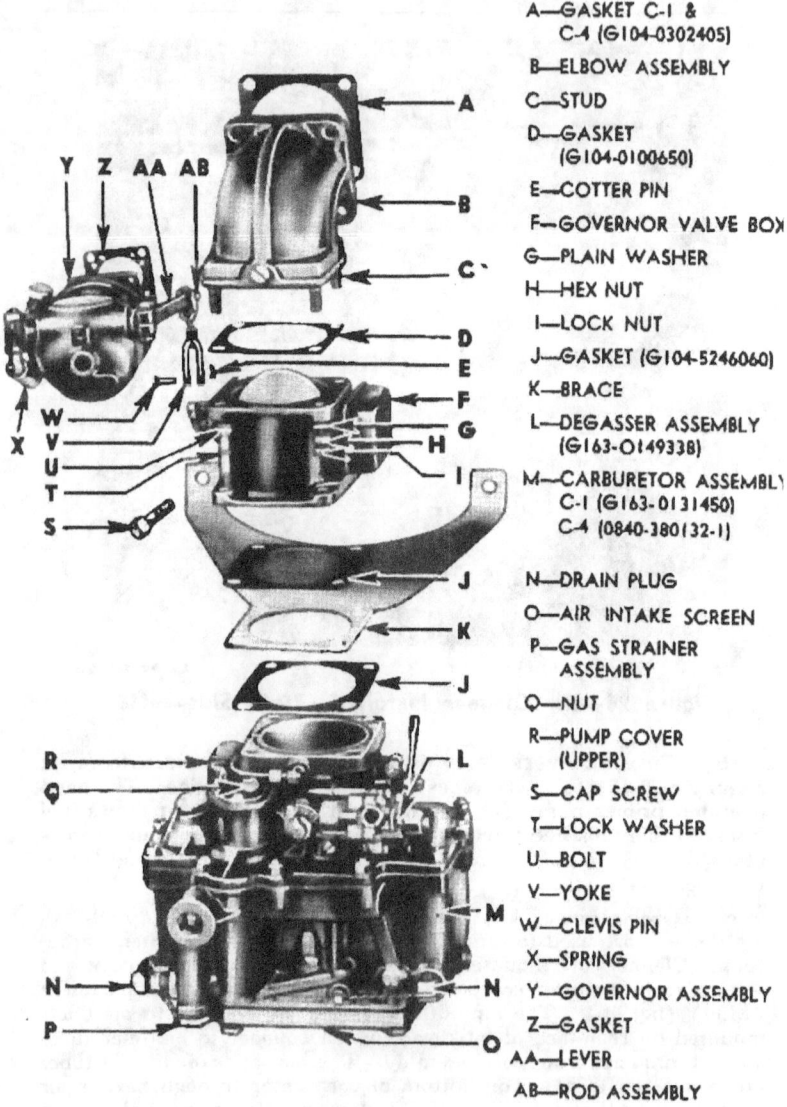

A—GASKET C-1 & C-4 (G104-0302405)
B—ELBOW ASSEMBLY
C—STUD
D—GASKET (G104-0100650)
E—COTTER PIN
F—GOVERNOR VALVE BOX
G—PLAIN WASHER
H—HEX NUT
I—LOCK NUT
J—GASKET (G104-5246060)
K—BRACE
L—DEGASSER ASSEMBLY (G163-0149338)
M—CARBURETOR ASSEMBLY C-1 (G163-0131450) C-4 (0840-380132-1)
N—DRAIN PLUG
O—AIR INTAKE SCREEN
P—GAS STRAINER ASSEMBLY
Q—NUT
R—PUMP COVER (UPPER)
S—CAP SCREW
T—LOCK WASHER
U—BOLT
V—YOKE
W—CLEVIS PIN
X—SPRING
Y—GOVERNOR ASSEMBLY
Z—GASKET
AA—LEVER
AB—ROD ASSEMBLY

RA PD 340493

Figure 97—Carburetor, Governor, Governor Valve Box, and Carburetor Elbow

above carburetor further controls volume of fuel-air mixture to limit top speed of engine.

(1) CARBURETOR ASSEMBLY (fig. 97). The Bendix-Stromberg updraft carburetor assembly is supported on engine and connected to distribution chamber in rear crankcase by governor valve box and an elbow.

(2) THROTTLE LINKAGE (fig. 13). The throttle in carburetor is actuated by a throttle cross shaft mounted on engine. Throttle cross shaft is connected by rods to accelerator pedal cross shaft in driving compartment. The accelerator pedals and the hand throttle control are described in paragraph 14 e.

(3) GOVERNOR AND GOVERNOR VALVE BOX (fig. 97 and 106). A flyball-type governor is mounted on rear end of engine oil pump and is driven from pump shaft. It is connected by levers and an adjustable rod and yoke to valve in governor valve box which is located between carburetor and carburetor elbow attached to engine rear crankcase. The function of governor and governor valve is to limit top speed of engine to a safe maximum under all operating conditions.

e. Description of Exhaust System (fig. 116). The exhaust system consists of a flexible joint and connector assembly and two muffler and tail pipe assemblies. The joint and connector assembly provides a flexible connection between two outlet pipes of exhaust manifold and two mufflers which are mounted in brackets across rear end of the hull, in engine compartment. The tail pipes, which are integral with the mufflers, direct exhaust gas upward through grille covered openings in hull roof. The exhaust system parts are secured to each other by saddle-type clamps.

f. Data.

Fuel tank capacity	Right, 90 gal; left, 75 gal
Fuel tank vent	Through filler cap
Fuel tank drain plug	Reach by removing plug in hull floor
Fuel tank pumps	2
Fuel tank pump pressure	5 to 7 lb
Engine fuel pump	1
Engine fuel pump pressure, fuel cold	5 to 7 lb
Fuel filter location	In each tank
Fuel screen location	In each tank around pump
Gas strainer location	In carburetor
Air cleaners, number and type	2, heavy-duty oil-bath
Air cleaner cup oil capacity	8 qt
Governed speed of engine, full load	2,400 rpm
Carburetor connections	½ in. std. pipe
Primer pipe connections	⅛ in. std. pipe

92. AIR CLEANERS.

a. Cleaner Action (fig. 99). Air enters cleaner through louvers at upper corners of body and is drawn downward through internal passages until it reaches cup and disk where its direction is reversed as it strikes the oil. The centrifugal action created as air changes

TM 9-755
92

Part Three—Maintenance Instructions

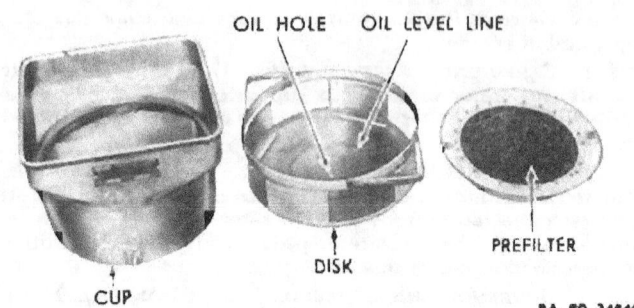

Figure 98—Parts To Remove When Changing Oil in Air Cleaner

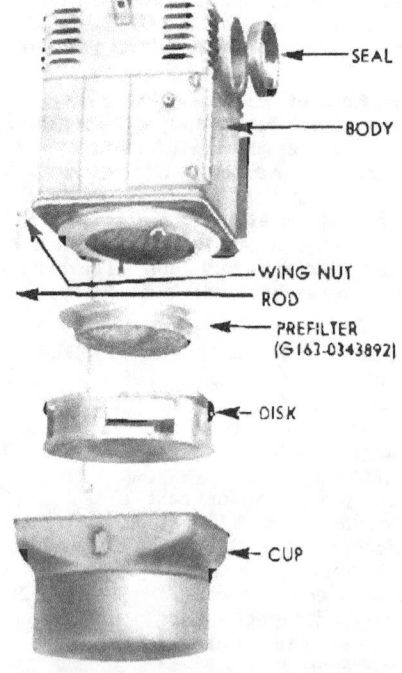

Figure 99—Air Cleaner—Disassembled

direction causes dirt particles in air to be thrown into and trapped by the oil. Air is then drawn upward through prefilter above disk and through filter element in cleaner body where additional particles of dirt are removed. The air then passes out of cleaner into air

Fuel and Air Intake and Exhaust Systems

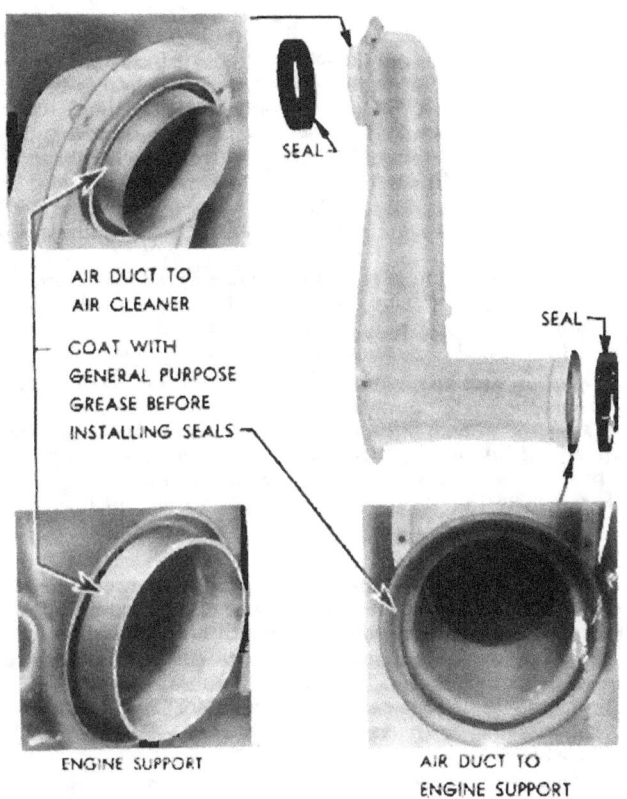

Figure 100—Air Duct Seals

duct leading to carburetor. The oil becomes laden with dirt and must be changed, and prefilter and filter must be cleaned periodically, in order to maintain efficiency of air cleaner.

b. **Changing Oil** (fig. 98). Open hull subfloor doors on M18 vehicle, or open rear seat and covers on M39 vehicle. Loosen wing nuts and unhook rods from cup while supporting cup to prevent it from falling, then remove cup and disk from cleaner. Remove the prefilter which is attached to cleaner body by two wing nuts. Drain oil from cup, wash all parts in dry-cleaning solvent, and dry with an air stream. Fill cup with specified oil (par. 38 d (1)) until it is level with line on disk, with disk installed in cup. Install prefilter on cleaner body and tighten wing nuts. Install cup and disk on cleaner body, attach rods to cup, and tighten wing nuts securely to prevent air leaks. Close subfloor doors (M18) or rear seat end covers (M39).

c. Removal.

(1) Remove cup, disk, and prefilter (subpar. b above).

(2) On M39 vehicle only, disconnect end of rear seat back pad which is attached with flat-head screws and finish washers. Remove rear seat back end plate which is attached with eight ⅜-inch, cap screws plain washers, and lock washers. When removing left air cleaner remove tachometer shaft shield which is attached to shaft and hull with three 5/16-inch cap screws and lock washers.

(3) When removing right air cleaner, remove cap screw and lock washer which attaches engine generator to filter box conduit to cleaner body.

(4) Remove safety nuts from studs which attach air cleaner body to bulkhead, pull top of body forward to disengage outlet from air duct and lift body out of hull. NOTE: *Some air cleaner bodies are attached with 7/16-inch cap screws and special flat washers.*

d. Cleaning and Inspection (fig. 99). Drain oil from cup, thoroughly wash all parts of air cleaner with dry-cleaning solvent, and dry with air stream. Clean filter element in cleaner body by sloshing body up and down in dry-cleaning solvent and allow it to drain. Inspect prefilter and filter element for damage. Inspect cup to make sure it is not distorted so that it will not fit airtight in lower end of body. Repair or replace damaged parts before installation.

e. Installation.

(1) Coat surface of air duct with general purpose grease and install a new seal (fig. 100). Coat surface of air cleaner outlet port with general purpose grease so that seal will enter smoothly.

(2) Lift air cleaner body into position on studs on bulkhead, using care to avoid damaging seal, and install four safety nuts (7/16 in.—14) tightened to 25-30 foot-pounds tension. NOTE: *Some air cleaner bodies are attached with cap screws (7/16 in.—14 x 1¼ in.) and special flat washers.*

(3) Attach engine generator to filter box conduit to right air cleaner body with one cap screw and lock washer through support clip.

(4) On M39 vehicle only, install tachometer shaft shield to shaft and hull with three cap screws (5/16 in.—24 x ⅝ in.) and lock washers. Attach rear seat back end plate with eight cap screws (⅜ in.—24 x 1 in.), plain washers, and lock washers. Attach seat back pad with flat-head screws (¼ in.—28 x ¾ in.) and finish washers.

(5) Install prefilter, fill cup and install cup and disk (subpar. b above).

93. CARBURETOR ASSEMBLY.

a. Description. The main and economizer jets in the carburetor are of the fixed-orifice type so that no adjustment is required for part and full throttle operation. An idle mixture adjusting lever is provided for adjustment of carburetor at idle speed. A throttle stop containing a stop screw is mounted on throttle shaft for purpose of setting minimum idling speed of engine. The carburetor assembly includes a degasser assembly which contains a diaphragm actuated by

Fuel and Air Intake and Exhaust Systems

engine intake vacuum. When throttle is closed to reduce engine speed, the high intake vacuum thus created causes diaphragm to move a plunger which cuts off supply of fuel to carburetor idle jets, thus reducing tendency of engine to backfire while decelerating. The degasser assembly contains a solenoid controlled by "Idle Fuel Cut-off" switch on instrument panel. When switch is depressed, battery current energizes solenoid, causing it to actuate a plunger in carburetor which stops flow of fuel to idle jets, thus stopping engine.

    b. **Non-interchangeability of Carburetors.** The use of an R975-C1 carburetor on an R975-C4 engine will give too rich a fuel mixture; resulting in poor acceleration, low power, fouling of spark plugs, and other mechanical difficulties. The use of an R975-C4 carburetor on an R975-C1 engine will result in too lean a fuel mixture due to its smaller jets; this will cause overheating and detonation which may seriously damage engine. Consequently, it is important that these two carburetors, which are alike in design and appearance, should not be interchanged between these two engines.

    c. **Identification.** R975-C4 engine carburetors have been identified by painting a large white "C4" on the side of main body; however, all carburetors were not thus identified. The CWR part number was stamped on the flange as a means of identification, but it was discovered that many of R975-C4 carburetors produced were stamped in error with the R975-C1 carburetor part number CWR 202199. Therefore, it is necessary, for the purpose of distinguishing between these two carburetors, that identification be made only from parts list number stamped on carburetor specification plate.

    (1) R975-C1 ENGINE CARBURETOR. NAR9G carburetors for R975-C1 engine are stamped on specification plate with one of the following; the last parts list issue number (-5) shown provides a carburetor with latest recommended jet combination.

                Parts List No. 380113—1
                 Parts List No. 380113—2
                 Parts List No. 380113—3
                 Parts List No. 380113—4
                 Parts List No. 380113—5

    (2) R975-C4 ENGINE CARBURETOR. NAR9G carburetors for R975-C4 engine are stamped on the specification plate with one of the following.

                 Parts List No. 380118—1
                 Parts List No. 380118—2
                 Parts List No. 380118—3
                 Parts List No. 380118—4
                 Parts List No. 380132—1

    d. **Carburetor Idle Adjustment** (fig. 101). Start and warm up engine until it has reached a normal operating temperature (par. 17 b and c). Remove carburetor inspection hole cover (par. 183 a). Fully release hand throttle control and adjacent throttle stop screw on carburetor as required to give engine idle speed of 700 revolutions per minute. Slowly move the idle mixture adjusting lever on carburetor to

TM 9-755
93

Part Three—Maintenance Instructions

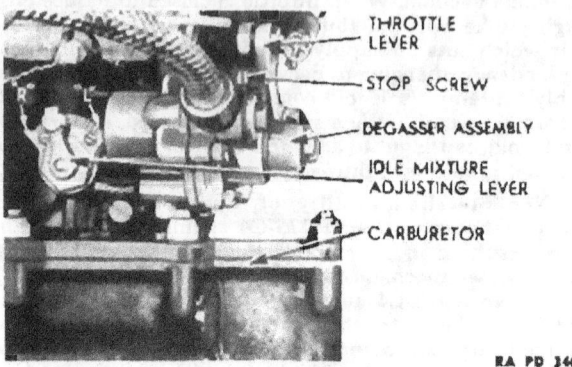

Figure 101—Carburetor Idle Adjustment and Throttle Stop Screw

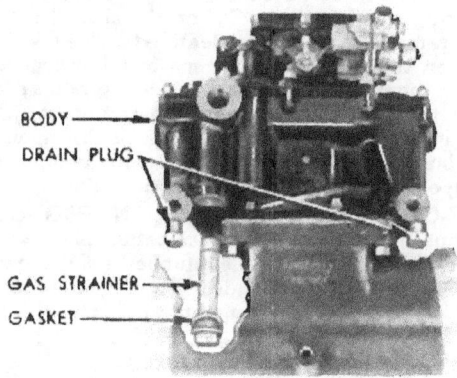

Figure 102—Carburetor Drain Plugs and Gas Strainer

right toward "L" (lean) side of quadrant until engine begins to run unevenly, then move lever back to left toward "R" (rich) side of quadrant one notch at a time until engine operates evenly. If engine idle speed has changed when adjustment is completed, adjust throttle stop screw again to give engine idle speed of 700 revolutions per minute. Install inspection hole cover on hull rear door (par. 183 n).

e. **Draining Carburetor and Cleaning Gas Strainer** (fig. 102). Cut lock wires and remove two square head plugs at lower rear corners of carburetor main body to drain gas, water and sediment from float chamber. Remove gas strainer and gasket which are located in lower left corner of carburetor main body. Wash strainer in dry-cleaning solvent and dry out with air stream. Place a new basket over strainer, coat threads of strainer and drain plugs with joint and thread com-

242

TM 9-755
93

Fuel and Air Intake and Exhaust Systems

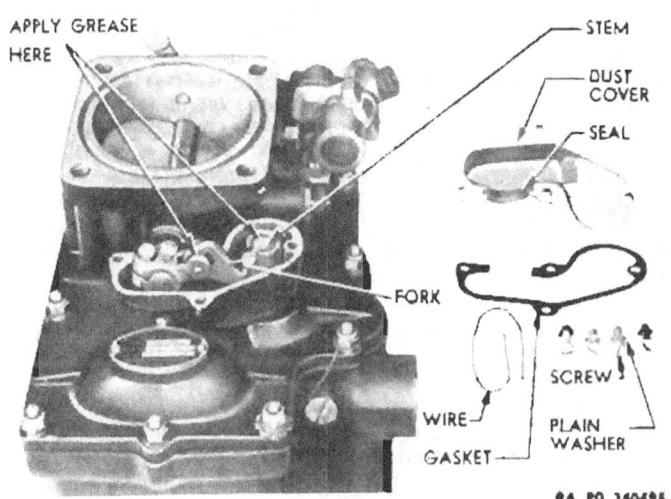

Figure 103—Accelerating Pump Stem and Fork

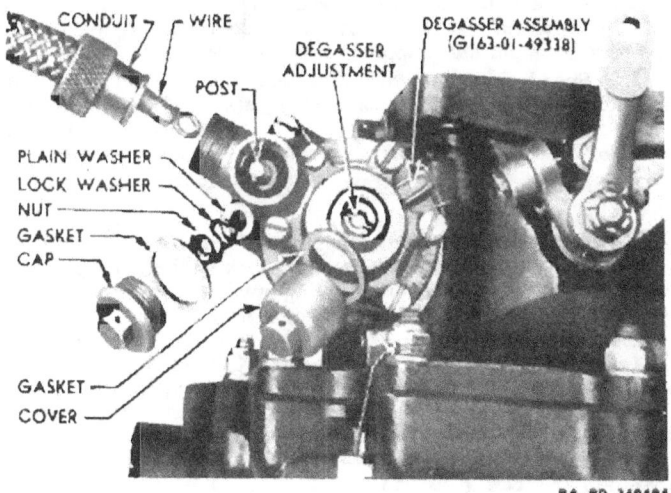

Figure 104—Degasser Adjustment and Wire Connection

pound, type A and install parts in main body. Tighten screen and plugs firmly and install lock wires.

f. **Correction for Sticking Accelerating Pump Stem** (fig. 103). Accelerating pump stem and fork are packed with grease when carbu-

TM 9-755
93

Part Three—Maintenance Instructions

retor is assembled; however, the stem may become dry and stick in its bushing after some service. A sticking pump stem will cause the throttle to stick and may cause engine to spit when accelerating. If there is indication that pump stem is sticking it should be freed up and lubricated. Cut lock wires, remove four cover attaching screws and plain washers, and remove pump upper dust cover using care to avoid damaging rubber seal over throttle shaft. Apply a liberal quantity of general purpose grease on fork and around pump stem while working throttle back and forth. When pump stem and fork are free and well lubricated, install dust cover with a new gasket, attach cover with four screws and plain washers, and install lock wire through screw heads.

g. **Degasser Adjustment** (fig. 104). No adjustment is provided for idle fuel cut-off section of degasser assembly; if unit fails to cut off and stop engine, (refer to paragraph 51 h). If engine backfires on deceleration, adjust degasser as follows:

(1) Start engine and warm up to normal operating temperature (par. 17 b, and c). Check carburetor idle adjustment and set engine idle speed at 700 rpm (subpar. d above).

(2) Cut lock wire and remove degasser adjusting screw cover and gasket. With a small screwdriver, turn degasser adjusting screw clockwise as far as it will go, then slowly turn adjusting screw counterclockwise, one notch at a time, until engine just begins to surge or roll. Finally, turn adjusting screw clockwise six to eight notches. The engine should idle smoothly with this setting.

(3) Speed engine up to 1,800 rpm, then close throttle quickly and allow engine to settle back to idling speed. If engine rolls excessively, or stops, turn adjusting screw clockwise one or two notches. Again open and close throttle quickly to check adjustment.

(4) Install adjusting screw cover and gasket and install lock wire.

h. **Degasser Replacement** (fig. 104). Cut lock wire and remove terminal cap and gasket. Remove nut, lock washer, and plain washer from terminal post. Unscrew coupling nut and disconnect conduit and wire. Cut lock wire and remove three screws and lock washers which attach degasser assembly to carburetor throttle body. To install, attach degasser assembly with a new gasket (G104-1593998) to throttle body with three screws and lock washers secured with lock wire. Connect wire conduit to degasser and secure wire on terminal post with a plain washer, lock washer and nut. Install gasket and terminal cap and secure with a lock wire. Start engine and test degasser operation (subpar. g above).

i. **Carburetor Removal** (fig. 97).

(1) Close the three fuel shut off valves (par. 14 a). Open hull rear door (par. 181 a).

(2) Cut lock wire and remove terminal cap and gasket from degasser. Remove nut, lock washer, and plain washer from terminal post (fig. 104). Unscrew coupling nut and disconnect conduit and

Fuel and Air Intake and Exhaust Systems

wire from degasser. Remove safety nut which attaches conduit support clip to left rear corner of throttle body.

(3) Disconnect link from throttle lever. Loosen clamp and disconnect fuel supply hose (fig. 108) from elbow on carburetor.

(4) Remove clamps from inner ends of air inlet flexible joints and turn rubber seals back over ends of air scoop, and pull joints from air scoop.

(5) Cut lock wire between plug in carburetor elbow and plug in throttle body. While supporting carburetor, remove safety nuts and plain washers from bolts which attach carburetor to governor valve box, remove bolts and remove carburetor from engine (fig. 97).

(6) Cut lock wires and remove four screws and plain washers which attach air scoop to carburetor and remove scoop and air intake screen.

j. Carburetor Installation (fig. 97).

(1) Attach air intake screen and air scoop to carburetor main body with four screws and plain washers; install lock wires through screws. Examine rubber seals on ends of air scoop and if damaged install new seals.

(2) Place carburetor in position with new gasket, insert attaching bolts from above and install plain washers and safety nuts on lower ends of bolts (fig. 97). Tighten nuts evenly and firmly. Install lock wire between plug in throttle body and plug in carburetor elbow.

(3) Push flexible joints over ends of air scoop, turn rubber seals from air scoop over the ends of the joints, install and tighten flexible joint clamps.

(4) With throttle valve fully closed, throttle lever must be one notch forward of vertical position. Remove lever from shaft, if necessary, set in required position and tighten attaching nut. Attach throttle control link to the throttle lever with lock washer and nut. Check throttle linkage for full opening and closing of throttle valve and adjust linkage as required (par. 94).

(5) Connect wire and conduit to degasser and secure wire on terminal post with a plain washer, lock washer and nut (fig. 104). Install gasket and terminal cap, and secure with a lock wire. Attach conduit support clip to carburetor by means of left rear throttle body attaching safety nut.

(6) Start engine and warm up to normal operating temperature (par. 17 b and c). Adjust carburetor (subpar. d above). Check fuel feed hose connection for leaks.

(7) Close hull rear door (par. 181 b).

94. ACCELERATOR PEDAL TO THROTTLE VALVE LINKAGE, AND HAND THROTTLE CONTROL.

a. General. The accelerator pedal to carburetor throttle valve linkage must be adjusted properly to insure smooth control of the engine throughout the entire range from fully closed to wide open throttle; otherwise, maximum engine and vehicle performance will

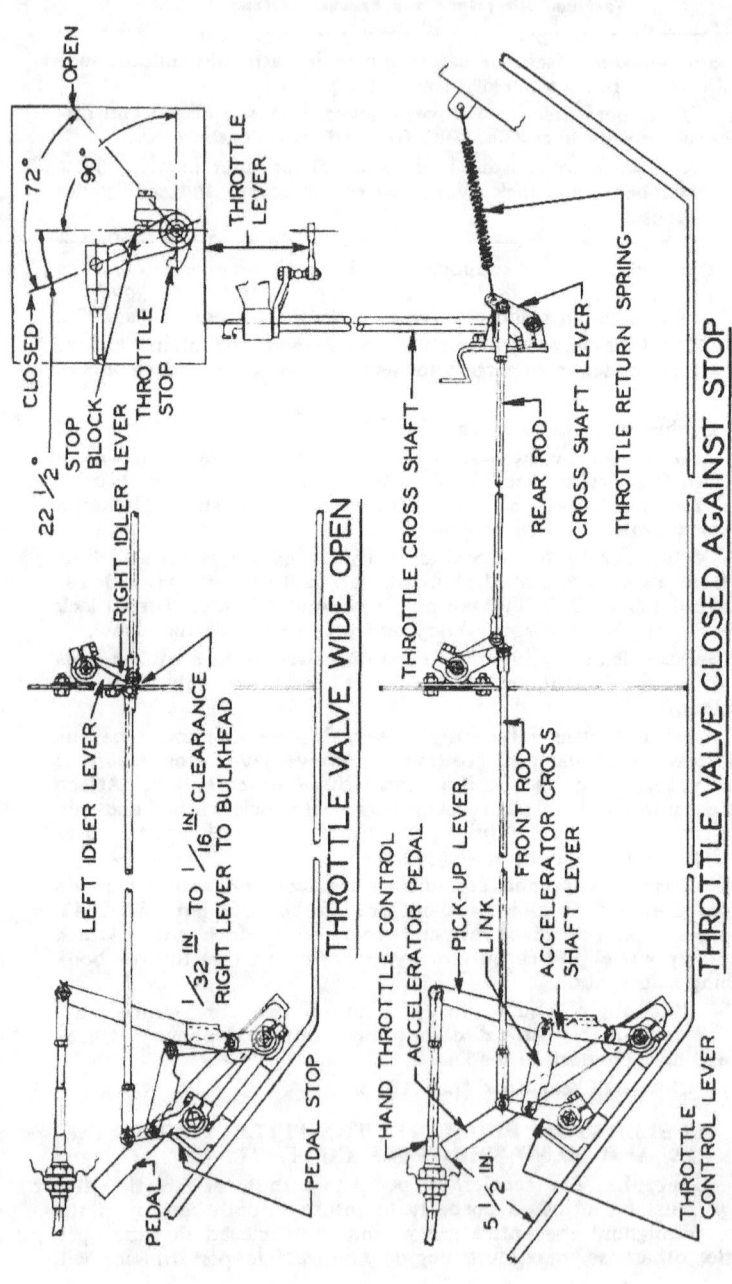

Figure 105—Accelerator Pedal and Throttle Linkage Adjustments

Fuel and Air Intake and Exhaust Systems

not be secured. The linkage, and hand throttle control, must permit throttle valve to close against slow idle stop on carburetor when accelerator pedal is released, and it must also permit throttle to be fully opened against wide open stop when accelerator pedal is pushed down against the stop under pedal.

b. **Operating Inspection.**

(1) Warm up engine and adjust idle speed to 700 revolutions per minute as described in paragraph 93 d, then stop engine (par. 17 i).

(2) With hand throttle control and accelerator pedals in fully released positions check to determine whether throttle stop screw is in contact with throttle stop block on carburetor throttle body.

(3) Have another man push accelerator pedal down against stop on hull front plate and check to determine whether throttle valve is wide open, with 0.010-inch to 0.020-inch clearance between throttle stop on throttle shaft and throttle stop block on throttle body. Make sure that pedal contacts pedal stop.

(4) Operate accelerator pedal five or six times through entire range, noting whether there is any sticking or binding at any point, whether there is any evidence of lost motion in linkage, and whether return spring moves throttle and linkage to closed position when pedal is released.

(5) Operate hand throttle control to determine whether it operates throttle linkage satisfactorily.

(6) If linkage does not operate smoothly, without lost motion, and give complete opening and closing of throttle valve, make a complete adjustment of linkage as described in subparagraph c below.

c. **Adjustment** (fig. 105). The linkage must be adjusted in the following order to obtain proper throttle operation.

(1) Open hull rear door (par. 181 d) and disconnect throttle operating rod from throttle lever on carburetor.

(2) Remove siren switch junction box and accelerator pedal linkage guard which are attached to mounting bracket by four $\frac{5}{16}$-inch cap screws and lock washers.

(3) Tighten all nuts at accelerator pedal cross shaft flanges. Tighten attaching lock nuts and clamp screws on cross shaft lever, throttle control lever, and throttle cross shaft on engine. Inspect all rod end pins for excessive wear and replace worn pins.

(4) Fully release hand throttle control. Move accelerator pedal through full range and if any binding or sticking exists, check for cause, and lubricate cross shaft bearings. (L, M, N, fig. 36), idler shaft bracket (U, fig. 37), and throttle cross shaft bearings (Y, Z, fig. 38). When pedal is released, throttle return spring must move cross shaft lever back into contact with hand throttle control pick-up lever; replace return spring if it is weak.

(5) Measure distance between hull front plate and top front edge of accelerator pedal. If distance is not $5\frac{1}{2}$ inches, disconnect hand throttle control from pick-up lever by removing rod end pin,

adjust yoke on control to secure this distance, then install pin with plain washer and cotter pin and tighten lock nut against yoke.

(6) Disconnect throttle control front rod from throttle control lever by removing rod end pin. Hold accelerator pedal down against pedal stop, pull rod forward until idler lever strikes bulkhead, and adjust yoke on rod until pin will just go through yoke and lever. Lengthen the rod by one turn of the yoke, to provide clearance of $\frac{1}{32}$-inch to $\frac{1}{16}$-inch between idler lever and bulkhead, then install rod and pin with a plain washer and cotter pin, and tighten lock nut against yoke.

(7) Move throttle lever through full range and check for evidence of sticking. If sticking exists free up accelerating pump stem (par. 93 f).

(8) Close carburetor throttle and check position of throttle lever. Lever should be 22½ degrees forward of vertical position, which is equal to one serration on throttle shaft. If lever is not properly set, remove lever from shaft, change its position as required, and secure lever on shaft with nut and cotter pin. Connect operating rod to throttle with nut and lock washer.

(9) Recheck position of accelerator pedal; if distance to hull front plate is now less than 5½ inches as set in step (5) above, throttle control rear rod is too long. Hold pedal down against pedal stop. If clearance between throttle stop and stop block on carburetor throttle body is more than 0.010 inch to 0.020 inch, rear rod is too long; if throttle stop strikes stop block before pedal contacts pedal stop, rod is too short.

(10) Disconnect throttle control rear rod from throttle cross shaft lever by removing rod end pin. Adjust yoke, as required, to secure distance of 5½ inches between pedal and hull plate when throttle is fully closed, and clearance of 0.010 inch to 0.020 inch between throttle stop and stop block when pedal is pushed down against pedal stop. In some cases it may be necessary to change the position of lever on throttle shaft by one serration in order to secure required settings. Connect rod to lever with rod and pin, plain washer and cotter pin, and tighten lock nut against yoke.

(11) Install linkage guard and siren switch junction box, attaching them to mounting bracket with four cap screws ($\frac{5}{16}$ in.—24 x ¾ in.) and external-tooth lock washers. Close hull rear door (par. 181 b).

95. GOVERNOR.

a. Description (fig. 106). The governor is a "fly-ball" type in which the centrifugal force of rotating weights is opposed by a calibrated governor spring. When engine is not running, governor spring holds valve in governor valve box wide open. When engine is running, centrifugal force of rotating governor weights tends to overcome governor spring force and move governor valve toward closed position and slow down engine. At point of balance between these opposing forces engine is held to nearly constant maximum speed when carburetor throttle valve is wide open. The action of governor is transmitted to governor valve by means of governor throttle lever which

Fuel and Air Intake and Exhaust Systems

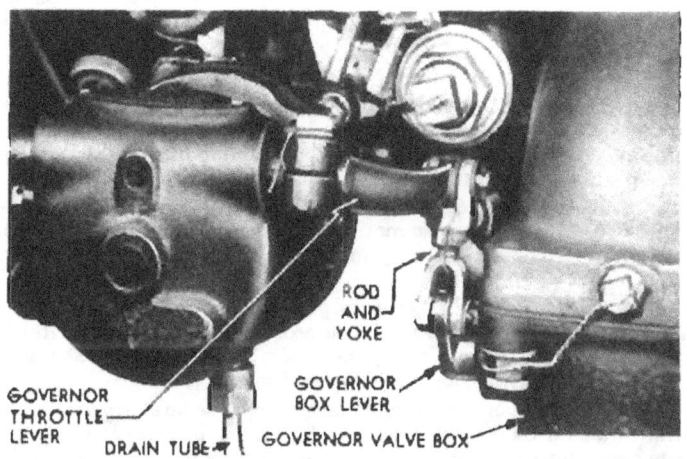

Figure 106—*Governor and Connections to Governor Valve Box*

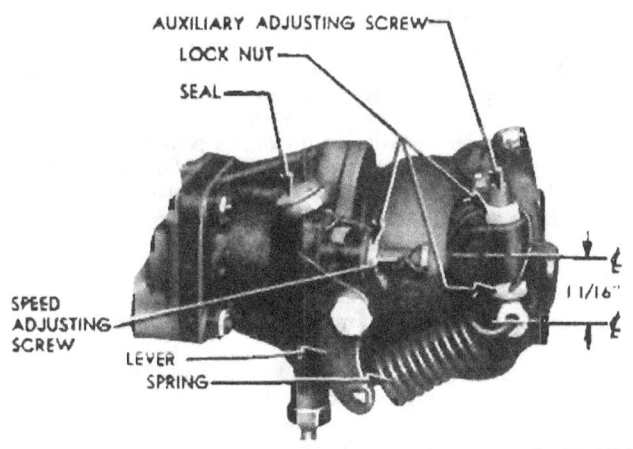

Figure 107—*Governor Adjustments*

is connected to governor box lever by an adjustable rod and yoke assembly.

b. Removal (fig. 106.) Open hull rear door (par. 181 a). Remove cotter pin, plain washer, and yoke end pin which connects governor throttle lever to rod and yoke assembly. Unscrew coupling

Part Three—Maintenance Instructions

nut and disconnect oil drain tube. Remove four lock nuts, hexagon nuts and plain washers from mounting studs and remove governor from oil pump.

c. **Installation** (fig. 106).

(1) Place a new gasket over mounting studs. Install governor on oil pump with throttle lever towards governor valve box, and install plain washers, hexagon nuts, and lock nuts on mounting studs. Connect oil drain tube and tighten coupling nut.

(2) Check governor throttle lever to make sure that it is held all the way up without lost motion by governor spring, then connect it to rod and yoke assembly by means of yoke end pin; do not install cotter pin.

(3) Check clearance between governor box lever and stop pin on governor box; this should be approximately $1/64$ inch. Remove lock wire and adjust rod and yoke assembly, if necessary, to secure this clearance, then tighten lock nut and install lock wire.

(4) Move governor throttle lever down several times to make certain that no binding exists at the rod and yoke, then install plain washer and cotter pin on yoke end pin. Lubricate yoke end pin with oil can.

(5) Check governor action and adjust, if necessary, as described in subparagraph d below.

d. **Governor Adjustment.** Governors on R975-C1 engines are to be set at 2,400 revolutions per minute with the transfer case clutch engaged, the transmission shift lever in neutral and carburetor throttle wide open; governors on R975-C4 engines are to be set at 2,500 revolutions per minute under same conditions. This requires resetting a new governor after installation on engine, since original setting was accomplished at factory with no load. The following procedure must be followed when checking a governor and adjusting it if necessary.

(1) Check throttle linkage to make certain that carburetor throttle valve is wide open when accelerator pedal is pushed down against its stop (par. 94).

(2) Make certain that transfer case clutch is engaged, then start engine (par. 17 b and c). Warm engine up for 15 minutes, with transmission in neutral.

(3) Push accelerator pedal down against stop to open throttle wide, and note maximum speed reached by engine as shown on tachometer. If maximum speed is other than 2,400 revolutions per minute on R975-C1 engine, or 2,500 revolutions per minute on R975-C4 engine, or if the governor causes the engine to "surge," adjust as described in the following steps.

(4) Open hull rear door (par. 181 a).

(5) Remove lock wire and seal from speed adjusting screw and lock nut, and loosen lock nut (fig. 107). With carburetor throttle valve wide open, turn adjusting screw as required to obtain engine speed of 2,400 revolutions per minute on R975-C1 engine, or 2,500 revolutions per minute on R975-C4 engine. Turning adjusting screw

Figure 108—Engine Fuel Pump and Feed Pipe Check Valve

clockwise increases engine speed; turning screw counterclockwise decreases speed. Tighten lock nut when proper speed is obtained, and fasten lock nut and adjusting screw securely with lock wire and a new seal.

(6) If the governor causes engine to "surge," check distance from center of rocker shaft to center of eye in auxiliary adjusting screw. If other than $1\tfrac{3}{16}$ inch, cut lock wire and adjust the screw to this dimension.

(7) Make certain that there is no binding or excessive friction in governor to valve box linkage. Recheck to determine that carburetor throttle valve opens to wide open position.

(8) If "surge" is still present, lengthen the auxiliary adjusting screw approximately $\tfrac{1}{16}$ inch, or a turn at a time, until surging is eliminated. Tighten lock nuts and install lock wire.

(9) Close hull rear door (par. 181 h).

96. ENGINE FUEL PUMP.

a. Removal (fig. 108). Close fuel tank shut-off valves (par. 14 a). Open hull rear door (par. 181 a). Loosen hose clamps and disconnect coupling hoses from feed pipe check valve and fuel pump. Remove two ¼-inch nuts which attach check valve to brace. Unscrew coupling nuts and disconnect two oil drain pipes from tee on front side

TM 9-755

Part Three—Maintenance Instructions

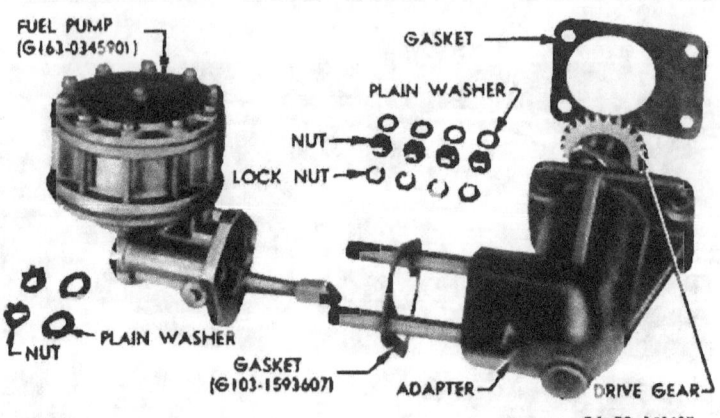

Figure 109—Engine Fuel Pump and Drive Shaft Adapter

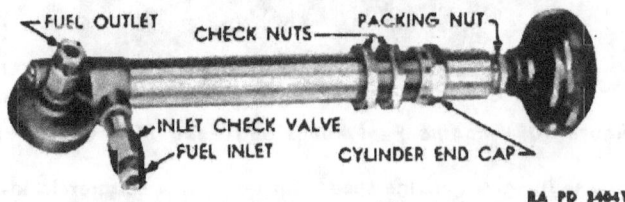

Figure 110—Primer Pump Assembly

of fuel pump drive shaft adapter. Remove four lock nuts, hexagon nuts, and plain washers which attach drive shaft adapter to accessory drive housing and remove pump and adapter, making sure that drive gear is removed with adapter. Cut lock wire, remove two nuts and plain washers from studs, and separate pump from adapter. Remove outlet elbow and feed pipe check valve from fuel pump.

b. **Installation** (fig. 109). Coat threads of outlet elbow and pump to check valve pipe and elbow with joint and thread compound, type A, and install outlet elbow and check valve on pump. Place a new gasket over studs on adapter housing, install pump with plain washers and slotted nuts, tighten nuts and install lock wire. Place a new gasket over studs on accessory drive housing, install adapter and pump assembly and anchor the adapter with four plain washers, hexagon nuts and lock nuts. Connect two oil drain pipes to tee on front side of adapter and tighten coupling nuts. Attach feed pipe check valve to brace, using spacers as required, lock washers, and two nuts (¼ in. — 28). Connect feed pipe coupling hoses to check valve, connect carburetor coupling hose to fuel pump, and tighten all hose clamps. Start engine (par. 17 b and c) and check for fuel leaks at hose connections, and oil leaks at joints of drive shaft adapter. Close hull rear door (par. 181 b).

97. PRIMER PUMP.

a. **Description** (fig. 110). When left fuel tank shut off valve is open and fuel tank pump is running, fuel is supplied to primer pump under pressure. Pulling out on pump plunger causes a charge of fuel to be drawn into pump through a check valve on inlet side. Pushing plunger in causes inlet check valve to close, and forces fuel out of pump through a check valve on outlet side which is connected by pipes to primer pipe distributor on engine (fig. 95). Mechanism inside pump prevents fuel from flowing through except when plunger is being operated; however, if pump is incorrectly assembled or valves are held open by dirt, fuel can leak through primer system into intake pipes. This will cause an excessive richness which cannot be compensated for by idle adjustment of carburetor. If this condition occurs, replace primer pump (subpar. b and c below).

b. **Removal.** Close the three fuel shut-off valves (par. 14 a). Loosen compression nuts and disconnect pipes from primer pump. Hold front end of pump and unscrew end cap from pump cylinder. Remove plunger, end cap, gasket, and bushing assembly intact without disturbing packing nut. Remove the 3/8-inch cap screw, internal-tooth lock washer and nut from mounting slip at front end of pump cylinder. Remove rear check nut and slide pump cylinder forward through hole in pump mounting bracket. Install check nut, plunger and attached parts on cylinder.

c. **Installation.** Remove plunger, and cap, gasket and bushing assembly from replacement pump cylinder. Remove rear check nut, push end of barrel through hole in mounting bracket and install check nut; do not tighten nut. Install plunger, end cap, gasket, and bushing assembly on cylinder using care to avoid damaging plunger leather. Connect fuel pipes and tighten compression nuts. Install mounting clip at front end of barrel and attach it to hull with one cap screw (3/8 in.—24 x 7/8 in.) internal-tooth lock washer, and nut. Hold pump barrel while tightening check nuts against mounting bracket to avoid straining fuel pipe connections. Turn on left fuel tank shut-off valve and fuel pump to check pipes and pump for leaks.

98. FUEL TANK PUMPS.

a. **Description.** The fuel tank pumps (G163-0345900) are centrifugal pusher-type electric units containing 24-volt shunt wound motors. The pumps are fully submerged in fuel, which circulates under pressure through motors to cool them and to provide a practically constant viscosity bearing lubricant under all temperature conditions. CAUTION: *Since the pumps depend on fuel for cooling and lubrication, they should never be operated when fuel tanks are empty.*

TM 9-755
98

Part Three—Maintenance Instructions

Figure 111—Removing Fuel Pump and Screen

b. **Removal.**

(1) Move engine out upon hull rear door (par. 75).

(2) Remove plug from hull floor under tank, remove tank drain plug and drain fuel tank. After tank is drained, install both plugs securely.

(3) If right pump is being removed, remove outlet elbow (Q, fig. 114). If left pump is being removed, disconnect primer pipe at tee and remove tee (V, fig. 113). Disconnect throttle control rod from idler lever at front end.

(4) Remove the sheet metal cover which is attached to tank insulator with six $\frac{3}{8}$-inch cap screws and lock washers (W, fig. 113; R, fig. 114). Remove felt insulator which is behind cover.

(5) Disconnect fuel shut-off valve operating rod from valve by removing cotter pin, plain washer, and rod end pin. Disconnect operating rod from cross shaft lever by removing U-shaped retaining washer and springs. Remove rod assembly.

(6) Unscrew captive screws to disconnect conduit housing from pump cable terminal. Disconnect capacitor lead and pump wire from terminal. Tape the wire to prevent accidental shorting.

(7) Remove nine $\frac{3}{8}$-inch cap screws and internal-tooth lock washers which attach pump cover to fuel tank. Remove pump and cover assembly from tank, and remove pump screen which is held in position in tank by pump cover (fig. 111).

Fuel and Air Intake and Exhaust Systems

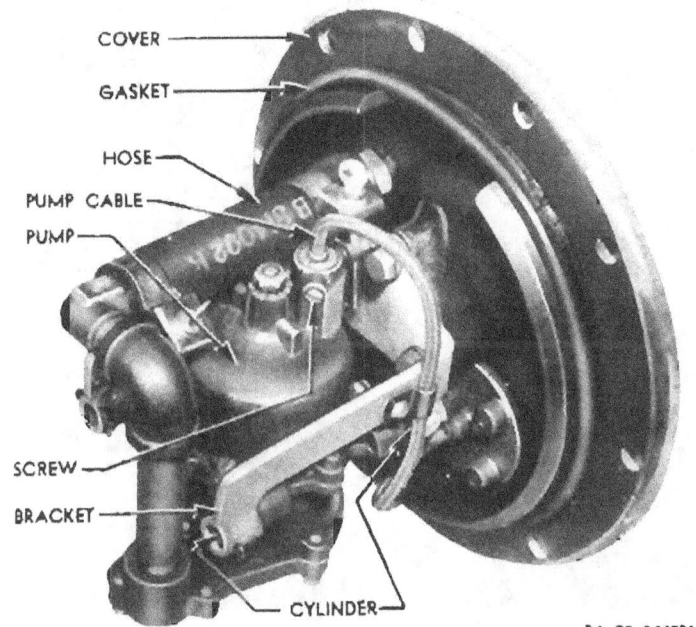

Figure 112—Fuel Tank Fuel Pump Mounted on Cover

(8) Remove set screw and unscrew pump cable connection from top of pump. Loosen hose clamp at elbow if connection to pump is a coupling hose, or unscrew coupling nuts and remove pipe if a pipe connection is used.

(9) Remove cotter pins on ends of pump mounting bracket, remove rubber cylinders which attach pump to bracket, and remove pump.

c. **Cleaning.** When cleaning a fuel pump that has been removed from tank, wash it in dry-cleaning solvent and dry with compressed air. If pump appearance indicates presence of gum, remove the gum with alcohol, acetone, or a 50-50 mixture of both. CAUTION: *Do not allow alcohol or acetone to contact the brushes, windings of field coil, or armature, because the cleaning solution will damage the insulation of the windings.* Thoroughly clean dirt and lint from pump screen and small screen on pump assembly.

d. **Installation** (fig. 112).

(1) Clean joint surfaces of pump cover and install a new cover gasket, using gasket cement. Clean joint surface of fuel tank.

(2) Place rubber cylinders in sleeves of pump mounting bracket nearest pump cover. Push outlet pipe elbow into coupling hose if

TM 9-755

Part Three—Maintenance Instructions

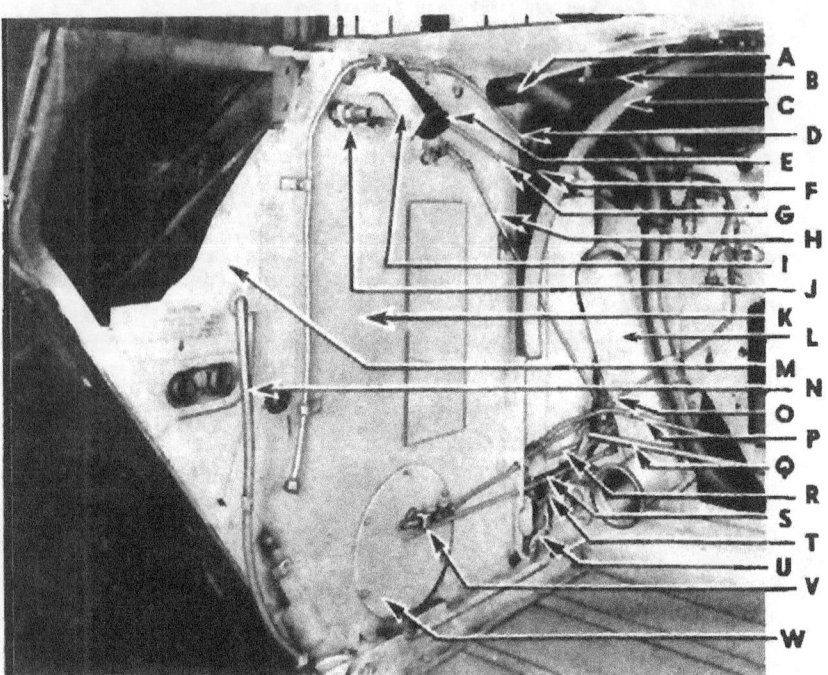

A—AIR TUBE
B—CONDUIT
C—BAFFLE
D—PIPE
E—HORN
F—GROMMET
G—DILUTION PIPE (OUTLET)
H—PRIMER PIPE
I—HORN SUPPORT
J—CONNECTOR
K—INSULATOR
L—AIR DUCT
M—INSPECTION PLATE
N—STARTER CONDUIT & WIRE
O—DILUTION PIPE (OUTLET)
P—BALANCE PIPE
Q—CROSS SHAFT
R—SHUT OFF VALVE OPERATING ROD
S—THROTTLE ROD
T—PIPE
U—DRAIN VALVE LEVER
V—TEE
W—COVER

RA PD 340498

Figure 113—Items Affected When Removing Left Fuel Tank

TM 9-755
98

Fuel and Air Intake and Exhaust Systems

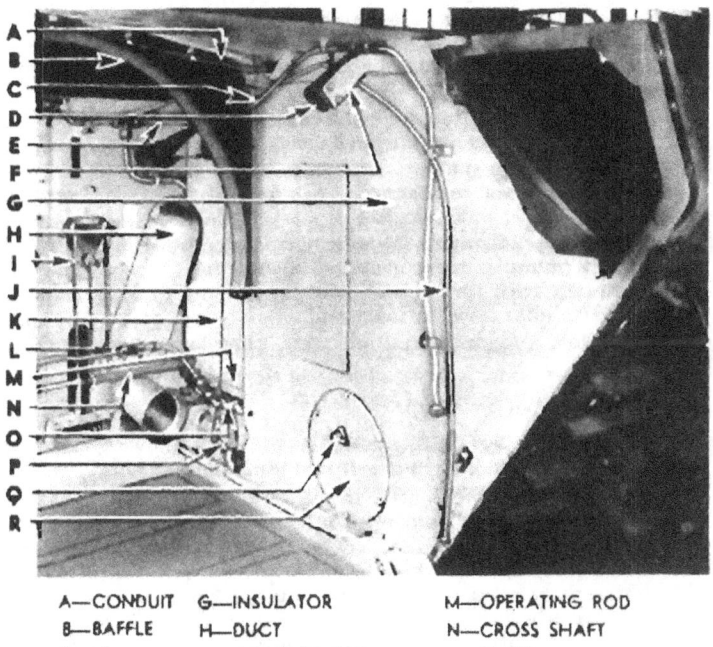

A—CONDUIT G—INSULATOR M—OPERATING ROD
B—BAFFLE   H—DUCT   N—CROSS SHAFT
C—PIPE   I—CONTROL ROD O—VALVE
D—HORN   J—TACHOMETER SHAFT P—DRAIN VALVE LEVER
E—PIPE   K—DRAIN VALVE CABLE Q—ELBOW
F—SUPPORT L—BALANCE PIPE R—COVER

Figure 114—Items Affected When Removing Right Fuel Tank

hose connection is used, and place pump over rubber cylinders in bracket. Insert rubber cylinders into sleeves and pump at ends of mounting bracket and install cotter pins to hold cylinders in place. Tighten hose clamp; however, if outlet connection is a pipe, install pipe and tighten coupling nuts securely.

(3) Screw pump cable connection into top of pump, install set screw and tighten securely.

(4) Place pump screen in opening in fuel tank and place pump and cover assembly in position with shut-off valve towards top of tank. Attach pump cover to fuel tank with nine cap screws ($\frac{3}{8}$ in.— 16 x 1 in.) and internal-tooth lock washers tightened to 20-25 foot-pounds tension.

(5) Connect pump wire and capacitor lead to pump cable terminal and tighten terminal screw firmly. Connect conduit housing to terminal and tighten the two captive screws.

TM 9-755
98-99

Part Three—Maintenance Instructions

---

(6) Place fuel shut-off valve operating rod in position and connect front end to cross shaft lever with a spring on each side of lever trunnion and a U-shaped retaining washer snapped into groove in end of rod. Connect rod to shut-off valve with a rod end pin, plain washer, and cotter pin.

(7) Make certain that fuel shut-off valve is closed and fill tank with fuel. Turn on fuel pump and check for fuel leaks.

(8) Place felt insulator over pump outlet and install the sheet metal cover with six cap screws ($\frac{3}{16}$ in.—24 x $\frac{5}{8}$ in.) and lock washers (W, fig. 113; R, fig. 114).

(9) If right fuel pump is being installed, install elbow (Q, fig. 114); if left pump is being installed, install tee (V, fig. 113) after coating threads with thread and joint compound, type A. Connect throttle rod to idler lever at front end.

(10) Move engine into hull and complete the installation (par. 76).

99. FUEL TANKS.

a. Removal (figs. 113 and 114). The following procedure covers removal of either right or left fuel tank, the procedure being the same on either side except where otherwise specified.

(1) Remove engine from vehicle (par. 75).

(2) Drain both fuel tanks by removing plugs from hull floor under tanks and removing drain plugs from tanks. For removal of left tank, also drain engine oil tank and leave drain plug out.

(3) Unscrew coupling nuts to disconnect pipes at front fire extinguisher horn. Detach clips which support the center and rear fire extinguisher pipes to tank, roof, and tank insulator. Detach rear fire extinguisher horn from support and remove horn with pipes attached. Remove front nozzle and support, and remove horn to tee pipe. If right tank is being removed, detach tachometer shaft support clips, pull shaft through engine air baffle and coil it up to avoid damage.

(4) Remove engine air baffle which is attached to tank insulator and roof support with eight $\frac{3}{16}$-inch cap screws and lock nuts. When left baffle is removed also disconnect and remove heater air tube if installed on vehicle.

(5) For removal of left tank, disconnect throttle rod at idler lever, but do not remove rod from tank insulator. Disconnect and remove two primer pipes and the oil dilutions valve pipe. Remove hose fitting from oil tank inlet connection. Remove vent and filler from engine oil tank.

(6) Remove elbow or tee from fuel tank pump outlet and remove the sheet metal cover over pump, which is attached to tank insulator with six $\frac{3}{16}$-inch cap screws and lock washers. Remove felt insulator which is behind the cover.

(7) For removal of left tank, remove rear junction box inspection plate which is attached with five $\frac{3}{16}$-inch cap screws, special flat washers, and two lock nuts. Slide plate out far enough on starter cable conduit to place plate in sponson at rear of oil tank. Remove tank in-

TM 9-755

Fuel and Air Intake and Exhaust Systems

sulator which is attached with 10 $\frac{3}{16}$-inch cap screws and flat washers. For removal of right tank, remove tank insulator which is attached with 14 cap screws and special flat washers.

(8) Disconnect ground strap which is attached to rear end of tank with a $\frac{3}{16}$-inch cap screw and two lock washers. Disconnect wire from fuel gage tank unit. Unscrew captive screws to disconnect conduit housing from fuel pump cable terminal, disconnect capacitor lead and pump wire from terminal, and detach conduit support clips from fuel tank.

(9) Remove clevis pin which attaches hull drain valve operating lever to mounting bracket, pull cable and lever up and lay parts on hull roof.

(10) Loosen hose clamps to disconnect and remove balance pipe from both tanks.

(11) Disconnect control handle rods from fuel valve cross shaft on side from which tank is being removed. Disconnect operating rod from fuel shut-off valve (1 left, 2 right) by removing cotter pin, plain washer and rod end pin. Remove cap screws which attach cross shaft brackets to mounting brackets and remove shaft with operating rods attached.

(12) Remove air cleaner (par. 92 c). Remove air duct which is attached to rear side of bulkhead with six $\frac{3}{16}$-inch cap screws and lock washers.

(13) Remove fuel tank filler cap. Loosen screws in filler neck top washer to relieve pressure on filler neck grommets, insert screwdriver under washer and work washer and grommet assembly out of hull opening and from filler neck (fig. 115). Remove filler neck and gasket which are attached with six round head screws. Lift filter out of tank. Remove filler neck adapter and lower washer which are attached to tank with six flathead screws.

(14) Remove lock nut and clip from fuel tank anchor bolt at bottom of tank (N, M, fig. 94). Loosen lock nuts on tee bolts which anchor the fuel tank support straps at both ends of tank and remove support straps (D, F, fig. 94). Move tank out into engine compartment and remove from vehicle.

(15) Remove fuel tank gage unit (par. 170 d). Remove fuel tank pump and cover assembly which is attached to tank with nine $\frac{3}{8}$-inch cap screws and lock washers. Remove pump screen from tank.

b. **Installation** (figs. 113 and 114). The following procedure covers installation of either right or left fuel tank, procedure being the same on either side except where otherwise specified.

(1) Inspect interior of tank and blow out any dirt with an air stream. Clean all joint surfaces. Make sure that support pads are in good condition and securely riveted to tank.

(2) Install fuel pump and screen (par. 98 d (1) through (4)). Install fuel tank gage unit (par. 170 d).

(3) Move tank into position in hull. Engage support straps with hooks on sponson bottom plate, engage support strap tee bolts with hooks on hull side plate and turn lock nuts down snug but not tight.

TM 9-755
99

Part Three—Maintenance Instructions

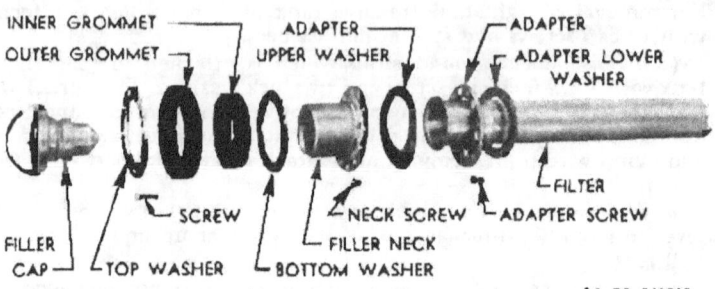

Figure 115—Filler Cap, Filler Neck, Filter, and Attaching Parts

Place clip over anchor bolt at bottom of tank, install lock nut and tighten securely. Tighten support strap tee bolt nuts securely.

(4) Place a new lower washer and the filler neck adapter on tank filler flange, with hole in adapter centered as near as possible with opening in hull roof plate, and attach to flange with six flathead screws (¼ in.—20 x ⅝ in.) (fig. 115). Place new washer and filler neck in position on adapter, with holes in alinement and neck centered as near as possible in opening in hull roof plate, and attach neck to adapter with six round screws (¼ in.—28 x ⅜ in.) Push assembled washers and filler neck grommets into roof plate opening, with screw heads up and hole in inner grommet alined with filler neck. If grommet cannot be alined with neck without strain, remove assembly, remove screws, and turn eccentric inner grommet in the eccentric outer grommet until a position is obtained which will permit alinement with filler neck. Install screws, install washer and grommet assembly, and tighten screws evenly to compress grommets against filler neck and roof plate.

(5) Attach air duct to rear side of bulkhead with six cap screws (⁵⁄₁₆ in.—24 x 1 in.), plain washers and lock washers. Install air cleaner (par. 92 d).

(6) Place fuel valve cross shaft in position and anchor the brackets to mounting brackets with four cap screws (⁵⁄₁₆ in.—24 x 1 in.). Connect control handle rods to levers on cross shaft and place handles in "OFF" positions. Place operating rod yoke in position in shut-off valve stem and check alinement of pin holes; adjust yoke if necessary so that rod end pin will go freely through holes in yoke and stem. Secure the pin with a plain washer and cotter pin, and tighten lock nut on rod against yoke.

(7) Place balance pipe in support clips and connect coupling hoses to fittings on both tanks; tighten hose clamps.

(8) Attach hull drain valve operating lever to mounting bracket on hull floor with clevis pin secured with a cotter pin.

(9) Connect wire to fuel gage tank unit. Clean surface of tank and connect ground strap, placing one internal-external toothed lock

TM 9-755
99-100

Fuel and Air Intake and Exhaust Systems

washer on both sides of strap and attaching strap with one cap screw (⁵⁄₁₆ in.—24 x ⅝ in.). Connect fuel pump wire and capacitor lead to pump cable terminal and tighten terminal screw firmly. Connect conduit housing to terminal and tighten two captive screws. Attach conduit support clips (3 right, 2 left) to tank with cap screws (⁵⁄₁₆ in.—24 x ⅝ in.) and lock washers.

(10) Install drain plug in fuel tank and fill tank with fuel. If left tank is being installed, install drain plug in engine oil tank and fill tank with oil. Check for fuel and oil leaks.

(11) Place round felt insulators over inlet and outlet fittings on oil tank, if left tank is being installed. Install tank insulator, attaching top edge with cap screws (⁵⁄₁₆ in.—24 x 1⅛ in.) and square washers, and bottom and rear edge with cap screws (⁵⁄₁₆ in.—24 x ¾ in.) and round washers. On left tank installation, install rear junction box inspection plate, attaching it with five cap screws (⁵⁄₁₆ in.—24 x ⅝ in.) and square washers, with lock nuts on two top screws. Place felt insulator over pump outlet and install the sheet metal cover with six cap screws (⁵⁄₁₆ in.—24 x ⅝ in.) and lock washers. Install elbow (right) or tee (left) in pump outlet.

(12) If left tank is being installed, install oil dilution valve pipe and two primer pipes and tighten hose clamps. Attach throttle rod to idler lever with rod end pin, plain washer and cotter pin. Install vent and filler in engine oil tank.

(13) Install engine air baffle, attaching it to tank insulator with five cap screws (⁵⁄₁₆ in.—24 x ¾ in.) inserted from rear and secured with plain washer and lock nuts. Attach baffle to roof support with one cap screw and lock nut at center, and with two cap screws external-tooth lock washers and lock nuts which also attach the conduit support clips.

(14) Connect pipe to front fire extinguisher horn and tee, attach horn support to tank with two cap screws (⁵⁄₁₆ in.—24 x ⅝ in.), external-tooth lock washers and nuts. Place rear fire extinguisher horn, with pipes attached, in position and connect center pipe to front horn. Tighten pipe coupling nuts securely, then attach horn to support on tank insulator with two cap screws (⁵⁄₁₆ in.—24 x ⅝ in.) and external-tooth lock washers. Attach center pipe support clips to tank and roof plate with cap screws and lock washers, and attach rear pipe support clips to tank insulator with cap screws and lock washers. If right tank is being installed, push tachometer shaft through grommet in air baffle, attach one support clip with one insulator top attaching screw, and attach rear support clip at rear fire extinguisher pipe upper clip.

(15) Install engine in vehicle and complete the installation (par. 76).

100. MUFFLERS AND CONNECTOR.

a. Removal (fig. 116).

(1) Remove the three air outlet grilles above mufflers (par. 183 c).

(2) Apply a liberal quantity of penetrating oil to muffler joints and clamps.

TM 9-755
100-101

Part Three—Maintenance Instructions

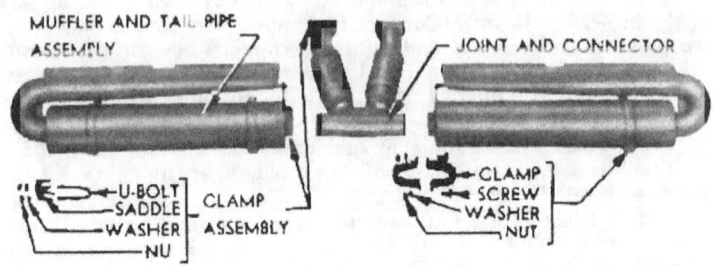

Figure 116—Muffler and Tail Pipe Assemblies, and Joint and Connector

(3) Remove nuts and lock washers from ⅜-inch cap screws which attach four muffler clamps to mounting brackets on hull. Loosen nuts on saddle clamp U-bolts, separate mufflers from connector and remove mufflers from hull.

(4) If flexible joint and connector is to be removed, open hull rear roof door and tie it so it cannot fall. Loosen nuts on saddle clamp U-bolts and remove connector from exhaust manifold outlet pipes.

b. Installation (fig. 116).

(1) If flexible joint and connector was removed, place saddle clamps over front ends of connector and place ends of connector over outlet pipes of exhaust manifold; do not tighten clamps.

(2) Place saddle clamps over rear ends of connector and place mufflers in position with ends engaging ends of connector.

(3) Install the four clamps which anchor the mufflers to mounting brackets on hull and tighten clamp bolts securely.

(4) Turn clamps at mufflers so that saddles are on top and tighten U-bolts. Turn clamps at exhaust manifold so that saddles are on top, position connector on manifold pipes so that flexible sections are in a neutral position, and tighten U-bolts.

(5) Close hull rear roof door and anchor it with three cap screws (½ in.—20 x 1½ in.) tightened to 50-60 foot-pounds tension. Install the three air outlet grilles (par. 183 e).

Section XXII
TRANSFER CASE ASSEMBLY (REAR)

101. DESCRIPTION AND DATA.

a. Description. The transfer case assembly (rear) transmits power from engine to propeller shaft which extends forward to transmission at a lower level than engine crankshaft. It also provides a means of disconnecting engine from power train when this is desirable. The unit consists of a train of three gears enclosed in a

TM 9-755

101

Transfer Case Assembly (Rear)

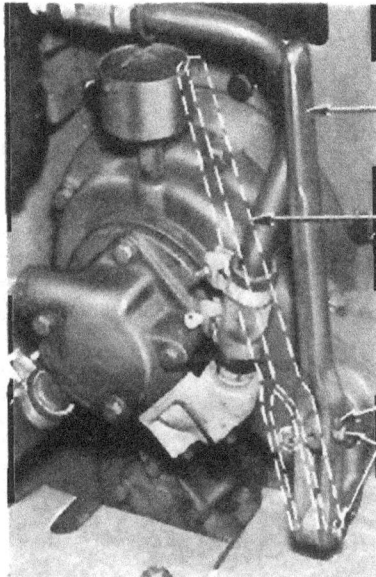

Figure 117—Transfer Case Shifter Lever Installed

case which is mounted on bulkhead forward of engine. A sliding clutch, splined to input shaft and actuated by a shifter fork and shifter lever (par. 102), engages input or upper gear to transmit power to gear train. When clutch is moved out of engagement with input gear by pulling shifter lever away from case, power cannot be transmitted through transfer case to power train. A spring-loaded poppet ball engages notches in shifter rod to hold clutch in either engaged or disengaged position. Lubricant for gears and bearings within case is circulated by differential oil cooler oil pump and the transfer case oil pump which are mounted on front side of case and are driven by input and idler shafts respectively. A splined yoke installed on splined output shaft to which it is secured by a nut, provides a mounting for propeller shaft rear universal joint.

b. Data.

Number of teeth on input shaft gear	40
Number of teeth on idler shaft gear	35
Number of teeth on output shaft gear	40
Gear ratio, input to output	1 to 1
Gears and shafts supported by	Ball and roller bearings
How lubricated	Oil circulated by pump
How coupled with engine	Splined universal-joint flange
How coupled with propeller shaft	Splined universal-joint yoke
How engaged and disengaged	Manual lever
How vented	Breather

TM 9-755
102

Part Three—Maintenance Instructions

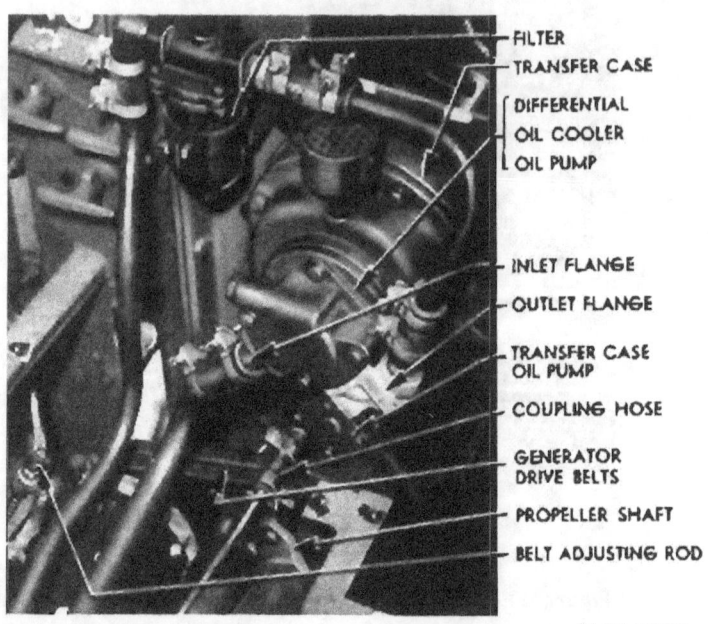

Figure 118—Connections to Transfer Case

102. TRANSFER CASE SHIFTER LEVER.

a. Purpose. The shifter lever is to be used only when it is necessary to disconnect engine from power train during maintenance operations. It is not necessary to disconnect engine from power train when starting engine in sub-zero temperatures.

b. Installation. The shifter lever with rod end pins and cotter pins, is stowed in the tool stowage box. To install lever, connect it to transfer case shifter rod and to shifter fork cap by means of rod end pins and cotter pins as shown in figure 117.

c. Operation. Disengage transfer case clutch by pulling shifter lever away from bulkhead; engage clutch by pushing lever toward bulkhead. A spring-loaded poppet ball engages notches in shifter rod to hold transfer case clutch in either position. CAUTION: *Do not operate shifter lever while engine is running as gear damage will result.* If clutch cannot be engaged because gear teeth do not mesh, turn engine with hand crank (magneto switch off) until engagement is secured.

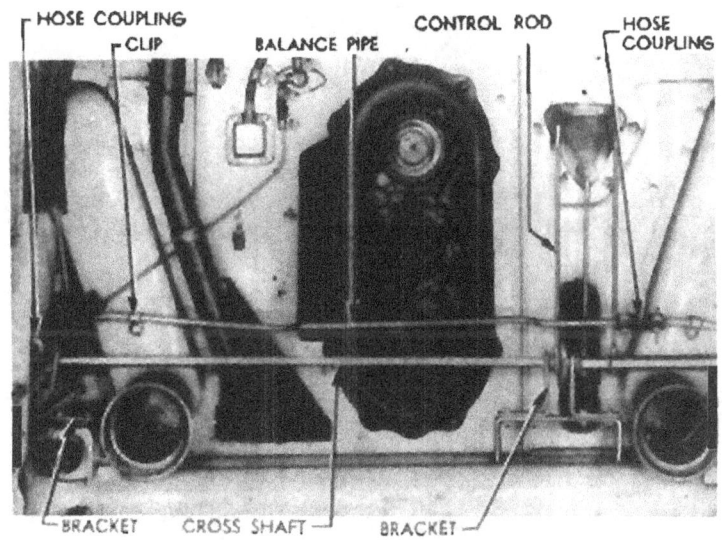

Figure 119—Fuel Tank Balance Pipe and Fuel Valve Cross Shaft Connections

103. TRANSFER CASE REMOVAL.

a. **Removal of Parts in Crew or Fighting Compartment.**

(1) On M18 vehicle, remove right and left rear subfloor plates (par. 184 c). On M39 vehicle, open rear seat center cover and remove rear seat back with pad attached (par. 185 c).

(2) Disconnect inlet and outlet flanges which are attached to differential oil cooler oil pump by $\frac{3}{16}$-inch cap screws and lock washers.

(3) Loosen hose clamps and remove coupling hose on oil return pipe from outlet fitting on transfer case pump (fig. 118). NOTE: *In M18 vehicles having serial number below 1351, return pipe is connected to pump by a Sealflex fitting which is disconnected by unscrewing compression nut (fig. 148).*

(4) Disengage clutch in transfer case (par. 102 c) so that propeller shaft may be turned by hand.

(5) Flatten bent up tongues of lock plates and remove four cap screws which attach propeller shaft to rear universal joint. Push propeller shaft forward as far as possible.

(6) Unscrew safety nut on upper end of belt adjusting rod (fig. 191) far enough to permit removal of generator drive belts from pulleys.

Part Three—Maintenance Instructions

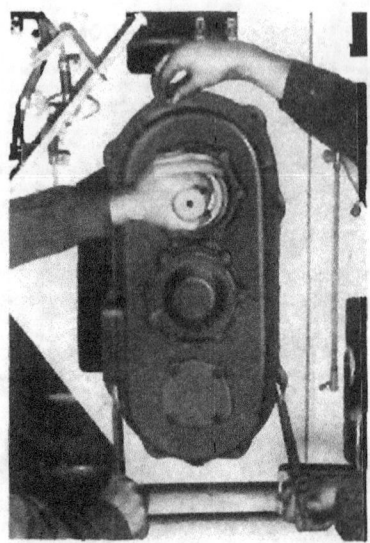

Figure 120—Removing Transfer Case

(7) Remove breather by unscrewing it from transfer case, and remove shifter lever, if it is installed.

b. **Removal of Parts in Engine Compartment.**

(1) Disconnect and move engine out upon rails on hull rear door (par. 75).

(2) Remove drain plug from hull floor, remove drain plug from left fuel tank and drain tank (capacity 75 gal.). Install drain plugs and tighten securely.

(3) Loosen hose clamps and pull coupling hoses from both ends of left fuel tank balance pipe (fig. 119), push pipe out of clip on bulkhead and remove pipe. NOTE: *In M18 vehicles having serial numbers below 1351, the balance pipe is connected by Sealflex unions which are disconnected by unscrewing the compression nuts (fig. 148).*

(4) Disconnect control handle rod from right end of fuel valve cross shaft. Remove two $\frac{5}{16}$-inch cap screws which secure each cross shaft bracket to mounting brackets and tie right end of shaft up out of way, using care not to damage fuel valve operating rod attached to left end of cross shaft.

(5) Loosen the $\frac{1}{2}$-inch safety nuts on the eight cap screws which attach the transfer case to bulkhead and remove all but the two upper nuts and cap screws. Insert large punches, or other suitable tools ($\frac{1}{2}$ in. diam.), through the second from the bottom screw holes

Transfer Case Assembly (Rear)

to support transfer case while removing the two upper nuts and cap screws. With two men using punches as handles, lift transfer case out of opening in bulkhead and lower it to hull floor (fig. 120).

(6) Attach sling to transfer case and lift it out of vehicle.

(7) Flatten bent up tongues of lock plates under screw heads on rear side of drive pulley and remove four $\frac{3}{8}$-inch cap screws with lock plates which attach universal joint and pulley to transfer case yoke.

(8) Remove cotter pin and slotted nut which secures universal joint yoke to transfer case output shaft. Remove yoke and pulley, and install nut on output shaft to protect threads.

104. TRANSFER CASE INSTALLATION.

a. **Installation of Parts in Engine Compartment.**

(1) Place pulley over yoke, install yoke on transfer case output shaft and install nut ($1\frac{1}{4}$ in.—18). Tighten nut securely and install cotter pin ($\frac{1}{8}$ in. x $2\frac{1}{2}$ in.).

(2) Inspect mating surfaces of yoke and universal joint to make sure they are clean and free of burs. Attach universal joint and pulley to yoke with two lock plates and four cap screws ($\frac{3}{8}$ in.—24 x $1\frac{1}{4}$ in.) installed through inner side of pulley. Tighten screws to 28-33 foot-pounds tension and bend tongues of lock plates up against flats on screw heads.

(3) Attach sling to transfer case and hoist it into engine compartment.

(4) Have one man in fighting compartment to install transfer case attaching cap screws and two men in engine compartment to lift transfer case into position. Insert large punches, or other suitable tools, ($\frac{1}{2}$ in. diam.) through the second from the bottom screw holes in case to serve as handles. Lift transfer case into position in bulkhead opening, entering forward ends of punches into screw holes in bulkhead to support transfer case while installing two cap screws ($\frac{1}{2}$ in.—20 x $1\frac{3}{4}$ in.) and safety nuts. Install six remaining cap screws and safety nuts and tighten all nuts to 80-100 foot-pounds tension.

(5) Install fuel valve cross shaft by attaching each bracket to mounting brackets with two cap screws ($\frac{5}{16}$ in.—24 x 1 in.), position cross shaft so that left fuel valve operating rod is not bent, and tighten bracket screws securely. Attach control handle rod to right end of shaft with nut ($\frac{1}{4}$ in.—28) and lock washer.

(6) Place left fuel tank balance pipe in support clip on bulkhead, push ends of balance pipe into coupling hoses and tighten hose clamps. NOTE: *If Sealflex unions are used to connect pipes adjust compressions nuts as described in paragraph 125 b.*

(7) Move engine into position, make all connections, and close hull rear door (par. 76).

Part Three—Maintenance Instructions

h. **Installation of Parts in Crew or Fighting Compartment.**

(1) Place generator drive belts on pulleys and adjust tension (par. 143 h). Install breather.

(2) Connect propeller shaft to universal joint with two lock plates and four cap screws (⅜ in.—24 x $^{27}/_{32}$ in.) tightened to 28-33 foot-pounds tension. Bend tongues of lock plates up against flats on screw heads.

(3) Push oil return pipe into coupling hose on outlet fitting on transfer case pump and tighten clamp securely. NOTE: *If Sealflex fitting is used, adjust compression nut as described in paragraph* 125 h.

(4) Fill differential oil cooler oil pump inlet pipe with SAE 50 engine oil to eliminate air pocket and inject oil into pump inlet port to insure good seal. Wipe joint surfaces of pump and flanges to make sure they are clean.

(5) Coat new gaskets (fibre, $^{3}/_{16}$ in. thick) with gasket cement and insert them between both flanges and the pump. Attach inlet flange to pump with two cap screws ($^{5}/_{16}$ in.—18 x ¾ in.) and lock washers. Attach outlet flange to pump with one cap screw ($^{5}/_{16}$ in.—18 x 2¼ in.) and one cap screw ($^{5}/_{16}$ in.—18 x 2¾ in.), both with lock washers. Tighten all flange attaching screws to 18-22 foot-pounds tension.

(6) Install breather on transfer case, engage transfer case clutch (par. 102).

(7) On M18 vehicle, install right and left rear subfloor plates (par. 184 c). On M39 vehicle, install rear seat back and close rear seat center cover (par. 185 d).

Section XXIII
PROPELLER SHAFT AND UNIVERSAL JOINT

105. DESCRIPTION.

a. **Description (fig. 121).** The propeller shaft, which transmits power from transfer case to transmission, is a steel tube with a yoke welded to each end, to which a universal joint is attached by cap screws. The rear universal joint is attached to yoke on transfer case output shaft by cap screws which also attach generator drive pulley to yoke. The front universal joint is attached to flange on transmission transfer case input shaft by cap screws. Separate cap screws attach blower drive pulley to flange.

106. REMOVAL.

a. **Propeller Shaft Removal—M18 Vehicle.**

(1) Remove center front subfloor plate and right rear subfloor plate (par. 184 c).

(2) Remove attaching screws and lock washers from center sub-

Propeller Shaft and Universal Joint

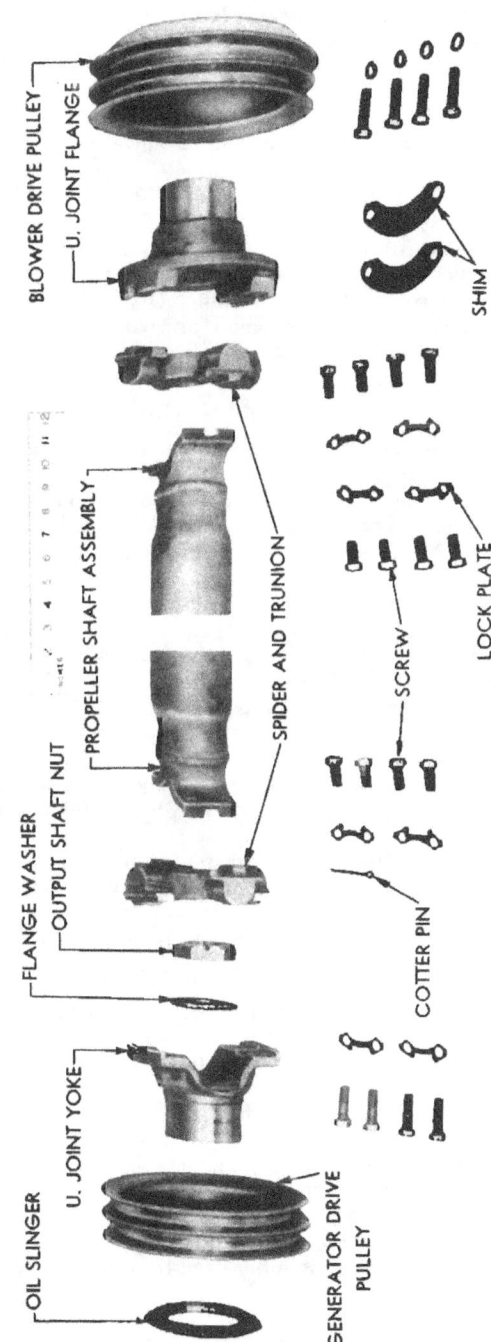

Figure 121—Propeller Shaft and Universal Joints—Disassembled

floor plate, move plate and attached slip ring box to the right, and block it up as far as possible.

(3) Remove oil cooler inlet duct cover which is secured to inlet duct with fourteen $\frac{5}{16}$-inch cap screws, and to blower housing with six $\frac{5}{16}$-inch self-tapping cap screws all having plain and lock washers.

(4) Disengage clutch in transfer case (par. 102) so that propeller shaft may be turned by hand.

(5) Flatten bent-up tongues of lock plates and remove four $\frac{3}{8}$-inch cap screws which attach propeller shaft to universal joint bearings at each end, and remove bearings from spider (fig. 122).

(6) Lift front end of propeller shaft and carefully work shaft out from under center subfloor plate, extending shaft into left driving compartment as far as required to accomplish removal.

b. **Propeller Shaft Removal—M39 Vehicle.**

(1) Remove batteries (par. 142 g) and battery box (par. 188 a).

(2) Remove oil cooler inlet duct cover which is secured to inlet duct with fourteen $\frac{5}{16}$-inch cap screws and to blower housing with six $\frac{5}{16}$-inch self-tapping cap screws, all having plain and lock washers.

(3) Detach rear edge of front seat pad, remove hinge pins and remove front seat cover.

(4) Remove four flathead screws, nuts, and lock washers which attach the two oil pipe brackets to propeller shaft guard. Remove guard which is anchored at bottom and both ends by fourteen $\frac{3}{8}$-inch cap screws with plain and lock washers.

(5) Remove front seat center support rail (fig. 143) which is attached by a $\frac{3}{8}$-inch cap screw, plain and lock washers at rear end and by a $\frac{1}{2}$-inch cap screw, plain and lock washers at front end.

(6) Remove support brackets from oil pipes, loosen clamps at hose couplings on both ends of oil pipes over propeller shaft, and remove pipes.

(7) Flatten bent up tongues of lock plates and remove four $\frac{3}{8}$-inch cap screws which attach propeller shaft to universal joint bearings at each end and remove bearings from spider (fig. 123).

(8) Lift front end of propeller shaft and remove shaft.

c. **Front Universal Joint Removal—M18 Vehicle** (fig. 122).

(1) Remove covering parts and disconnect propeller shaft from universal joints at both ends as described in steps (1), (3), (4) and (5) of subparagraph a above. Do not remove center subfloor plate (step (2)).

(2) Remove blower belt guard which is anchored on left end by a $\frac{3}{8}$-inch cap screw and loosen blower belts (par. 122 a).

(3) Remove blower driven pulley, and shims between pulley and hub, which are attached to hub with four $\frac{3}{8}$-inch cap screws and lock washers. Do not lose shims.

(4) Lower rear end of propeller shaft, block up front end, and pull universal joint and flange assembly from transmission input shaft.

Propeller Shaft and Universal Joint

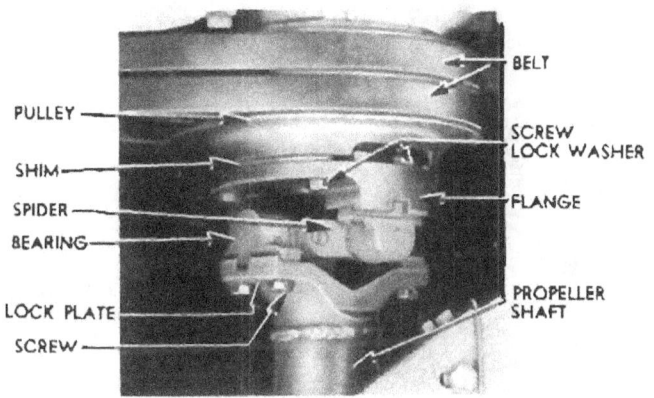

Figure 122—Front Universal Joint

(5) Remove blower drive pulley, and shims between pulley and flange, which are secured to flange with four 3/8-inch cap screws and lock washers. NOTE: *Tie each set of shims together to avoid losing or mixing them.*

(6) Flatten bent up tongues of lock plates and remove four 3/8-inch cap screws with lock plates which attach universal joint bearings to flange.

d. **Front Universal Joint Removal—M39 Vehicle** (fig. 122).

(1) Remove left battery (par. 142 g) and battery floor pad. Remove inspection hole cover from opening under pad.

(2) Disengage clutch in transfer case (par. 102) so that propeller shaft may be turned by hand.

(3) Remove blower belt guard which is attached with two 3/8-inch cap screws, nuts and lock washers, and loosen blower belts (par. 122 a).

(4) Remove blower driven pulley, and shims between pulley and hub, which are attached to hub with four 3/8-inch cap screws and lock washers. Do not lose shims.

(5) Flatten bent up tongues of lock plates and remove four 3/8-inch cap screws which attach propeller shaft to universal joint bearings at each end and remove bearings from spider (fig. 123). Move propeller shaft to rear as far as possible.

(6) Remove four 3/8-inch cap screws and lock washers which attach the blower drive pulley and shims to flange on transmission input shaft. NOTE: *Tie each set of shims together to avoid losing or mixing them.*

TM 9-755
106-107

Part Three—Maintenance Instructions

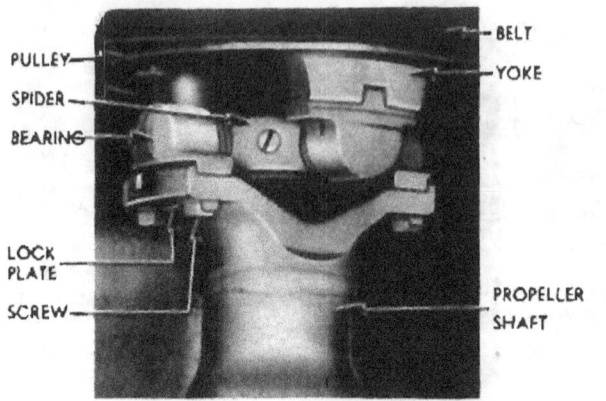

Figure 123—Rear Universal Joint

(7) Remove flange from transmission input shaft, and remove pulley from flange and lift flange out through inspection hole.

(8) Flatten bent up tongues of lock plates, remove four $\frac{3}{8}$-inch cap screws and remove universal joint spider and bearings from flange.

 e. **Rear Universal Joint Removal** (fig. 123).

(1) On M18 vehicle, remove left and right rear subfloor plates (par. 184 c). On M39 vehicle, open rear seat center cover and remove rear seat back with pad attached (par. 185 c).

(2) Disengage clutch in transfer case (par. 102) so that propeller shaft may be turned by hand.

(3) Flatten bent up tongues of lock plates and remove four $\frac{3}{8}$-inch cap screws with lock plates which attach propeller shaft to rear universal joint bearings, and remove bearings from spider. Push shaft forward as far as possible.

(4) Flatten bent up tongues of lock plates under screw heads on rear side of drive pulley, unscrew four $\frac{3}{8}$-inch cap screws which attach universal joint bearings to transfer case yoke, and remove universal joint.

107. INSTALLATION.

 a. **Propeller Shaft Installation—M18 Vehicle.** Both ends of propeller shaft are identical; either end may be placed to the rear.

(1) Move propeller shaft from fighting compartment into driving compartment to left of transmission far enough to get rear end of shaft under center subfloor plate; then carefully work shaft down into position between universal joints.

(2) Inspect mating surfaces of universal joint bearings and pro-

Propeller Shaft and Universal Joint

peller shaft yokes to make sure they are clean and free of burs. Place bearings on spiders and connect propeller shaft to each bearing with one lock plate and two cap screws ($3/8$ in.—24 x $27/32$ in.) tightened to 28-32 foot-pounds tension. Bend tongues of lock plates up against flats on screw heads.

(3) Rotate propeller shaft by hand to determine whether blower drive and driven pulleys are in alinement. If belts rub on one side of grooves in pulleys, and propeller shaft is hard to turn, pulleys are not in alinement. Remove cap screws one at a time and add, or remove, shims between pulley and flange as required to secure alinement, making sure to use shims of same thickness on both sides of pulley.
NOTE: *Shims are available in thicknesses of $1/16$-inch and $1/8$-inch.*

(4) Lubricate universal joints.

(5) Install oil cooler inlet duct cover and secure it to inlet duct with 14 cap screws ($3/16$ in.—24 x $3/4$ in.) and to blower housing with 6 selftapping cap screws ($3/16$ in.—24 x $5/8$ in.). Use plain and lock washers with all cap screws.

(6) Move center subfloor plate into position and secure it with six cadmium plated cap screws ($3/8$ in.—24 x 1 in.) and external-tooth lock washers tightened to 28-33 foot pounds tension.

(7) Install right rear subfloor plate with eight cap screws ($3/8$ in.—24 x 1 in.) and plain washers. Install center front subfloor plate and secure it to blower housing and floor support with nine cap screws ($3/8$ in.—24 x 1 in.) and plain washers, and to blower belt guard with two cap screws ($3/8$ in.—24 x 1 in.) with plain and lock washers. Tighten all cap screws to 20-25 foot-pounds tension.

(8) Engage transfer case clutch (par. 102).

b. **Propeller Shaft Installation—M39 Vehicle.** Both ends of propeller shaft are identical; either end may be placed to the rear.

(1) Inspect mating surfaces of universal joint bearings and propeller shaft yokes to make sure they are clean and free of burs. Set propeller shaft in position, install bearings on spiders, and attach propeller shaft to each bearing with a lock plate, two cap screws ($3/8$ in.—24 x $27/32$ in.) tightened to 28-33 foot-pounds tension. Bend tongues of lock plates up against flats on screw heads.

(2) Lubricate universal joints.

(3) Rotate propeller shaft by hand to determine whether blower drive and driven pulleys are in alinement. If belts rub on one side of grooves in pulleys, and propeller shaft is hard to turn, pulleys are not in alinement. Remove cap screws one at a time and add, or remove, shims between pulley and flange as required to secure alinement, making sure to use shims of same thickness on both sides of pulley.
NOTE: *Shims are available in thicknesses of $1/16$ inch and $1/8$ inch.*

(4) Place oil pipes in position, push ends into hose couplings and tighten hose clamps securely. Install support brackets on pipes.

(5) Install front seat center support rail by attaching it to seat frame at rear end with one cap screw ($3/8$ in.—24 x $3/4$ in.) and lock

TM 9-755
107

Part Three—Maintenance Instructions

washer, and to front support with one cap screw (½ in.—20 x 1 in.) and lock washer.

(6) Install propeller shaft guard by anchoring it at lower edges and to front and rear seat supports using 14 cap screws (⅜ in.—24 x ¾ in.) with plain washers and lock washers. Tighten screws to 25-30 foot-pounds tension. Attach each oil pipe support bracket to guard with two flat head screws (⁵⁄₁₆ in.—18 x ¾ in.), lock washers and nuts.

(7) Install front seat cover, install hinge pins and secure them with cotter pins in each end. Attach seat pad to seat support with flat head screws and finish washers.

(8) Install oil cooler inlet duct cover and secure it to inlet duct with 14 cap screws (⁵⁄₁₆ in.—24 x ¾ in.) and to blower housing with 6 self-tapping cap screws (⁵⁄₁₆ in.—24 x ⅝ in.). Use plain and lock washers with all cap screws.

(9) Install battery box (par. 188 h) and batteries (par. 142 h).

c. Front Universal Joint Installation—M18 Vehicle (fig. 122).

(1) Inspect mating surfaces of flange and universal joint bearings to make sure they are clean and free of burs. Place bearings on opposite ends of spider and attach bearings to flange with two lock plates and four cap screws (⅜ in.—24 x 1¹⁵⁄₁₆ in.) tightened to 28-33 foot-pounds tension. Bend tongues of lock plates up against flats on screw heads.

(2) Attach blower drive pulley and shims to flange with four cap screws (⅜ in.—24 x 1⅜ in.) and lock washers tightened to 28-33 foot-pounds. NOTE: *Make sure that shims on each side of pulley are of equal thickness.*

(3) Install flange on transmission input shaft, using care not to damage oil seal in transmission. Place blower belts over drive pulley.

(4) Inspect mating surfaces of universal joints, bearings, and propeller shaft yokes to make sure they are clean and free of burs. Place bearings on spiders and connect propeller shaft to each bearing with one lock plate and two cap screws (⅜ in.—24 x ²⁷⁄₃₂ in.) tightened to 28-32 foot-pounds tension. Bend tongues of lock plates up against flats on screw heads.

(5) Place belts in grooves of drive and driven pulleys and install driven pulley on hub, placing original number of shims between pulley and hub, and attach pulley with four cap screws (⅜ in.—24 x 1¼ in.) and lock washers. Adjust blower belts to proper tension (par. 122 a).

(6) Rotate propeller shaft by hand to determine whether blower drive and driven pulleys are in alinement. If belts rub on one side of grooves in pulleys, and propeller shaft is hard to turn, pulleys are not in alinement. Remove driven pulley and add, or remove, shims as required to secure alinement.

(7) Install blower belt guard by anchoring left end to tapping block with one cap screw (⅜ in.—24 x 1 in.) and lock washer.

Propeller Shaft and Universal Joint

(8) Complete installation of parts as described in steps (4), (5), (7), and (8) of subparagraph *a* above.

d. **Front Universal Joint Installation—M39 Vehicle** (fig. 122).

(1) Inspect mating surfaces of flange and universal joint bearings to make sure they are clean and free of burs. Place bearings on opposite ends of spider and attach bearings to flange with two lock plates and four cap screws (3/8 in.—24 x 1 7/16 in.) tightened to 28-33 foot-pounds tension. Bend tongues of lock plates up against flats on screw heads.

(2) Install flange on transmission input shaft, using care not to damage oil seal in transmission. Place blower belts over drive pulley.

(3) Attach blower drive pulley and shims to flange with four cap screws (3/8 in.—24 x 1 3/8 in.) and lock washers tightened to 28-33 foot-pounds. NOTE: *Make sure that shims on each side of pulley are of equal thickness.*

(4) Inspect mating surfaces of universal joints, bearings, and propeller shaft yokes to make sure they are clean and free of burs. Place bearings in spiders and connect propeller shaft to each bearing with one lock plate and two cap screws (3/8 in.—24 x 27/32 in.) tightened to 28-32 foot pounds tension. Bend tongues of lock plates up against flats on screw heads.

(5) Place belts in grooves of drive and driven pulleys and install driven pulley on hub, placing original number of shims between pulley and hub, and attach pulley with four cap screws (3/8 in.—24 x 1 1/4 in.) and lock washers. Adjust blower belts to proper tension (par. 122 a).

(6) Rotate propeller shaft by hand to determine whether blower drive and driven pulleys are in alinement. If belts rub on one side of grooves in pulleys, and propeller shaft is hard to turn, pulleys are not in alinement. Remove driven pulley and add, or remove, shims as required to secure alinement.

(7) Install blower belt guard and anchor it with two cap screws (3/8 in.—24 x 3/4 in.), lock washers, and nuts.

(8) Install inspection hole cover and battery floor pad. Install battery (par. 142 h).

e. **Rear Universal Joint Installation** (fig. 123).

(1) Inspect mating surfaces of transfer case yoke and universal joint to make sure they are clean and free of burs.

(2) Attach universal joint and pulley to transfer case yoke with two lock plates and four cap screws (3/8 in.—24 x 1 1/4 in.) installed through rear side of pulley. NOTE: *Bend ends of lock plate tongues slightly to facilitate insertion of chisel for final bending after screws are tightened.* Tighten screws to 28-32 foot-pounds tension and bend tongues of lock plates up against flats on screw heads.

(3) Place generator belts on pulleys and adjust to proper tension (par. 143 a).

(4) Connect propeller shaft to universal joint with two lock plates and four cap screws (3/8 in.—24 x 27/32 in.) tightened to 28-32 foot-

TM 9-755
107-108

Part Three—Maintenance Instructions

pounds tension. Bend tongues of lock plates up against flats on screw heads.

(5) Engage transfer case clutch (par. 102).

(6) Lubricate universal joint.

(7) On M18 vehicle, install left and right rear subfloor plates (par. 184 c). On M39 vehicles, install rear seat back and close rear seat center cover (par. 185 d).

Section XXIV
TORQMATIC TRANSMISSION ASSEMBLY

108. DESCRIPTION AND DATA.

a. **Description.** The Torqmatic transmission assembly is a manually controlled, hydraulically operated unit designed to prevent possibility of stalling the engine under heavy loads, and to automatically vary the torque applied to the driving sprockets in accordance with changing operating requirements while the engine is held fairly constant at its most efficient speed. The transmission is a constant mesh planetary type which provides neutral, reverse, and three forward speed ratios. Any one of these positions may be selected by the driver through positioning of a shift lever, without manual operation of a clutch. The assembly consists of three principal sections which are joined together into one compact unit completely enclosed and thoroughly lubricated by a pressure oiling system (par. 118 a). The transmission assembly is attached to the differential carrier and transmits power to the differential ring gear through a bevel pinion gear on the transmission output shaft. The principal sections of the transmission assembly (fig. 124) are as follows:

(1) TRANSFER CASE. The transfer case, located on input end of assembly, contains a train of three gears which transmits power from under-slung propeller shaft to torque converter which is located at a higher level.

(2) TORQUE CONVERTER. The torque converter is located between transfer case and transmission. It functions as a fluid clutch and also multiplies the torque output of engine in accordance with power requirements of vehicle. Under light driving loads and high speeds, engine torque is not increased; under heavy driving loads and low speeds engine torque may be increased approximately 4½ times by torque converter.

(3) TRANSMISSION. The transmission case houses planetary transmission gears which are controlled by two hydraulically operated clutches and two hydraulically operated servo bands to produce various speed ranges. Power from torque converter is transmitted through planetary gears to output shaft in accordance with speed range or gear ratio selected by operator.

TM 9-755
108

Torqmatic Transmission Assembly

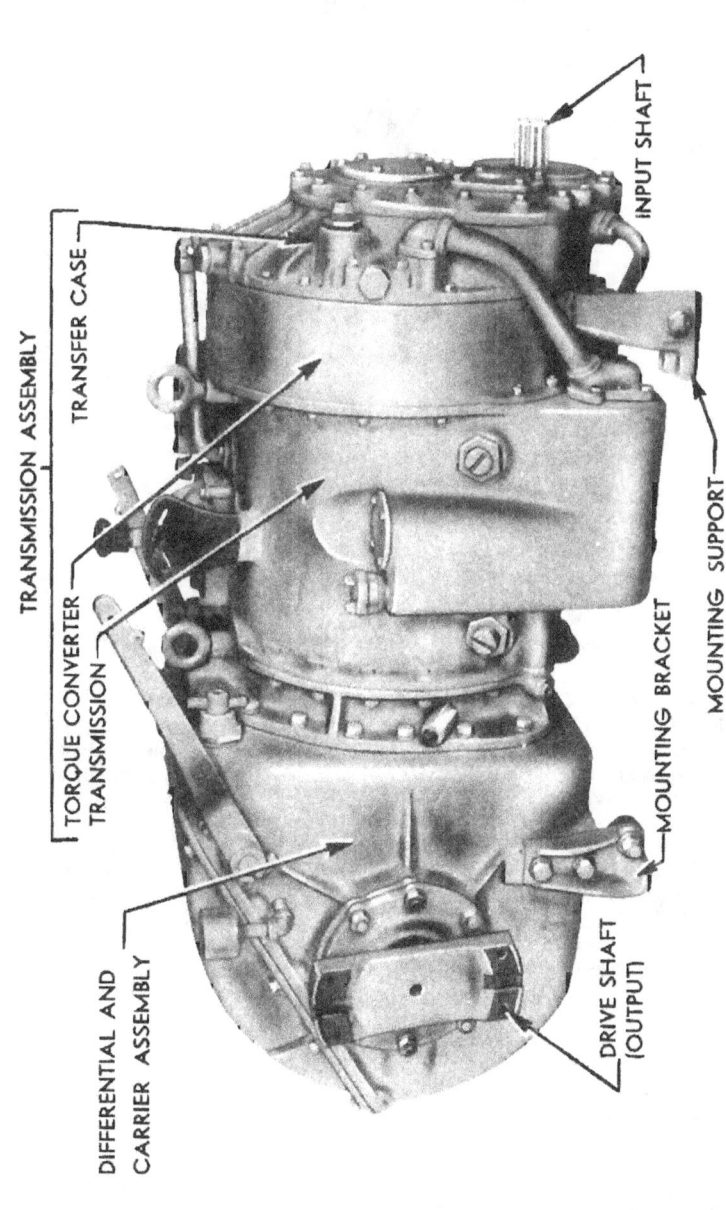

Figure 124—Transmission and Differential Assembly

Part Three—Maintenance Instructions

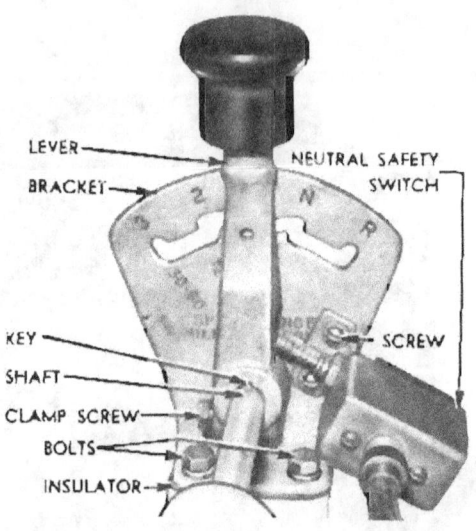

Figure 125—Manual Shift Lever, Bracket, and Neutral Safety Switch

b. Data.

Manufacturer	Detroit Transmission Division
Type	Torqmatic—hydraulic
Type gears used	Constant mesh planetary
Torque converter type	3-stage

Transmission ratios:
Reverse (R)	1.0 to 1.322 overdrive
Low (1)	1.0 to 1.0 direct drive
Intermediate (2)	1.0 to 2.337 overdrive
High (3)	1.0 to 4.105 overdrive

Converter torque multiplication:
Low range at stall (vehicle at stand still)	4.8 to 1
Intermediate range, normal operation	2.0 to 0.43
High range, normal operation	1.1 to 0.24
Number of gears in transfer case	3
Transfer case gear ratio	1.29 to 1
Transmission operation	Hydraulic
How controlled	Manual shift lever
Number of oil pumps	2
Pressure in control system	100 lb per sq. in.
Pressure in lubrication system	15 lb per sq. in.
Maximum oil flow through cooler	40 to 45 gal. per min.
Weight of transmission (dry)	1,347 lb
Weight of transmission and differential assembly (dry)	2,512 lb

Torqmatic Transmission Assembly

109. MANUAL SHIFT LEVER AND BRACKET.

a. Removal (fig. 125).

(1) Remove starter neutral safety switch which is attached to shift lever bracket by two round head machine screws, external toothed lock washers and plain washers.

(2) Remove shift lever bracket and bracket insulator which are attached to transmission case by two ⅜-inch bolts and lock washers.

(3) Loosen clamp screw on shift lever and tap lever off valve shaft.

b. Installation (fig. 125).

(1) Aline keyway in lever with key in valve shaft, with clamp screw towards front of vehicle, and tap lever on shaft until shaft projects through lever about ¼ inch.

(2) Place insulator on transmission case and place bracket over insulator with bolt flange under shift lever. Attach these parts with two bolts (⅜ in.—16 x 1 in.) and lock washers tightened to 28-33 foot-pounds tension.

(3) Tap lever towards bracket with latch pin in slot in bracket until lever just clears bracket without binding in any position; then tighten lever clamp screw to 20-25 foot-pounds tension.

(4) Install starter neutral safety switch on shift lever bracket with key on switch bracket engaged in keyway in lever bracket and secure with two round-head machine screws (10-32 x ½ in.) provided with plain washers and external toothed lock washers. Adjust position of switch for correct timing as described in paragraph 149 c.

110. TRANSMISSION BAND ADJUSTMENT.

a. Transmissions which have been rebuilt or overhauled but have not been run on a dynamometer should have band adjustment checked after 3 to 4 miles of operation in the vehicle. High spots on lining will have worn off during these few miles of operation and a proper adjustment can then be made. Adjustment procedure for each band is as follows:

(1) Remove left shield which is attached to transmission by two lifting eye bolts, and by a cap screw at front and rear ends.

(2) Straighten edges of lock washer which are bent up against lock nut and down against a flat spot in transmission case (fig. 126). NOTE: *Some early production transmissions were not equipped with the grooved screw and lock washer, the screw being anchored in position by the lock nut only.*

(3) Place servo band adjusting wrench (41-W-490-250) over back nut with blade engaged in adjusting screw slot. Using box wrench (41-W-576) and flexible handle (41-H-1502) on the adjusting wrench (fig. 127), loosen adjusting screw lock nut approximately 1½ turns to free adjusting screw.

(4) Tighten adjusting screw to 50 foot-pounds tension with blade and handle, then loosen screw (counterclockwise) three-quarters of a turn. Hold screw at this point and tighten lock nut securely. CAUTION: *To avoid disconnecting adjusting screw from band, do*

TM 9-755

Part Three—Maintenance Instructions

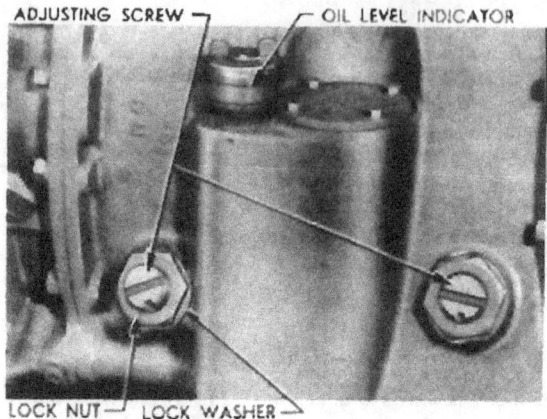

Figure 126—Transmission Band Adjusting Screws

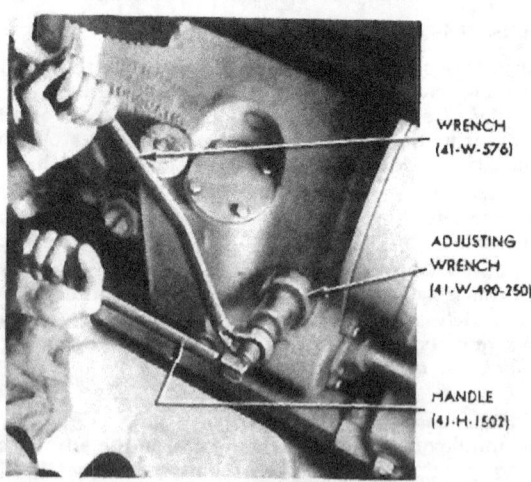

Figure 127—Adjusting Transmission Band With Wrench 41-W-490-250

not back out screw more than ¼ inch. If screw is disconnected from band, the transmission will have to be disassembled.

(5) Bend one edge of lock washer up against flat on nut and another edge down against flat on boss of transmission case.

(6) Install left transmission shield and secure it with two lifting eye bolts, one output end cover cap screw (½ in.—13 x 1½ in.) and one torque converter housing cap screw (⅜ in.—16 x 1⅛ in.) with lock washers.

111. TRANSMISSION REMOVAL.

a. **Coordination With Higher Authority.** Replacement of this major assembly with a new or rebuilt unit is normally a third echelon operation, but may be performed in an emergency by second echelon, provided authority for performing this replacement is obtained from the appropriate commander. Tools needed for the operation which are not carried in second echelon may be obtained from a higher echelon of maintenance.

b. **Removal of Transmission and Differential Assembly (fig. 124).** NOTE: *The transmission and differential must be replaced as a unit assembly and not separately. When joining these separate assemblies together special gages are required to obtain proper adjustment of ring gear with pinion gear.*

(1) Remove three drain plugs in hull floor, remove drain plugs in bottom of torque converter housing, transmission case, and differential carrier, and allow oil to drain from these units. After oil is drained, install all plugs and tighten securely.

(2) Remove hull front door (par. 180 a).

(3) Disconnect speedometer shaft and remove eight ⁵⁄₁₆-inch bolts with plain and lock washers which secure the universal joint center guard to inner and outer guards (fig. 152).

(4) Cut safety wires, remove four ½-inch cap screws which attach each universal joint to its differential drive shaft, and push universal joints and final drive pinion shafts away from differential as far as possible.

(5) Remove left shield which is attached to transmission by two lifting eye bolts, one ½-inch output end cover cap screw and one ⅜-inch torque converter housing cap screw.

(6) Remove manual shift lever and bracket and tie starter neutral safety switch up out of way (par. 109 a).

(7) Remove cotter pins and rod end pins which connect links to control levers on differential. Unhook return springs from control levers.

(8) Turn both brake adjusting shafts (fig. 131) counterclockwise until brake control levers are low enough to clear hull when assembly is removed.

TM 9-755
111

Part Three—Maintenance Instructions

Figure 128—Extension Rails Installed

(9) Completely loosen blower belts by loosening clamp screw and moving belt-tightener lever towards transmission (fig. 144).

(10) On M18 vehicles, remove center front subfloor plate (par. 184 c). On M39 vehicles, remove left battery (par. 142 g), remove battery floor pad and lift out inspection hole cover which is attached by a chain. Those operations will provide access to universal joint flange when transmission is installed.

(11) Remove transmission transfer case breather (H, fig. 35).

(12) Remove two screws which secure conduit housing to transmission oil temperature gage unit and remove conduit and wire from unit.

(13) Unscrew compression nuts at elbows on torque converter and move ends of oil pipes clear of unions.

(14) Loosen hose clamps on coupling hoses on the three oil pipes at right rear side of transmission and pull hoses from the front pipes. Remove the upper oil pipe which is connected to elbow at top of differential carrier, and remove elbow from carrier. NOTE: *In M18 vehicles having serial numbers below 1351, these pipes are connected by Sealflex fittings (fig. 148) which are disconnected by unscrewing the compression nuts.*

(15) Remove two ⅝-inch cotter pins, nuts and bolts which anchor the differential carrier to mounting rails. Remove two ⅝-inch cotter pins, nuts, and bolts which anchor the transmission to mounting rails.

Torqmatic Transmission Assembly

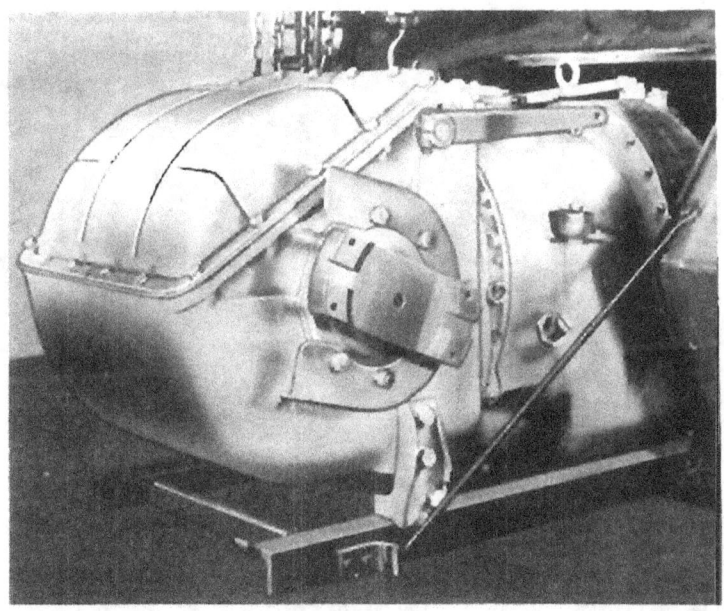

Figure 129—*Transmission and Differential on Extension Rails*

(16) Assemble extension rail cross bar (41-B-90), extension rails (41-R-38 and 41-R-38-10), and brackets (41-B-1926-250 and 41-B-1926-255) with bolts as shown in figure 128. Rest ends of rails on hull plate in line with differential mounting rails and anchor with hull front door bolts (½ in.—20 x 1¼ in.). Support front ends of rails with two hooks (41-H-2737) engaging rail brackets and attached to hull with front door bolts.

(17) Using vehicle crowbar to start mounting bearings out of recesses in mounting rails, move transmission and differential assembly out upon extension rails (fig. 129). NOTE: *Shims may be found between mounting rails and transmission on differential support brackets; mark or tag shims so that they will be placed in same location during installation.*

(18) Install the lifting eye bolts (41-B-1586-100) in transmission case, attach hoist and lift the assembly from extension rails.

(19) If the transmission and differential assembly is to be replaced by another assembly, remove oil temperature gage unit, remove right transmission shield, install the elbow in top of differential and connect oil pipe to this elbow. Plug all openings to prevent entrance of dirt.

TM 9-755
112

Part Three—Maintenance Instructions

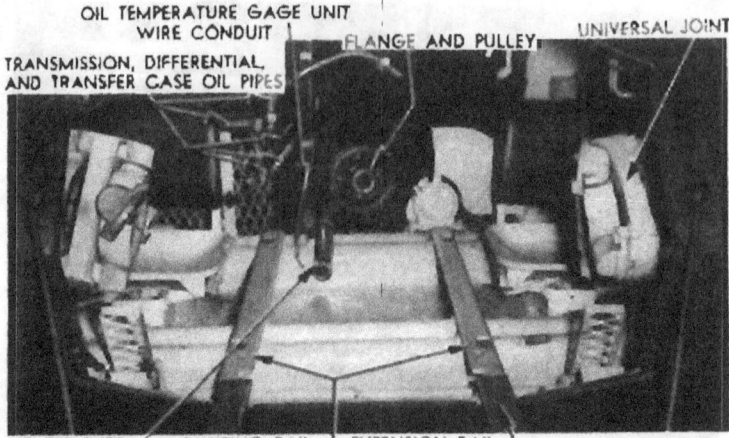

Figure 130—Driving Compartment With Transmission and Differential Removed

112. TRANSMISSION INSTALLATION.

a. Replacement Assembly. A replacement transmission and differential assembly includes the lifting eye bolts, brake control levers, oil strainer and attached oil pipes and transfer case oil return pipe attached to differential carrier. Refer to paragraph 111 a for authority for replacement of assembly.

b. Installation of Transmission and Differential Assembly.

(1) If replacement assembly is being installed, install oil temperature gage unit in elbow on right side of torque converter; install right shield between transmission and the differential oil strainer and secure it to transmission with one valve body cap screw and one oil outlet elbow cap screw; remove oil pipe connected to elbow on top of differential carrier and remove elbow.

(2) Install extension rails if they were removed from vehicle (par. 111 b (16)).

(3) Attach hoist to lifting eyebolts in transmission and place transmission and differential assembly on extension rails (fig. 129).

(4) Remove lifting eyebolts, manual shift lever and bracket and transfer case breather if installed, and turn both brake adjusting shafts (fig. 131) counterclockwise until brake control levers are low enough to go through hull door opening.

(5) Slowly roll the assembly into hull upon mounting rails. As the assembly nears final position, have one man hold flange and pulley on

Torqmatic Transmission Assembly

propeller shaft so that flange will engage the splined transmission input shaft. Have another man hold the three oil pipes on right side so the coupling hoses will engage oil pipes supported by bracket on transmission (fig. 130).

(6) When the assembly is in final position, aline bolt holes in support brackets and mounting rails. Install shims between brackets and rails as required to provide a firm and even support, with holes in shims alined with other bolt holes. Shims are furnished in $1/16$ and $1/8$ inch thicknesses.

(7) Install two bolts ($5/8$ in.—18 x $2\,5/8$ in.) through mounting rails and differential mounting brackets, with castle nuts on upper ends. Install two bolts ($5/8$ in.—18 x $2\,3/8$ in.) through rails and transmission supports, with nuts on upper ends. Tighten all securely and install cotter pins ($1/8$ in. x $1\,1/4$ in.) through nuts and bolts.

(8) Install elbow in top of differential carrier, connect oil pipe to elbow, and anchor rear end of pipe with clip which secures two other pipes to bracket at right rear side of transmission. Push coupling hoses on rear pipes over ends of pipes attached to bracket and tighten hose clamps securely. NOTE: *In M18 vehicles having serial numbers below 1351, these pipes are connected with slip-joint type Sealflex unions. Refer to paragraph 125 b for procedure for connecting these unions.*

(9) Connect two oil pipes to elbows on torque converter and adjust tension on compression nuts and seals as described in paragraph 125 h.

(10) Press wire into terminal in transmission oil temperature gage unit and attach conduit to unit by installing conduit housing with two screws.

(11) Install transmission transfer case breather (H, fig. 35).

(12) On M18 vehicles only, install center front sub-floor plate and secure it to blower housing and floor support with nine cap screws ($3/8$ in.—24 x 1 in.) and plain washers, and to blower belt guard with two cap screws ($3/8$ in.—24 x 1 in.) with plain and lock washers. Tighten screws 20-25 foot-pounds tension.

(13) On M39 vehicles only, place cover over inspection hole, install battery floor pad and left battery (par. 142 h).

(14) Adjust blower belts to correct tension (par. 122 a).

(15) Connect return springs to brake control levers, connect links to control levers (par. 116 c) and adjust brakes (par. 114).

(16) Install manual shift lever, bracket, and starter neutral safety switch (par. 109 h), and time safety switch (par. 149 c).

(17) Install left transmission shield and secure it with two lifting eye bolts, one output end cover cap screw ($1/2$ in.—13 x $1\,1/2$ in.) and one torque converter housing cap screw ($3/8$ in.—16 x $1\,1/8$ in.) with lock washers.

(18) Connect each final drive universal joint to its differential drive shaft with four cap screws ($1/2$ in.—20 x $1\,1/2$ in.) tightened to 75-100 foot-pounds tension. Install safety wire through heads of adjacent screws.

(19) Install both universal joint center guards and attach each one to inner and outer guards with eight bolts (9/16 in.—24 x 5/8 in.) provided with plain and lock washers (fig. 152). When installing left guard, anchor the speedometer shaft with one guard inner attaching bolt and one additional bolt. Connect speedometer shaft.

(20) Remove extension rails and install hull front door (par. 180 b).

c. Record of Replacement. Record replacement of transmission and differential assembly on W.D. A.G.O., Form No. 478, M.W.O. and Major Unit Assembly Replacement Record.

Section XXV
CONTROLLED DIFFERENTIAL ASSEMBLY

113. DESCRIPTION AND DATA.

a. Description. The differential assembly is attached to transmission to form a compact unit assembly (fig. 124). It is called a controlled differential because it not only functions as a differential to transmit power equally to both tracks under changing road conditions but also may be controlled for steering and braking by manually operated brakes within assembly. The differential assembly mounted inside carrier (housing) is driven from transmission output shaft through a bevel pinion gear which meshes with a bevel ring gear mounted on differential case cover. Steering and braking is accomplished by two sets of brake shoes which surround drums which are connected through gears to each differential drive (output) shaft. Each set of shoes may be applied to its drum by external brake hand levers. When right hand brake shoes are applied, right hand drive shaft is slowed down which will cause vehicle to turn to right; when left hand brake shoes are applied, vehicle will turn to left. When both sets of brake shoes are applied equally, vehicle will be slowed down or stopped without turning to either side. The differential assembly is completely enclosed, and is lubricated by cooled oil as described in paragraph 118 b.

b. Data.

Maximum steering ratio (one drive shaft to other)	1.61 to 1
Ring gear and pinion type	Spiral bevel
Number of teeth—ring gear	47
Number of teeth—pinion	15
Gear ratio	3.133 to 1
Number of steering brake drums	2
Drum diameter	18 in.
Drum width	4¼ in.
Number of steering brake shoes (3 per drum)	6
Brake shoe lining width	4 in.

Controlled Differential Assembly

Figure 131—Steering Brake Shoe Adjustment

Brake shoe lining thickness	¼ in.
Weight of differential assembly (dry)	1,165 lb
Weight of transmission and differential assembly (dry)	2,512 lb

114. STEERING BRAKE ADJUSTMENT.

a. **Setting of Brake Controls.** Push right hand (auxiliary) brake hand levers forward into latch springs (fig. 14). Push left hand (driver's) brake hand levers forward and apply locking levers to engage locking pawls in first notches in quadrants. The control levers on differential must be up against stop pins on differential carrier. If levers are not against stop pins, adjust links as described in paragraph 116 c.

b. **Adjust Brake Shoes** (fig. 131). With brake controls properly set (supar. a above), turn both brake adjusting shafts clockwise by hand until tight; then turn both shafts counterclockwise exactly one complete turn. If adjustment is correctly made, the brake shoes will be fully applied when the brake hand levers are slightly back of vertical position, with both levers of each set having equal travel.

115. STEERING BRAKE SHOE REPLACEMENT.

a. **Removal of Brake Shoes.** Three brake shoes (upper, center, and lower) are joined together around each brake drum. The procedure for removal of either set of shoes is as follows:

(1) Remove hull front door (par. 180 a).

(2) Remove 12 bolts and 6 nuts, with lock washers, which anchor cover to differential carrier. Raise cover slightly and remove tapered dowels which are located on six cover studs. Remove cover and gasket.

TM 9-755
115

Part Three—Maintenance Instructions

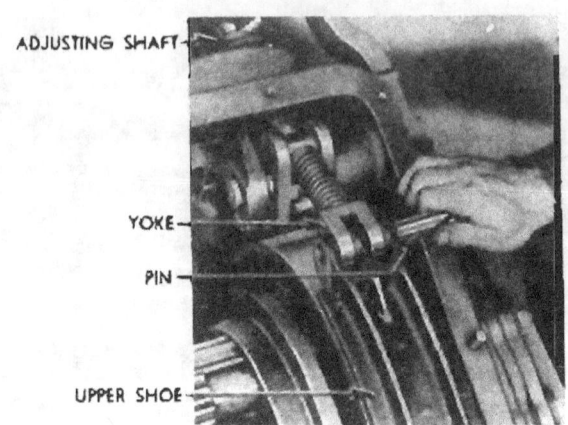

Figure 132—Disconnecting Upper Brake Shoe From Drum

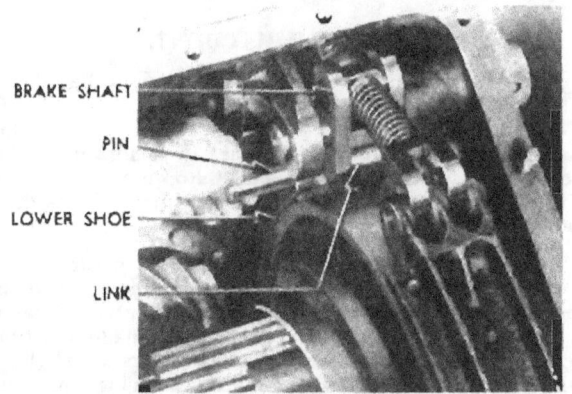

Figure 133—Disconnecting Lower Brake Shoe From Link

(3) Turn brake adjusting shaft counterclockwise as far as it will go to relieve all tension on brake shoes (fig. 131).

(4) Remove cotter pin from inner end of yoke to upper shoe pin and remove pin (fig. 132). Remove cotter pin from outer end of link to lower shoe pin and remove pin (fig. 133).

(5) Raise upper shoe from drum and lift it up (fig. 134) while

Controlled Differential Assembly

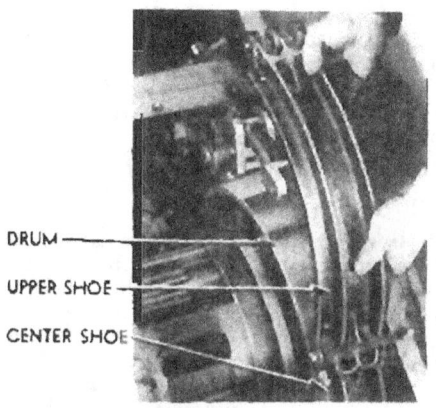

Figure 134—Removing Brake Shoes

working the other two shoes around underneath the drum until all are removed.

(6) Remove cotter pins from pins which join the shoes together and remove the brake shoe pins.

b. **Inspection of Brake Drums.** While shoes are removed, carefully examine brake drum for scores, cracks, glazed spots or other conditions which would warrant replacement. Should inspection indicate replacement necessary, notify higher authority.

c. **Installation of Brake Shoes.** The lower and center brake shoes are identical. The upper end of the upper shoe has an offset boss which is drilled for the yoke-to-shoe pin (¾ in. diameter).

(1) Connect the upper, center, and lower shoes together with shoe pins (⅝ in. x 4½ in.). Install cotter pins (³⁄₃₂ in. x 1 in.) in both ends of shoe pins.

(2) Pass a wire around under the drums and attach it to free end of lower shoe. Start lower shoe into differential carrier and under the drum; then pull the set of shoes into position around drum with the wire. Remove wire.

(3) Mate the upper end of lower shoe with the brake shaft link and install brake shoe pin (⅝ in. x 4½ in.) through shoe and link (fig. 133). Install cotter pins (³⁄₃₂ in. x 1 in.) in both ends of shoe pin.

(4) Mate the upper end of upper shoe with the adjusting rod to the yoke and install shoe pin (¾ in. x 3⅝ in.) (fig. 132). Install cotter pins (³⁄₁₆ in. x 1¼ in.) in both ends of shoe pin.

(5) Adjust brake shoes as described in paragraph 114.

(6) Clean off surfaces of differential carrier and cover and place a

TM 9-755
115-116

Part Three—Maintenance Instructions

new cover gasket on the carrier. Place cover over the gasket using care to prevent gasket getting out of place.

(7) Place the tapered, split dowels, small end down, over the six cover studs and install lock washers and nuts (½ in.—20) on the studs. Install 12 bolts (½ in.—13 x 1⅜ in.) and lock washers which secure cover to differential carrier. Tighten bolts and nuts uniformly to 54-64 foot-pounds tension.

(8) Install hull front door (par. 180 h).

116. STEERING BRAKE CONTROLS.

a. Description (fig. 14). The steering brakes are operated by duel controls which permit steering and braking by either driver or assistant driver as described in paragraph 14 f. Each set of steering brake shoes is actuated by a brake control lever keyed to a brake shaft which extends through side of the differential carrier. Each control lever is connected by an adjustable link to a short lever which is keyed to one of two brake cross shafts supported in brackets attached to hull above the differential. Hand levers mounted on both ends of each cross shaft provide means of rotating shaft and applying brake shoes controlled by that shaft. The hand levers actuate cross shaft through toothed collars keyed to shaft and so constructed that either lever may be used while other is latched forward in a non-operating position. The two hand levers on cross shaft that controls right brake shoes are located to right of each driver's seat; the two hand levers on shaft that controls left brakes are located to the left of each driver's seat. Left hand (driver's) set of hand levers contain locking pawls which are moved by locking levers into engagement with notched quadrants mounted on cross shaft brackets to lock brakes in applied position for parking (fig. 135). Right hand (auxiliary) set of hand levers cannot be locked for parking. Brakes are applied by pulling rearward on hand levers and are released by return springs connected to control levers. Each cross shaft has a universal joint in the middle to facilitate alinement and prevent binding. A cam secured to each cross shaft operates a stop light switch (par. 150 h (3)).

b. Adjustment of Brake Hand Lever Locking Pawl (fig. 135). Proper adjustment of brake hand lever locking pawl is very important, as improper adjustment may permit levers to drop down from non-operating position and lock brakes while vehicle is being steered from assistant driver's seat. When pawls are applied by moving locking levers to left, pawls must fully engage notches in quadrants and locking levers must bear against hand levers to lock pawls securely. When hand levers are then pulled rearward, pawls in moving out of notches must cause locking levers to swing to right to fully released position so that pawls cannot engage next notch in quadrant. If locking pawls are improperly adjusted, loosen lock nut and turn rod as required to secure adjustment described above, then tighten lock nut securely.

c. Adjustment of Links and Control Levers (fig. 14). When brake hand levers are forward in non-operating position and return

Controlled Differential Assembly

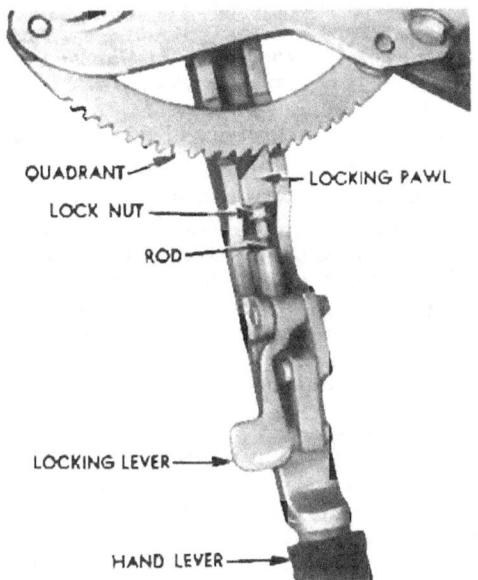

Figure 135—Brake Locking Pawl, Locking Lever, and Quadrant

springs are connected to brake control levers, control levers should be in contact with stop pins on each side of differential carrier. If either lever does not contact its stop pin, adjust as follows:

(1) Remove cotter pin and rod end pin which connects link to control lever. Return spring then should pull control lever up against stop pin.

(2) Loosen lock nut and turn lower clevis so that rod end pin will pass through clevis and control lever, with lever against stop pin. NOTE: *Lower clevis is offset and must be turned so that it engages control lever without binding.*

(3) Tighten lock nut, and install cotter pin (⅛ in. x 1 in.) in rod end pin.

d. **Removal of Brake Controls.** Hand levers or quadrants may be removed from cross shafts while in vehicle; however, for other replacements it is advisable to remove brake controls from vehicle as an assembly.

(1) Remove cotter pins and rod end pins which connect links to control levers on differential assembly. Unhook return springs from control levers and cross shaft brackets.

(2) Remove two ½-inch bolts and lock washers which attach

each cross shaft bracket to hull front plate, while shafts are supported to prevent falling. NOTE: *Shims may be installed between one or more brackets and hull to aline shafts and prevent binding. Mark or tag all shims so that they may be placed in original location upon installation of parts.*

(3) Work assembled controls out of driving compartment through a driver's door hatch, being careful not to damage door rubber seal.

e. **Disassembly of Brake Controls** (fig. 136).

(1) Remove the ½-inch cap screws and lock washers from hand lever collars on both ends of both cross shafts, drive collars from cross shafts. NOTE: *Screws engage grooves in cross shafts; therefore, screws must be removed before removing collars.*

(2) Remove hand levers, spring washers, and brackets from cross shafts.

(3) Loosen set screws and remove stop light switch operating cams from cross shafts.

(4) Remove cotter pins and rod end pins which attach the links to cross shaft levers.

(5) Remove the ½-inch bolts and lock washers from cross shaft levers and drive levers from shafts. NOTE: *Bolts engage grooves in cross shafts.*

(6) Remove cross shaft lever keys from cross shafts.

(7) If a quadrant is worn and requires replacement, remove the two ½-inch cap screws and lock washers which attach quadrant to bracket.

f. **Assembly of Brake Control Parts** (fig. 136).

(1) The left cross shaft assembly (G) is approximately one inch longer than the right cross shaft assembly (O). Place left shaft on bench with end having two rounded bracket bearing journals to the left; place right shaft on bench with the similar end to the right.

(2) Drive cross shaft lever keys (Q) centrally into keyways adjacent to universal joints in both shafts.

(3) The left cross shaft lever (E) is longer than the right cross shaft lever (M). Install left lever over left end of left cross shaft with small end pointing away from near edge of bench; install right lever over right end of right cross shaft with small end pointing away from near edge of bench.

(4) Aline keyways with keys and drive both levers into position so that clamp bolts will go through grooves in shafts. Install self-locking clamp bolts (½ in.—20 x 1½ in.) and tighten to 80-100 foot-pounds tension.

(5) The bore of left cam (D) is smaller than the bore of right cam (P). Install left cam over right end of left cross shaft with set screw up and cam pointing towards near edge of bench. Install right cam over left end of right cross shaft with set screw up and cam pointing towards near edge of bench. Both cams will be positioned and tightened on cross shafts after installation in vehicle.

Controlled Differential Assembly

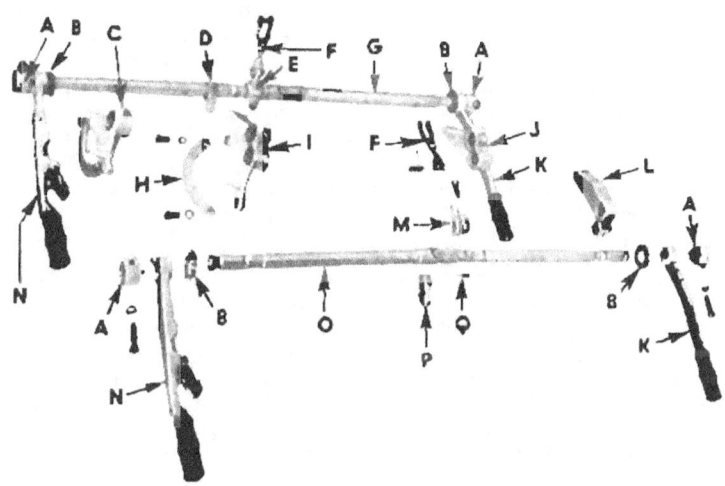

A—HAND LEVER COLLAR
B—SPRING WASHER
C—OUTER LEFT BRACKET
D—LEFT STOP LIGHT SWITCH OPERATING CAM
E—LEFT CROSS SHAFT LEVER
F—CONTROL LEVER LINK
G—LEFT CROSS SHAFT
H—HAND LEVER QUADRANT
I—CENTER LEFT BRACKET
J—CENTER RIGHT BRACKET
K—AUXILIARY BRAKE HAND LEVER
L—OUTER RIGHT BRACKET
M—RIGHT CROSS SHAFT LEVER
N—BRAKE HAND LEVER
O—RIGHT CROSS SHAFT
P—RIGHT STOP LIGHT SWITCH OPERATING CAM
Q—LEVER TO CROSS SHAFT KEY

RA PD 340372

Figure 136—Brake Control Parts—Disassembled

(6) Install quadrants (H) on left cross shaft brackets (C) and (I) with notches slanting away from flanged ends of brackets and secure each quadrant with two cap screws (½ in.—20 x 1 in.) and lock washers tightened to 80-100 foot-pounds tension.

(7) Install center left bracket (I) and center right bracket (J) over their respective ends of both cross shafts, with bolt flanges to the right.

(8) Install outer left bracket (C) and outer right bracket (L) over outer ends of left and right cross shafts with bolt flanges to the right.

(9) Install spring washers (B) over each end of both cross shafts with pronged sides outward.

(10) Install driver's hand levers (N) on left ends of both cross

shafts with notches in hubs outward. Install auxiliary hand levers (K) on right ends of both cross shafts with notches in hubs outward.

(11) Install hand lever collars (A) on each end of both cross shafts with notches inward. NOTE: *One double width serration in collar and on shaft permits installation of collar on shaft in one position only; also, a mark on end of shaft alines with slot in collar.*

(12) Drive collars on shafts until clamp screw holes aline with grooves in shafts then install cap screws (½ in.—20 x 2 in.) with lock washers and tighten to 80-100 foot-pounds tension.

g. Installation of Brake Controls.

(1) Work assembled controls into driving compartment through a driver's door hatch with driver's hand levers (N) on left side of vehicle, being careful not to damage door rubber seal.

(2) Lift assembly into place and attach each cross shaft bracket to tapping blocks on hull front plate with two bolts (½ in.—20 x 1¼ in.) and lock washers tightened to 80-100 foot-pounds tension. NOTE: *If shims were found on removal on controls (subpar. d (2) note above) install shims in original position to assure proper alinement of cross shafts.*

(3) Attach return springs to cotter pin on each brake control lever and to drilled boss on center cross shaft bracket.

(4) Adjust and connect links as described in subparagraph c above.

(5) Place right cam over hub of universal joint with end of cam centered on right stop light switch lever. Place left cam so it is centered on left stop light switch lever. Adjust cams as described in paragraph 154 c (2).

117. DIFFERENTIAL REPLACEMENT.

a. The transmission and differential must be replaced as a unit assembly and not separately. When joining these separate assemblies together special gages are required to obtain proper adjustment of ring gear with pinion gear. Refer to paragraph 111 for information covering authority for replacement of transmission and differential assembly, replacement procedure, and record of replacement.

Section XXVI

TRANSMISSION, DIFFERENTIAL, AND TRANSFER CASE LUBRICATION SYSTEM

118. DESCRIPTION AND DATA.

a. Transmission Lubrication System (fig. 137). Two oil pumps incorporated in transmission assembly supply oil under pressure to lubricate all moving parts, operate hydraulic system which controls gear changes, and maintain proper oil level in torque converter. The

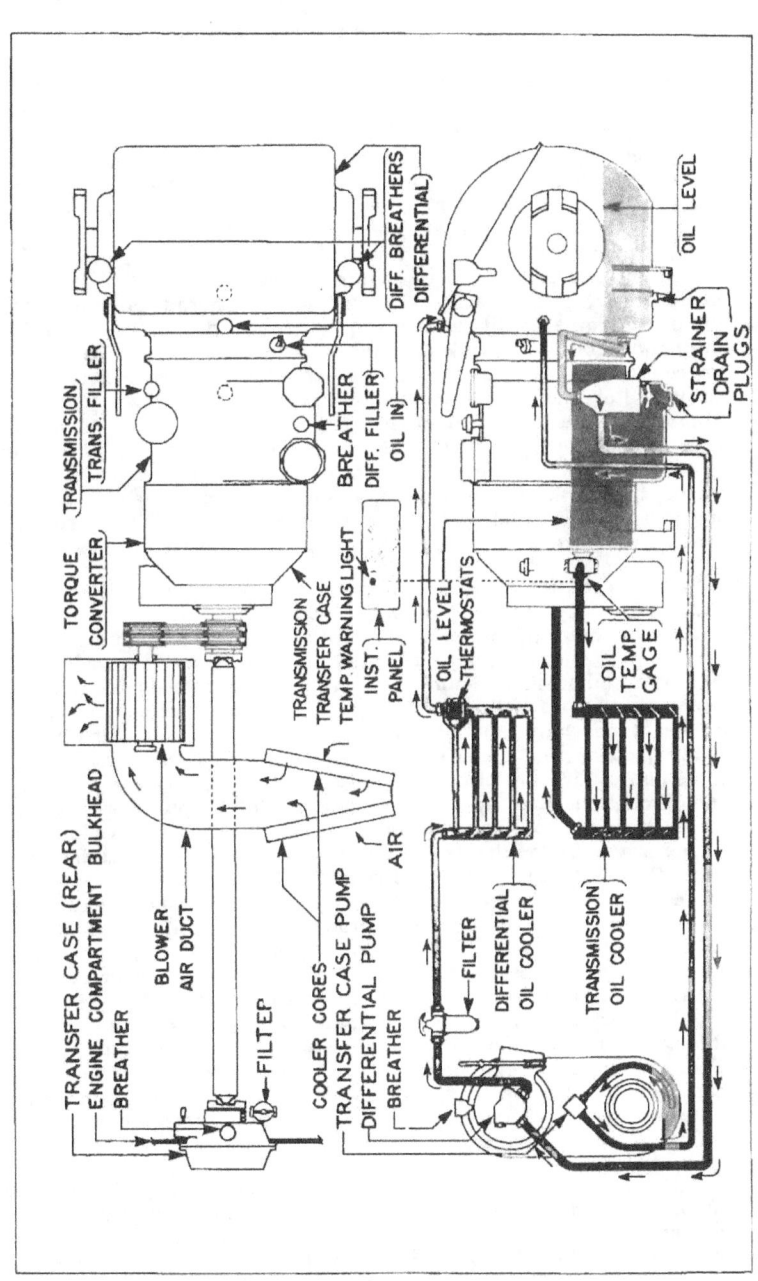

Figure. 137—Transmission, Differential, and Transfer Case Lubrication System

Part Three—Maintenance Instructions

front pump, mounted in transfer case and driven by converter front rotor shaft, supplies oil pressure whenever engine is running with rear transfer case clutch engaged. The rear pump, mounted in output end cover and driven by transmission output shaft, supplies oil pressure only when vehicle is in forward motion. In normal operation rear pump supplements oil delivery from front pump; however, in downhill operation with vehicle "over-running" engine, or when vehicle is being towed, rear pump alone supplies oil pressure for lubrication and control. Both pumps draw oil from sump in transmission case through a screen (J, fig. 35). The output pressure from pumps is regulated by valves to 15 pounds per square inch for lubrication, 100-110 pounds per square inch for control and 75 pounds per square inch entering the torque converter near its center. The turbine rotor increases converter oil pressure to approximately 110 pounds per square inch at outer edge of turbine housing where oil is led through pipes to the external cooler core and returned to center of turbine housing. Whenever converter oil temperature reaches high limit of 280°-290° F, a temperature sending unit located in converter oil outlet elbow causes a warning light on instrument panel to burn red.

b. **Differential and Transfer Case Lubrication System** (fig. 137 and 138). Differential and transfer case lubrication system consists of following units: a differential oil cooler oil pump and a transfer case oil pump, both mounted on and driven by transfer case (rear); an oil strainer mounted on right side of transmission; an oil filter mounted on bulkhead to right of transfer case; an oil cooler located under hull subfloor (M18); or front seat and battery box (M39); connecting pipes and fittings. The larger portion of oil is contained in differential carrier. Oil is drawn from bottom of carrier and through strainer by differential oil cooler oil pump which then forces it through oil filter and oil cooler back into top of differential carrier where oil discharges at point of contact between differential ring and pinion gears. When oil is cold, a thermostatically controlled valve in cooler permits most of the oil to by-pass cooler core while a small amount of oil circulates through core. As oil warms up, the valve starts closing, until, at operating temperature of approximately 150° F, valve is closed and all oil is forced through cooler core. A bleeder hole in differential oil pump shaft permits oil to feed into a drilled passage in transfer case input shaft, from which it circulates through oil passages to lubricate gears and bearings in transfer case. The flow of oil is regulated by an oil check rod located in drilled passage in input shaft. Surplus oil which accumulates in lower end of transfer case is withdrawn and returned to differential carrier by transfer case oil pump.

c. **Transmission and Differential Oil Cooler and Blower** (fig. 138). A transmission and differential oil cooler, consisting of a V-shaped air inlet duct and two cooler cores, is located under hull subfloor beneath the auxiliary generator (M18) or under front seat and battery box (M39). The larger torque converter oil cooler core

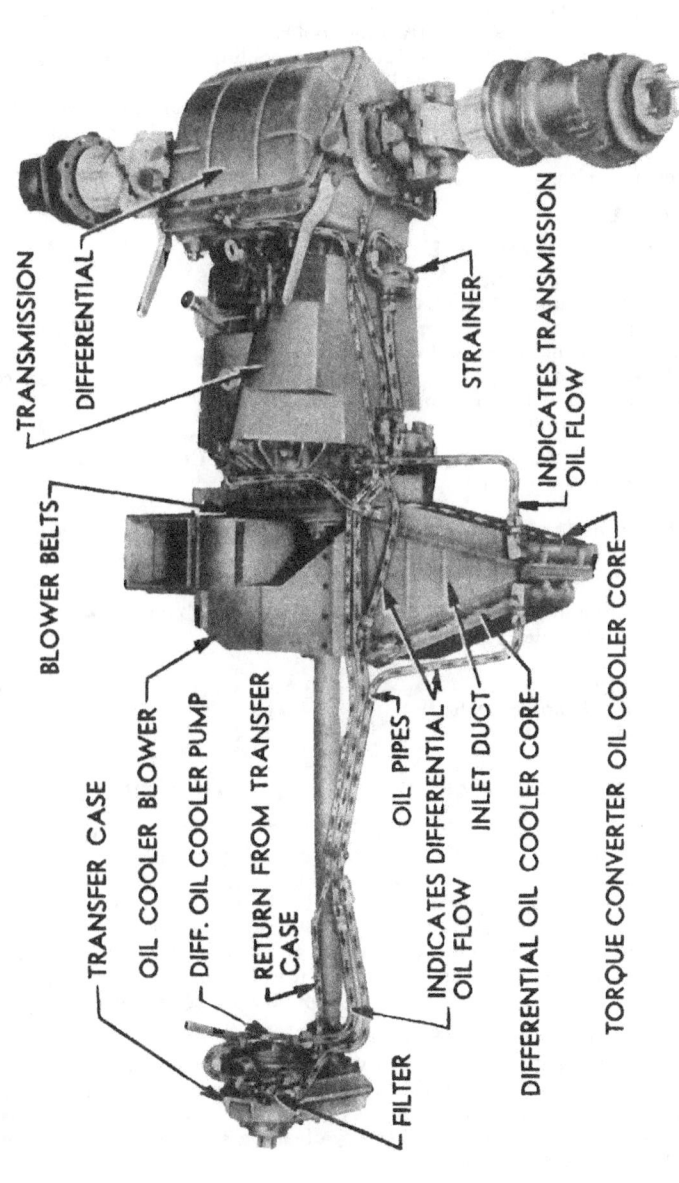

Figure 138—Lubrication System Components

is mounted on front side of inlet duct; the smaller differential oil cooler core is mounted on rear side of inlet duct. A removable cover on top surface of inlet duct permits inspection and cleaning of cores and provides access to attaching bolts. A sirocco-type blower connected to left end of inlet duct draws cooling air through cores and discharges it through a grille covered opening in hull roof. The blower is driven by two V-belts from pulleys on the transmission transfer case input shaft. A belt tightener (fig. 144) located at rear of left driving compartment is provided to maintain proper belt tension.

d. Data.

(1) TRANSMISSION LUBRICATION SYSTEM.

Number of oil pumps	2
Pressure in lubrication passages	15 psi
Pressure in control system	100-110 psi
Pressure in converter and oil cooler	75 psi
Oil cooler type	Air cooled core
Oil capacity—transmission and cooler	48 qt
Maximum allowable oil temperature	280-290° F

(2) DIFFERENTIAL AND TRANSFER CASE LUBRICATION SYSTEM.

Number of oil pumps	2
Strainer type	Removable screen
Filter type	Purolator
Oil cooler type	Air cooled core
Capacity—differential, transfer case and cooler	20 qt

(3) TRANSMISSION AND DIFFERENTIAL OIL COOLER AND BLOWER.

Number of cooler cores	2
Cooler core type	Plate
Blower type	Sirocco
Number of blower belts	2
Belt adjustment	Yes

119. DIFFERENTIAL OIL COOLER OIL PUMP.

a. Description. The differential oil cooler oil pump, which is mounted on and driven by the transfer case (fig. 118), is an internal-external rotor type having a maximum capacity of 18 gallons per minute at 2,400 revolutions per minute. Its function is to circulate oil through cooler system and supply oil to both differential and transfer case. The pump contains a spring-loaded relief valve which opens to bypass oil from outlet side to inlet side of pump whenever pressure on outlet side exceeds 30 pounds per square inch.

b. Relief Valve and Spring. If a large particle of dirt should enter pump and lodge on relief valve seat, oil will be bypassed rather than circulated through the system. If this occurs, remove valve cover, gasket, spring and valve as shown in figure 139. Clean parts and valve seat thoroughly with dry-cleaning solvent, install parts and tighten valve cover securely.

Transmission, Differential, and Transfer Case Lubrication System

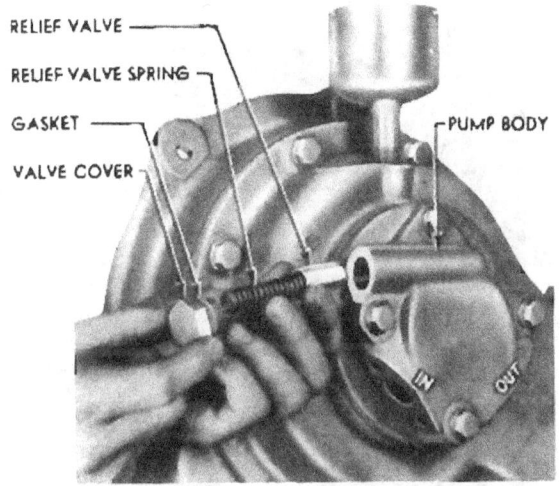

Figure 139—Oil Pump Relief Valve

c. **Oil Pump Removal** (fig. 140).

(1) On M39 vehicle only, open rear seat center cover and remove rear seat back (par. 185 c).

(2) Disconnect inlet and outlet flanges which are attached to pump with 5/16-inch cap screws and lock washers.

(3) Remove four 3/8-inch cap screws and lock washers which attach oil pump to transfer case. Pull pump straight out, then remove gasket and oil check rod which is located in drilled passage in transfer case input shaft.

d. **Oil Pump Installation** (fig. 140).

(1) Insert the oil check rod into drilled passage in transfer case input shaft.

(2) Fill inlet pipe with SAE 50 engine oil to eliminate air pocket and inject engine oil into pump inlet port while turning shaft to form oil seal around rotors and assure good suction when pump is started.

(3) Place oil pump in position with a new gasket (vellumoid 0.033 in. thick). Anchor pump to transfer case with one cap screw (3/8 in.—16 x 2 in.) and three cap screws (3/8 in.—16 x 3 1/4 in.) all provided with lock washers and tightened uniformly to 20-25 foot-pounds tension.

(4) Be sure that joint surfaces of flanges and pump are clean. Coat new gaskets (fibre, 1/18 in. thick) with gasket cement and insert them between both flanges and the pump.

TM 9-755
119

Part Three—Maintenance Instructions

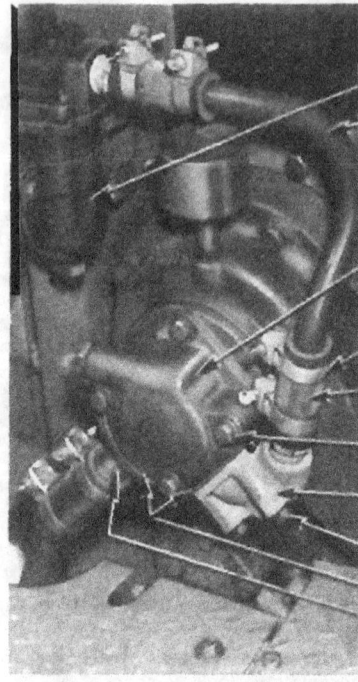

Figure 140—Differential Oil Cooler Oil Pump Connections

(5) Attach inlet flange to pump with two cap screws ($\frac{3}{16}$ in.—18 x $\frac{3}{4}$ in.), and lock washers. Attach outlet flange to pump with one cap screw ($\frac{3}{16}$ in.—18 x 2$\frac{1}{4}$ in.) and one cap screw ($\frac{3}{16}$ in.—18 x 2$\frac{3}{4}$ in.), both with lock washers. Tighten all flange attaching screws to 18-22 foot-pounds tension.

(6) On M39 vehicle only, install rear seat back with pad attached (par. 185 d), and close rear seat center cover.

c. **Priming the Oil Pump.** An air leak at any joint or connection between differential and inlet side of oil pump, or air entering oil system when oil pipes are disconnected for any reason, will cause oil pump to lose its prime and fail to pump oil. This condition can be detected after a few minutes of vehicle operation by placing a hand on oil return pipe at top of differential. If this pipe is not warm, it indicates that oil pump is not functioning and differential and transfer case are not being lubricated. Check all joints and connections for signs of oil leakage and make necessary corrections; if no leaks are

TM 9-755
119-120

Transmission, Differential, and Transfer Case Lubrication System

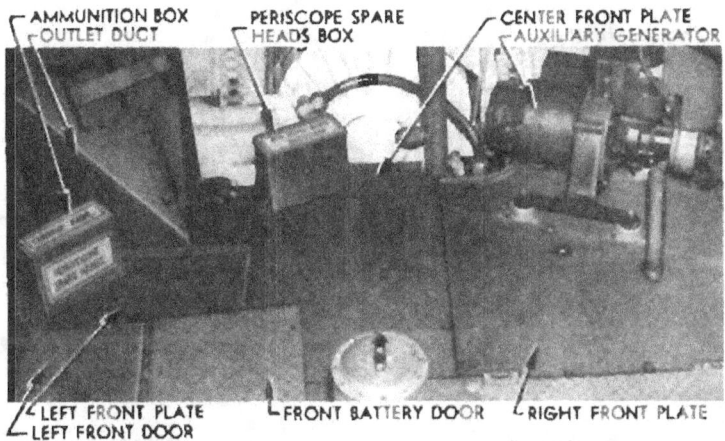

Figure 141—Hull Subfloor Plates and Auxiliary Generator—M18

found, prime oil pump. To prime pump, disconnect inlet flange (subpar. c (2), above), fill inlet pipe with prescribed differential lubricant and inject lubricant into pump. Connect inlet flange (subpar. d (5), above) and check operation as previously described.

120. TRANSFER CASE OIL PUMP.

a. Description. The transfer case oil pump, which is mounted on and driven by transfer case (fig. 118) is an internal-external rotor type. Its sole function is to withdraw surplus oil from bottom of transfer case and return it to differential. The inlet side of pump is connected by an oil suction pipe to a flange at lower end of transfer case. The outlet side of pump is connected by piping to differential carrier.

b. Oil Pump Removal (fig. 118).

(1) On M18 vehicle, remove right rear subfloor plate (par. 184 c). On M39 vehicle, open rear seat center cover and remove rear seat back with pad attached (par. 185 c).

(2) Loosen hose clamps and remove coupling hoses from inlet and outlet fittings on pump. NOTE: *In M18 vehicles having serial numbers below 1351, the suction pipe and return pipe are connected to pump by Sealflex fittings which are disconnected by unscrewing the compression nuts (fig. 148).*

(3) Remove the four ⅜-inch cap screws and lock washers which attach pump to transfer case and pull the pump straight out. Remove oil pump gasket.

TM 9-755
120-121

Part Three—Maintenance Instructions

c. Oil Pump Installation (fig. 118).

(1) Inject SAE 50 engine oil into pump while turning shaft to form oil seal around rotors. Clean joint surfaces of oil pump and transfer case.

(2) Place pump in position with a new gasket (vellumoid, 0.033 in. thick) and attach pump to transfer case with four cap screws (3/8 in.—16 x 1 1/4 in.) and lock washers tightened to 20-25 foot-pounds tension.

(3) Install coupling hoses over inlet and outlet fittings and tighten hose clamps securely. NOTE: *Refer to paragraph 125 b for information on proper tightening of Sealflex fittings which were used in M18 vehicles having serial numbers below 1351.*

(4) On M18 vehicle only, install right rear subfloor plate and secure it with eight cap screws (3/8 in.—24 x 1 in.) and plain washers tightened to 28-33 foot-pounds tension. On M39 vehicle only, install rear seat back, with pad attached (par. 185 d), and close rear seat center cover.

121. TORQUE CONVERTER AND DIFFERENTIAL OIL COOLER CORES.

a. Torque Converter Oil Cooler Core Removal, M18 Vehicle (fig. 142).

(1) Remove auxiliary generator (par. 144 h).

(2) Remove center front subfloor plate (fig. 141) which is anchored by eleven 3/8-inch cap screws and plain washers.

(3) Loosen clamp screws and hose clamp on heater air tube tee clamp under propeller shaft. Push lower generator heater air tube to right as far as it will go.

(4) Remove right front subfloor plate which is anchored by ten 3/8-inch cap screws and plain washers. Remove lower generator heater air tube from upper tube on floor plate.

(5) Remove drain plug in hull floor under torque converter (third plug from front) remove 1/4-inch pipe plug from converter housing and drain oil from housing; then install both plugs and tighten securely.

(6) Unscrew compression nuts on oil pipe fittings at both ends of cooler core, loosen compression nuts at other ends of both oil pipes and carefully move oil pipes out of the way.

(7) Remove service cover and gasket which is attached to top of inlet duct by 10 round head machine screws with plain and lock washers.

(8) From inside inlet duct, remove seven self-tapping cap screws with plain and lock washers which attach lower edge of core to inlet duct. NOTE: *In M18 vehicles having serial numbers below 1368, the lower edge was attached with machine screws and nuts like the upper edge.*

TM 9-755
121

Transmission, Differential, and Transfer Case Lubrication System

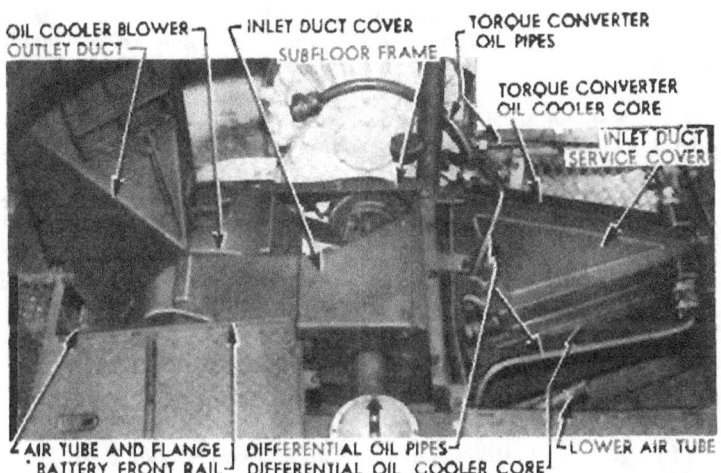

Figure 142—Oil Cooler, Blower, and Oil Pipe Connections—M18

(9) Remove seven round head machine screws with plain and lock washers and nuts which attach upper edge of core to inlet duct.

(10) Remove eight $\frac{5}{16}$-inch cap screws with plain and lock washers which extend through the brackets and attach core to inlet duct at each end.

(11) Lift cooler core out of hull and drain oil.

b. **Torque Converter Oil Cooler Core Removal, M39 Vehicle** (fig. 143).

(1) Remove batteries (par. 142 g) and battery box (par. 188 a), and open front seat cover.

(2) Remove drain plug in hull floor under torque converter (third plug from front), remove ¼-inch pipe plug from converter housing and drain oil; then install and tighten both plugs securely.

(3) Disconnect oil pipes at fittings on torque converter by unscrewing compression nut next to fitting.

(4) Remove core by performing steps (7) through (11) of subparagraph a above.

(5) Unscrew fittings from core with oil pipes attached.

c. **Differential Oil Cooler Core Removal, M18 Vehicle** (fig. 142).

(1) Remove auxiliary generator, center front and right front subfloor plates, and lower generator heater air tube (steps (1) through (4), subpar. a above).

TM 9-755
121

Part Three—Maintenance Instructions

(2) Loosen hose clamp and remove oil pipe coupling hoses from fittings at both ends of cooler core. NOTE: *In M18 vehicles having serial numbers below 1351, the oil pipes are connected by Soalflex fittings which are disconnected by unscrewing the compression nut* (fig. 148).

(3) Remove service cover and gasket which is attached to top of inlet duct by 10 round-head machine screws with plain and lock washers.

(4) Remove 14 round-head machine screws with plain and lock washers and nuts which attach upper and lower edges of cooler core to inlet duct.

(5) Remove six $\frac{5}{16}$-inch cap screws with plain and lock washers which extend through bracket and attach core to inlet duct at each end.

(6) Lift cooler core out of hull and drain oil.

d. **Differential Oil Cooler Core Removal, M39 Vehicle** (fig. 143).

(1) Remove batteries (par. 142 g) and battery box (par. 188 a) and open front seat cover.

(2) Remove core by performing steps (2) through (6) in subparagraph c above.

e. **Torque Converter Oil Cooler Core Installation, M18 Vehicle** (fig. 142).

(1) Attach cooler core to inlet duct with eight cap screws ($\frac{5}{16}$ in.—24 x 1 in.) having plain and lock washers, through brackets at each end.

(2) Secure upper edge of core to duct with seven round-head machine screws (No. 12—24 x $\frac{5}{8}$ in.) having plain and lock washers.

(3) Secure lower edge of core to duct with seven self-tapping cap screws (No. 12—24 x $\frac{5}{8}$ in.) having plain and lock washers, installed from inside inlet duct. NOTE: *In M18 vehicles having serial numbers below 1368, lower edge is secured with seven machine screws, washers, and nuts like upper edge.*

(4) Install service cover and gasket and secure them to inlet duct with 10 round-head machine screws (No. 12—24 x $\frac{5}{8}$ in.) having plain and lock washers.

(5) Connect oil pipes to core and adjust tension of compression nuts as described in paragraph 125 b.

(6) Push end of lower generator heater air tube, which does not have hose clamp, over lower end of upper tube on right front subfloor plate. As floor plate is lowered into position, guide other end of lower air tube over outlet of tee clamp under propeller shaft. Tighten hose clamp and clamp screws.

(7) Attach right front subfloor plate with ten cap screws ($\frac{3}{8}$ in. —24 x 1 in.) and plain washers tightened to 20-25 foot-pounds tension.

TM 9-755
121

Transmission, Differential, and Transfer Case Lubrication System

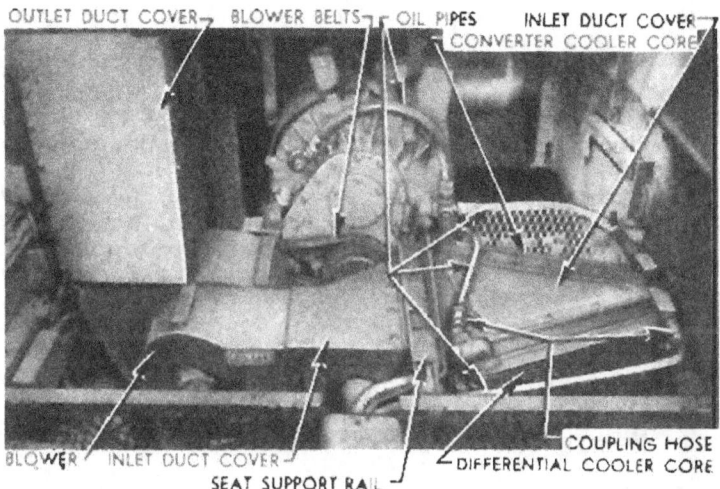

Figure 143—Oil Cooler, Blower, and Oil Pipe Connections—M39

(8) Install center front subfloor plate and secure it to blower housing and floor support with nine cap screws (⅜ in.—24 x 1 in.) and plain washers, and to blower belt guard with two cap screws (⅜ in.—24 x 1 in.) with plain and lock washers. Tighten screws to 20-25 foot-pounds tension.

(9) Install auxiliary generator (par. 144 i).

(10) Fill transmission to proper level with oil.

f. **Torque Converter Oil Cooler Core Installation, M39 Vehicle** (fig. 143).

(1) Coat threads of oil pipe fittings with thread and joint compound and install oil pipes on core, placing pipe with elbows at left end of core.

(2) Install core on inlet duct by performing steps (1) through (4) in subparagraph e above.

(3) Connect oil pipes to fittings in torque converter.

(4) Install battery box (par. 188 h) and batteries (par. 142 h). Close front seat cover.

(5) Check and fill transmission (par. 38).

g. **Differential Oil Cooler Core Installation, M18 Vehicle** (fig. 142).

(1) Attach cooler core bracket to inlet duct with six cap screws (⁵⁄₁₆ in.—24 x 1 in.) having plain and lock washers which extend through brackets at each end.

305

TM 9-755
121-122

Part Three—Maintenance Instructions

(2) Secure upper and lower edges of core to inlet duct with 14 round-head machine screws (No. 12—24 x ⅝ in.) having plain washers, lock washers and nuts.

(3) Install service cover and gasket and secure them to inlet duct with 10 round-head machine screws (No. 12—24 x ⅝ in.) having plain and lock washers.

(4) Inspect oil pipe coupling hoses and replace them if not in good condition.

(5) Connect oil pipes to fittings on cooler core by means of coupling hoses and tighten hose clamps securely. NOTE: *In M18 vehicles having serial numbers below 1351, the oil pipes are connected by Sealflex fittings. Connect these and adjust compression nuts as described in paragraph 125 b.*

(6) Install lower generator air tube, right front and center front subfloor plates and generator (steps (6) through (9), subpar. c above).

(7) Check and fill differential.

h. **Differential Oil Cooler Core Installation, M39 Vehicle** (fig. 143).

(1) Install core on inlet duct by performing steps (1) through (5) of subparagraph g above.

(2) Install battery box (par. 188 b) and batteries (par. 142 h). Close front seat cover.

(3) Check and fill differential.

122. TRANSMISSION AND DIFFERENTIAL OIL COOLER BLOWER BELTS AND PULLEYS.

a. **Blower Belts Adjustment.** The two V-belts which drive the blower are adjusted to proper tension by the belt tightener located at rear of driving compartment to left of transmission. The belts are correctly adjusted when they may be depressed ½ inch by finger pressure applied to top sides midway between pulleys (fig. 144). To adjust tension, loosen clamp screw on belt tightener mounting bracket, move lever away from transmission to tighten belts or toward transmission to loosen belts, until correct tension is obtained; then tighten clamp screw securely.

b. **Blower Belt Removal.**

(1) On M18 vehicle, remove center front subfloor plate (fig. 260). On M39 vehicle, remove left battery (par. 142 g), battery floor pad, and inspection hole cover under pad.

(2) Disengage transfer case clutch (par. 102) so that propeller shaft can be turned by hand.

(3) Loosen belts (subpar. a above) and remove belt guard.

(4) Remove blower belt driven pulley, and shims between pulley and hub, which are attached to hub with four ⅜-inch cap screws and lock washers. Do not lose shims.

TM 9-755
122

Transmission, Differential, and Transfer Case Lubrication System

Figure 144—Blower Belt Tightener and Correct Belt Adjustment

(5) Flatten bent up tongues of lock plates and remove four ⅜-inch cap screws which attach propeller shaft to front universal joint bearings and remove bearings from spider (fig. 122).

(6) Remove belts by working them through between end of propeller shaft and universal joint spider.

c. **Blower Belt Installation.**

(1) Work new belts (G163-0118253) through between end of propeller shaft and universal joint spider.

(2) Place bearings on universal joint spider and attach propeller shaft to each bearing with one lock plate and two cap screws (⅜ in.—24 x ²⁷⁄₃₂ in.) tightened to 28-33 foot-pounds tension. Bend tongues of lock plates up against flats on screw heads.

(3) Place belts in grooves of drive and driven pulleys and install blower driven pulley on hub, placing original number of shims between pulley and hub, and attach pulley with four cap screws (⅜ in.—24 x 1¼ in.) and lock washers.

(4) Adjust belts to proper tension (subpar. *a* above) and install belt guard.

(5) Engage transfer case clutch (par. 102).

(6) On M18 vehicle only, install center front subfloor plate and secure it to blower housing and floor support with nine cap screws (⅜ in.—24 x 1 in.) and plain washers, and to blower belt guard with two cap screws (⅜ in.—24 x 1 in.) with plain and lock washers. Tighten all cap screws to 20-25 foot-pounds tension.

(7) On M39 vehicle only, install inspection hole cover and battery floor pad. Install battery (par. 142 h).

d. **Blower Drive or Driven Pulley Replacement.** On M18 vehicle, removal of drive or driven pulley is covered in paragraph 106

307

TM 9-755
122-123

Part Three—Maintenance Instructions

c and installation is covered in paragraph 107 c. On M39 vehicle, removal of pulleys is covered in paragraph 106 d and installation is covered in paragraph 107 d.

123. TRANSMISSION AND DIFFERENTIAL OIL COOLER BLOWER.

a. **Welding Blower Fan to Shaft, M18** (fig. 145). On M18 vehicles below serial No. 293 the blower fan was secured to shaft by set screws. In some cases fan became loose on its shaft, permitting fan to move and strike housing; this condition often recurred after set screws had been securely tightened. Beginning with vehicle serial No. 293, this condition was corrected in production by assembling fan on shaft with a press fit. All M18 vehicles bearing serial numbers lower than 293 will be changed according to the following procedure:

(1) Remove center front subfloor plate (fig. 260) by removing eleven $\frac{3}{8}$-inch cap screws.

(2) Remove oil cooler inlet duct cover which is secured to inlet duct with fourteen $\frac{3}{16}$-inch cap screws having plain and lock washers, and to blower housing with six $\frac{3}{16}$-inch self-tapping cap screws having plain and lock washers.

(3) Check clearance between fan rim and inlet cone with a steel scale or other suitable tool. This clearance should be $\frac{1}{4}$ inch, plus or minus $\frac{1}{16}$ inch. Loosen set screws and tap fan along shaft to secure this clearance, if necessary; then tighten set screws securely.

(4) Disengage clutch in rear transfer case (par. 102) to permit turning propeller shaft by hand.

(5) Arc weld fan hub to shaft, working through blower inlet. Weld completely around the shaft, turning propeller shaft as required to turn blower shaft. The completed weld should have a $\frac{1}{8}$-inch to $\frac{3}{16}$-inch continuous fillet. CAUTION: *Under no circumstance should this operation be done with an acetylene torch as high temperatures would distort the shaft.*

(6) Carefully inspect fan and housing and remove any tools, welding rod ends, or other foreign material; then install inlet duct cover and secure it to inlet duct with 14 cap screws ($\frac{3}{16}$ in.—24 x $\frac{3}{4}$ in.) having plain and lock washers, and to blower housing with 6 self-tapping cap screws ($\frac{3}{16}$ in.—24 x $\frac{5}{8}$ in.) having plain and lock washers.

(7) Paint a white "X" on inlet duct cover to show that fan has been welded to shaft.

(8) Install center front subfloor plate and secure it to blower housing and floor support with nine cap screws ($\frac{3}{8}$ in.—24 x 1 in.) and plain washers and to blower belt guard with two cap screws ($\frac{3}{8}$ in.—24 x 1 in.) with plain and lock washers. Tighten all cap screws to 20-25 foot-pounds tension.

(9) Engage transfer case clutch (par. 102).

b. **Blower Removal, M18 Vehicle** (fig. 142).

(1) Remove left front and center front subfloor plates (fig. 260).

TM 9-755
123

Transmission, Differential, and Transfer Case Lubrication System

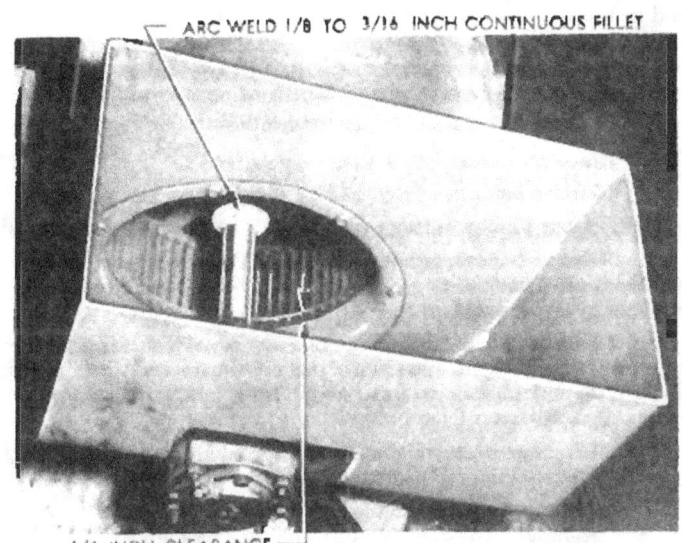

Figure 145—Fan Hub Welded to Shaft

(2) Remove front sub-floor frame which is anchored at rear end by two 3/8-inch cap screws with lock washers and nuts, to bracket on roof support by one 3/8-inch cap screw and lock washer, and to right front subfloor plate by two cap screws and plain washers.

(3) Remove subfloor battery front rail which is attached to floor frame at front end of battery box by four 3/8-inch cap screws with lock washers and nuts.

(4) Remove blower to battery heater air tube and flange which is attached to blower by two 3/8-inch cap screws with plain and lock washers and nuts.

(5) Remove outlet duct and door assembly which is attached by nineteen 3/16-inch cap screws with plain and lock washers.

(6) Loosen belts (par. 122 a) and remove belt guard.

(7) Remove blower belt driven pulley, and shims between pulley and hub, which are attached to hub with four 3/8-inch cap screws and lock washers. Do not lose shims.

(8) Remove inlet duct cover which is secured to inlet duct by fourteen 3/16-inch cap screws having plain and lock washers, and to blower housing by six 3/16-inch self-tapping cap screws having plain and lock washers.

TM 9-755
123
Part Three—Maintenance Instructions

(9) Remove four 5/16-inch self-tapping cap screws having plain and lock washers which secure inlet air duct to blower housing.

(10) Remove two 7/16-inch bolts and lock washers at rear and right front mounting support brackets, and two 3/8-inch bolts having plain washers and lock washers at left front mounting bracket.

(11) Lift blower assembly out of hull.

c. **Blower Removal, M39 Vehicle** (fig. 143).

(1) Remove batteries (par. 142 g) and battery box (par. 188 a).

(2) Loosen blower belts (par. 122 a) and remove belt guard.

(3) Remove blower driven pulley, and shims between pulley and hub, which are attached to hub with four 3/8-inch cap screws and lock washers. Do not lose shims.

(4) Remove outlet duct and door assembly which is anchored by eighteen 5/16-inch cap screws with plain washers and lock washers, and by four 3/8-inch cap screws having nuts, lock washers and plain washers. Pull bottom of duct rearward to remove it.

(5) Disconnect blower housing from inlet duct by removing ten 5/16-inch self-tapping cap screws having plain washers and lock washers.

(6) Remove two 7/16-inch bolts and lock washers at rear and right front support brackets, and remove two 3/8-inch bolts with plain washers and lock washers at left front mounting bracket.

(7) Move blower to left, raise outlet end until housing is separated from inlet duct, and remove blower from vehicle.

d. **Oil Cooler Blower Installation, M18 Vehicle.**

(1) Lower blower into position in hull and anchor it by two bolts (7/16 in.—20 x 1 in.) and lock washers at right front and rear mounting support brackets, and by two bolts (3/8 in.—24 x 3/4 in.) with plain washers and lock washers at left front mounting brackets.

(2) Attach inlet air duct to blower housing by four self-tapping cap screws (5/16 in.—24 x 5/8 in.) with plain washers and lock washers.

(3) Attach inlet air duct to blower housing by four self-tapping cap screws (5/16 in.—24 x 5/8 in.) with plain washers and lock washers.

(4) Install inlet duct cover and attach it to inlet duct by 14 cap screws (5/16 in.—24 x 3/4 in.) with plain and lock washers, and to blower housing by six self-tapping cap screws (5/16 in.—24 x 5/8 in.) with plain and lock washers.

(5) Install and secure outlet duct and door assembly by eight cap screws (5/16 in.—24 x 3/4 in.) on front edge, six cap screws (5/16 in.—24 x 1 1/2 in.) on rear edge, and five cap screws (5/16 in.—24 x 1/2 in.) on top edge, using plain and lock washers on all screws.

(6) Place blower belts in grooves of drive and driven pulleys and install driven pulley on hub, placing original shims between pulley

Transmission, Differential, and Transfer Case Lubrication System

and hub, and attach pulley with four cap screws ($3/8$ in.—24 x 1 in.) and lock washers. Adjust belts to proper tension (par. 122 a).

(7) Rotate propeller shaft by hand to determine whether blower pulleys are in alinement. If belts rub on one side of grooves in pulleys, and propeller shaft is hard to turn, pulleys are not in alinement. Remove driven pulley and add, or remove, shims as required to secure alinement.

(8) Install blower belt guard with one cap screw ($3/8$ in.—24 x 1 in.) and lock washer.

(9) Attach cooler to battery air tube and flange to blower housing by two cap screws ($3/8$ in.—24 x 1 in.) with plain washers, lock washers, and nuts.

(10) Attach subfloor battery front rail to floor frame at front end of battery box by four cap screws ($3/8$ in.—24 x 1 in.) with lock washers and nuts.

(11) Install and secure left front subfloor plate by six cap screws ($3/8$ in.—24 x 1 in.) and plain washers. Attach carbine ammunition box over floor plate by two cap screws ($5/16$ in.—24 x $5/8$ in.) and lock washers.

(12) Install and secure front subfloor frame to subfloor frame at rear end by two cap screws ($3/8$ in.—24 x 1 in.) with lock washers and nuts, to bracket on roof support by one cap screw ($3/8$ in.—24 x 1 in.) and lock washer, and to right front subfloor plate by two cap screws ($3/8$ in.—24 x 1 in.) and plain washers.

(13) Install center front subfloor plate and secure it to blower housing and floor support with nine cap screws ($3/8$ in.—24 x 1 in.) and plain washers and to blower belt guard with two cap screws ($3/8$ in.—24 x 1 in.) with plain and lock washers. Tighten all cap screws to 20-25 foot-pounds tension.

c. **Blower Installation, M39 Vehicle** (fig. 143).

(1) Examine the felt seal which extends from under outlet end of inlet duct to make sure it is in good condition and securely cemented to hull floor plate.

(2) Lower blower into place, entering blower housing into inlet duct, using care not to damage felt seal.

(3) Attach blower housing to inlet duct with 10 self-tapping cap screws ($5/16$ in.—24. x $5/8$ in.) having plain washers and lock washers. Leave screws loose.

(4) Install bolt ($7/16$ in.—20 x 1 in.) and lock washer at rear and right front mounting brackets, and install two bolts ($7/16$ in.— 20 x 1 in.) with plain washers and lock washers at left front mounting bracket. Leave bolts loose.

(5) Place outlet duct and door assembly in position and attach

TM 9-755
123

Part Three—Maintenance Instructions

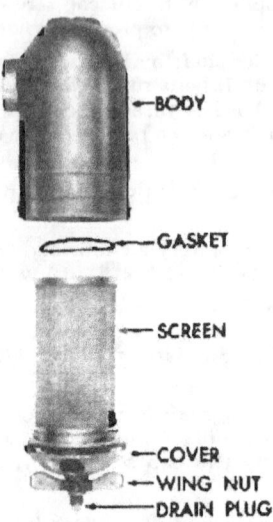

Figure 146—Differential Oil Cooler Oil Strainer—Disassembled

top and left rear side with 15 cap screws ($\frac{7}{16}$ in.—24 x $\frac{7}{8}$ in.), having plain washers and lock washers; attach front side with four cap screws ($\frac{3}{8}$ in.—24 x $\frac{7}{8}$ in.) having lock washers and nuts, and three cap screws ($\frac{7}{16}$ in.—24 x $\frac{5}{8}$ in.) having plain washers and lock washers.

(6) Tighten the $\frac{7}{16}$-inch mounting bolts to 55-65 foot-pounds tension. Tighten the $\frac{3}{8}$-inch mounting bolts to 28-30 foot-pounds tension. Tighten all $\frac{7}{16}$-inch cap screws to 10-12 foot-pounds tension, and all other $\frac{3}{8}$-inch cap screws to 20-25 foot-pounds tension.

(7) Place blower belts in grooves of drive and driven pulleys and install driven pulley on hub, placing original number of shims between pulley and hub, and attach pulley with four cap screws ($\frac{3}{8}$ in.—24 x 1$\frac{1}{4}$ in.) and lock washers. Adjust belts to proper tension (par. 122 a).

(8) Rotate propeller shaft by hand to determine whether blower drive and driven pulleys are in alinement. If belts rub on one side of grooves in pulley, and propeller shaft is hard to turn, pulleys are not in alinement. Remove driven pulley and add, or remove, shims as required to secure alinement.

(9) Attach belt guard to front plate with one cap screw ($\frac{3}{8}$ in.—24 x $\frac{3}{4}$ in.), lock washers and nut tightened to 20-25 foot-pounds tension.

(10) Install battery box (par. 188 b) and batteries (par. 142 h).

Transmission, Differential, and Transfer Case Lubrication System

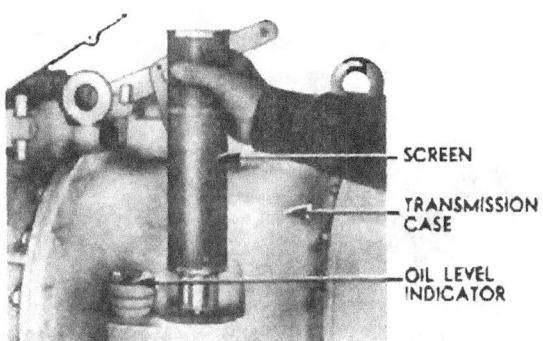

Figure 147—Removing Transmission Oil Screen

124. STRAINER, FILTER, AND SCREEN.

a. **Cleaning Differential Oil Cooler Oil Strainer** (fig. 146). The differential oil cooler oil strainer is mounted on right side of transmission so that the screen can be removed for cleaning without removing strainer assembly. Remove 1/8-inch pipe plug and drain oil from strainer. Cut lock wire and unscrew wing nut, then turn bottom cover one-third turn and remove screen from strainer body. Wash screen in dry-cleaning solvent and blow it out with an air stream. When screen is installed, make sure that gasket is in good condition and that bottom cover is turned one-third turn to fully engage strainer body; then tighten wing nut securely and anchor it with lock wire.

b. **Cleaning Differential Oil Cooler Oil Filter** (fig. 118). The oil cooler oil filter mounted on the bulkhead to right of the transfer case is the same as used on the engine. It is serviced as described in paragraph 80.

c. **Cleaning Transmission Oil Screen** (fig. 147). The transmission oil screen is located under a cover on left side of transmission core to rear of the oil level indicator. Remove left shield which is attached to transmission by the two lifting eye bolts and a cap screw at front and rear ends. Remove cover which is retained by four cap screws (9/16 in.—18 x 7/8 in.) and lock washers, and lift screen out of transmission case. Wash screen in dry-cleaning solvent and blow it out with an air stream. When screen is installed, make sure cover gasket is in good condition and tighten cover cap screws to 18-22 foot-pounds tension. Install left shield.

125. OIL PIPES AND FITTINGS.

a. **Description.** The oil pipes are made of steel tubing formed to shape. All pipes except those connected to torque converter (trans-

Part Three—Maintenance Instructions

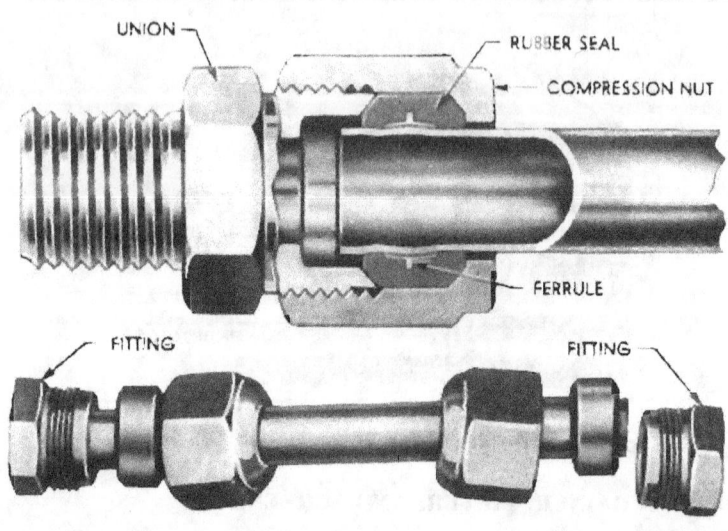

Figure 148—Sealflex Oil Pipe Connections

mission) are joined together and to fittings on the various assemblies by coupling hoses secured with hose clamps (fig. 142). The torque converter oil pipes are provided with Sealflex connections at both ends. These connections (fig. 148) consist of a synthetic rubber seal fitted over a metal ferrule swaged to pipe, and a compression nut which compresses seal against pipe, ferrule, and union or other fitting. This type of connection provides a flexible but oiltight coupling. In early production M18 vehicles, Sealflex connections are used where coupling hoses are located on later vehicles. At some points in early production installations, slip-joint type Sealflex connections are used also. In these connections ferrules are not swaged on pipe but are free to slide, thus providing for expansion and contraction of oil pipes. At slip joints pipes are marked with white paint which does not show beyond compression nuts when joints are properly assembled.

b. **Connecting and Tightening Sealflex Fitting.** (fig. 148). Inspect fittings and seal to make sure they are clean and seal is in good condition. Insert end of oil pipe squarely into union or other fitting and screw compression nut on fitting until seal is firmly compressed, using a standard open end wrench of proper size. Then loosen compression nut two flats or one-third turn; the connection will then have a firm "feel" but will have some flexibility. If the connection does not feel firm, seal is damaged or it is improperly assembled on the ferrule.

c. **Inspection and Replacement of Oil Pipes.** If inspection of oil pipes reveals that they are contacting and chafing against other parts, they should be repositioned to provide proper clearance or should be insulated to prevent damage. All supporting clips and brackets must be in place and securely mounted. Crushed or restricted oil pipes should be replaced. When oil pipes are installed make certain that they are clean internally, and avoid bending to facilitate installation. Operate vehicle and test for oil leaks at all connections before installing covering parts.

Section XXVII

FINAL DRIVE

126. DESCRIPTION AND DATA.

a. Description (fig. 149). The two final drive assemblies function both as a means of transmitting power from differential to tracks and as track compensating devices. Power from differential is transmitted to each final drive assembly through universal joints (fig. 152) which connect differential drive shafts to final drive pinion shafts. The final drive pinion shaft turns a pinion which meshes with a gear splined to wheel spindle upon which final drive sprockets and hub are mounted. The shaft, pinion, gear, and spindle are supported by ball bearings in a wheel carrier which is supported in bearings in a wheel carrier support bolted to hull side plate. Compensating links connect wheel carriers to front track wheel support arms and cause the carriers and drive sprockets to swing forward and backward on supports as track wheels rise and fall, thus compensating for varying track tension as vehicle moves over uneven ground.

b. Data.

Number of final drive universal joints 2
Number of spider and trunnion assemblies per joint 2
Number of final drive assemblies per vehicle 2
Number of teeth on pinion 17
Number of teeth on gear 37
Reduction ratio in final drive assembly .... 2.175 to 1
Number of sprocket hubs per vehicle 2
Number of drive sprockets on each hub 2

127. FINAL DRIVE SPROCKETS AND HUB.

a. Removal. Since one sprocket is bolted against inner side of hub flange it cannot be removed with hub on wheel spindle; therefore, sprockets and hub must be removed from vehicle as an assembly.

(1) Remove front end guard and raise side guards (par. 131).

(2) Disconnect track by removing track link pin in front of No. 1 (front) track wheel (par. 132 d). Release parking brakes and roll track back on support rollers by turning sprocket with crow bar.

TM 9-755
127

Part Three—Maintenance Instructions

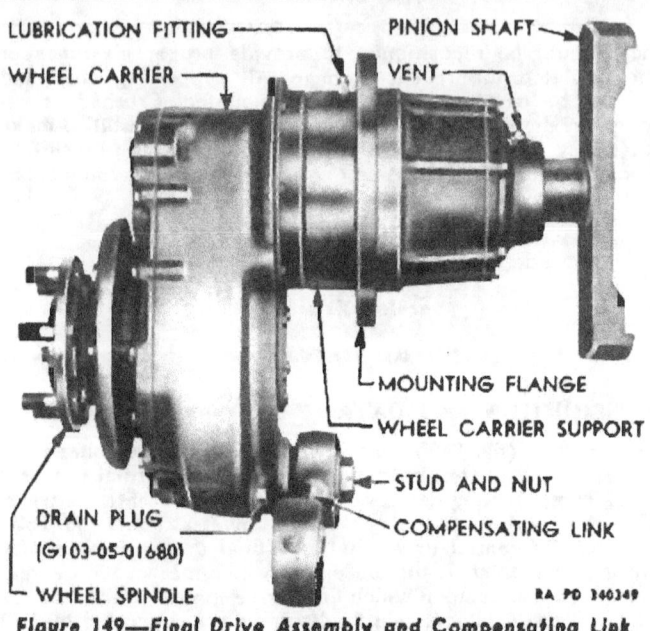

Figure 149—Final Drive Assembly and Compensating Link

Figure 150—Removing Sprockets and Hub

TM 9-755
127-128

Final Drive

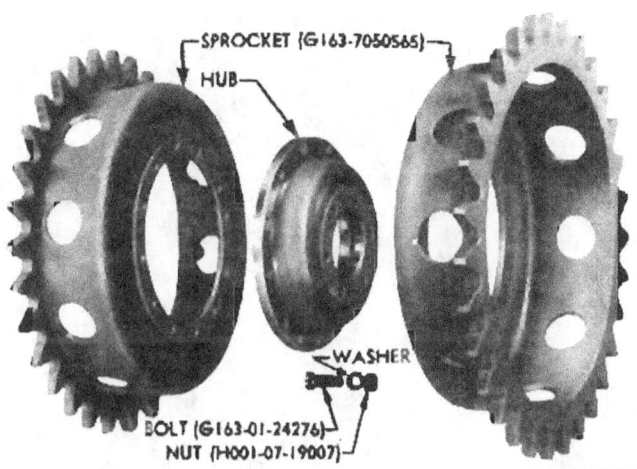

Figure 151—Final Drive Sprockets and Hub—Disassembled

(3) Remove six ¾-inch nuts and lock washers which anchor sprocket hub to final drive wheel spindle. Attach chain hoist to outer sprocket and use crow bar to keep sprockets from tilting as the assembly is removed from wheel spindle (fig. 150) and lowered to ground.

(4) Remove twelve ⅝-inch bolts, nuts, and lock washers which attach sprockets to hub (fig. 151).

b. Installation.

(1) Place sprocket hub between flanges of final drive sprockets and install 12 bolts (⅝ in.—18 x 2⁷⁄₁₆ in.) from concave side of hub (fig. 151). Install lock washers and nuts on bolts and tighten to 130-150 foot-pounds tension.

(2) Attach chain hoist to sprocket on convex side of hub and lift hub and sprocket assembly into place on wheel spindle, using a crow bar to keep sprockets from tilting (fig. 150). Install lock washers and nuts (¾ in.—16) on spindle bolts and tighten nuts to 200-230 foot-pounds tension.

(3) Roll track over sprockets and take up slack by turning sprockets with crow bar. Connect track (par. 132 f) and adjust tension (par. 132 h). Install front end guard, lower side guards, and secure all guards (par. 131).

128. FINAL DRIVE UNIVERSAL JOINTS.

a. Description (fig. 153). Each final drive universal joint consists of two spider and trunnion assemblies anchored to one coupling

TM 9-755
128

Part Three—Maintenance Instructions

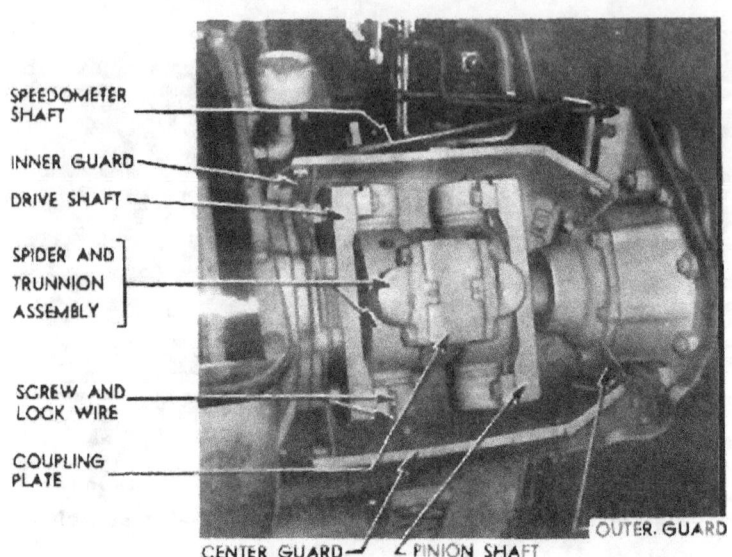

Figure 152—Final Drive Universal Joint and Guards

plate by eight cap screws. The universal joints are covered by guards in driving compartment.

b. **Removal.**

(1) Remove hull front door (par. 180 a).

(2) Remove eight $\frac{5}{16}$-inch bolts with plain and lock washers which secure the universal joint center guard to inner and outer guards (fig. 152).

(3) Cut lock wires, remove four $\frac{1}{2}$-inch cap screws which attach universal joint to final drive pinion shaft, and push shaft outward as far as it will go.

(4) Support universal joint to prevent falling while removing the four $\frac{1}{2}$-inch cap screws which attach universal joint to differential drive shaft, and remove universal joint.

(5) Cut lock wires, remove four $\frac{1}{2}$-inch cap screws which attach each spider and trunnion assembly to the coupling plate and tap assemblies from plate.

c. **Installation.**

(1) Inspect mating surfaces of coupling plate and universal joint to make sure they are clean and free of burs.

(2) Place each spider and trunnion assembly on coupling plate

Final Drive

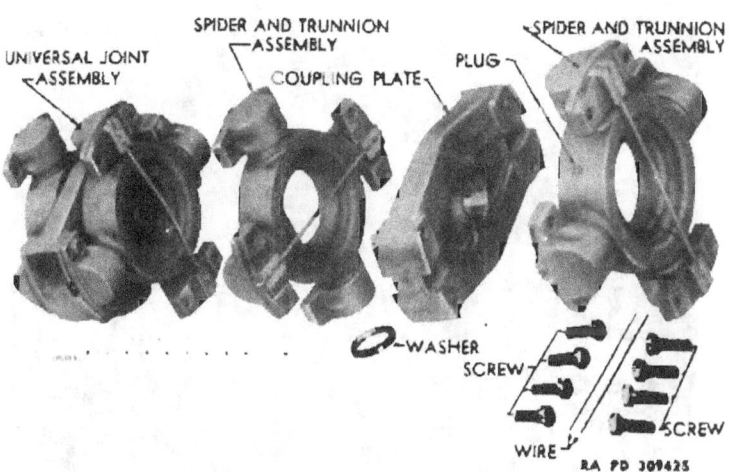

Figure 153—Final Drive Universal Joint—Disassembled

and secure it with four cap screws (½ in.—20 x 1½ in.) tightened to 75-100 foot-pounds tension. Anchor each pair of cap screws with a lock wire.

(3) Attach universal joint to differential drive shaft with four cap screws (½ in.—20 x 1½ in.) tightened to 75-100 foot-pounds tension.

(4) Attach universal joint to final drive pinion shaft with four cap screws (½ in.—20 x 1½ in.) tightened to 75-100 foot-pounds tension.

(5) Anchor each pair of attaching screws with a lock wire.

(6) Install universal joint center guard and attach it to inner and outer guards with eight bolts (⁹⁄₁₆ in.—24 x ⅝ in.) provided with plain and lock washers. When installing left guard, anchor the speedometer shaft with one guard inner attaching bolt and one additional bolt.

(7) Lubricate universal joint.

(8) Install hull front door (par. 180 b).

129. FINAL DRIVE ASSEMBLY.

a. **Removal.** The following procedure covers removal of either left or right final drive assembly.

(1) Disconnect track and remove sprockets and hub (par. 127 a). Disconnect compensating link at wheel support arm by removing the 1⅛-inch lock bolt and lock washer.

(2) Remove hull front door and final drive universal joint (par. 128 b). Remove outer end guard which is secured by four bearing retainer bolts on final drive assembly.

TM 9-755
129

Part Three—Maintenance Instructions

Figure 154—Removing Final Drive Assembly, Using Fixtures 41-F-2994-3 and 41-F-2994-8

(3) Remove four ⅜-inch upper dirt shedder-to-carrier cap screws and lock washers and attach final drive lifting fixture (41-F-2994-3). Install vehicle crow bar in lifting fixture (41-F-2994-8), attach chain hoist to eye of fixture, and insert hooked end in fixture (41-F-2994-3) on final drive assembly (fig. 154). Adjust chain hoist so that final drive can be supported by a man holding a crow bar.

(4) Remove eight ¾-inch bolts, nuts and lock washers which anchor final drive wheel carrier support to hull side plate.

(5) While one man holds crow bar to support final drive assembly, install four bolts (¾ in.—10) having 2 inches of thread in tapped holes in flange of carrier support; then tighten bolts evenly to force support out of hull side plate. NOTE: *Early production M18 carrier supports which were not tapped for bolts may have to be forced out with a jack inside the hull.* Lift final drive assembly away from hull, leaving pinion shaft attached to universal.

(6) Remove 1¼-inch safety nut from stud and tap compensating link from stud. Install nut on stud to protect threads.

(7) Remove lifting fixture (41-F-2994-3) and install cap screws and lock washers.

Final Drive

(8) Cut lock wires and remove four ½-inch cap screws which attach pinion shaft to universal joint and remove shaft.

b. Installation (fig. 154).

(1) Remove pinion shaft from final drive assembly and attach it to universal joint with four cap screws (½ in.—20 x 1½ in.) tightened to 75-100 foot-pounds tension. Anchor each pair of screws with lock wire.

(2) Install compensating link and safety nut (1¼ in.—12) on stud, tighten smaller outer stud nut to 225-275 foot-pounds tension and larger inner nut to 250-300 foot-pounds tension.

(3) Remove four upper dirt shedder-to-carrier cap screws (⅜ in. —16 x ¾ in.) and lock washers and attach final drive lifting fixture (41-F-2994-3).

(4) Lightly coat surface of hull side plate around hole with joint sealing compound (51-C-1616).

(5) Install vehicle crow bar in lifting fixture (41-F-2994-8), attach chain hoist to eye of fixture, insert hooked end of fixture in lifting fixture (41-F-2994 3) on final drive assembly and lift assembly into position. With one man holding final drive assembly in position another man must insert pinion shaft into support and guide support into hole in hull side plate, using care to engage splines on shaft with splines in pinion inside the assembly.

(6) Aline bolt holes with a punch and install eight bolts (¾ in.— 16 x 3 in.) through support and hull plate from the outside. Install lock washers and nuts on these bolts inside hull and tighten nuts to 200-230 foot-pounds tension.

(7) Remove lifting fixture (41-F-2994-3) from final drive and install shedder-to-carrier cap screws and lock washers tightened to 15-20 foot-pounds tension.

(8) Remove four pinion shaft bearing retainer bolts (⁷⁄₁₆ in.— 24 x 1½ in.) and lock washers, install universal joint outer end guard in alinement with inner end guard and tighten retainer bolts to 56-64 foot-pounds tension.

(9) Install universal joint, center guard, and hull front door (par. 128 e).

(10) Connect compensating link to wheel support arm with link bolt (1⅜ in.—16 x 3 in.) and lock washer, tightened to 200-250 foot-pounds tension.

(11) Lubricate final drive assembly.

(12) Install sprockets and hub and connect track (par. 127 b).

c. Record of Replacement. Record replacement of final drive assembly on W.D. A.G.O., Form No. 478, M.W.O. and Major Unit Assembly Replacement Record.

Part Three—Maintenance Instructions

Section XXVIII

TRACKS AND SUSPENSION

130. DESCRIPTION AND DATA.

a. Description (fig. 155).

(1) TRACKS. Two individually driven steel tracks, 12 inches wide, provide the necessary traction to propel the vehicle. Each complete track is composed of 83 separate steel track links with center guides, connected together with straight link pins which are anchored to one link and carried in rubber bushings in the adjoining link (fig. 162). The links are of interlocking design to eliminate vibration and wear that occurs when track wheels pass over opening between track links. Two drive sprockets, one on each side, pull the tracks forward over the support rollers and lay them down in the path of the advancing track wheels. Hinged track guards cover the upper portion of the tracks at front and rear ends of the vehicle (fig. 156).

(2) TRACK WHEELS, SUPPORT ARMS, AND TORSION BARS (fig. 155). Ten dual, rubber-tired track wheels, five on each side, are carried on individual support arms attached to independent torsion bars which function as springs. Each track wheel consists of a hub and two wheel disks bolted together (fig. 170). The support arms are splined and welded to tubular axle shafts which are mounted in roller bearings in housings bolted to the lower corner of hull. The axle shafts are splined to solid steel torsion bars, which extend across the hull through protective housings to be anchored in splined retainers located in the opposite axle shaft housings (fig. 179). The torsion bars for opposite wheels are staggered to permit location of all axle shafts at the same level.

(3) SHOCK ABSORBERS AND SPRING BUMPERS (fig. 155). Double-acting, heavy-duty truck type, hydraulic shock absorbers mounted on hull side plates are connected by steel links to each support arm except No. 3 right and left (fig. 185). These shock absorbers control the rate of movement upward and downward of the track wheels when traveling over rough ground. Volute spring bumpers mounted on hull side plates are provided to limit the upward travel of all support arms under extremely rough ground conditions.

(4) COMPENSATING WHEELS AND COMPENSATING LINKS (fig. 155). An adjustable dual compensating wheel supports each track at rear end of vehicle and provides a means of adjusting track tension. Each compensating wheel, consisting of a hub and two disk wheels bolted together, is carried on a support which is mounted on a support spindle bolted to hull side plate (fig. 169). An eye bolt attached to wheel support and connected by an adjusting nut to a bracket bolted to support spindle is provided to move compensating wheel and support rearward to take up slack in track. Compensating links connect final drive carriers to front track wheel support arms and cause carriers and drive sprockets to swing foward and backward as front track

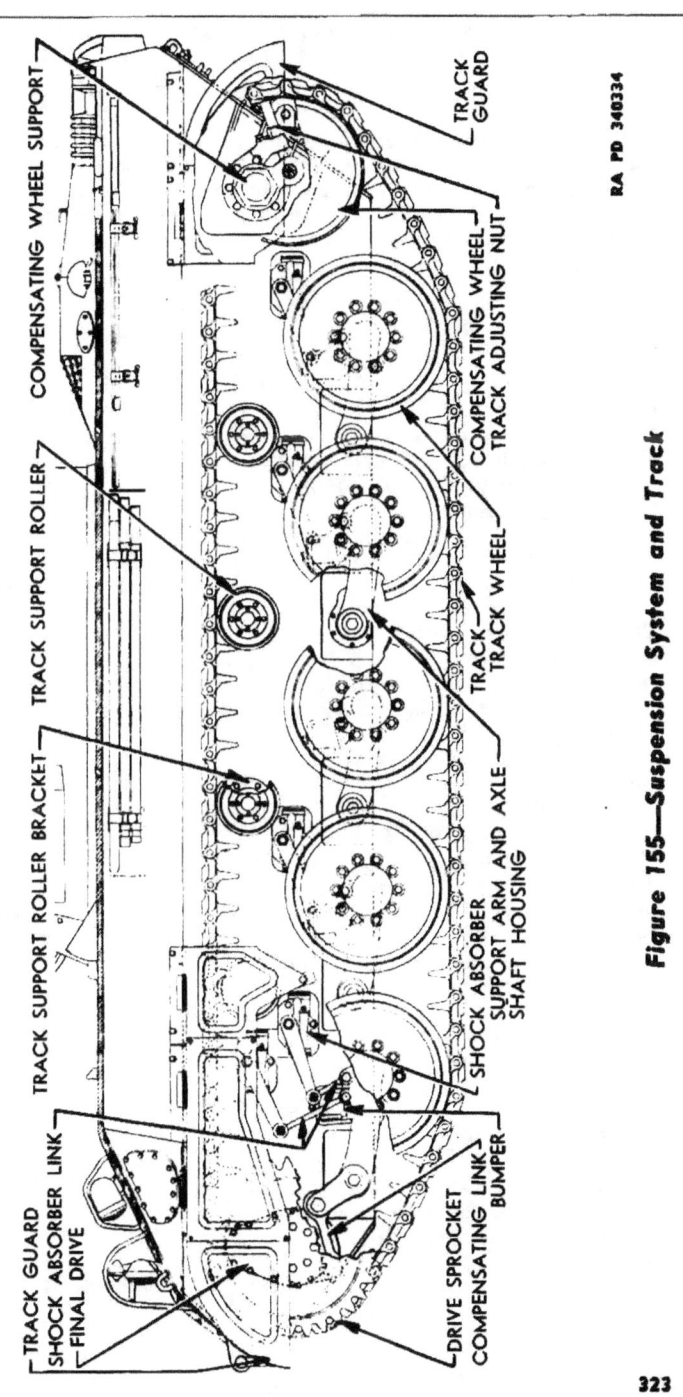

Figure 155—Suspension System and Track

TM 9-755
130

Part Three—Maintenance Instructions

wheels rise and fall, thus compensating for varying track tension as vehicle moves over uneven ground.

(5) TRACK SUPPORT ROLLERS (fig. 168). Four dual, rubber-tired track support rollers are carried on spindles of support brackets which are bolted to hull side plates. Each roller consists of a hub and two roller disks bolted together. These support rollers support the upper sides of the tracks between the compensating wheels and the drive sprockets.

b. **Data.**

(1) TRACKS.
Number of tracks ... 2
Width of track ... 12 in.
Length of track ... Approximately 37 ft
Number of links per track ... 83
Type of track link ... Steel, with center guide and rubber bushings

(2) TRACK WHEELS AND HUBS.
Number per vehicle ... 10
Type ... Dual, demountable
Disks per wheel (or hub) ... 2
Disk type ... Steel, rubber-tired
Outside diameter of tire (new) ... 26 in.
Thickness of tire (new) ... 1¼ in.
Bearings per wheel hub ... 2
Bearing type ... Tapered roller

(3) SUPPORT ARMS AND AXLE SHAFT HOUSINGS.
Number per vehicle ... 10
Bearings per arm or axle shaft ... 2
Bearing type ... Roller
Type of springing ... Torsion bar

(4) TORSION BARS.
Number per vehicle ... 10
Type ... Solid steel
Length ... 73⅛ in.
Interchangeability ... No, see fig. 177

(5) SHOCK ABSORBERS AND LINKS.
Number per vehicle ... 10
Absorber type ... Hydraulic, double-acting
Absorber fluid capacity ... 496 to 516
Link type ... Self alining
Link bushing material ... Self-lubricating fabric

(6) SUPPORT ARM SPRING BUMPERS.
Number per vehicle ... 10
Type ... Volute spring

(7) COMPENSATING WHEELS AND HUBS.
Number of wheel hubs per vehicle ... 2
Number of wheels per hub ... 2
Wheel type ... Steel disk
Wheel bearings and type ... Same as track wheels

324

Figure 156—Track Guards Raised

(8) COMPENSATING LINKS.
Number per vehicle	2
Type	Solid steel
How connected	Through self-alining roller bearings

(9) SUPPORT ROLLERS AND BRACKETS.
Number per vehicle	10
Type	Dual, demountable
Disks per hub (or bracket)	2
Disk type	Steel, rubber-tired
Diameter of tire (new)	10 in.
Thickness of tire (new)	½ in.
Number of bearings per hub	2
Bearing type	Tapered roller

131. TRACK GUARDS.

a. **Description (fig. 156).** The track guards are attached to sponsons and are hinged so that they may be raised when necessary, or they may be removed from vehicle individually by removing attaching screws. The front and rear end guards, and front and rear side guards may be raised without disturbing adjoining guards; however, the front side guards must be raised to permit raising the idler guards.

b. **Front End Guards.** Each front end guard is attached to hull on inner edge by two cap screws (⅜ in.—24 x ⅞ in.) having plain and lock washers, to upper hinge section on outer edge by one cap screw (⅜ in.—24 x 1 in.) having plain and external toothed lock washers, and to side guard at lower edge by one cap screw (⅜ in.—24 x ⅝ in.) having plain and external toothed lock washers. These screws must be removed in order to raise guard. The upper section of each guard is attached to hull by one cap screw (⅜ in.—24 x ⅞ in.) having plain and lock washers, and to sponson extension plate by

TM 9-755
131

Part Three—Maintenance Instructions

---

five cap screws (3/8 in.—24 x 7/8 in.) having a plain washer under screw head on top and a plain washer and safety nut under extension plate. These screws must be removed, in addition to other attaching screws, in order to remove guard from vehicle.

c. Front Side Guards. Each front side guard is anchored to upper or hinge plate section by three cap screws (3/8 in.—24 x 1 in.) having plain and external-toothed lock washers, and to adjoining guards at lower edge by two cap screws (3/8 in.—24 x 5/8 in.) having plain and external-toothed lock washers. These screws must be removed in order to raise the guard. The upper section is attached to sponson by two cap screws (1/2 in.—20 x 1 in.) having plain and lock washers, and to sponson extension plate by two cap screws (1/2 in.—20 x 1 1/4 in.) having plain washers under heads on top and plain washers and safety nuts under extension plate. These screws must be removed, in addition to other attaching screws, in order to remove guard from vehicle.

d. Front Idler Guard. Each front idler guard is attached to upper or hinge plate section by one cap screw (3/8 in.—24 x 1 in.) having plain and external-toothed lock washers, and to brace mounted on hull by one cap screw (5/8 in.—18 x 2 in.) having a plain washer under head on outside and safety nut on inside. These screws must be removed, and side guard raised (subpar. b, above), in order to raise this guard. The upper section is attached to sponson by two cap screws (1/2 in.—20 x 1 in.) having plain and lock washers, which must be removed in addition to other attaching screws in order to remove guard from vehicle.

e. Rear Side Guard. Each rear side guard is anchored to the upper or hinge plate section by two cap screws (3/8 in.—24 x 1 in.) and to rear end guard by one cap screw (3/8 in.—24 x 5/8 in.) all having plain and external-toothed lock washers, and to brace mounted on hull by one cap screw (5/8 in.—18 x 2 in.) having a plain washer under head on outside and safety nut on inside. These screws must be removed in order to raise the guard. The upper section is attached to sponson by two cap screws (1/2 in.—20 x 1 in.) having plain and lock washers which must be removed, in addition to attaching screws, in order to remove guard from vehicle.

f. Rear End Guard. Each rear end guard is anchored to the upper or hinge plate section by two cap screws (3/8 in.—24 x 1 in.) having plain and external-toothed lock washers, to side guard by one cap screw (3/8 in.—24 x 5/8 in.) having plain and external-toothed lock washers, and to brackets on hull by two cap screws (3/8 in.—24 x 3/4 in.) having plain and lock washers. These screws must be removed in order to raise guard. The upper section is attached to side of sponson by two cap screws (1/2 in.—20 x 1 in.) having plain and lock washers, and to bracket on rear end of sponson by three cap screws (3/8 in.—24 x 3/4 in.) having plain and external-toothed lock washers. These screws must be removed in addition to other attaching screws, in order to remove guard from vehicle.

TM 9-755
132

Tracks and Suspension

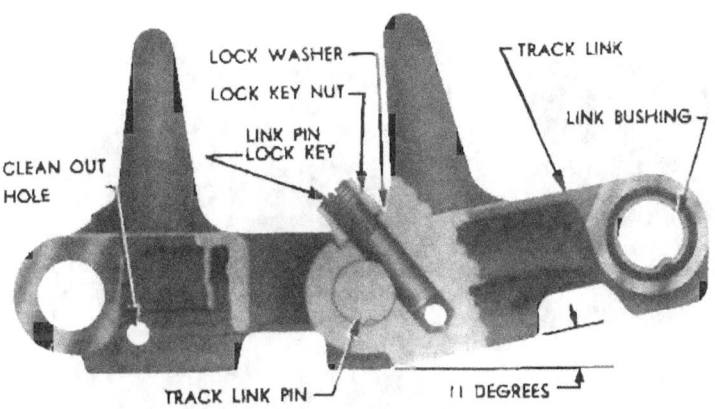

Figure 157—Track Links Assembled, Showing Positions of Parts

132. TRACKS.

a. Description (fig. 157). Each track consists of 83 track link assemblies joined together by link pins. Each track link is a heavy forged steel plate bored to receive two link bushings in one end and a link pin and lock key in other end. The road contact surface of link has webs and hard-surfaced bosses or grousers to provide traction, and opposite surface has an integral lug projecting from it which passes between dual track wheel disks, thereby functioning as a guide to hold track in position on turns. An arrow, and the words "Forward on Ground" are embossed on top surface of link to indicate direction in which assembled track is to be installed on vehicle. A bushing assembly, consisting of a steel sleeve on which rubber bushings are vulcanized, is pressed into each of two large holes bored laterally in link. A longitudinal key formed in steel sleeve engages a groove in link pin and prevents pin from turning in sleeve, so that hinge action is accomplished by flexing of rubber bushings. The solid steel link pin has a groove cut lengthwise for engagement with key in track bushings, and has a flat spot machined at the middle for engagement with lock key. The round steel lock key has a flat surface machined on one side at an angle to center line of key so that it acts as a wedge when drawn up against flat spot on link pin by a nut on threaded outer end of lock key. A screwdriver slot cut off center in threaded outer end of lock key is used to locate wedge surface in proper position when key is installed in link. When two links are connected together by link pin and lock key, they meet at an angle of 11 degrees which helps to equalize the flexing of rubber bushings in both directions as links roll around sprockets or over obstacles on ground.

227

TM 9-755

Part Three—Maintenance Instructions

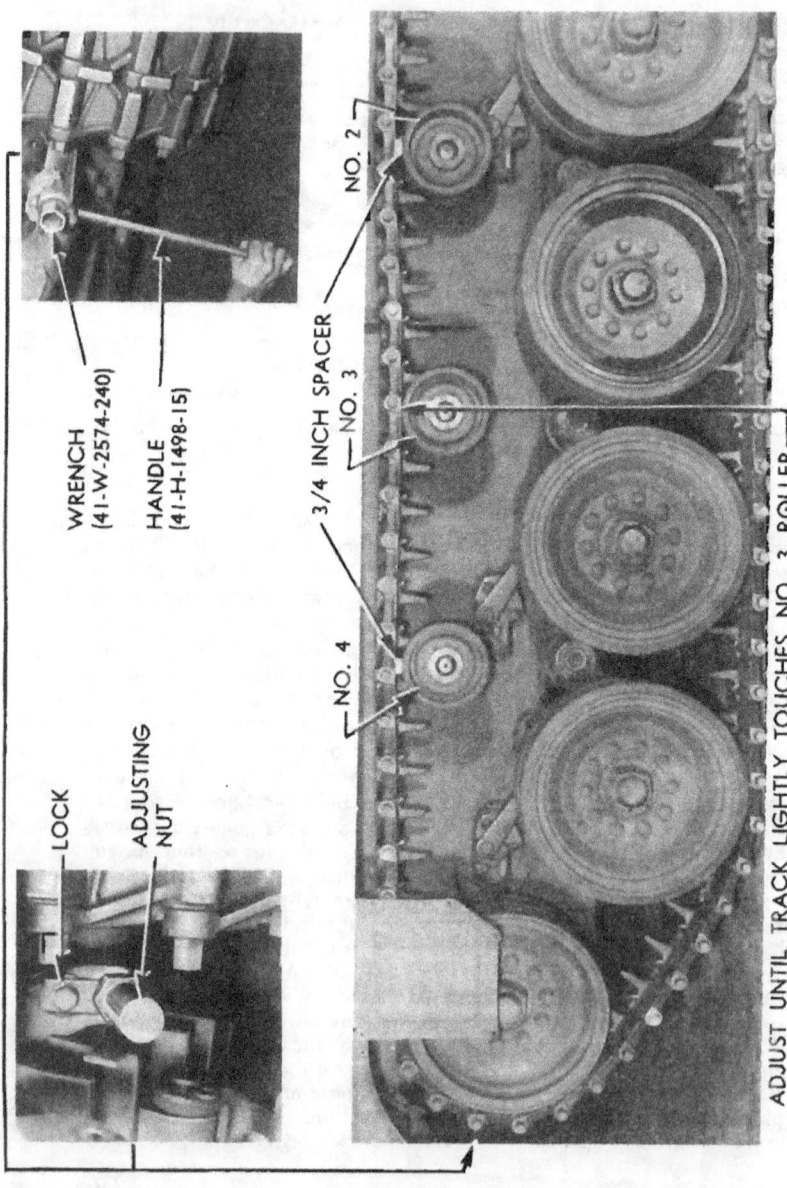

Figure 158—Adjustment of Track

Tracks and Suspension

b. **Track Adjustment.** It is very important that both tracks are kept properly adjusted, to prevent unnecessary wear and breakage. If both tracks do not have equal tension, the vehicle may lead to side on which the track is tighter, making it necessary to use opposite brake excessively for steering. Check tension of both tracks, and adjust to equal tension if necessary, in the following manner:

(1) Release parking brakes so that vehicle is free on tracks.

(2) Raise track and place a spacer ¾ inch thick between track and tires of both disks of No. 2 and No. 4 support rollers (fig. 158).

(3) The track should just touch tires of No. 3 support roller disks with light pressure so that disks can be turned by hand. If track presses on No. 3 support roller disks so they cannot be turned by hand, track is too loose and must be tightened; if track does not contact disks, track is too tight and must be loosened.

(4) Raise rear end guard (par. 131 f) and remove adjusting nut lock which is attached to adjusting bracket by one ¾-inch cap screw and lock washer.

(5) Using a track adjusting wrench (41-W-2574-240) and a handle (41-H-1498-15), turn track compensating wheel adjusting nut on-eyebolt in clockwise direction to tighten track, or in counterclockwise direction to loosen track until tension specified in step (3) is obtained.

(6) Remove spacers (step (2), above). Install adjusting nut lock with one cap screw (¾ in.—16 x 1½ in.) and lock washer tightened to 180-200 foot-pounds tension. Lower rear end guard and secure it with attaching screws (par. 131 f).

c. **Inspection and Classification of Tracks.** Since it is desirable to secure as much wear as possible from tracks, more frequent inspections should be made, as allowable wear limits are approached, before removing from vehicles. Tracks to be considered serviceable for overseas use or domestic use will meet specifications listed below:

(1) The contact pads, called grousers, on bottom surface of track links are 1 1/16 inch high when new. Tracks classified as *serviceable for overseas use* will be those with grouser height of more than 9/16 inch.

(2) Tracks classified as *serviceable for domestic use only* will be those with grouser height of less than 9/16 inch but more than ½ inch.

(3) Track links must not be broken or cracked.

(4) Link guides must not be bent or worn so thin that there is possibility of breaking off in service. Guides worn to a knife edge are not serviceable.

(5) Link bushings must not be broken. Broken bushings can be detected by extreme looseness and sagging when link is between two support rollers on vehicle.

(6) Track link pins must not be worn so that they will not make proper contact with teeth of drive sprockets.

d. **Track Removal.** Removal of a track is accomplished by removing a link pin between two links near an end track wheel, and

TM 9-755

Part Three—Maintenance Instructions

Figure 159—Starting Removal of Track Link Pin

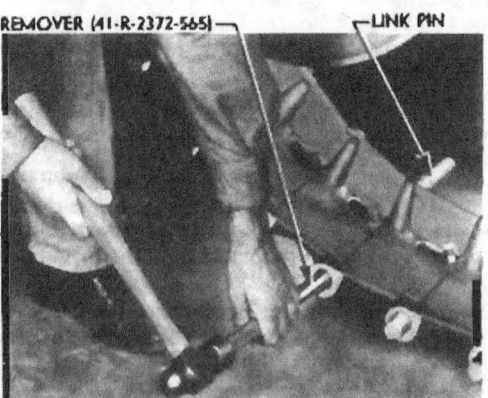

Figure 160—Removing Link Pin With Remover 41-R-2372-565

rolling the free upper section out upon the ground. The track may be laid out either in front of or in back of the vehicle; it is preferable to lay it out in front if space is available.

(1) Remove track end guards, raise side guards (par. 131 e), and completely loosen track (par. h above).

Tracks and Suspension

(2) If track is to be laid out in front of vehicle, select for removal a track link pin behind the rear track wheel; if track is to be laid out at rear of vehicle, select a link pin ahead of the front track wheel.

(3) Loosen the lock key nut (fig. 157) above link pin to be removed and back nut off key approximately half the threads. Clean mud out of hole below lock key, working through clean-out hole in link, drive key down until it is loose, remove nut and lock washer, and continue driving key down until it will free the lock pin.

(4) Place sledge hammer or other heavy weight against inner edge of track link, start link pin out with a sledge hammer (41-S-3726) (fig. 159), then drive pin out with link pin remover (41-R-2372-565) (fig. 160). NOTE: *If mud has worked in around link pin sufficiently to cause pin to stick, work track links up and down with crow bar while driving on link pin.*

(5) Release parking brakes and turn drive sprocket with crow bar (41-B-175) to roll free section of track over support rollers (fig. 161). As free end of track leaves support rollers and sprockets, or compensating wheel, it must be supported by two men holding a crowbar underneath, to prevent track from falling. Continue to support free end of track with bar as it is rolled out on ground.

(6) If individual links in old track are to be replaced, proceed with subparagraph e, below. If a new track is to be installed, disconnect free section of old track by removing a link pin near the end track wheel, remove this section, and proceed with subparagraph f, below.

e. **Track Link Replacement (fig. 162).** With track laying upon the ground (subpar. d (6), above), individual links may be replaced as required by disconnecting them from adjoining links.

(1) Remove the connecting track link pins (subpar. d (3) and (4), above) and remove link.

(2) Set replacement link in position and start link pin through adjoining links, with groove in pin engaging keys in bushings, and flat spot on middle of pin facing toward lock key hole.

(3) Install lock key in links, hold it down with screwdriver so that smaller section formed by off-center slot is toward link pin, and drive link pin in until centered in link.

(4) Lubricate threads of lock key with white lead and oil. Install lock washer and nut. While tightening nut, work link up and down slightly to allow flat surface of lock key to seat squarely against flat spot on link pin.

(5) Tighten lock key nut to 120-130 foot-pounds tension, and leave flat side of nut parallel to centerline of track assembly (fig. 163). This position is necessary to prevent corner of nut from contacting track wheels and support rollers.

(6) Install other links and pins in the same manner, as required.

f. **Track Installation.** The track assembly of 83 links may be

TM 9-755

Part Three—Maintenance Instructions

Figure 161—Removing Track by Turning Sprockets

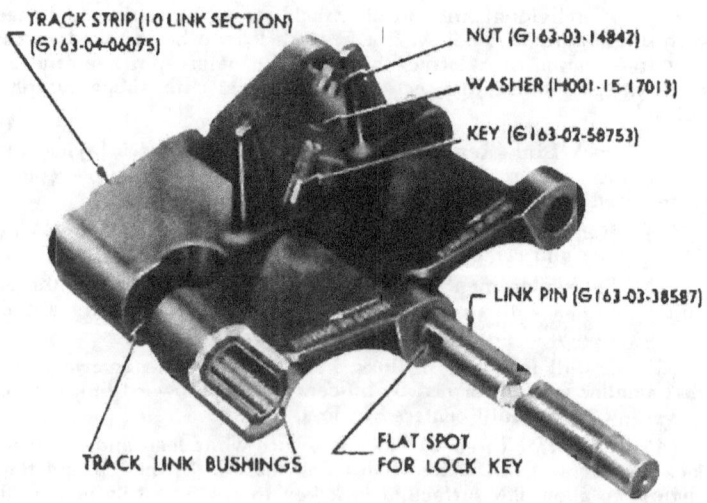

Figure 162—Track Parts

laid out either in front of or behind vehicle, depending on space available.

(1) Lay track upon ground with guides up and in line with space between track wheel disks. If track is in front of vehicle, arrows in

TM 9-755
132
Tracks and Suspension

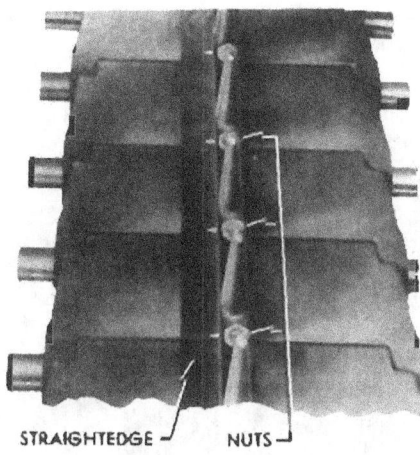

Figure 163—Alinement of Lock Key Nuts

Figure 164—Installing Track on Sprocket

links (fig. 162) must point away from vehicle; if it is behind vehicle, arrows must point toward vehicle. When track is finally installed, arrows on links on ground must point to front of vehicle.

(2) If a section of old track is under track wheels (subpar. d (6), above), temporarily connect new track to old by installing a link pin. If track wheels are on the ground, place a block at end of track or dig a shallow trench under end of track so that track wheels will roll over end of track without pushing track out of position.

TM 9-755
132

Part Three—Maintenance Instructions

Figure 165—*Installing Track Over Support Rollers*

(3) Tow vehicle upon new track until rearward track wheel rests upon fourth link from end of track.

(4) If track is laid out in front of vehicle, the forward end must be brought back and placed on final drive sprockets by two men supporting track with crowbar under seventh or eighth link. A third man must then turn sprockets with crowbar to roll track back upon support rollers, while the other two men continue to support track with crow bar to prevent fouling (fig. 164). When end of track reaches first support roller, insert track adjusting wrench handle (41-H-1498-15) in holes in first link and pull track rearward over support rollers and compensating wheel until track end links are together (fig. 165).

(5) If track is laid out behind vehicle, the rear end of track must be brought forward over compensating wheel, support rollers, and drive sprockets in manner described in step (4) above until track end links are together.

(6) Turn sprockets with crowbar to take up slack until end links can be joined together. While holding end links together by means of crowbar placed under track (fig. 166) install and lock track link pin as described in subparagraph *e*, above.

(7) Adjust track tension (subpar. b, above) and install track guards (par. 131).

TM 9-755
132-133

Tracks and Suspension

RA PD 301383

Figure 166—Installing Track Link Pin

g. Record of Replacement. Record replacement of track assembly on W.D. A.G.O., Form No. 478, M.W.O. and Major Unit Assembly Replacement Record.

133. COMPENSATING LINKS.

a. Description (fig. 149). Each compensating link has a self-alining roller bearing installed in each end and retained by snap rings. Bearing spacers on both sides of each bearing are surrounded by oil seals pressed and staked into the link to retain lubricant and exclude dirt and water from the bearing. Each link is attached through the bearings to a front wheel support arm by a large bolt and to a final drive assembly by a stud mounted in the final drive wheel carrier.

b. Link Replacement. Each compensating link must be removed and installed with the final drive assembly to which it is connected. Procedure for removal and installation is given in paragraph 129.

c. Staking Oil Seals in Link. Effective with M18 vehicle serial number 253, bearing oil seals (G163-03-82854) are staked in place during production. The staking operations, rather than a press fit alone

TM 9-755
133-134

Part Three—Maintenance Instructions

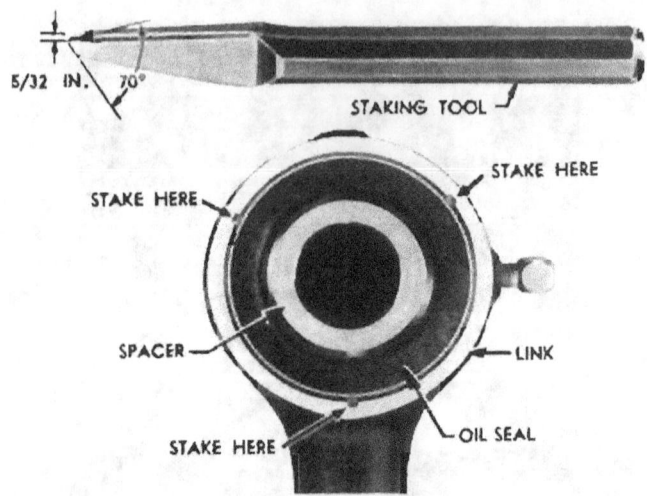

Figure 167—Oil Seal Staked in Link—Staking Tool Required

provides a more positive means of retaining the seals in position. When servicing vehicles in the field built prior to this change in production, or installing a link canabalized from an early production vehicle, all oil seals must be hand staked at three points. A suitable tool for staking can be easily made by grinding a cold chisel to dimensions shown in figure 167. Staking with a center punch has been found to be unsatisfactory.

134. TRACK SUPPORT ROLLER ASSEMBLY.

a. **Description** (fig. 168). Each track support roller consists of two rubber tired steel disks installed on a hub so that the track link guides can pass between them. The rubber tires are vulcanized to the disks, and the disk assemblies must be discarded when the tires are worn out, broken, or separated from disks. The hub rotates upon two tapered roller bearings mounted on a solid steel spindle pressed and welded into a cone-shaped bracket which is bolted to the hull side plate. An oil seal pressed into the inner end of hub, a dirt slinger pressed on the hub, and a dirt shedder on the spindle are provided to retain lubricant and exclude dirt and water from bearings. The outer end of hub is closed by a cap and gasket attached by four cap screws. NOTE: *On vehicles having serial numbers below 489 the hub cap is screwed into hub.*

b. **Removal.** Raise rear end track guard (par. 131 f) turn

Tracks and Suspension

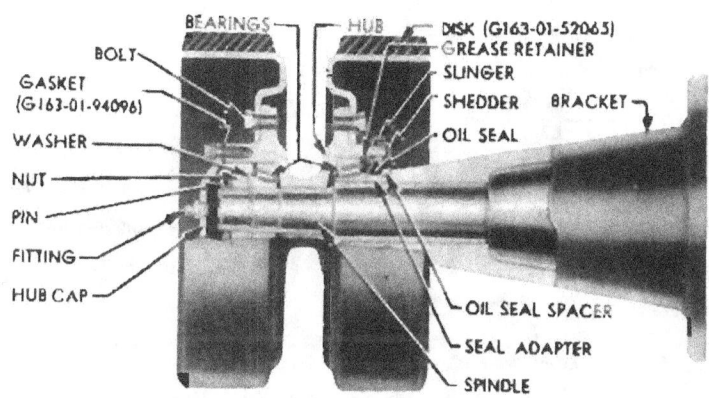

Figure 168—Track Support Roller Assembly—Sectional View

compensating wheel adjusting nut to obtain maximum looseness of track (par. 132 h) and block track up clear of support roller. Remove the six ½-inch bolts and lock washers which attach the support roller bracket to hull side plate, and remove support roller assembly.

c. Installation. With track loose and blocked up, attach support roller bracket to hull side plate with six bolts (½ in.—20 x 2 in.) and lock washers tightened to 75-84 foot-pounds tension. Lubricate support roller bearings (par. 37). Remove blocks from under track, adjust track tension (par. 132 h) and anchor rear end guard (par. 131 f).

135. COMPENSATING WHEELS, BEARINGS, AND SEALS.

a. Description (fig. 169). Two compensating wheels are installed back-to-back on a wheel hub so that track link guides can pass between them. The wheel hub rotates on two tapered roller bearings mounted on a spindle pressed and swaged into compensating wheel support. An oil seal pressed into inner end of hub, a dirt slinger pressed on hub, and a dirt shedder on spindle are provided to retain lubricant and exclude dirt and water from bearings. The outer end of hub is closed by a cap and gasket attached by four cap screws. The wheel support is mounted upon ball and roller bearings on a support spindle bolted to hull side plate. The support is rotated on spindle to move wheel rearward and tighten track tension by means of an eye bolt and adjusting nut.

b. Removal.

(1) REMOVE WHEELS. Raise rear end and side guards (par.

TM 9-755
135

Part Three—Maintenance Instructions

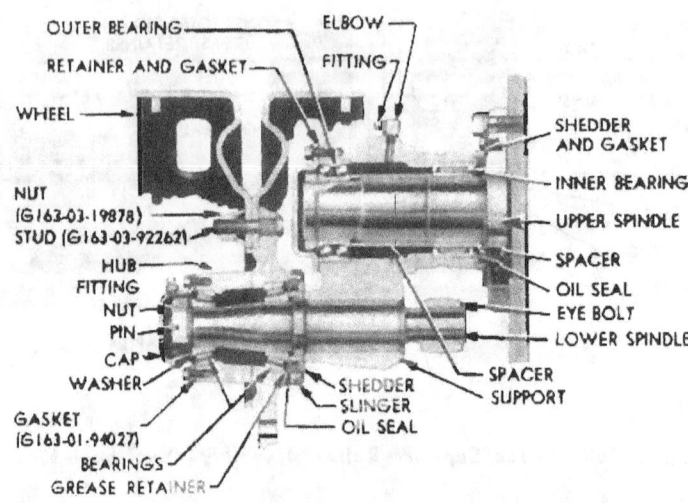

Figure 169—Compensating Wheel, Hub, and Support—Sectional View

131 e and f) and remove track guard brace which is anchored to tapping plate on hull with two ½-inch cap screws and lock washers. Disconnect track by removing link pin behind rear track wheel, release parking brake and roll track forward on support rollers to clear compensating wheel (par. 132 d). Remove ten ¾-inch special flanged nuts from hub studs, place crowbar (41-B-175) between compensating wheels to separate them and remove both wheels from hub.

(2) REMOVE WHEEL HUB, BEARINGS, AND SEALS. The compensating wheel hub, bearings, and oil seal are identical with track wheel parts and are removed from spindle in same manner as described in paragraph 136 b.

c. Installation.

(1) INSTALL OIL SEAL, BEARINGS, AND WHEEL HUB. These parts are identical with track wheel parts and are installed in same manner as described in paragraph 136 c.

(2) INSTALL WHEELS. Place two compensating wheels back to back on hub, install 10 special flanged nuts (¾ in.—16) on studs and tighten evenly to 275-300 foot-pounds tension. Connect track (par. 132 f) and adjust tension (par. 132 h). Attach track guard brace to tapping plate on hull with two cap screws (½ in.—20 x 1¼ in.) and lock washers tightened to 75-100 foot-pounds tension. Anchor rear side and end guards (par. 131 e and f).

Tracks and Suspension

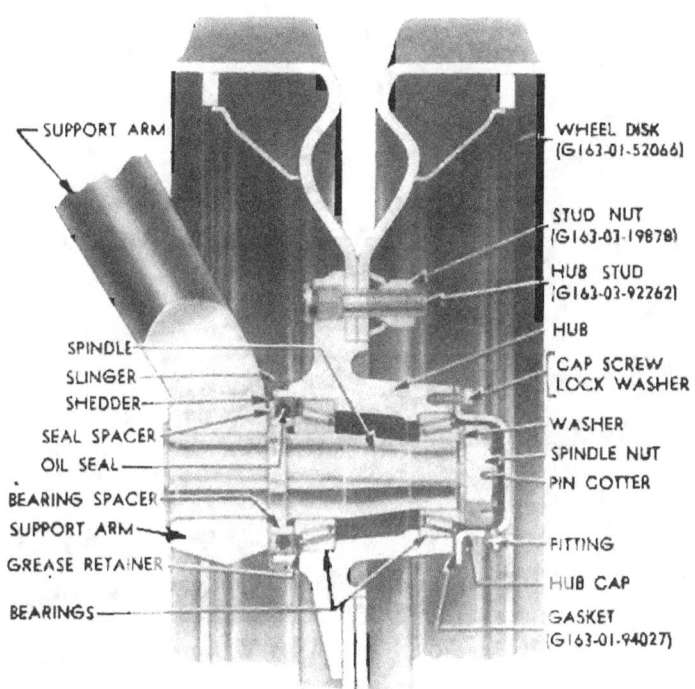

Figure 170—Track Wheel Disks, Hub, Bearings, and Seal—Sectional View

136. TRACK WHEELS, BEARINGS, AND SEALS.

a. Description (fig. 170). Two rubber-tired track wheel disks are installed back-to-back on a wheel hub so that track link guides can pass between them. The rubber tires are vulcanized to wheel disks, and disk assemblies must be discarded when tires are worn out, broken, or separated from disks. The wheel hub rotates on two tapered roller bearings mounted on a spindle pressed and swaged into a wheel support arm (par. 138). An oil seal pressed into inner end of hub, a dirt slinger pressed on hub, and a dirt shedder on spindle are provided to retain lubricant and exclude dirt and water from bearings. Outer end of hub is closed by a cap and gasket attached by four cap screws.

b. Removal.

(1) REMOVE WHEEL DISKS. Remove nut and lock washer from shock absorber link stud with offset box wrench (41-W-576) and pull

TM 9-755
136

Part Three—Maintenance Instructions

Figure 171—Track Wheel Lifter 41-L-1379 in Position To Lift Wheel

stud from wheel support arm with remover (41-R-2366-975) (fig. 184). Attach arm of track wheel lifter (41-L-1379) to support arm at stud hole, and rest lower end of lifter on lifter block placed over track link guides to rear of track wheel (fig. 171). Start engine and slowly reverse vehicle to force track wheel up over lifter block, stopping vehicle and setting parking brakes when lifter is vertical and wheel is clear of lifter block (fig. 172). Shut engine off. Remove 10 flanged nuts (¾ in.—16) from hub studs, place crow bar between wheel disks to separate them and remove disks from hub.

(2) REMOVE WHEEL HUB. Remove four ⅜-inch cap screws, lock washers, hub cap and gasket from hub. Remove cotter pin, nut and washer from wheel spindle. Place crow bar between hub and support arm, pry out with steady pressure to push outer bearing off spindle, then remove hub and inner bearing from spindle. CAUTION: *Use care to avoid damage to oil seal on spindle.*

(3) REMOVE BEARING RACES. If wheel bearings and races are chipped or scored, the outer races must be removed from the hub. Carefully drive races out of hub, using a brass drift and hammer. The

TM 9-755

Tracks and Suspension

Figure 172—Track Wheel Raised by Wheel Lifter 41-L-1379

grease retainer will be removed as inner bearing race is driven from hub.

(4) REMOVE OIL SEAL. A worn or damaged oil seal must be replaced to prevent entrance of dirt and water into wheel bearings. Pry oil seal off bearing spacer on spindle, using chisel between oil seal and oil seal spacer. Use care to avoid damaging dirt shedder. If oil seal cannot be removed from bearing spacer, drive against inner side of dirt shedder to remove shedder, oil seal spacer, oil seal, and bearing spacer from spindle. NOTE: *The bearing spacer is a drive fit on spindle; therefore, shedder and spacer will probably be damaged and require replacement.*

c. Installation.

(1) INSTALL OIL SEAL. If dirt shedder and bearing spacer were removed, place new shedder over wheel spindle with cupped edge pointing outward and drive it tight against support arm. Install a new bearing spacer over spindle and drive it tight against dirt shedder. Place oil seal spacer over bearing spacer. Install oil seal on bearing spacer with spring side towards dirt shedder and drive it tight against oil seal spacer with seal replacer (41-R-2383-950) (fig. 173).

(2) INSTALL BEARING RACES. Place new bearing race in hub

TM 9-755
136-137

Part Three—Maintenance Instructions

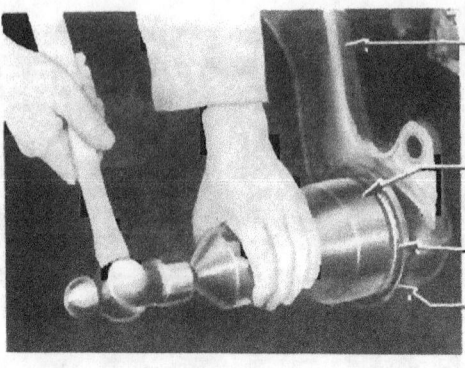

Figure 173—Installing Oil Seal With Replacer 41-R-2383-950

with thick edge inward and drive it down against shoulder in hub, using a brass drift or hardwood block, being careful to prevent cocking race in hub. Place a new grease retainer in replacer (41-R-2390-450) with flanged side inward and position replacer on hub (fig. 174). Drive retainer squarely into place and remove tool.

(3) INSTALL WHEEL HUB AND ADJUST BEARINGS. Pack inner bearing cone with grease, seasonal grade (par. 38), and install on wheel spindle with small end outward. Pack inside of hub with grease and install hub on wheel spindle, being careful not to damage oil seal. Pack outer bearing cone with grease and push it on spindle and into race in hub, centering hub on spindle. Install keyed washer and spindle nut. Turn spindle nut up tight while rotating hub (do not use wrench longer than 10 inches), then loosen nut one flat and aline nearest slot with hole in spindle; then install cotter pin ($3/_{16}$ in. x 2½ in.). Install hub cap and new gasket with four cap screws (⅜ in.—24 x ¾ in.) and lock washers tightened to 20-24 foot-pounds tension. Force grease into fitting on hub cap until it just starts to come out at seal.

(4) INSTALL WHEEL DISKS. Place two wheel disks back-to-back on hub, install ten flanged nuts (½ in.—16) on studs and tighten evenly to 275-300 foot-pounds tension. Start engine and slowly move vehicle forward until track wheel is down on track and track wheel lifter is free. Remove lifter block and disconnect wheel lifter from wheel support arm. Connect shock absorber link to support arm with lockwasher and nut (¾ in.—16) and tighten securely with offset box wrench (41-W-576).

137. TORSION BARS.

a. *Description and Identification* (fig. 179). Torsion bars are solid steel shafts of high carbon alloy steel serrated at each end. One

342

Tracks and Suspension

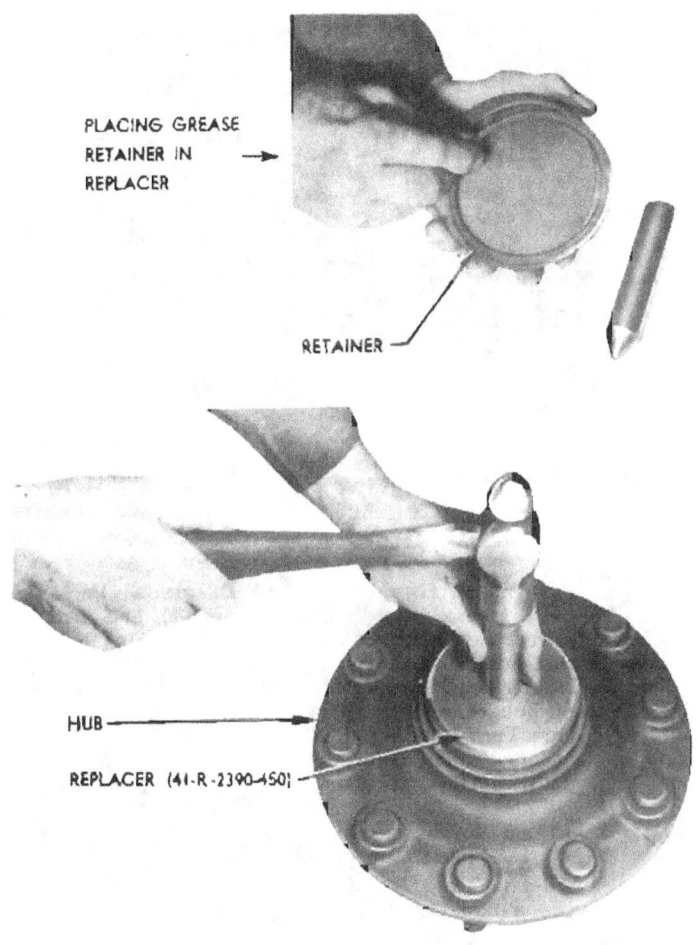

Figure 174—Installing Grease Retainer With Replacer 41-R-2390-450

end engages in the internally serrated axle shaft which is integral with the support arm and other end is anchored to opposite axle shaft housing through engagement with internally serrated torsion bar retainer inclosed in housing. As track wheel moves upward, when going over an obstruction, support arm pivots on axle shaft which imparts a twist to torsion bar. The bar resists this twisting action and there-

Part Three—Maintenance Instructions

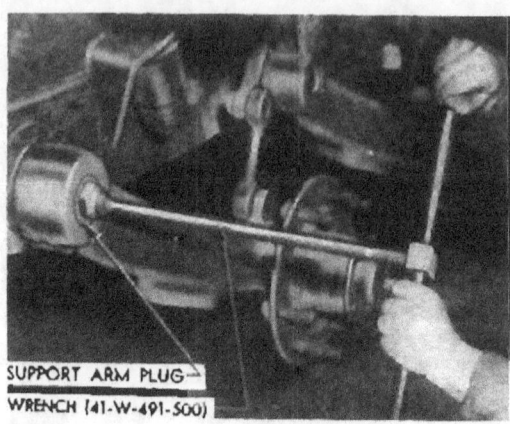

Figure 175—Removing Support Arm Plug, Using Wrench 41-W-491-500

Figure 176—Removing Torsion Bar, Using Tongs 41-T-2723

fore functions as a spring. In production, each bar is given a definite twist or set in direction bar will twist in supporting vehicle. Therefore, torsion bars are not interchangeable from left to right and if installed on wrong side will fail very quickly. Bars are also of different strengths from front to rear and are distinguished by letters A, B, C, D following a common part number (fig. 177). Each bar is marked

Tracks and Suspension

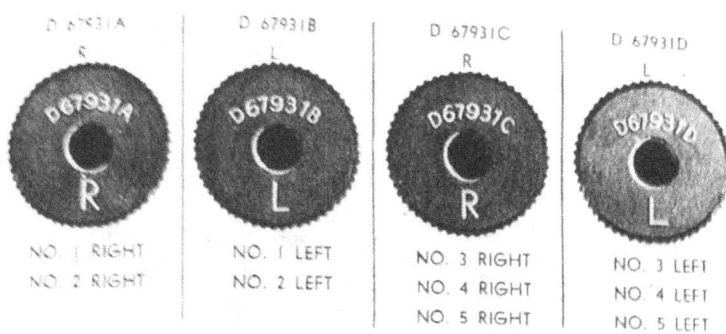

Figure 177—Torsion Bar Identification Marks

on one end with its part number and a letter "L" or "R" to indicate on which side of vehicle it is to be installed, and this end must engage the axle shaft when installed in vehicle. These markings may be observed by removing support arm plug in end of axle shaft.

b. Removal.

(1) Jack up vehicle and support it on suitable blocks about 23 inches high placed under the jack pads welded to bottom of hull at each corner.

(2) Disconnect shock absorber link (par. 139 b) and completely loosen track (par. 132 b).

(3) If No. 1 (front) torsion bar is to be removed, remove final drive sprockets and hubs from both sides of vehicle (par. 127 a).

(4) Remove support arm plug from outer end of axle shaft housing, using a plug wrench (41-W-491-500) (fig. 175) and remove ½-inch pipe plug from axle shaft housing cap on opposite side of vehicle.

(5) Attach hoist to wheel or use a crowbar to raise track wheel slightly until movement can be noted between serrations in support arm and those on torsion bar. While moving support arm up and down slightly, have second man push on opposite end of torsion bar through plug hole in housing cap. When torsion bar comes out of axle shaft far enough to clear serrations, grasp bar with lifting tongs (41-T-2723) (fig. 176) and remove bar.

c. Installation. Before the torsion bar is installed, make certain that it is correct part for location, as shown by part number and letter stamped on one end (fig. 177).

(1) Coat both serrated ends of torsion bar with rust-proofing oil.

(2) Start unmarked end of torsion bar through axle shaft, grasp outer serrated end with lifting tongs (41-T-2723) (fig. 176) and raise inner end of bar so it will start into torsion bar retainer in opposite

TM 9-755
137-138

Part Three—Maintenance Instructions

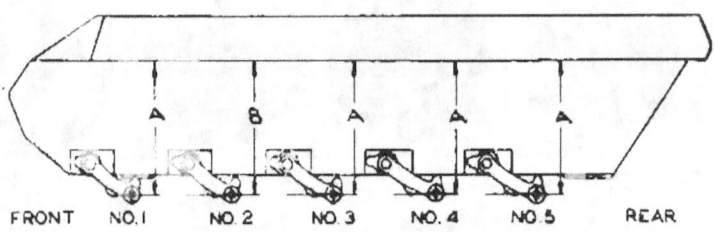

Figure 178—Locating Support Arms To Install Torsion Bars

axle shaft housing. Push bar into retainer until serrations are just ready to engage.

(3) Lift track wheel to position so that center of wheel spindle is the correct distance from bottom of sponson as shown in figure 178 (wheel and hub on spindle), engage serrations by moving support arm up or down slightly and push torsion bar fully into place. Recheck setting with wheel and arm freely supported by torsion bar only. NOTE: *Dimensions given in figure 178 are nominal, with tolerances of plus or minus ½ inch. When setting cannot be made to exact nominal dimension, setting to larger dimension is preferred, that is, the given dimension plus tolerance of ½ inch.*

(4) Install ½-inch pipe plug in housing cap (on opposite side) and install support arm plug in axle shaft housing using plug wrench (41-W-491-500) (fig. 175).

(5) Install final drive sprockets and hubs, and connect track (par. 127 h) after installation of No. 1 torsion bar, if this was removed.

(6) Connect shock absorber link (par. 139 c).

(7) Remove blocks from under vehicle and adjust track (par. 132 b).

138. WHEEL SUPPORT ARM AND AXLE SHAFT HOUSING.

a. **Description** (fig. 179.) The support arm assembly consists of a heavy steel forging into which a tubular axle shaft is splined and welded at front end and a solid steel wheel spindle is pressed and swaged into rear end. The axle shaft is supported in axle shaft housing by two large diameter roller bearings separated by a tubular spacer. A cap, bolted to outer end of housing and a retaining nut screwed into inner end of housing position and hold bearings in place.

Tracks and Suspension

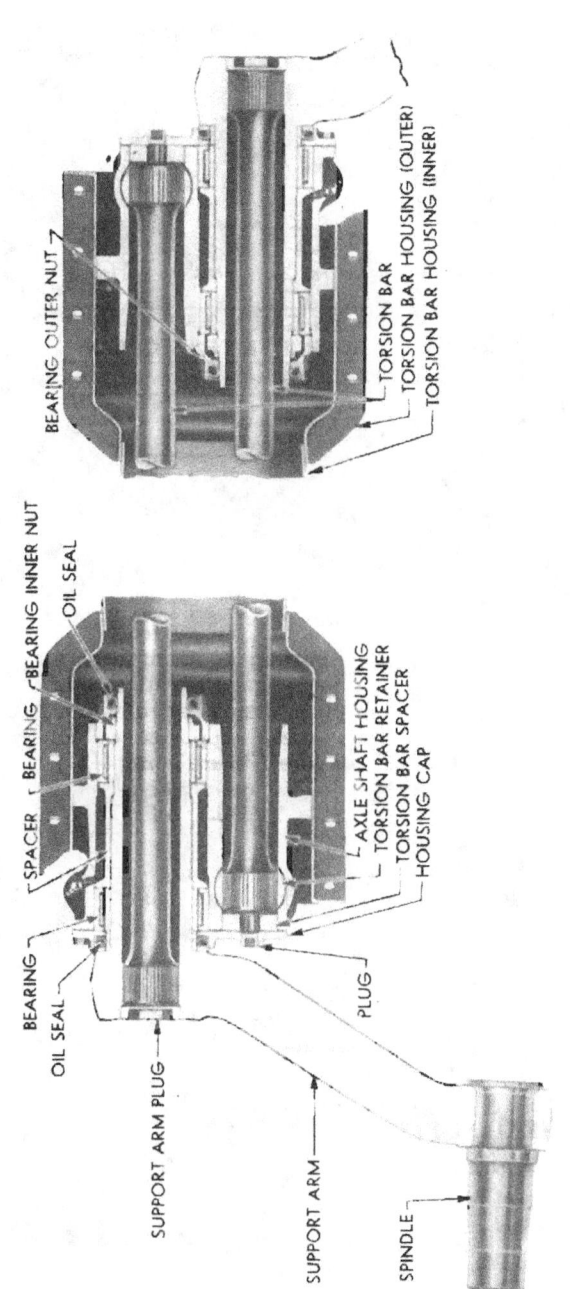

Figure 179—Wheel Support Arms, Housings, and Torsion Bars—Sectional View

BEFORE WELDING

AFTER WELDING

Figure 180—Before and After Welding Joint Between Hull and Torsion Bar Outer Housing

TM 9-755
138

Tracks and Suspension

The axle shaft is retained in housing by a nut at its inner end which bears against inner bearing race. Oil seals are located in housing cap and bearing outer retaining nut to retain lubricant and exclude dirt and water from bearings. The housing incloses a splined torsion bar retainer into which splined end of the torsion bar from opposite axle shaft is engaged and anchored. The flange of axle shaft housing contains grooves in which seals are installed (fig. 183) to secure a water tight joint between housing and hull. The flange of housing also contains three blind tapped holes and eight through bolt holes for attachment to hull.

b. **Modification of Axle Housing Mounting—M18.** In M18 vehicles having serial numbers 1 through 81, the axle shaft housings were attached to hull with $\frac{1}{2}$-inch bolts. Under severe operating conditions, the front housing attaching bolts become loose, causing bolt holes to become enlarged and the bolts eventually shear off. Beginning with M18 vehicle number 82, bolts are $\frac{9}{16}$-inch diameter and all new housings are drilled for $\frac{9}{16}$-inch bolts. In M18 vehicles having serial numbers below 201, joint between hull and torsion bar outer housing requires strengthening by welding (fig. 180). M18 vehicles in field having serial numbers below those specified will be modified as follows; however, it will not be necessary to modify any mountings other than front unless failure elsewhere is evident.

(1) If old housing is in good condition and the three inside tapped holes have good threads, it will not be necessary to replace housing. Remove outside bolts and nuts, one at a time, and drill or ream holes in housing and hull for $\frac{9}{16}$-inch bolts. Install new bolts ($\frac{9}{16}$ in.—18 x 2$\frac{1}{4}$ in.), with nuts and lock washers inside hull and tighten to 100-120 foot-pounds tension. Continue to use $\frac{1}{2}$-inch bolts in inside tapped holes; retapping these holes to $\frac{9}{16}$ inch is not practicable.

(2) When replacing old housing (subpar. c and d, below,) with new housing having $\frac{9}{16}$-inch bolt holes, bolt new housing in place with two $\frac{1}{2}$-inch bolts in opposite holes. Shift housing on hull until all other bolt holes are centered on holes in hull and tighten attaching bolts. Drill or ream bolt holes in hull using housings as a guide, and install bolts ($\frac{9}{16}$ in.—18 x 2$\frac{1}{4}$ in.), with nuts and lock washers inside. Remove $\frac{1}{2}$-inch bolts, drill these holes and install large bolts. Drill or ream three upper center holes in hull and install bolts ($\frac{9}{16}$ in.—18 x 1$\frac{1}{2}$ in.) with lock washers inside hull. Tighten all bolts to 100-120 foot-pounds tension.

(3) On vehicles with serial numbers 1 through 200, remove axle shaft housing (subpar. c, below) and weld the joint between hull and torsion bar outer housing at front, top, and rear as shown in figure 180. Install housing (subpar. d, below).

c. **Removal.**

(1) If No. 1 (front) support arm and axle shaft housing is to be removed, remove final drive sprockets and hubs from both sides of vehicle (par. 127 a) and disconnect compensating link from the support arm that is to be removed.

349

TM 9-755

Part Three—Maintenance Instructions

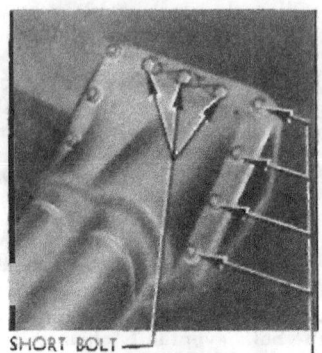 

SHORT BOLT
NUT (WELDED TO HULL)
VIEW INSIDE HULL

LONG BOLT   SAFETY WIRE
VIEW OUTSIDE HULL

RA PD 140339

Figure 181—Axle Shaft Housing to Hull Mounting

HOUSING AND SUPPORT ARM   TORSION BAR

RA PD 140400

Figure 182—Removing Support Arm and Axle Shaft Housing Assembly

(2) If No. 5 (rear) support arm and axle shaft housing is to be removed, remove engine (par. 75) and fuel tank (par. 99 a) in order to reach inside attaching bolts.

(3) Jack up vehicle and support it on suitable blocks about 23 inches high placed under jack pads welded to bottom of hull at each corner.

Tracks and Suspension

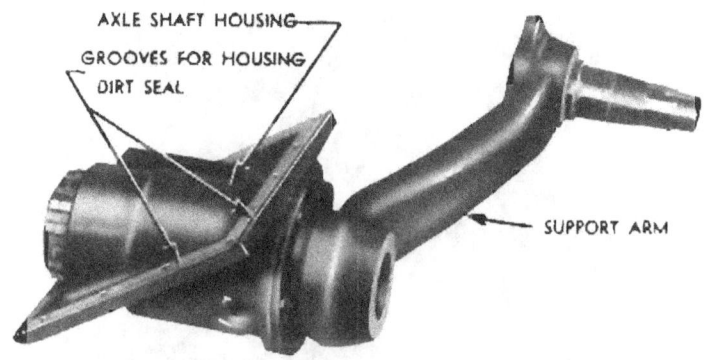

Figure 183—Wheel Support Arms With Housing Assembly

(4) Disconnect shock absorber link (par. 139 b) and completely loosen track (par. 132 b).

(5) Remove track wheel disks and hub from spindle of support arm (par. 136 b). Cover spindle with tape to protect threads.

(6) Remove right and left torsion bars (par. 137 b).

(7) Cut lock wires and remove three ⁹⁄₁₆-inch bolts inside hull and four ⁹⁄₁₆-inch center side bolts outside hull, leaving corner bolts to support the housing temporarily (fig. 181). NOTE: *In M18 vehicle having serial numbers below 1,258, the nuts for outside bolts were loose and secured by lock washers instead of being welded to hull.*

(8) Insert serrated end of torsion bar into serrated section of axle shaft and attach chain hoist to bar (fig. 182). Support arm and housing assembly by holding outer end of torsion bar down while removing four corner bolts, then remove assembly from vehicle.

d. **Installation.** On M18 vehicles having serial numbers below 31, refer to subparagraph b above for information on changes in mounting bolts. On M18 vehicles having serial numbers below 201, refer to subparagraph b (3) above for instructions on welding torsion bar outer housing.

(1) Thoroughly clean surfaces of hull and axle shaft housing to insure a water tight joint. Place dirt seals in grooves in flange of housing (fig. 183), making sure to get tight joints at corners, and coat seals thoroughly with joint sealing compound (51-C-1616).

(2) Insert serrated end of torsion bar into serrated section of axle shaft, attach chain hoist to bar, and lift arm and housing assembly into place while holding outer end of torsion bar down to balance weight of assembly (fig. 182).

(3) Install eight bolts (⁹⁄₁₆ in.—18 x 2⅛ in.) through housing from outside of hull, and three bolts (⁹⁄₁₆ in.—18 x 1⅜ in.) from

TM 9-755
138

Part Three—Maintenance Instructions

Figure 184—Removing Shock Absorber Link, Using Remover 41-R-2366-975

inside hull (fig. 181). Tighten all bolts to 100-120 foot-pounds tension and install lock wires through heads of adjacent pairs of bolts. NOTE: *In M18 vehicles having serial numbers below 1258, nuts for outside bolts are not welded to hull. The bolts ($\%_{16}$ in.—18 x 2¼ in.) are secured inside the hull by free nuts and lock washers. The inside bolts ($\%_{16}$ in.—18 x 1½ in.) are provided with lock washers. Safety wires are not used.*

(4) Install right and left torsion bars as described in paragraph 137 c (1) through (4), setting the support arm so that center of spindle is the correct distance from bottom of sponson as shown in figure 178 (wheel off hub).

(5) Connect shock absorber links (par. 139 c) and compensating link if disconnected.

(6) Install track wheel hub and disks (par. 136 c).

(7) Remove blocks and lower vehicle to ground.

(8) Install final drive hubs and sprockets and connect track (par. 127 b), if removed.

(9) Adjust track (par. 132 b).

(10) Install fuel tank (par. 98) and engine (par. 76), if removed.

e. *Record of Replacement.* Record replacement of support arm and axle shaft housing assembly on W.D. A.G.O., Form 478, M.W.O. and Major Unit Assembly Replacement Record.

Tracks and Suspension

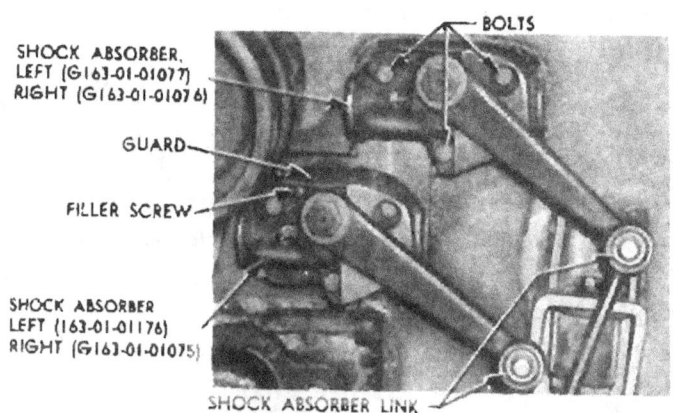

Figure 185—Shock Absorbers and Guards Installed

139. SHOCK ABSORBERS AND LINKS.

a. **Description** (fig. 185). The track wheel shock absorbers, which are bolted to the hull side plates and protected by guards, are heavy duty double-acting units. The hollow upper portion of the shock absorber body which serves as a reservoir for fluid, is filled through openings sealed by filler screw plugs and soft metal gaskets. The links which connect shock absorber arms to wheel support arms are steel rods having large eyes in each end into which stud and bushing assemblies are pressed and staked in place. The bushings contain fabric cores saturated with life time lubricant.

b. **Removal.** Remove nut and lock washer from stud at each end of shock absorber link, using offset box wrench (41-W-576). Pull studs from shock absorber and support arms, using remover (41-R-2366-975) (fig. 184). Remove shock absorber and guard which are attached to hull side plate with three ⅝-inch bolts (⅝ in.) and lock washers.

c. **Installation** (fig. 185). Place shock absorber in position on hull with arm pointing to front of vehicle, place guard over front of absorber body, and install three bolts (⅝ in.—18 x 3¹¹⁄₁₆ in.) and lock washers. Tighten bolts to 130-149 foot-pounds tension. NOTE: *Early production guards do not have a hole through which to install front upper bolt. Install and tighten this bolt first and then install guard and two remaining bolts.* Remove filler plug screw, move arm slowly through full travel several times to work out all air from fluid and then check fluid level, which must be even with bottom of filler hole. Add specified shock absorber fluid, if necessary, and install filler plug screw. Place studs of link assembly in support arm and absorber arm, install lock washers and nuts (¾ in.—16) and tighten to 180-200 foot pounds tension with offset box wrench (41-W-576).

140. WHEEL SUPPORT ARM SPRING BUMPER.

a. Description (fig. 168). Each wheel support arm spring bumper consists of a volute spring welded to a steel plate which is threaded for bolt by which assembly is attached to a bracket on hull side plate. A steel button is pressed into small diameter lower end of spring at point of contact with wheel support arm.

b. Replacement. A spring bumper is detached from its mounting bracket on hull by removing one bolt (⅝ in.—18 x 1¼ in.) and lock washer. When a spring bumper is installed on vehicle, tighten attaching bolt to 100-120 foot-pounds tension.

Section XXIX
BATTERIES AND CHARGING SYSTEM

141. DESCRIPTION, CIRCUITS, AND DATA.

a. Description. Batteries and charging system consists of the following components.

(1) BATTERIES. Two 12-volt, 168 ampere hour, 6-cell storage batteries are used to supply current for all electrical circuits in vehicle. Batteries in M18 vehicle are located under two subfloor doors to left of turret slip ring box (fig. 260). Batteries in the M39 vehicle are located in a battery box at front end of crew compartment.

(2) RECEPTACLE (OUTLET) (fig. 15). A receptacle is mounted on master switch box to provide an outlet to battery circuit for plugging in a slave battery, or a battery charger, when vehicle batteries are too low to function properly.

(3) ENGINE GENERATOR (fig. 191). A 26-volt 50 ampere, shunt-wound generator is used to provide current during operation of vehicle and to keep batteries charged. The generator is driven by two V-belts from pulleys mounted on universal joint to transfer case yoke; therefore, it is in operation whenever engine is running, with transfer case clutch engaged. The generator is supported by a hinge pin on a mounting plate attached to hull floor by studs through shock absorbing bushings. An adjusting rod is provided to move generator on hinge pin to adjust tension of drive belts. In the M18 vehicle, generator is reached by removing right rear subfloor plate. In the M39 vehicle, generator is reached by raising rear seat center cover which is hinged and secured by two flush fasteners under rear corners of seat pad.

(4) AUXILIARY GENERATOR—M18 VEHICLE (fig. 24). A Homelite Model HRUH-28 auxiliary generator is mounted on subfloor in right front corner of fighting compartment in the M18 vehicle. Auxiliary generator is an integral direct current power plant with a

capacity of 1,500 watts at 30 volts. It consists of an electric generator with control box attached, directly coupled to and driven by a 2-cycle gasoline engine. Auxiliary generator is provided to charge batteries when engine generator is not operating, or when engine generator output is not sufficient to meet unusual operating requirements.

(5) GENERATOR REGULATOR ASSEMBLIES (fig. 196). The generator regulator assemblies are mounted with main filter box on a plate attached to vehicle bulkhead through shock absorbing bushings. Engine generator regulator is located on left side of filter box, and auxiliary generator regulator is located on right side of filter box. Each regulator assembly contains a current regulator, a voltage regulator, and a cut-out relay. The current regulator functions to control maximum output of generator. The voltage regulator functions to regulate generator output to proportionate requirements of load; this is dependent upon condition of batteries and number of current-consuming devices that are in operation. For example, if batteries are in a nearly discharged condition or load requirements are increased, voltage regulator will cause charging rate to increase. As batteries approach full charge or load is decreased, charging rate will become proportionately lower to prevent over-charging. The cut-out relay functions to automatically open circuit when generator is not operating, thus preventing batteries from discharging through generator to ground.

(6) CIRCUIT BREAKERS. An automatic circuit breaker is located in main filter box (fig. 196) and is connected in series with engine generator circuit to protect generator and wiring against overload which might result from inadvertent closing of cut-out relay from gunfire shock when generator is not operating.

(7) WIRES, CONDUITS, AND JUNCTION BOXES. All wires are enclosed in shielded conduits for protection of wires and to effect adequate suppression of radio interference. Junction boxes are used for convenience in making connections of wires in separate conduits. The junction boxes in battery circuit are battery junction box and master switch box. The additional junction boxes in charging circuits are auxiliary generator junction box, main filter box, and starter junction box.

b. Battery and Charging Circuits (fig. 186).

(1) BATTERY CIRCUIT. The batteries are connected to provide single-wire with ground-return circuits of both 12 and 24 volts. The negative terminal of one battery is grounded to vehicle and positive terminal of this battery is connected to negative terminal of other battery by a jumper cable. A feed wire connected to negative terminal of grounded battery supplies 12-volt current since this current is taken from only one battery. Another feed wire connected to positive terminal of other battery supplies 24-volt current since both batteries are connected in series. Both feed wires run to battery junction box (fig. 230) and from there to master switch box (figs. 240 or 242), where 24-volt wire is connected to upper (24-volt) switch and 12-volt wire is connected to lower (12-volt) switch. From master switch box,

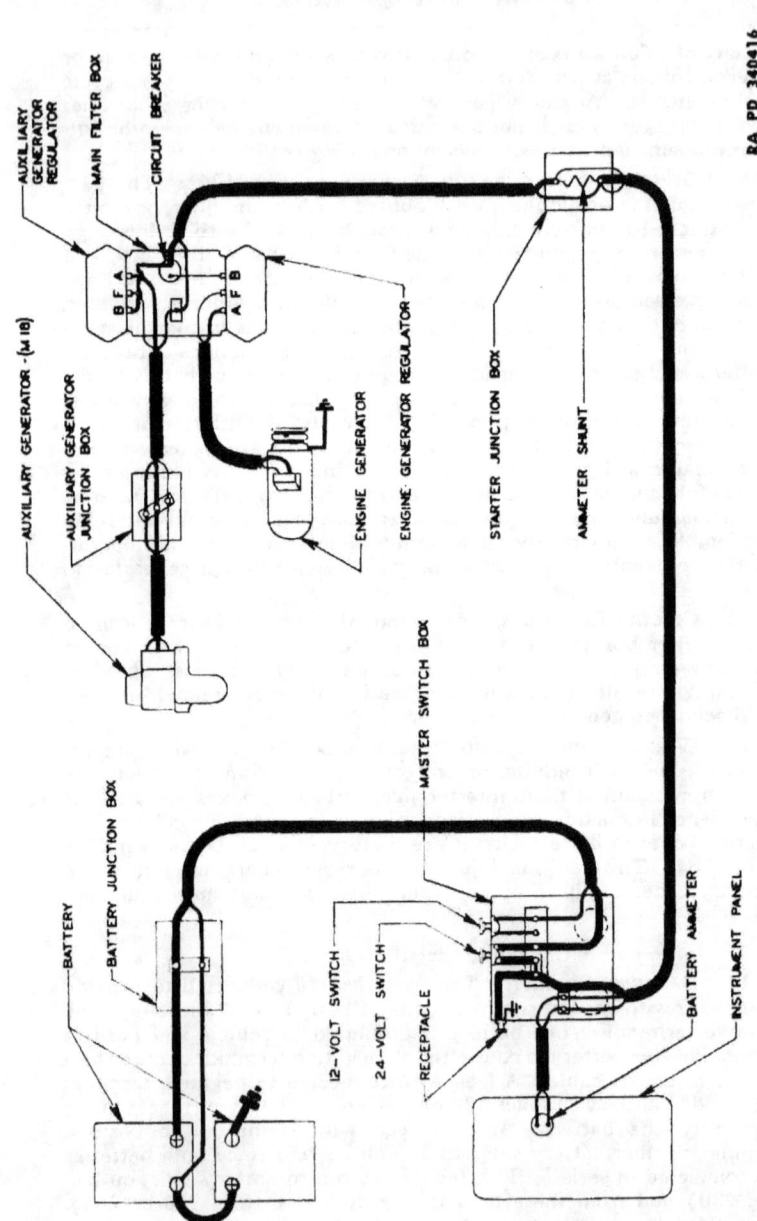

Figure 186—Battery and Charging Circuits

TM 9-755
141

Batteries and Charging System

12-volt current is distributed to radio, and 24-volt current is distributed to all other electrical units in vehicle. The ammeter (N, fig. 206) on instrument panel does not indicate current discharge from batteries to electrical circuits.

(2) ENGINE GENERATOR CIRCUIT. Generator is connected to a single-wire with ground-return circuit. Generator is grounded to vehicle by a jumper wire (fig. 221). A conduit containing armature and field wires connects generator to main filter box (fig. 196) on vehicle bulkhead, where wires connect to engine generator regulator. Wires carry generator current from regulator to circuit breaker in filter box, from breaker to starter junction box (fig. 232) and from junction box to master switch box (fig. 240 or 242) where connection is made to 24-volt master switch battery circuit (subpar. b (1), above). An ammeter shunt in starter junction box is connected by two separate wires to ammeter (N, fig. 206) on instrument panel, which indicates charge going into batteries from generator.

(3) AUXILIARY GENERATOR CIRCUIT. Auxiliary generator is grounded to hull by a strap (fig. 220). A field wire, an armature wire, and a starting wire inclosed in a conduit, connect terminals in generator control box to terminals in auxiliary generator junction box, from which connections are made to auxiliary generator regulator and terminal block in main filter box. A jumper wire connects battery (output) terminal of regulator to upper or output side of circuit breaker in filter box, where auxiliary generator circuit joins engine generator circuit (subpar. b (2) above). Automatic circuit breaker in filter box is not in auxiliary generator circuit, since a separate manual reset circuit breaker is contained in auxiliary generator control box. The starting wire connects at terminal block in main filter box to a wire from starter junction box to furnish battery current for starting auxiliary generator engine. The charging rate of auxiliary generator is indicated by ammeter in auxiliary generator control box as well as by instrument panel ammeter.

c. Data.
(1) BATTERIES.
Number of batteries 2
Type   6 cell
Voltage—each   12
Voltage circuits provided   12 and 24
Ampere-hour capacity   168
(2) ENGINE GENERATOR.
Make   Delco Remy
Model No.   1117303
Winding   4-pole, shunt
Rotation, drive end view   Clockwise
Generator speed to engine speed   1.52 to 1
(3) AUXILIARY GENERATOR.
Make   Homelite
Model No.   HRUH-28
Rating   1,500 watts, 30 volts, D-C

TM 9-755
141-142

Part Three—Maintenance Instructions

Generator winding	4-pole, shunt
Engine type	Single cylinder, air cooled, 2-cycle
Engine speed	3,400 to 3,700 rpm
Carburetor	Tillotson, float-feed type
Fuel consumption, under full load	½ gal per hr
Fuel	80 octane gasoline
Ignition	Built-in magneto
Spark plug type	Champion HO-14S
Governor	Automatic, built-in, non-adjustable

(4) GENERATOR REGULATORS.

Number	2
Make	Delco-Remy
Model No., auxiliary generator	1118492
Model No., engine generator	1118478
Type	3 unit, 24-volt

142. BATTERIES.

a. **General.** The service which using arms can perform on batteries includes checking specific gravity, adding water, cleaning, replacing cables, recharging or replacement.

b. **Checking Specific Gravity.** Specific gravity must be checked before adding water, as water does not mix immediately and a true reading will not be obtained. When checking specific gravity with a hydrometer a true reading will be obtained only when electrolyte temperature is normal, or 80° F. When electrolyte temperature is above or below 80° F, a correction must be made to hydrometer reading to obtain true specific gravity. Remove battery filler caps and take temperature of electrolyte with a thermometer, take specific gravity with a hydrometer and then correct the hydrometer reading to obtain true specific gravity by means of scale shown in figure 187, which includes an example to show method of figuring actual specific gravity. A fully charged battery will have an actual specific gravity of 1.275 to 1.300. In normal temperatures, battery must be recharged if specific gravity is 1.200 or below. Refer to paragraph 27 e (7) for requirements in freezing temperatures, and to paragraph 30 for requirements in torrid zones.

c. **Adding Water.** Add distilled water to each battery cell as required to bring fluid level up to ½ inch above the top of plates, as a general rule; however, some batteries used have filling instructions on filler caps and these instructions must be observed. When water is added in freezing weather, operate auxiliary generator or vehicle for a short time to flow current into batteries, which will cause water and electrolyte to mix.

d. **Charging.** When specific gravity is found to be below 1.200 (subpar. b above), batteries must be recharged. This may be accomplished with batteries in vehicle by plugging in a charger at receptacle on master switch box (fig. 15). NOTE: *Observe charging instructions which may be stamped on filler caps.* If batteries cannot

TM 9-755
142

Batteries and Charging System

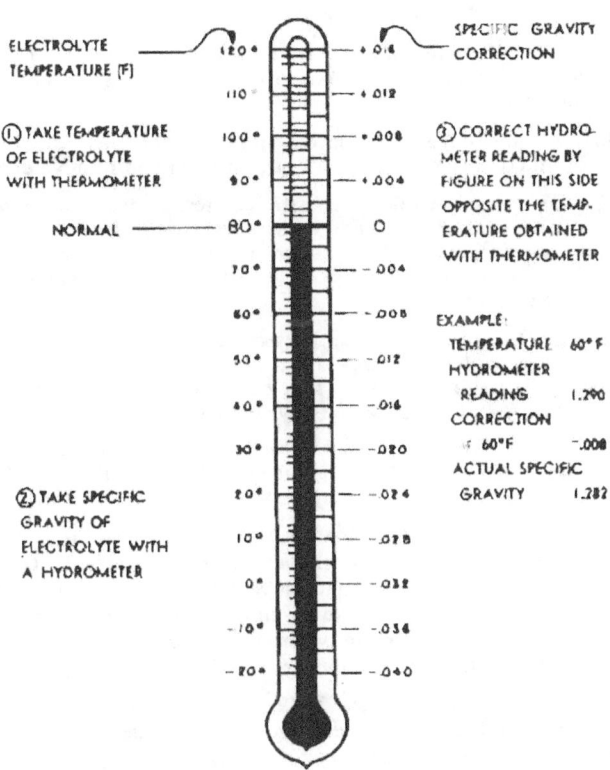

Figure 187—*Electrolyte Specific Gravity Correction Chart*

be brought up to properly charged condition, they must be replaced for service by higher authority.

e. **Spilled Electrolyte and Corroded Terminals.** Battery electrolyte will cause corrosion of all metal parts touched, such as terminals and cables, battery box, or hull components. If spilling or overflowing of electrolyte occurs it must be immediately cleaned from affected areas. Remove batteries (subpar. g, below) and wash battery box, cable terminals, and other affected hull areas with an alkaline solution. Open hull escape door to allow solution to drain from hull. Wash batteries and terminal posts with alkaline solution, using care not to get solution into battery cells. Clean all corrosion from terminal posts and scrape posts and interior of cable terminals until bright. Install batteries (subpar. h, below).

Part Three—Maintenance Instructions

f. **Battery Cable Replacement.**

(1) Turn both master switch box switches off (fig. 15).

(2) On M18 vehicle, open front and rear battery doors (fig. 260). On M39 vehicle, open battery box cover and remove battery box left side cover (fig. 262).

(3) Remove cover from battery junction box, disconnect battery cables in junction box and at battery terminals and remove cables.

(4) Before new cables are installed, scrape terminals lightly to provide good contact and coat with petrolatum. Install cables and connect them in junction box as shown in figure 230, and to batteries as shown in figure 189. Tighten all attaching nuts securely and install junction box cover.

(5) On M18 vehicle, close battery doors and anchor each door with two cap screws ($\frac{3}{8}$ in.—24 x 1 in.). On M39 vehicle, install battery box left side cover with four cap screws ($\frac{3}{8}$ in.—24 x $\frac{3}{4}$ in.) and lock washers, then close battery box cover.

g. **Battery Removal.** Before removal, make certain that both master switch box switches are turned off (fig. 15). CAUTION: *Secure assistance to handle batteries; do not tilt or handle roughly.*

(1) REMOVE BATTERIES, M18 VEHICLE. Open front and rear battery doors (fig. 260). Loosen cable terminal clamp bolts, remove terminals from battery posts, and move cables clear of batteries. Remove the three battery hold-down clamps which are anchored with $\frac{7}{16}$-inch safety nuts and plain washers. Lift each battery out of box by means of handles and remove from vehicle.

(2) REMOVE BATTERIES, M39 VEHICLE. Open battery box center cover, and remove left and right covers which are anchored by eight $\frac{3}{8}$-inch cap screws and lock washers (fig. 262). Remove seat back pad which is attached with 11 flat-headed screws and finish washers. Remove the $\frac{7}{16}$-inch battery hold-down clamp safety nuts and washers from studs, remove ten $\frac{3}{8}$-inch cap screws which anchor battery box rear plate and lower plate and cover to horizontal position (fig. 188). Loosen cable terminal clamp bolts, remove terminals from battery posts, and remove hold-down clamps. Slide batteries to rear and remove from vehicle.

h. **Battery Installation.**

(1) GENERAL. Remove battery floor pads and inspect battery box to make sure it is in good condition and free of corrosion. If electrolyte has spilled into battery box, wash off with alkaline solution and open hull escape door to allow solution to drain from hull. Dry surfaces thoroughly, scrape off corrosion and paint bare areas with acid-resisting paint. Replace battery floor pads and hold-down clamp pads if in bad condition. On M39 vehicle, make sure that inspection hole cover is

Batteries and Charging System

Figure 188—Battery Box Opened for Removal of Batteries—M39

in place under left battery floor pad. Wash batteries with alkaline solution, if necessary, to remove all acid from outside of case and terminal posts, using care not to get solution into battery cells. Scrape any corrosion from battery posts and battery terminals to secure bright metal surfaces, and coat surfaces with petrolatum.

(2) BATTERY INSTALLATION, M18 VEHICLE. Place batteries (H-15-500907) in battery box with terminal posts together at middle of box. Place hold-down clamps over studs, install plain washers and safety nuts (7/16 in.—20) on studs and tighten until clamp pads are firmly compressed. Connect ground cable to negative (—) post of front battery, connect 24-volt cable to positive (+) post of rear battery, and connect battery-to-battery cable to remaining posts with 12-volt cable attached to rear battery negative (—) terminal, as shown in figure 189. Close battery doors and anchor with cap screws (3/8 in.—24 x 1 in.) tightened to 20-25 foot-pounds tension.

(3) BATTERY INSTALLATION, M39 VEHICLE. Place batteries (H-15-500907) in battery box with terminal posts together at middle of box. Raise battery box rear plate and cover, and attach rear plate with 10 cap screws (3/8 in.—24 x 3/4 in.) and lock washers tightened to 25-30 foot-pounds tension. Place hold-down clamps over studs, install plain washers and safety nuts (7/16 in.—20) on studs and tighten

TM 9-755
142-143

Part Three—Maintenance Instructions

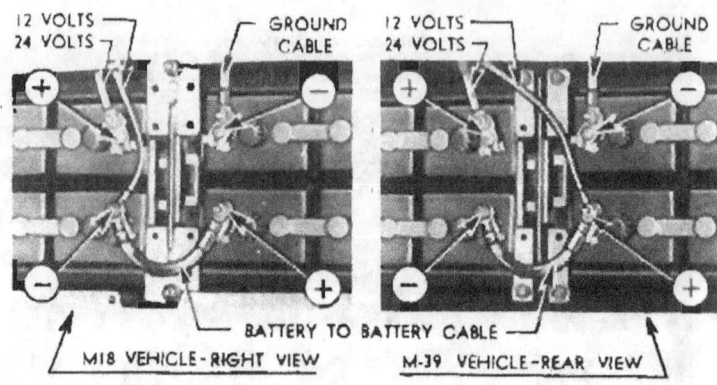

Figure 189—Cable Connections to Batteries

until clamp pads are firmly compressed. Connect ground cable to negative (—) post of right battery, connect 24-volt cable to positive (+) post of left battery, and connect the battery to battery cable to remaining posts with 12-volt cable attached to right battery positive (+) post, as shown in figure 189. Install right and left battery box covers and attach each cover with four cap screws ($\frac{3}{8}$ in.—24 x $\frac{3}{4}$ in.) and lock washers tightened to 25-30 foot-pounds tension. Close and latch center cover. Install front seat back pad with 11 flat head screws ($\frac{1}{4}$ in.—28 x $\frac{5}{8}$ in.) and finish washers.

143. ENGINE GENERATOR.

a. **Drive Belt Adjustment** (fig. 190). On M18 vehicle, remove right rear subfloor plate (fig. 260); on M39 vehicle, open rear seat center cover. Back off lower adjusting rod nut and tighten safety nut on end of rod until belts can be depressed $\frac{1}{2}$ inch midway between pulleys when pressed firmly with fingers. Tighten lower adjusting rod nut against upper bracket. Install subfloor plate (M18) or close rear seat center cover (M39).

b. **Drive Belt Removal.** Disengage transfer case clutch so that propeller shaft can be turned by hand (par. 102). On M18 vehicle, remove right rear sub-floor plate (fig. 260); on M39 vehicle, open rear seat center cover. Flatten bent up tongues of lock plates and remove four $\frac{3}{8}$-inch cap screws with lock plates which attach propeller shaft to rear universal joint (fig. 123). Push propeller shaft forward and remove universal joint trunnion bearings from spider. Unscrew safety nut on upper end of belt adjusting rod far enough to remove belts from generator pulley. Work belts out between end of propeller shaft and universal joint, pushing shaft forward and turning it as required to secure clearance.

TM 9-755
143

Batteries and Charging System

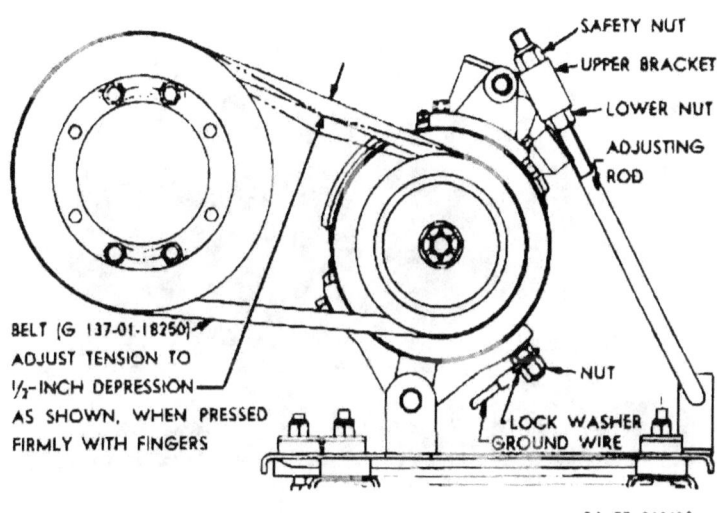

Figure 190—Generator Drive Belt Adjustment

c. **Drive Belt Installation.** Work belts through between end of propeller shaft and universal joint, pushing shaft forward and turning it as required to secure sufficient clearance. Place belts over drive and driven pulleys and adjust tension (subpar. a above). Place universal joint trunnion bearings on spider and connect them to propeller shaft with two lock plates and four cap screws ($\frac{3}{8}$ in.—24 x 1 in.) tightened to 28-33 foot-pounds tension (fig. 123). Bend tongues of lock plates up against flats on screw heads. Install subfloor plate (M18), or close rear seat center cover (M39). Engage transfer case clutch (par. 102). NOTE: *Check tension of belts after vehicle has been driven 25 miles, and readjust if necessary.*

d. **Commutator and Brushes.** If generator output is below normal, the commutator or brushes may require service. On M18 vehicle, remove right rear subfloor plate; on M39 vehicle, open rear seat center cover. Remove cover band on rear end of generator and inspect commutator and brushes. The commutator requires cleaning only when excessively carbonized or when too much arcing occurs. To clean, start engine, place a strip of 2/0 flint paper over thin flat piece of wood having a smooth square end and hold it against commutator until it is bright. CAUTION: *Do not use emery cloth.* Remove any grease or grit by holding rag moistened with dry-cleaning solvent against commutator. CAUTION: *Make certain that solvent is noninflammable.* If brushes are sticking in holders, are worn so that brush springs ride on brush holders, or if lead wires are loose, remove gen-

TM 9-755
143

Part Three—Maintenance Instructions

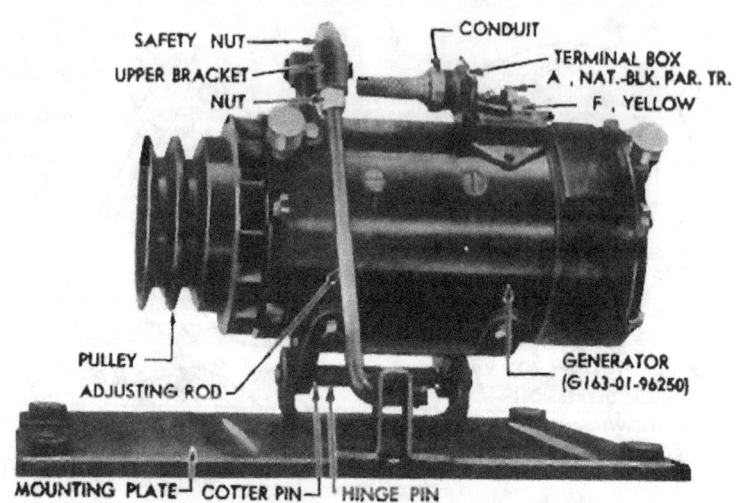

Figure 191—Engine Generator on Mounting Plate

erator (subpar. e and f, below) so that brushes can be freed up or replaced properly.

e. **Generator Removal** (fig. 191). On M18 vehicle, remove right rear subfloor plate and open right rear subfloor door; on M39 vehicle open rear seat center cover. Remove terminal cover, disconnect two wires in terminal box, unscrew conduit coupling nut and move conduit out of way. Remove safety nut from upper end of adjusting rod, swing generator to right until adjusting rod comes out of upper bracket, and move drive belts off of pulley. Disconnect ground wire which is attached to generator by a nut and lock washer on left front mounting bracket stud (fig. 190). Remove two cotter pins from hinge pin, support generator while removing hinge pin, and lift generator from mounting plate.

f. **Generator Installation** (fig. 191). Place generator in position on mounting plate, install hinge pin and secure it with two cotter pins (1/8 in. x 1 in.) Attach ground wire to left front stud of mounting bracket with nut and lock washer (fig. 190), making sure that contact surfaces are clean. Swing generator to right, insert upper end of adjusting rod into upper bracket, place drive belts on pulley, and install safety nut on adjusting rod. Adjust belt tension (subpar. a above). Connect conduit to terminal box, connect yellow wire to terminal stud marked (F), and connect other wire (Nat.-Blk, Par. Tr.) to terminal stud marked (A). Install terminal box cover and install lock wire through cover and terminal box screw heads. Start engine and check generator operation. Polarize generator, if necessary (subpar. g, below). Close right rear subfloor door and install right rear subfloor plate (M18), or close rear seat center cover (M39).

g. *Polarizing Generator.* When a new generator is installed it may be necessary to polarize it to establish residual magnetism and cause the current to flow in proper direction. To polarize generator, remove main filter box cover which is attached with six captive screws, disconnect wire at terminal marked "FIELD" or "F" on engine generator regulator and touch it for an instant to either terminal on the circuit breaker (fig. 196). Connect wire to regulator being sure to connect condenser wire also, and install filter box cover.

h. *Drive Pulley Replacement.* Remove rear universal joint (par. 106 e). Remove cotter pin and nut, and pull universal joint yoke from transfer case output shaft. Remove old pulley and place new pulley over yoke, install yoke on output shaft and install nut (1¼ in.—18). Engage transfer case clutch (par. 102), tighten nut securely and install cotter pin (⅛ in. x 2½ in.). Install rear universal joint (par. 107 e).

144. AUXILIARY GENERATOR, M18.

a. *Cleaning Spark Plug and Adapter.* The following information is contained on an instruction plate attached to fan housing: "Use only Champion HO-14S spark plug. Do not use standard aircraft plug as gap of approximately 0.012 inch is too close. This engine requires a gap setting of 0.025 inch. It is impractical to open up gap on standard aircraft plug."

(1) Unscrew nut and disconnect spark plug conduit. Remove cap and spring assembly. Remove spark plug and gasket using ⅞-inch hexagon socket wrench. Remove spark plug adapter and gasket from cylinder, using one inch hexagon socket wrench.

(2) Clean the spark plug, and if necessary, adjust point gap to 0.025 inch. If points are badly worn, replace with new Champion HO-14S plug (M001-01-09365).

(3) Scrape all carbon and lead deposits from all holes and both sides of the adapter baffle. If the baffle shows signs of erosion around the holes or at the edge, replace adapter.

(4) Install baffle with new gasket and tighten securely with one-inch hexagon socket wrench. Install spark plug with new gasket and tighten firmly, but not too tight, using ⅞-inch hexagon socket wrench. Install cap and spring assembly. Connect spark plug conduit to plug and tighten nut firmly.

b. *Magneto Test, Inspection, and Adjustment* (fig. 192).

(1) TEST. Disconnect conduit from spark plug and hold end of high tension wire ¼ inch from cylinder shield. Spin engine quickly with starting rope or, if connected to vehicle battery, by depressing control switch button (fig. 24). If no spark, or only a weak spark, is

TM 9-755
144

Part Three—Maintenance Instructions

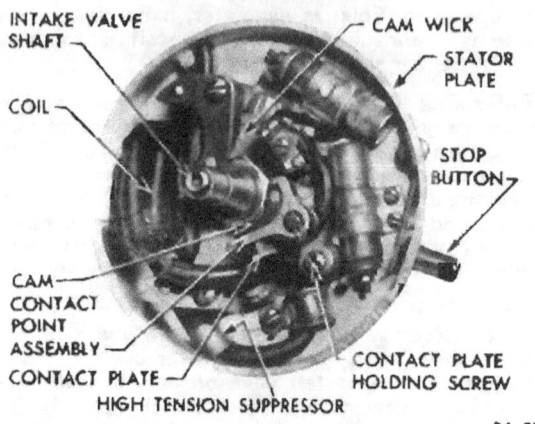

Figure 192—Magneto Stator Plate Assembly

obtained check contact point condition and adjustment (steps (2) and (3) below).

(2) INSPECTION. Remove magneto rotor by loosening rotor nut. NOTE: *Do not remove three screws holding stator plate to rotor.* Remove spark plug to relieve engine compression and permit turning the flywheel. Inspect contact point surfaces and if badly pitted, replace with a new set. Uneven or pitted contact points may be restored to a true, even condition with a contact point dressing tool. NOTE: *Do not use a file on contact point surfaces.* After dressing, wipe points with a clean, dry cloth to remove all dust particles. Tighten all wiring connections in magneto.

(3) ADJUSTMENT. Turn flywheel slowly counterclockwise until breaker arm fiber rests on highest point of cam, approximately 1/8 inch past breaking edge of cam. Check contact point gap with a feeler gage; the correct setting is 0.020 inch. CAUTION: *To prevent damage to points, separate points by hand to place gage between surfaces; separate points by hand to remove gage.* If point gap adjustment is required, *slightly* loosen contact plate holding screw. Move contact plate away from cam to increase gap, or toward cam to decrease gap. After adjusting points to 0.020 inch gap, tighten contact plate holding screw securely and recheck gap with gage to make certain that adjustment did not change. Install spark plug and attach conduit. Install magneto rotor and tighten nut securely.

c. Carburetor Adjustment (fig. 193). The auxiliary generator engine will operate at full speed even when the carburetor is set considerably too rich, but under this condition excessive carbon is formed and poor fuel economy obtained. For this reason the carburetor should

TM 9-755
144

Batteries and Charging System

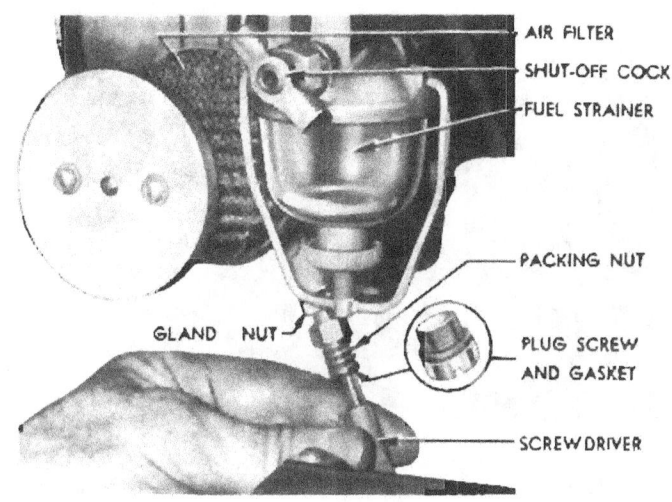

Figure 193—Adjusting Carburetor

be adjusted correctly. Proper adjustment of carburetor can be obtained only when the engine is warm and operating under full rated load.

(1) Remove plug screw and gasket from lower end of packing nut. Hold gland nut with wrench and loosen packing nut. The adjusting screw, which is concealed in the packing nut, can then be turned by inserting a small screwdriver in its slotted head. Turn clockwise for leaner mixture or counterclockwise for richer mixture.

(2) If the carburetor is completely out of adjustment, a setting for starting the engine can be obtained by closing adjusting screw lightly against its seat, then turning to counterclockwise 1¼ turns. CAUTION: *Do not turn adjusting screw hard against seat as this will damage both seat and needle end of screw.*

(3) With engine warm and operating under full rated load, turn adjusting screw clockwise until engine speed just begins to fall off, as noted by sound of the exhaust. Then turn adjusting screw back, counterclockwise, very gradually until engine reaches full speed. The proper setting is approximately one-eighth turn richer than the leanest point at which maximum speed is obtained. A slightly richer setting of one-eighth to one-quarter turn is advisable in extremely cold weather.

(4) After proper setting of adjusting screw is obtained, hold the screw by means of small screwdriver while tightening packing nut to make certain that adjustment is not changed. Install plug screw and gasket in lower end of packing nut.

TM 9-755
144

Part Three—Maintenance Instructions

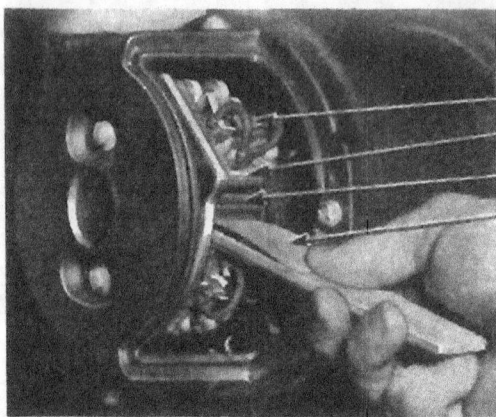

Figure 194—Cleaning the Generator Commutator

d. **Cleaning Fuel Strainer** (fig. 193). When dirt or water accumulates in the fuel strainer it must be cleaned out. Close fuel shut-off cock, unscrew knurled nut and swing bowl bail to one side, and remove bowl, gasket, and screen from strainer cover. Wash bowl and screen in dry-cleaning solvent and dry thoroughly. Install screen, gasket, bowl, and tighten bail nut securely. Open fuel shut-off cock.

e. **Cleaning Carburetor Air Filter** (fig. 193). Carburetor air filter must be removed to be cleaned, by removing two attaching screws and lock washers. Clean filter by rinsing in dry-cleaning solvent, then dip in engine oil and allow to drain. Wipe oil off end caps and install filter with two screws and lock washers.

f. **Cleaning Generator Commutator and Brushes.** Remove brush head cover plate and inspect the commutator and brushes. Commutator requires cleaning only when excessively carbonized or when too much arcing occurs. To clean, start engine, place strip of 2/0 flint paper over thin flat piece of wood having smooth square end and hold it against commutator until it is bright (fig. 194). CAUTION: *Do not use emery cloth.* If brushes do not seat properly, insert strip of 2/0 flint paper between commutator and brush with flint side against brush. While holding flint paper flat against commutator, rock engine back and forth until carbon shows across entire width of brush. If brush is worn so that brush spring rides on brush holder, or if lead wire is loose, replace brush. New brushes are shaped to commutator and do not require seating with flint paper.

g. **Auxiliary Generator Ammeter Replacement.** Unscrew button from starting switch, take out two attaching screws, and remove cover from control box. Disconnect two wires from ammeter. Remove

TM 9-755
144

Batteries and Charging System

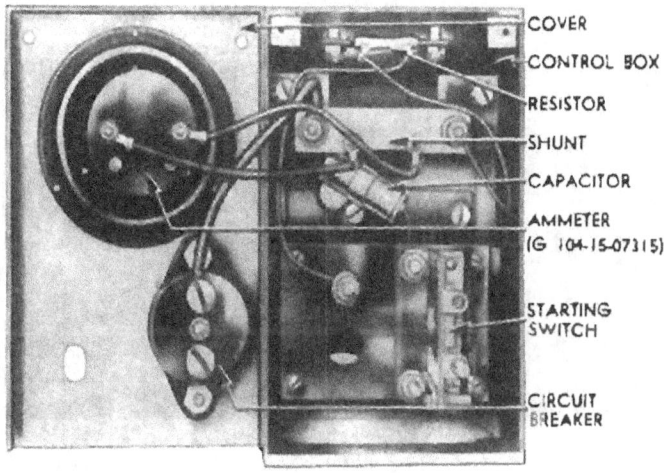

Figure 195—Interior of Control Box

three screws (No. 6—32 x 3/16 in.) which attach the ammeter mount to control box cover, and remove three screws (No. 4—36 x 3/8 in.) which attach the mount to ammeter. When ammeter is installed, be sure to connect wires as shown in figure 195.

h. **Auxiliary Generator Removal, M18 Vehicle.**

(1) Remove junction box cover and disconnect three wires from terminals in box (fig. 234). Disconnect conduit from junction box, and remove 3/16-inch cadmium-plated screw and lock washer which attaches conduit clip to hull side plate. Close shut-off cock at fuel tank (fig. 25), and disconnect fuel line at fuel strainer.

(2) Remove six cap screws (1/4 in.—28 x 5/8 in.) and lock washers which attach generator mountings to floor plate. Remove two cadmium-plated screws 1/4 in.—28 x 5/8 in.) and internal-external toothed lock washers which attach conduit clips and generator mountings to floor plate. Remove one cadmium-plated screw (1/4 in.—28 x 5/8 in.), plain washer and two internal-external toothed lock washers which attach ground strap to floor plate.

(3) Remove air outlet lower duct which is attached to upper duct by seven 3/16-inch cap screws and lock washers.

(4) Remove four 1/4-inch cap screws and lock washers which attach the exhaust pipe and gasket to engine. Remove generator assembly from vehicle.

(5) Examine exhaust pipe flexible coupling for cracks, using

TM 9-755
144

Part Three—Maintenance Instructions

Figure 196—Generator Regulators and Main Filter Box

care not to bend it, which might cause it to crack. If coupling is cracked, disconnect it from exhaust pipes.

i. **Auxiliary Generator Installation, M18.**

(1) If exhaust pipe flexible coupling was removed for replacement, connect new coupling to upper and lower exhaust pipes with cap screws (¼ in.—20 x ¾ in.) and lock washers, using new gaskets at joints.

(2) Place generator assembly in position on floor plate and attach exhaust pipe and new gasket to engine with four cap screws (¼ in.— 20 x ¾ in.) and lock washers, being careful not to strain and possibly crack exhaust pipe flexible coupling. Install air outlet lower duct and gasket with seven cap screws (⁵⁄₁₆ in.—24 x ¾ in.) and lock washers.

(3) Install six cap screws (¼ in.—28 x ⅝ in.) and lock washers through holes in rear mountings, and rear sides of front mountings, into subfloor plate. Attach ground strap to floor plate with one cadmium-plated screw (¼ in.—28 x ⅝ in.), placing one internal-external toothed lock washer between strap and plate, and one plain washer and one internal-external-toothed lock washer on top side of strap. Attach conduit and clips with two cadmium-plated screws (¼ in.—28 x ⅝ in.) and internal-external toothed lock washers through front mounting holes to floor plate.

(4) Attach conduit and clip to hull side plate with one cadmium-plated screw (⁵⁄₁₆ in.—24 x ½ in.) and external-toothed lock washer. Connect conduit to junction box, attach wires to terminal studs as shown in figure 234, and install junction box cover. Connect fuel line

to fuel strainer and open shut-off cock at fuel tank to check for tight connection.

145. GENERATOR REGULATORS.

　a. Removal (fig. 196). Turn master switches off (fig. 15). Remove main filter box cover which is attached with six captive screws. Disconnect wires and condensers from the three terminal studs of regulator to be removed. The engine generator regulator is on inner side, and auxiliary generator regulator is on outer side of filter box. Remove regulator which is attached to mounting plate by four $\frac{3}{16}$-inch cadmium-plated cap screws with plain and external-tooth lock washers.

　b. Installation (fig. 196). Install regulator on mounting plate with four cadmium-plated cap screws ($\frac{3}{16}$ in.—24 x 1¼ in.) having plain and external-toothed lock washers; attach ground strap with nearest attaching screw, placing an external-toothed lock washer between strap and regulator mounting lug and a plain washer and an external-toothed lock washer on top of strap. Attach condenser and other wires to terminal studs on regulator with lock washers and nuts, in accordance with wiring diagram on filter box cover (fig. 231). Make sure that wire terminals are clean and bright, and tighten nuts securely. Install filter box cover.

Section XXX

STARTER SYSTEM

146. DESCRIPTION, CIRCUIT AND DATA.

　a. Description. The starter system consists of starter, starter relay, starter control circuit breaker, starter switch, starter neutral safety switch, connecting wires in shielded conduits, and junction boxes. The starter is a 24-volt, 4-pole compound, intermittent duty motor mounted on the engine crankcase (fig. 198). It is equipped with hand crank mechanism through which the engine can be cranked by means of removable starting crank inserted through an opening in hull rear door (par. 183 c). The starter relay, which is a solenoid-type switch mounted in a box on front side of bulkhead (fig. 199), eliminates the necessity for large capacity starter switch. When energized by a relatively low amperage current it closes circuit which supplies the high-amperage current required by starter. Starter switch, located in the instrument panel (G, fig. 16), closes the circuit which energizes relay. Starter neutral safety switch, mounted on the transmission shift lever bracket (fig. 125) and actuated by shift lever, is closed to permit passage of energizing current to relay only when shift lever is in neutral. It prevents accidental starting of the engine when the transmission is in gear. The starter control circuit breaker

TM 9-755
146

Part Three—Maintenance Instructions

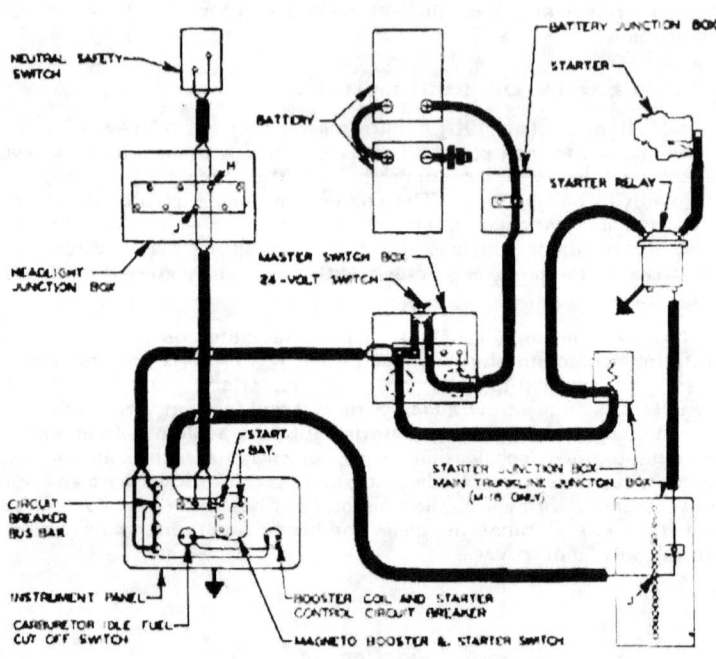

Figure 197—Starter Circuits

located in the instrument panel (A, fig. 16), is manual reset type breaker which protects starter control circuit against overload.

b. **Starter Circuit** (fig. 197). A single-wire system is used with ground-return. The circuit includes a low amperage control (energizing) circuit and a high-amperage power circuit. Wires in shielded conduit connect the 24-volt battery circuit in master switch box to common circuit breaker bus bar in the instrument panel. Wires conduct the low-amperage control current to the control circuit breaker, to starter switch, to neutral safety switch, and to starter relay, in order given. Outlet end of the relay energizing coil is grounded. Large capacity cables conduct high-amperage power current from master switch box to starter relay, and from relay to starter which is grounded to vehicle through engine. When starter switch is closed, with transmission shift lever in neutral, the control current energizes the relay which then closes power circuit to supply current to starter. If shift lever is not in neutral, control circuit is open at neutral safety switch so that no current can pass to energize relay.

Starter System

Figure 198—Starter Installed on Engine

c. Data.

Make	Delco-Remy
Model	1108685
Type	4-pole, compound, intermittent duty
Voltage	24
Control circuit breaker	30 amp.

147. STARTER.

a. Removal (fig. 198). Turn 24-volt master switch off (fig. 15). Open hull rear door (par. 181 a). Cut lock wire, unscrew cap from terminal shield and unscrew conduit coupling nut from shield. Remove ⅜-inch brass nut, internal-toothed lock washer, plain washer, and wire terminal from terminal stud in shield. Remove lock nuts and loosen large ⅜-inch nuts on the six starter flange studs, using crowfoot wrench (41-W-871-45). Support starter while removing nuts and plain washers, and remove starter. Remove gasket.

b. Installation (fig. 198). Place gasket over dowel pins on engine crankcase. Install starter over dowel pins with hand crank jaw straight down, and support starter while installing zinc-plated plain washers and nuts (⅜ in.—24) on all flange studs. Tighten nuts securely and install lock nuts (⅜ in.—24) using crowfoot wrench (41-W-871-45). Be careful not to strip lock nuts by excessive tightening. NOTE: *In some installations the starter attaching studs are drilled for use of lock wire (0.041 in. diam.) instead of lock nuts.* Place starter wire terminal on terminal stud through side opening in terminal shield and secure it with a zinc-plated plain washer, a cadmium-plated

TM 9-755
147-148

Part Three—Maintenance Instructions

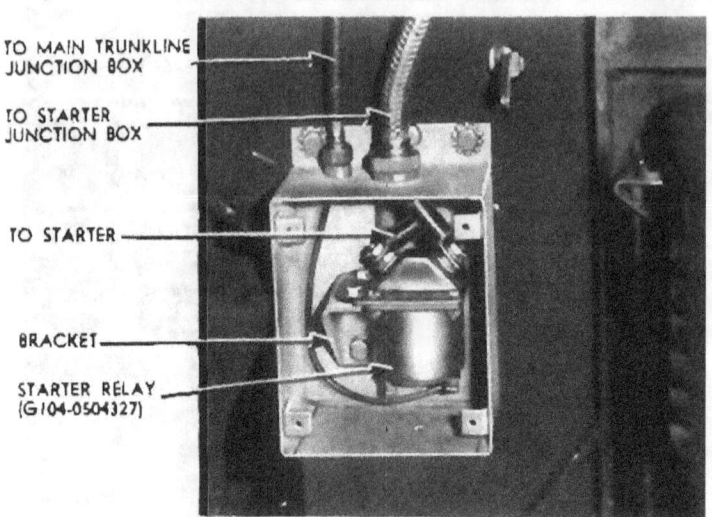

Figure 199—Starter Relay Installed in Box

external-toothed lock washer, and a brass nut ($3/8$ in.—16). Screw conduit coupling nut firmly on terminal shield. Install cap in shield and secure it with lock wire. Turn engine two complete revolutions by hand crank (50 turns of crank) to make certain that engine turns freely, then test starter installation by closing master switch and starter switch to crank engine several revolutions with starter. Close hull rear door (par. 181 b).

148. STARTER RELAY.

a. Removal (fig. 199).

(1) On M39 vehicle only, remove rear seat back (par. 185 e).

(2) Turn 24-volt master switch off (fig. 15).

(3) Remove starter relay junction box cover. Remove nuts and lock washers which attach two large wires and one small wire to relay. Remove two $5/16$-inch bracket attaching cap screws and lock washers and remove bracket and relay from box.

(4) Remove bracket which is attached to relay by two $1/4$-inch cap screws and lock washers.

b. Relay Installation (fig. 199).

(1) Install bracket on relay with two cadmium-plated cap screws ($1/4$ in.—28 x $3/8$ in.) and external-toothed lock washers.

(2) Install relay and bracket in relay junction box, attaching

Starter System

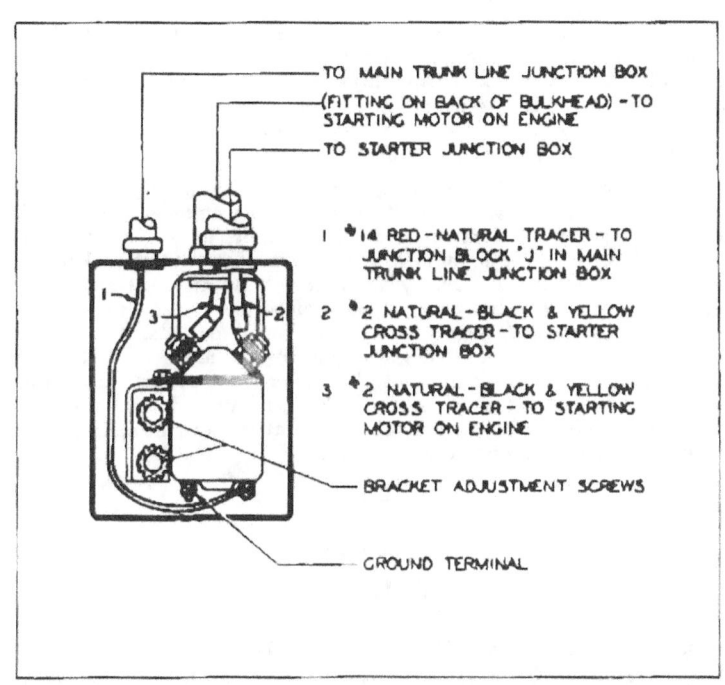

Figure 200—Starter Relay Junction Box Wiring Diagram

bracket with two cadmium-plated cap screws (5/16 in.—24 x 3/4 in.) and internal-external-toothed lock washers. Leave cap screws loose.

(3) Attach terminals of large wires to upper terminal studs, and terminal of small wire to lower terminal stud as shown on diagram on box cover (fig. 200). NOTE: *No wire is connected to ground terminal.* Install stud lock washers and nuts and tighten securely. Adjust position of bracket so there is ample clearance between lower terminals and relay box and tighten attaching screws. Install relay junction box cover.

(4) On M39 vehicle only, install rear seat back and pad (par. 185 d).

149. STARTER NEUTRAL SAFETY SWITCH.

a. Removal (fig. 125). Remove two round head screws, lock washers and plain washers which attach neutral safety switch to transmission shift lever bracket. Unscrew conduit coupling nut, remove switch cover, and disconnect the two wires from switch terminals.

TM 9-755
149-150

Part Three—Maintenance Instructions

b. Installation. Remove switch cover and place it over end of conduit. Connect two wires to switch terminals, install cover, attach conduit and tighten coupling nut firmly. Place switch on shift lever bracket, with key on switch mounting bracket engaged in keyway in lever bracket and secure it with two round head screws (10—32 x 42 in.) provided with plain washers and external-toothed lock washers. Check switch timing (subpar. c below).

c. Checking Switch Timing. Remove headlight junction box cover and disconnect wires to safety switch at terminals marked (H) and (J) (fig. 233). Connect leads of test light (fig. 47) to terminals on switch wires, plug light into instrument panel outlet socket, and close 24-volt master switch. Test light will burn with transmission shift lever in neutral but must go out before lever reaches reverse or first range positions. Slowly move shift lever towards reverse until test light just cuts off, then mark position of lever on shift lever bracket; repeat operation, moving shift lever towards first range. Shift position of switch on shift lever bracket as required until test light cuts off at points equi-distant from neutral in both directions, then tighten switch attaching screws securely.

Section XXXI
LIGHTING SYSTEM

150. DESCRIPTION, CIRCUITS, AND DATA.

a. Description. Refer to paragraph 15 for operating instructions on light controls. The vehicle lighting system consists of the following components.

(1) HEADLIGHT, BLACKOUT LIGHT AND TAILLIGHT SWITCH. This switch is described in paragraph 15 g. Service instructions are given in paragraph 160 b.

(2) STOP LIGHT SWITCH. Two stop light switches are mounted above transmission in position to be actuated by cams on brake cross shafts. They are connected in series so that stop lights will burn only when both steering brake band levers of either set are pulled rearward together. Service instructions are given in paragraph 154 c.

(3) HEADLIGHTS (fig. 203). Each headlight asembly consists of a body mounted upon a support bracket and inclosing a sealed beam lamp-unit which is retained in position by a door. The body is adjustable on support bracket to permit light to be properly aimed (par. 151 e). Blackout marker light is mounted upon top of body. A blackout light assembly, which is installed in place of left headlight assembly when blackout lights are required is identical in construction except that it contains sealed beam blackout lamp-unit instead of the headlight bright lamp-unit.

(4) HEADLIGHT BLACKOUT RESISTOR. A blackout resistor is connected in the circuit to the left headlight so that the current will be

Lighting System

reduced from 24 volts to 6 volts when the blackout light is installed in place of the left headlight assembly.

(5) TAIL AND STOP LIGHTS (fig. 204). Each tail and stop light assembly consists of two sealed beam lamp-units inclosed in a body and retained by a door. Upper lamp-unit in left light assembly is the service tail and stop light. Upper lamp-unit in right light assembly is blackout stop light. Lower lamp-units in both light assemblies are blackout taillights.

(6) TRAILER BLACKOUT TAIL AND STOP LIGHT RESISTOR, M39 (fig. 252). Blackout resistors are connected in circuit between lights switch on instrument panel and trailer receptacle, so that the current to trailer tail and stop lights will be reduced from 24 volts to either 6 volts or 12 volts as required by trailer lights.

(7) INSTRUMENT PANEL LIGHTS. Instruments in instrument panel are lighted by two lamps in sockets covered by removable caps in face plate (E and AE, fig. 206). Lights are turned on and varied in intensity by rheostat-type switch marked "PANEL LIGHTS" (M, fig. 206).

(8) DOME LIGHTS. Two dome lights are located on ceiling of driving compartment above transmission. In turret of M18 vehicle, dome light is located above turret wiring switch box and another is located at radio. All dome lights are operated by individual switches in lights.

(9) OUTLET SOCKETS. Outlet sockets into which accessory lights may be plugged are located in instrument panel (J and AD, fig. 206), right dome light connector box (fig. 201), rear junction box (fig. 235), and turret wiring switch box (fig. 18). All sockets except in dome light junction box are closed by removable plugs when not in use.

(10) CIRCUIT BREAKERS. Manual reset circuit breakers are connected in series in all lighting circuits to protect wiring against overload conditions. Headlight and taillight wiring is protected by circuit breaker on instrument panel marked "OUTSIDE LIGHTS". Instrument panel lights, and the hull dome lights and outlet sockets are protected by the circuit breaker in instrument panel marked "INSIDE LIGHTS" (fig. 206). The M18 turret dome lights and outlet socket in turret wiring switch box are protected by separate circuit breakers marked "DOME LAMP" and "TROUBLE LAMP" (fig. 18).

(11) WIRES AND CONDUITS. All light wires are enclosed in shielded conduits for protection of wires and to effect adequate suppression of radio interference. Conduits are fitted with multiple pin connectors to join conduits to equipment or junction boxes. The pins at ends of each wire are marked by key letters stamped on connectors to aid in identifying each wire in the conduit.

(12) JUNCTION BOXES. Junction boxes are used for convenience in making connections of wires in separate conduits. Junction boxes included in lighting system are: headlight junction box (fig. 233), connector box (for driving compartment dome lights), main trunkline junction box, M18 only (fig. 229), rear junction box (fig. 235), slip

TM 9-755
150

Part Three—Maintenance Instructions

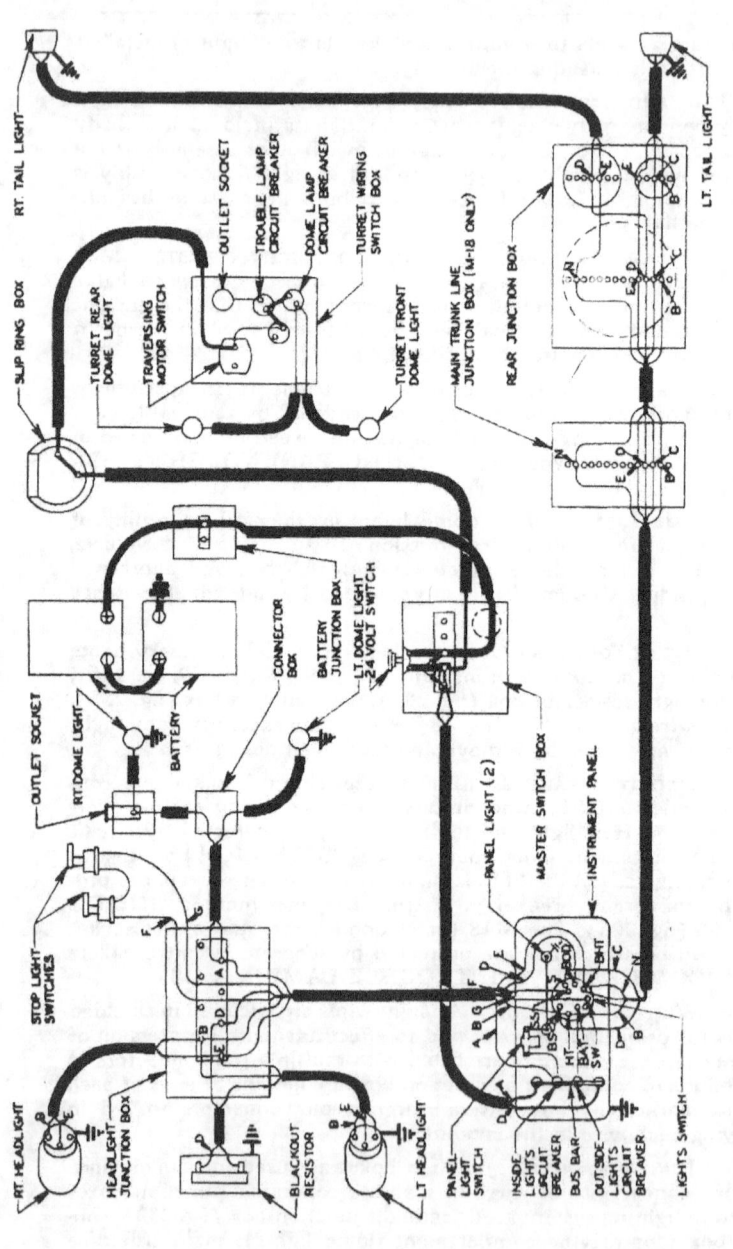

Figure 201—Lighting Circuits

Lighting System

ring box, M18 only (fig. 246) and turret wiring switch box, M18 only (fig. 18).

b. **Lighting Circuits** (fig. 201). All lights operate on 24-volt, single-wire with ground-return circuits. Wires in a conduit connect the 24-volt battery circuit in master switch box to common circuit breaker bus bar in instrument panel. A wire inside the panel connects the outside light circuit breaker to the "BAT" terminal of the light switch from which current is distributed to headlights and taillights. Wires inside instrument panel connect inside lights circuit breaker to the panel lights switch and lamps to instrument panel outlet socket. Inside lights circuit breaker is also connected in series with hull dome lights and outlet sockets. In the M18 vehicle, a separate circuit connects the 24-volt battery circuit in master switch box to slip ring box, from which connection is made to turret wiring switch box to supply current to turret dome lights and outlet socket. Figure 201 shows all units and connecting wires in lighting system. Individual unit wiring diagrams are contained in following units: instrument panel (fig. 207), headlight junction box (fig. 233), main trunk line junction box (fig. 229), and turret wiring switch box (fig. 245). Individual lighting circuits are as follows.

(1) HEADLIGHT CIRCUIT. Wires from light switch in instrument panel connect to a terminal block in headlight junction box, from which wires connect to both headlights and the headlight blackout resistor. When switch is in position "BO-MK", current is delivered to both headlight blackout markers. When switch is in position "HD-LT", current is delivered to both headlight lamp-units. When blackout light is installed in place of left headlight, and switch is in position "BO-DR", current is delivered to both headlight blackout markers and to left blackout headlight through blackout resistor, which reduces current to 6 volts.

(2) TAILLIGHT CIRCUIT. Wires from light switch in instrument panel connect through main trunkline junction box (M18 only) to rear junction box, from which connection is made to two taillights. In the M39 vehicle, wires go direct from instrument panel to rear junction box as a main trunkline junction box is not used. When lights switch is in position "BO-MK", current is delivered to both blackout taillights. When lights switch is in position "HD-LT", current is delivered to left taillight (bright). When lights switch is in position "BO-DR", current is delivered to both blackout taillights.

(3) STOP LIGHT CIRCUIT. Wires from lights switch in instrument panel go through headlight junction box to two stop light switches which are connected in series by a jumper wire. This circuit, which supplies current to stop lights, is open except when both stop light switches are closed by application of both brakes. From this circuit, wires connect to stop lamp-units in taillights through the same conduits and junction boxes described in taillight circuit. (Subpar. (3) above). When lights switch is in position "BO-MK", current is delivered to right blackout stop light when brakes are applied. When switch is in

positions "HD-LT" or "STOP-LT", current is delivered to left stop light (bright) when brakes are applied. When switch is in position "BO-DR", current is delivered to right blackout stop light when brakes are applied.

(4) TRAILER BLACKOUT TAIL AND STOP LIGHTS, M39 VEHICLE. Wires from lights switch on instrument panel lead to trailer blackout tail and stop light resistor (fig. 252). Wires from resistor lead to trailer electric receptacle at left rear corner of the hull. These wires pass through trailer brake resistor box.

(5) INSTRUMENT PANEL LIGHTS AND OUTLET SOCKETS. Wires inside instrument panel connect inside light circuit breaker to panel lights switch, from which wires deliver current to instrument panel lamps. Wires from inside lights circuit breaker also furnish current to outlet sockets in instrument panel, right dome light junction box, and rear junction box. Outlet socket in turret wiring switch box receives its current from supply line to switch box through trouble lamp circuit breaker.

(6) DOME LIGHTS. Wires from inside lights circuit breaker connect to dome lights in driving compartment through headlight junction box and connector box. Turret dome lights (M18) receive their current from supply line of turret wiring switch box through dome lamp circuit breaker.

c. Data.

(1) CIRCUIT BREAKERS.

Outside lights	20 amps
Inside lights	G163-0125880
Turret dome lamp	15 amps
Turret trouble lamp	G104-0500426

(2) LAMPS.

Blackout marker light	3 cp., single-
Dome light	contact
Instrument panel	(H104-190877)

(3) LAMP-UNITS.

Headlight (bright)	Sealed beam	(H104-193126)
Blackout headlight	Sealed beam	(M001-0107347)
Blackout stop	Sealed beam	(H004-504421)
Blackout taillight	Sealed beam	(H004-504418)
Service tail and stop	Sealed beam	(H004-504415)

151. HEADLIGHTS.

a. General. Headlights are carried in brackets in right side of driving compartment until their use is required. Headlight mounting sleeves in hull are weathersealed by plugs, which are chained to guards and carried in sockets on side of guards when headlights are installed.

b. Installation and Removal of Headlights. Unscrew headlight lock pins in front upper corners of hull (fig. 202), remove plugs and

TM 9-755
151

Lighting System

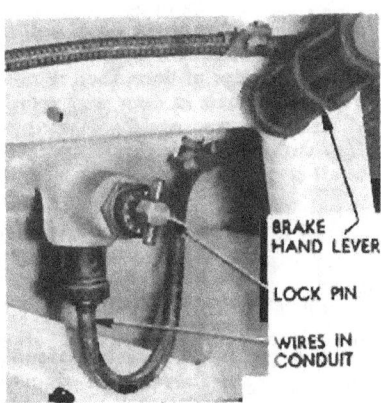

Figure 202—Headlight Lock Pin

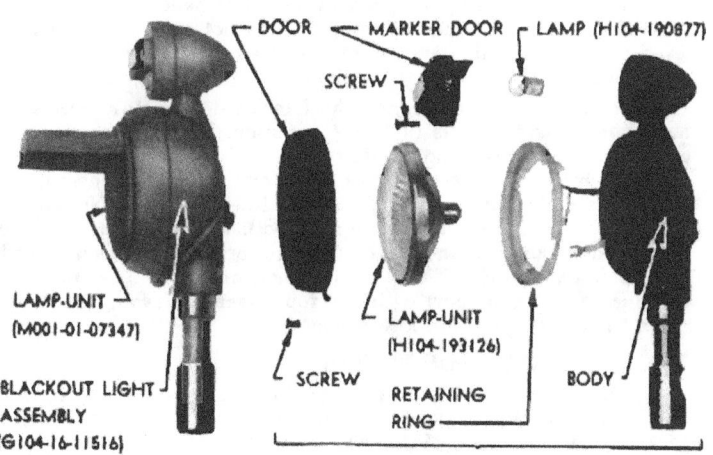

Figure 203—Blackout Light Assembly and Headlight Disassembled

place them in sockets on guards. Remove headlights from stowage brackets, install them in mounting sleeves and tighten lock pins securely. NOTE: *If blackout lights are to be used, install the blackout light in left mounting sleeve.* When headlights are removed after use, be sure to install plugs in mounting sleeves to prevent entrance of dirt and water.

c. **Replacement of Headlight Lamp-unit** (fig. 203). Remove attaching screw at lower edge of headlight door, pull door out and lift up from headlight body. Disconnect wires from lamp-unit and retaining ring. Unhook springs from edge of door, then remove retaining ring and lamp-unit. Place new lamp-unit in door with word "TOP" on lens at top of door. Install retaining ring and hook springs under edge of door. Attach orange (green-tracer) wire to lamp-unit and black wire to retaining ring. Install door on headlight body and secure it with screw.

d. **Replacement of Blackout Marker Lamp** (fig. 203). Remove attaching screw and lift off door. Press in on the lamp, turn counterclockwise and pull out. Insert new lamp into socket, press in and turn clockwise. Hook tip on top of door into slot in marker body, press door into position and install attaching screw.

e. **Adjustment of Headlights on Vehicle.** Headlights require no adjustment for aim unless they have been damaged or roughly handled. Park vehicle on a level area, install headlights, and turn light switch to "HD-LTS" position. Place a suitable screen 25 feet in front of headlights and at 90 degrees to centerline of vehicle. Mark a horizontal line on screen 5 1/4 inches lower than center of headlight lamp-units. Loosen three attaching screws and shift headlight body on support pad until upper outline of light beam hot spot is vertically centered on horizontal line and beam points straight ahead or parallel to centerline of vehicle; then securely tighten attaching screws.

f. **Replacement of Headlight Blackout Resistor.** Make certain that lights switch is in "OFF" position. Remove terminal cover and disconnect wires and conduit from resistor. Remove resistor assembly which is attached to hull with four screws. To install, attach resistor assembly to hull with four round-head screws (No. 10—32 x 1/4 in.) and cadmium-plated external-tooth lock washers. Attach conduit to cover support, and attach wires to terminal screws with cadmium-plated external-tooth lock washers and nuts (No. 10—32). Attach terminal cover to support with two round-head screws (No. 10—32 x 3/8 in.) and internal-tooth lock washers.

152. TAIL AND STOP LIGHTS.

a. **Replacement of Lamp-units.** Remove two attaching screws and remove taillight door. Pull lamp-unit to be replaced out of slip socket in taillight body. Push new lamp-unit into slip socket and install door. NOTE: *Refer to figure 204 for correct location of lamp-units.*

b. **Removal of Assembly.** Remove air outlet grille over light to be removed (par. 183 e). Remove cover and gasket which are attached to mounting bracket with four 3/8-inch bolts and lock washers. Disconnect wires from sockets by pushing in, turning counterclockwise, and pulling out on connector plugs. Remove safety nuts, lock washers, and flat washers from mounting studs and remove light assembly and seal from mounting bracket. Examine two rubber grommets and ground straps in mounting bracket and if they are in bad condition, push spacers out of grommets and push grommets out of mounting bracket.

TM 9-755
152-153

Lighting System

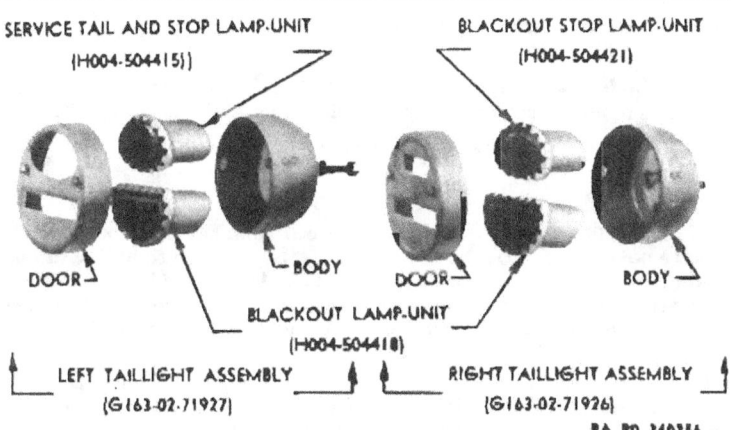

Figure 204—Tail and Stop Light Assemblies, Left and Right, Disassembled

c. **Installation of Assembly.** Push a rubber grommet through large hole in each ground strap so that end of strap is in groove in grommet. Install grommets in horizontal holes in mounting bracket, with ground straps forward, and push spacers into grommets. Install light assembly and seal through rear side of mounting bracket, making sure light is right side up and secure light with flat washer, cadmium-plated external-tooth lock washers, and safety nuts (¼ in.—20) on light studs. Connect wires by pushing connector plugs into sockets and turning clockwise. On right light, connect red wire to upper socket and orange wire to lower socket. On left light, connect double contact tan wires to upper socket and black (nat.-tracer) wire to lower socket. NOTE: *Pins on connector plugs, and slots in sockets are offset so that plugs can be installed in only one position; do not force plugs into sockets in wrong position.* Install gasket and cover on mounting bracket with four bolts (⅜ in.—24 x 1 in.) and lock washers. Install air outlet grille (par. 183 e).

153. DOME LIGHTS.

a. **Lamp Replacement.** Remove light cover attaching screws and remove cover. Press in on lamp, turn counterclockwise and pull out. Push new lamp into socket and turn clockwise. Install cover and attaching screws.

b. **Dome Light Replacement.** Remove light cover and lamp (subpar. a above). Disconnect wire from switch outer terminal. Remove two attaching screws and remove light from mounting bracket. Attach new light to mounting bracket with two round-head screws (No. 10—32 x ¾ in.), with plain and external-tooth lock washers. Connect wire to switch outer terminal. Install lamp and cover (subpar. a above).

TM 9-755
154-155

Part Three—Maintenance Instructions

154. LIGHT SWITCHES, INSTRUMENT LIGHTS, AND CIRCUIT BREAKERS.

a. **Light Switches on Instrument Panel.** Replacement of light switches mounted on instrument panel is covered in paragraph 160 c.

b. **Dome Light Switches.** Replacement of dome light switches is covered in paragraph 153.

c. **Stop Light Switches.**

(1) REPLACEMENT. Disconnect wires from switch to be replaced and remove switch from hull. To install, attach switch to hull with two cadmium-plated cap screws (¼ in.—28 x ½ in.) and lock washers. Connect two green wires to right hand switch, or one green jumper wire and one grey wire to left switch. Adjust operating cams (step (2) below).

(2) ADJUSTMENT OF OPERATING CAMS. Set both parking brake hand levers with pawls in fourth from front notches in quadrants. Place light switch on instrument panel in "STOP-LT" position. Loosen set screw on right operating cam, turn cam against switch lever until lever is fully depressed, then tighten set screw. Loosen set screw and move left operating cam against switch lever until service stop light (left, upper) just starts to burn, then tighten set screw. NOTE: *If 24-volt test lamp is available (fig. 47), disconnect one end of green jumper wire and connect lamp in series with jumper wire and switch terminal, then watch test lamp instead of stop light.* Loosen set screw and turn right operating cam away from switch lever, then back against lever until service stop light or test lamp just starts to burn and tighten set screw. After setting both operating cams, release brakes and test cam settings by pulling both levers back evenly until stop light burns, which should occur when pawls are at fourth or fifth notches in quadrants.

d. **Instrument Lights.** Replacement of instrument light lamps is covered in paragraph 162 a.

e. **Circuit Breakers.** Replacement of the lighting system circuit breakers contained in the instrument panel is covered in paragraph 161. Replacement of lighting system circuit breakers contained in the turret wiring switch box is covered in paragraph 168 d.

Section XXXII
INSTRUMENT PANEL ASSEMBLY

155. DESCRIPTION.

a. The instrument panel assembly is a metal box containing instruments, switches, circuit breakers, etc., described in paragraph 16. It is located in the sponson on left side of driving compartment and is mounted upon shock absorbing supports attached to sponson bottom plate. The face plate forms a panel which carries all equipment; case

Instrument Panel Assembly

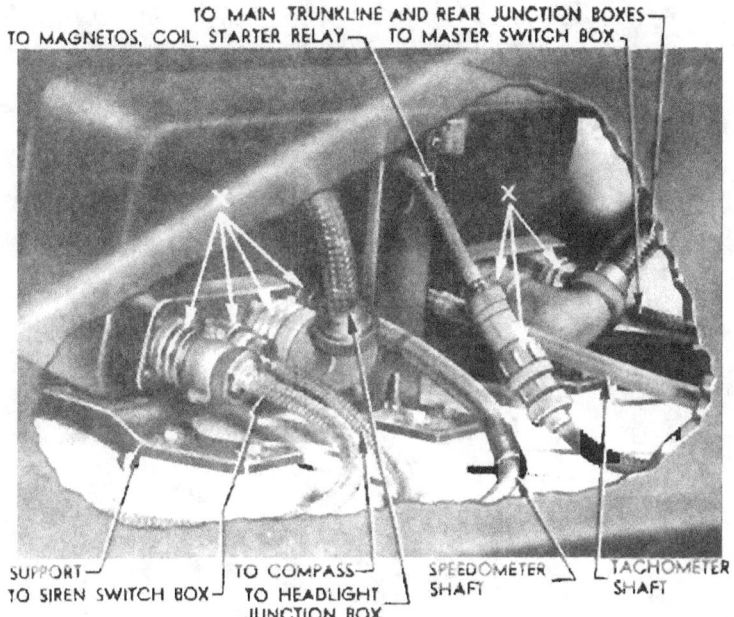

Figure 205—Connections to Instrument Panel

to which face plate is attached covers all equipment and connecting wires, and is attached to the mounting supports by means of studs which pass through rubber cushions in supports.

156. INSTRUMENT PANEL FACE PLATE.

a. Removal. It is not necessary to remove the instrument panel assembly in order to replace components. The instrument panel face plate upon which all components are mounted, is removed by the following procedure:

(1) Turn master switches off (fig. 15).

(2) Remove sponson side opening cover and gasket which are attached to sponson by 12 ½-inch bolts and lock washers.

(3) Disconnect the conduits attached to back of instrument panel by unscrewing the coupling nuts at points indicated by (X) in figure 205, and pull connector plugs from the receptacles in panel. CAUTION: *Do not attempt to disconnect at coupling nuts which attach conduits to connector plugs.*

(4) Disconnect speedometer and tachometer flexible shafts by unscrewing coupling nuts which attach them to instruments in panel.

TM 9-755

Part Three—Maintenance Instructions

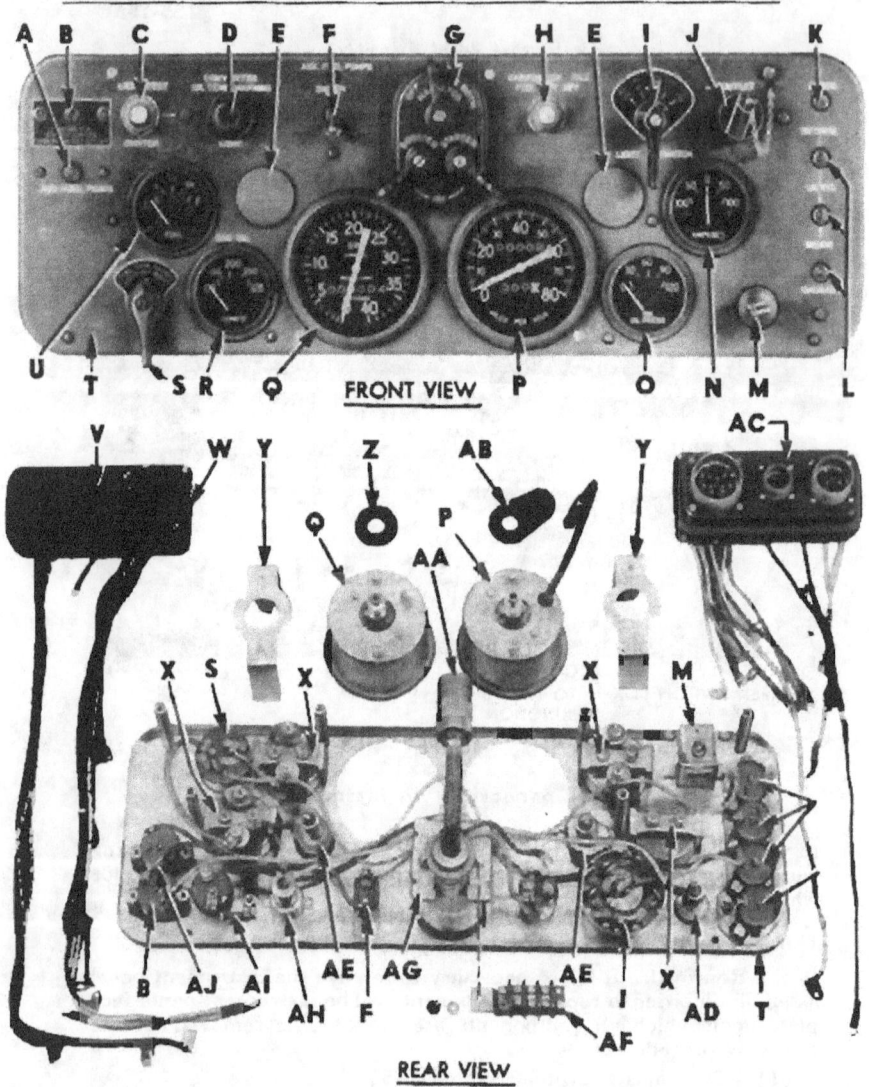

Figure 206—Instrument Panel Assembly—Front and Rear Views

Instrument Panel Assembly

- A   AUXILIARY FUEL PUMP CIRCUIT BREAKER (15 AMPS.)
- B   BOOSTER COIL, STARTER, IDLE CUT-OFF CIRCUIT BREAKER (30 AMPS.)
- C   LIGHT TEST SWITCH.
- D   CONVERTER OIL TEMPERATURE WARNING LAMP CAP.
- E   INSTRUMENT PANEL LAMP CAP.
- F   AUXILIARY FUEL PUMPS SWITCH.
- G   MAGNETO, BOOSTER AND STARTER SWITCH ASSEMBLY.
- H   FUEL CUT-OFF SWITCH.
- I   HEADLIGHT, BLACKOUT LIGHT AND TAILLIGHT SWITCH (G163-03-93788).
- J   OUTLET PLUG.
- K   SIREN CIRCUIT BREAKER (30 AMPS.)
- L   LIGHTS AND GAGES CIRCUIT BREAKERS (20 AMPS.)
- M   PANEL LIGHTS SWITCH.
- N   BATTERY AMMETER (M003-6199959).
- O   ENGINE OIL PRESSURE GAGE (G104-15-94117).
- P   SPEEDOMETER.
- Q   ENGINE TACHOMETER.
- R   ENGINE OIL TEMPERATURE GAGE (G104-2187858).
- S   FUEL GAGE SWITCH.
- T   INSTRUMENT PANEL FACE PLATE ASSEMBLY.
- U   FUEL GAGE (G104-1819406).
- V   AMPHENOL RECEPTACLE RETAINING PLATE (MAIN TRUNK AND MASTER SWITCH).
- W   RETAINING PLATE GASKET.
- X   MOUNTING CLAMP (FUEL, TEMPERATURE AND PRESSURE GAGES, AND AMMETER).
- Y   MOUNTING CLAMP (SPEEDOMETER AND TACHOMETER).
- Z   TACHOMETER TO INSTRUMENT PANEL GASKET.
- AA   MAGNETO, BOOSTER AND STARTER SWITCH CABLE ASSEMBLY.
- AB   SPEEDOMETER TO INSTRUMENT PANEL GASKET.
- AC   AMPHENOL RECEPTACLE RETAINING PLATE (COMPASS, HEADLIGHT, SIREN).
- AD   OUTLET SOCKET.
- AE   INSTRUMENT PANEL LIGHT SOCKET.
- AF   TERMINAL BLOCK.
- AG   MOUNTING CLAMP (MAGNETO, BOOSTER, STARTER SWITCH).
- AH   WARNING LAMP SOCKET.
- AI   LIGHT TEST SWITCH GROUND WIRE.
- AJ   BUS BAR.

RA PD 340340B

Legend for Figure 206—Instrument Panel Assembly—Front and Rear Views

Detach speedometer trip-set cable from support at bottom of panel by loosening nut.

(5) Remove the four round head screws which attach face plate, and remove face plate assembly from case.

b. Installation.

(1) Make sure that rubber gaskets are in place over amphenol receptacle retaining plates, speedometer, and tachometer (fig. 206).

(2) Place face plate assembly in instrument panel case and secure it to case with four round-head screws, using right upper screw to secure outlet plug chain.

(3) Attach speedometer trip-set cable to support at bottom of panel and tighten nut. Connect speedometer and tachometer shafts to their instruments and tighten coupling nuts securely.

(4) Starting at rear end of panel, connect wiring conduits to receptacles as shown in figure 205. When inserting connector plugs into receptacles, turn plugs so that their index grooves engage internal keys in receptacles, which will aline pins with proper sockets in receptacles and insure correct connections of circuits. Push plugs into receptacles, start coupling nuts and turn as tight as possible with fingers; then alternately push in on plugs and turn coupling nuts with fingers until plugs are fully seated and coupling nuts are finger-tight.

(5) Install sponson side opening cover and gasket with 12 bolts (½ in.—20 x 1¼ in.) and lock washers tightened to 75-100 foot-pounds tension.

157. BATTERY AMMETER (fig. 206).

a. Removal. Remove instrument panel face plate (par. 156 a). Remove three round-head screws and lock washers which attach amphenol retaining plate to posts on face plate over ammeter. Disconnect two wires from ammeter terminal studs, remove nuts, lock washers, fiber insulating plate, nuts and fiber washers from terminal studs, then remove mounting clamp and ammeter.

b. Installation. Place ammeter in face plate from front side, making sure it is right side up. Place mounting clamp over terminal studs and secure it with fiber washers and nuts on studs. Place insulating plate over terminal studs and install lock washers and nuts. Connect natural (red and black cross tracer) wire to negative terminal stud marker (—), and connect natural (green tracer) wire to positive terminal stud marker (+), securing each wire with a lock washer and nut (fig. 207 or 208). Attach amphenol retaining plate to posts on face plate with three round-head screws and lock washers. Install instrument panel face plate (par. 156 h).

158. GAGES IN INSTRUMENT PANEL (fig. 206).

a. Removal. Remove instrument panel face plate (par. 156 a). Remove the three round-head screws and lock washers that attach amphenol retaining plate to posts on face plate over gage to be removed. Disconnect wires from gage terminals, and remove mounting clamp which is secured to terminal studs by nuts and lock washers, and remove gage from face plate.

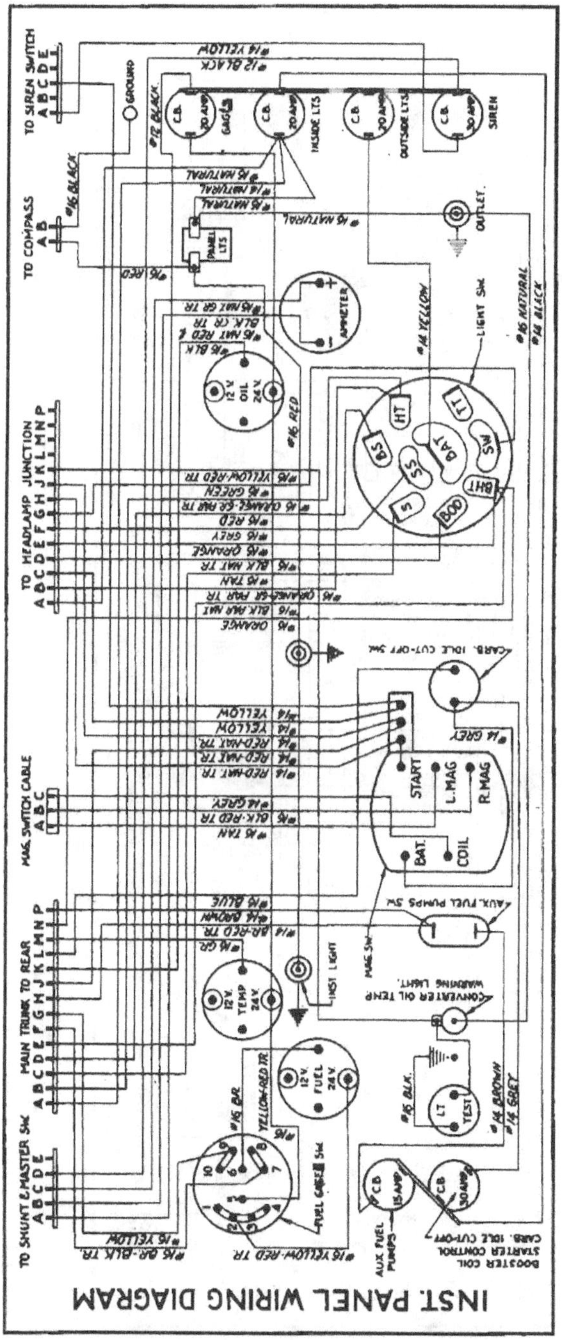

TM 9-755
158

Instrument Panel Assembly

Figure 207—Instrument Panel Wiring Diagram—M18

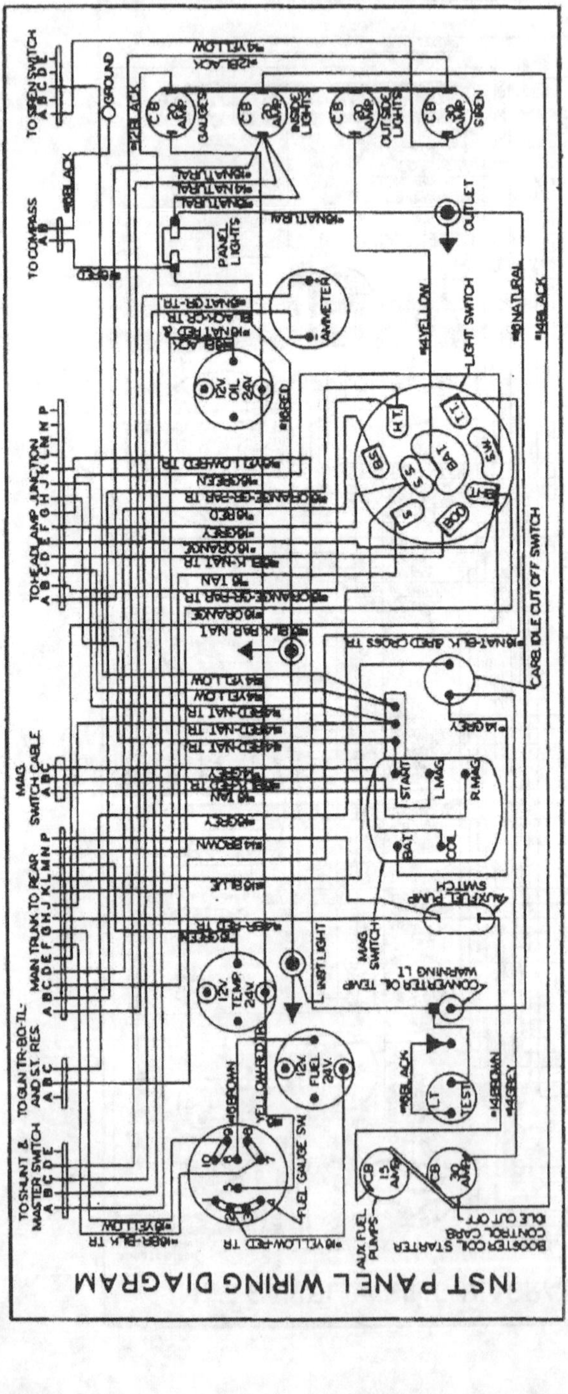

TM 9-755

Part Three—Maintenance Instructions

Figure 208—Instrument Panel Wiring Diagram—M39

Instrument Panel Assembly

b. **Installation.** Place gage in face plate from front side, making sure it is right side up. Place mounting clamp over terminal studs on gage and secure it with lock washers and nuts. Connect the yellow (red tracer) wire to the terminal marked "24V" on gage. Connect brown wire to terminal stud on fuel gage, or green wire to terminal stud on temperature gage, or black wire to terminal stud on pressure gage (fig. 207 or 208). Secure wires to terminals and studs with lock washers and nuts. Attach amphenol receptacle retaining plate to posts on face plate with three round-head screws and lock washers. Install instrument panel face plate (par. 156 b).

159. SPEEDOMETER OR TACHOMETER (fig. 206).

a. **Removal.** Remove instrument panel face plate (par. 156 a). Remove rubber gasket over mounting clamp and remove nuts and lock washers which secure mounting clamp to instrument to be removed. Remove instrument from front side of face plate.

b. **Installation.** Place instrument in face plate from front side, making sure it is right side up. Place mounting clamp over studs on instrument and secure it with lock washers and nuts. NOTE: *Attach black compass ground wire to lower stud when installing speedometer.* Place rubber gasket over mounting clamp around coupling threads. Install instrument panel face plate (par. 156 b).

160. SWITCHES IN INSTRUMENT PANEL.

a. **Replacement of Magneto, Starter, and Booster Switch Assembly** (fig. 206).

(1) REMOVAL. Remove instrument panel face plate (par. 156 a). Disconnect red (nat. tracer) wire from terminal block. Remove safety nuts which secure terminal block and mounting clamp to switch. Disconnect grey wire from carburetor idle fuel cut-off switch and remove switch assembly from face plate. Remove switch cover, disconnect wires from switch terminals and disconnect conduit from switch cover (fig. 209).

(2) INSTALLATION. Remove cover from replacement switch. Attach short grey wire to switch terminal marked "BAT" and attach short red (nat. tracer) wire to terminal marked "START". Attach conduit to cover and tighten coupling nut securely. Attach tan wire to switch terminal marked "L.MAG"; attach black (red tracer) wire to terminal marked "R.MAG"; attach grey wire to terminal marked "COIL". Make sure that wire terminals are turned so they will not contact the cover, and install cover on switch with two nuts and lock washers. Place switch assembly in face plate from front side, making sure it is right side up, and place mounting clamp over studs on switch. Connect grey wire to nearest terminal of carburetor idle fuel cut-off switch. Place terminal block over switch stud above fuel cut-off switch and secure block and mounting clamp with two safety nuts. Connect red (nat. tracer) wire to nearest terminal on terminal block. Install instrument panel face plate (par. 156 b).

b. **Replacement of Headlight, Blackout Light, and Taillight Switch** (fig. 206). Remove instrument panel face plate (par. 156 a).

TM 9-755
160

Part Three—Maintenance Instructions

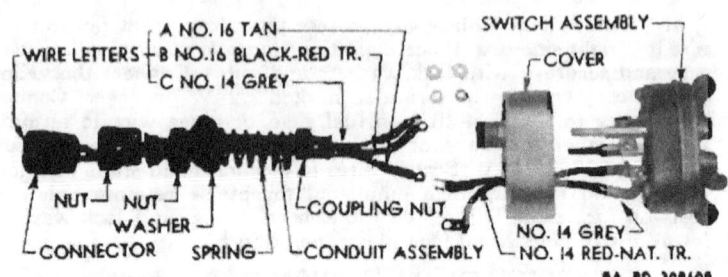

Figure 209—Magneto, Starter, Booster Switch, and Wires

Disconnect all wires from switch terminals. Remove oval-head screw which retains switch lever and carefully pry lever from switch shaft. Remove nut, lock washer and index plate and remove switch from face plate. To install, place switch in face plate from inner side and install index plate, lock washer and nut. Attach switch lever to shaft with oval-head screw. Connect colored wires to switch terminals marked as follows: red to "BS"; orange (green tracer) to "HT"; grey to "SS"; yellow to "BAT"; tan to "S"; black (nat. tracer) to "BOD"; green to "SW"; two orange and one black (nat. tracer) to "BHT" (fig. 207 or 208). NOTE: *In M39 instrument panel only, connect two grey wires to "SS", and one natural (black and red tracer) wire to "TT".* Install instrument panel face plate (par. 156 b).

c. **Replacement of Instrument Panel Light Switch** (fig. 206). Remove instrument panel face plate (par. 156 a). Remove amphenol retaining plate over switch and disconnect wires attached to switch terminals. Loosen lock screw in switch knob and unscrew knob from shaft. Remove special nut which secures switch to panel face plate. To install, place switch in face plate from inner side, install and tighten spanner nut securely. Screw knob on shaft until it has a slight clearance at spanner nut and tighten lock screw. Connect two red wires to inner terminal and two natural wires to outer terminal of switch (fig. 207 or 208). Attach amphenol retaining plate to posts on face plate with three round-head screws and lock washers.

d. **Replacement of Auxiliary Fuel Pump Switch** (fig. 206). Remove instrument panel face plate (par. 156 a). Disconnect wires and remove switch which is secured to panel face plate by two round-head screws. To install, place switch in face plate from inner side, with "ON" and "OFF" marks on switch in line with similar marks on face plate, and secure switch with two round-head screws. Connect one brown wire from circuit breaker to upper terminal, and connect brown and brown (red tracer) wires from amphenol receptacle to lower terminal of switch (fig. 207 or 208). Install instrument panel face plate (par. 156 b).

e. **Replacement of Fuel Gage Switch** (fig. 206). Remove instrument panel face plate (par. 156 a). Remove amphenol retaining plate over switch and disconnect wires attached to switch terminals being careful not to lose two small bus bars. Remove oval head screw in switch lever and carefully pry lever from switch shaft. Remove switch and index plate which are secured to panel face plate by hexagon nut and internal-tooth lock washer. To install, place switch in face plate from inner side, place index plate on outer side and secure them with an internal-tooth lock washer and hexagon nut. Install lever on switch shaft and secure it with an oval head screw. Place long bus bar over switch terminals "1" to "4"; place short bus bar over switch terminals "9", "6", "7". Connect colored wires to switch terminals marked as follows: yellow (red tracer) from fuel gage to "2"; yellow (red tracer) from temperature gage to "5"; brown to "6"; brown (black tracer) to "7"; yellow to "9" (fig. 207 or 208). Attach amphenol retaining plate to posts on face plate with three round head screws and lock washers. Install instrument panel face plate (par. 156 b).

f. **Replacement of Carburetor Idle Fuel Cut-off-Switch** (fig. 206). Remove instrument panel face plate (par. 156 a). Remove terminal block above switch and disconnect wires from switch. Remove rubber grommet which is installed around switch button, then remove switch which is secured to face plate with a hexagon nut and flat washer on the outside and an external-toothed lock washer and hexagon nut on the inside. To install, place hexagon nut and external-toothed lock washer on switch, place switch in face plate and secure it with a flat washer and nut. Install rubber grommet over switch button and engaged lip in large end in groove on switch body. Connect two grey wires to terminal nearest magneto switch and connect blue wire to opposite terminal (fig. 207 or 208). Place terminal block over stud on magneto switch and secure it with safety nut. Install instrument panel face plate (par. 156 b).

g. **Replacement of Light Test Switch** (fig. 206). This switch is attached to instrument panel face plate in same manner as the carburetor idle fuel cut-off switch, and is removed and installed in the manner described in subparagraph f above. When switch is installed, connect the black ground wire and the yellow (red tracer) wire to terminals (fig. 207 or 208).

161. CIRCUIT BREAKERS IN INSTRUMENT PANEL.

a. **Replacement of Lights, Gages, or Siren Circuit Breakers** (fig. 206). Remove instrument panel face plate (par. 156 a). Remove right amphenol retaining plate and disconnect all wires to four circuit breakers. Remove circuit-breakers-with-plate assembly which is attached to face plate with two nuts and lock washers. Disconnect bus bar and remove circuit breaker to be replaced, which is secured to mounting plate by two flathead screws. To install, attach circuit breaker (siren 30 amp., others 20 amp.) to mounting plate with two flat screws and connect to bus bar. Install circuit breakers with plate assembly on panel face plate and secure with two lock washers and nuts. Connect black wires from amphenol receptacle to each end of bus bar (fig. 207 or 208). Connect black wire from booster coil circuit breaker to inside lights circuit breaker at bus bar. Connect yellow wire from lights switch to inner terminal of outside lights circuit breaker. Connect yellow wire from amphenol receptacle to inner terminal of siren circuit breaker. Connect yellow (red tracer) wire from gages to inner terminal of gages circuit breaker. Connect three natural wires to inner terminal of inside lights circuit breaker. Attach amphenol retaining plate to posts on face plate with three round-head screws and lock washers. Install instrument panel face plate (par. 156 b).

b. **Replacement of Booster Coil and Starter Control Circuit Breaker** (fig. 206). Remove instrument panel face plate (par. 156 a). Disconnect wire and bus bar and remove circuit breaker, gasket, and name plate which are attached to panel face plate by two round-head screws. To install, place gasket and circuit breaker on inner side and name plate on outer side of face plate and secure with two round-head screws. Connect bus bar to lower terminal and, to other terminal, connect black wire from inside lights circuit breaker (fig. 207 or 208). Install instrument panel face plate (par. 156 b).

c. **Replacement of Auxiliary Fuel Pump Circuit Breaker** (fig. 206). Remove instrument panel face plate (par. 156 a). Disconnect wire and bus bar and remove circuit breaker and gasket which are attached to face plate with two round-head screws. To install, place gasket and circuit breaker on inner side of face plate and secure with two round-head screws Connect bus bar to upper terminal, and connect brown wire from fuel pumps switch to lower terminal (fig. 207 or 208). Install instrument panel face plate (par. 156 b).

162. PANEL LIGHTS AND OIL TEMPERATURE WARNING LIGHT.

a. **Replacement of Panel Light Lamp** (fig. 206). Carefully pry lamp cap from instrument panel face plate. Push in on lamp, turn counterclockwise and pull out. Push new lamp (3 cp.) into socket and turn clockwise. Push lamp cap into opening in panel face plate.

b. **Replacement of Converter Oil Temperature Warning Light Lamp** (fig. 206). Carefully pry warning lamp cap from panel face plate. Push in on lamp, turn counterclockwise and pull out. Push new lamp (3 cp.) into socket and turn clockwise. Push lamp cap into place over lamp.

Section XXXIII
ELECTRICAL EQUIPMENT

163. DESCRIPTION.

a. **General.** This section includes description and replacement instructions covering all electrical equipment not covered in previous sections, or in Section XXXXII which covers communication equipment. Approximate location of all electrical equipment in vehicles, except communications, is shown in figures 210 and 211 (M18) or figure 213 (M39).

Electrical Equipment

b. **Wiring.** All wiring is contained in shielded conduits for protection of wires and to effect adequate suppression of radio interference. Conduits are held in place by clamps or clips which form necessary bonding for suppression of radio interference. In most cases, a number of wires are run through one conduit, and many conduits are fitted with multiple-pin connector plugs at ends for easy attachment to junction boxes and equipment. Each pin in a connector plug is marked by a letter stamped on plug to identify circuit, and these letters are shown on pertinent wiring diagrams. Connectors provide accessible points for tests to locate grounds, shorts, or to check continuity of circuits. Identification of conduits, cables, and wires is given in paragraph 164 e.

c. **Circuit Breakers.** Either automatic or manual reset circuit breakers are used instead of fuses to protect all wiring circuits. These are contained in instrument panel, turret wiring switch box (M18), main filter box, auxiliary generator control box (M18), and accessory outlet box (M39). Replacement of circuit breakers is covered in paragraphs pertaining to equipment in which they are located.

d. **Ground Straps and Wires.** Ground straps or wires are installed at points where necessary to assure positive grounding of equipment to complete the electrical circuits, and to provide a low-resistance path to ground for any radio frequency circuits that may be generated within components to which they are attached. Ground straps and wires are identified in paragraph 165.

e. **Junction and Terminal Boxes.** Junction or terminal boxes are used where required to readily join conduits or wires in various circuits. These are identified in paragraph 166.

f. **Switch Boxes.** Master switch box (par. 167) serves as a junction box through which all circuits are connected to either the 12-volt or 24-volt battery current. It contains two master switches which control all current going into vehicle circuits. Turret wiring switch box (par. 168) contains switches and circuit breakers which control all current supplied to turret traversing electric motor, turret lights and outlet socket, and gun firing solenoid.

g. **Turret Slip Ring Box—M18 Vehicle.** Turret slip ring box (par. 169) provides a means of transmitting current to all electrical circuits in turret while turret is being rotated. It is mounted upon hull subfloor at the exact center of the turret, and is connected to the turret by a drag link which carries the wiring conduits leading to the turret wiring switch box and radio terminal box.

h. **Signal Sending Units.** Signal sending units are pressure, temperature, or fluid level measuring devices which operate electric gages and warning light on instrument panel.

(1) The transmission oil temperature sending unit, mounted in the oil outlet elbow on right side of torque converter, is a thermo switch which closes when oil temperature reaches highest allowable limit. Current then flows through unit, causing warning light on instrument panel to burn.

(2) Engine oil temperature gage unit, located on oil tank check

valve, is a resistor which varies current to gage on instrument panel in accordance with changes in temperature of oil.

(3) Oil pressure gage unit located forward of left magneto is a resistor which varies the current to oil pressure gage on instrument panel in accordance with changes in engine oil pressure.

(4) A fuel gage tank unit, mounted vertically in each fuel tank, is a resistance unit operated by a float within the assembly. As float rises or falls with change of level or fuel in the tank, it varies amount of current going through unit to fuel gage on instrument panel, so that gage registers amount of fuel in tank when fuel gage switch is closed.

i. **Siren and Siren Switch.** Siren is an electric signaling device in which sound is produced by a rotor having T-shaped blades which is rotated at high speed within stationary stator by small electric motor. It is mounted on front of hull and protected by a guard. Siren is operated by a push-type foot switch located in driving compartment conveniently placed for driver's left foot.

j. **Heater Control Box.** Heater control box, mounted in the driving compartment to left of driver's seat, controls winterization equipment, when installed in vehicle. The circuit breaker on rear end of box breaks circuits in case of overload. The box is equipped with clock operated heater switch which automatically shuts off current to limit time during which winterization equipment is in operation.

k. **Trailer Electric Brake Controller, M39 Vehicle.** This unit, located in left sponson forward of instrument panel, is operated by foot pedal in front of driver. It is a variable resistance unit which varies current to trailer electric brakes in accordance with pressure exerted on foot pedal. Knob on controller permits adjustment to vary current for given pedal pressure to suit road conditions.

l. **Trailer Electric Brake Resistor Box, M39 Vehicle.** This resistor box, located above instrument panel, contains a heavy duty 1.5 ohm resistor unit which reduces 24-volt battery current to 6 volts, permitting use of either two or four 6-volt brakes on trailer.

m. **Trailer Electric Receptacle, M39 Vehicle.** Electric receptacle, located on left rear corner of hull, provides quick detachable connection for trailer cable which carries current to trailer brakes and to trailer blackout tail and stop lights.

n. **Accessory Outlet Box, M39 Vehicle.** Accessory outlet box, mounted on left wall of crew compartment provides a 24-volt outlet for plugging in accessories not installed in vehicle. It is equipped with 15-ampere manual reset circuit breaker to protect circuits against overload.

164. WIRES, CABLES AND CONDUITS.

a. **Location.** Approximate location of all electrical equipment and their connecting wires, cables, or conduits is shown in figure 210 (M18 turret), figure 211 (M18 hull), and figure 213 (M39).

b. **Removal.** All electrical conduits are removed in essentially the same manner. If conduits are fitted with connector plugs, unscrew the coupling nuts and pull plugs from receptacles. Where connector

Electrical Equipment

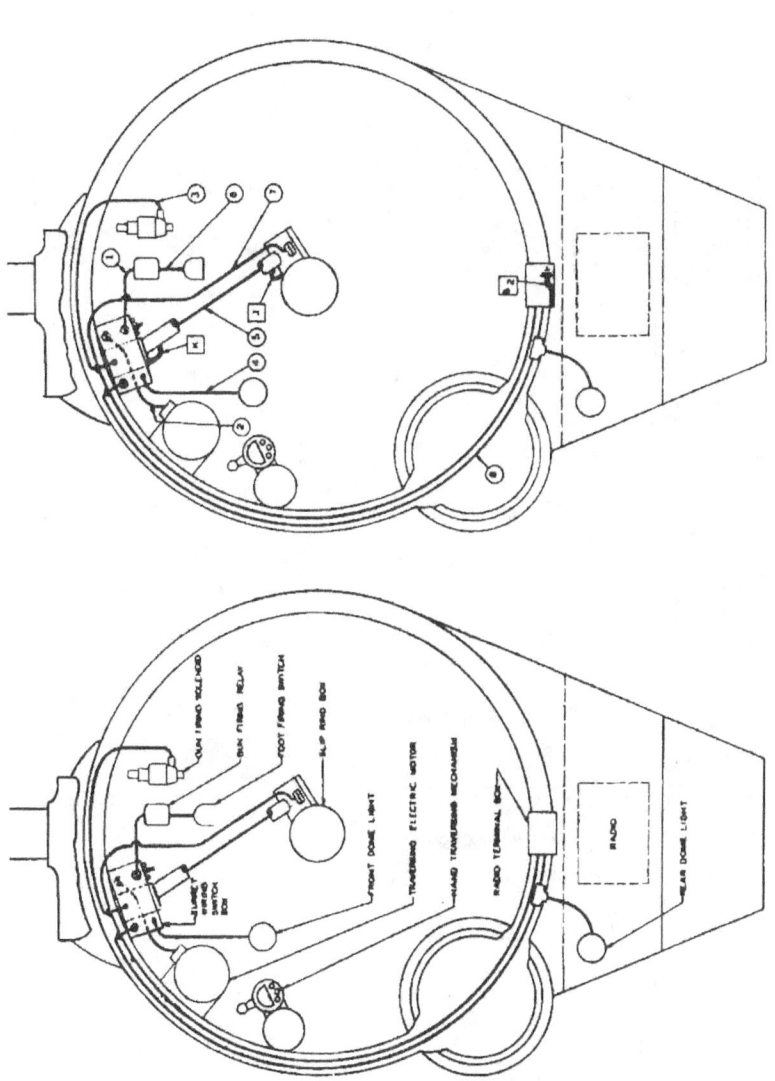

Figure 210—Turret Electrical Units, Connecting Wires, and Conduits—M18

TM 9-755
164

Part Three—Maintenance Instructions

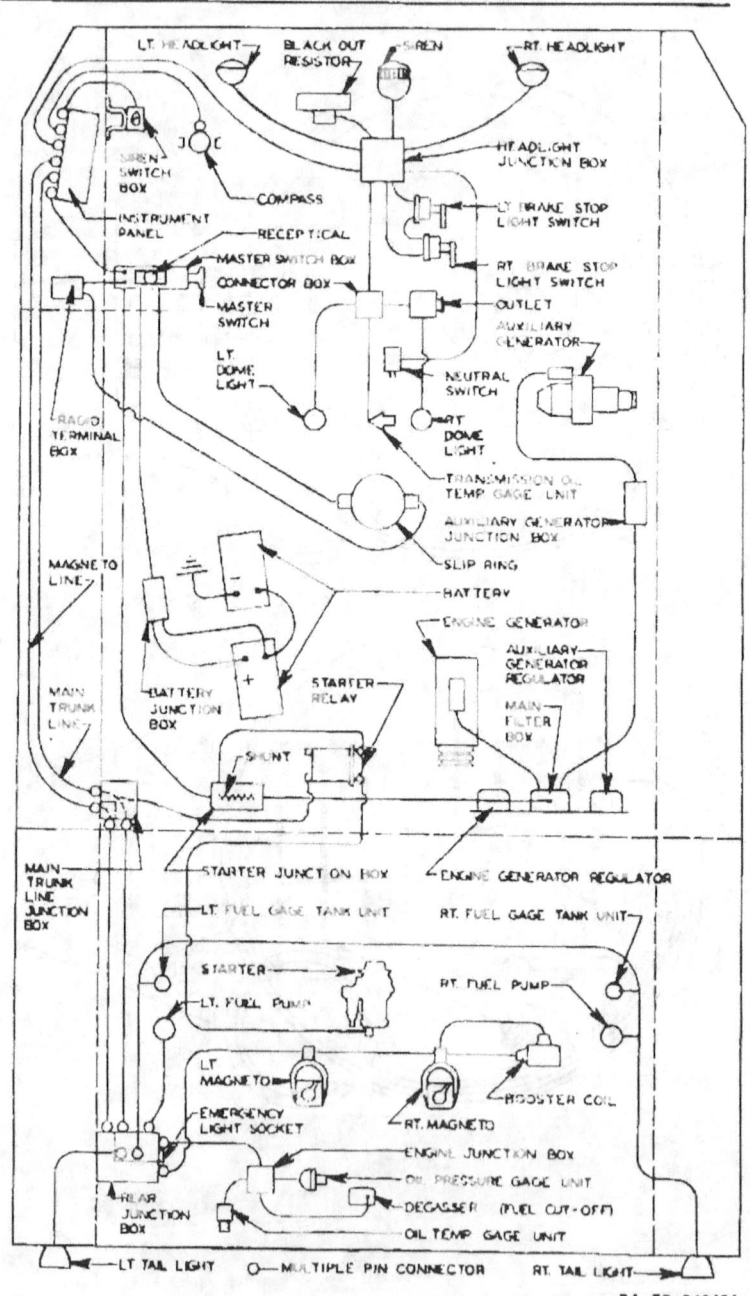

Figure 211—Hull Electrical Units—M18

Electrical Equipment

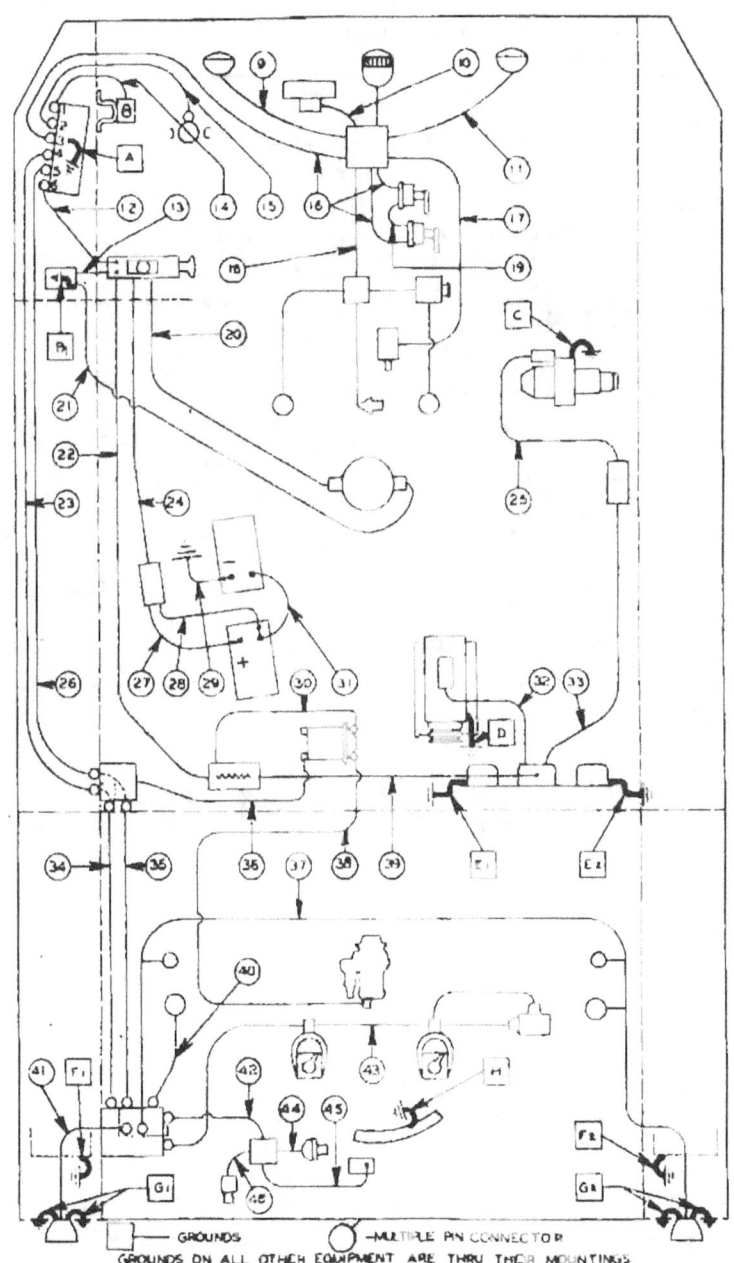

Figure 212—Hull Wires, Cables, and Conduits—M18

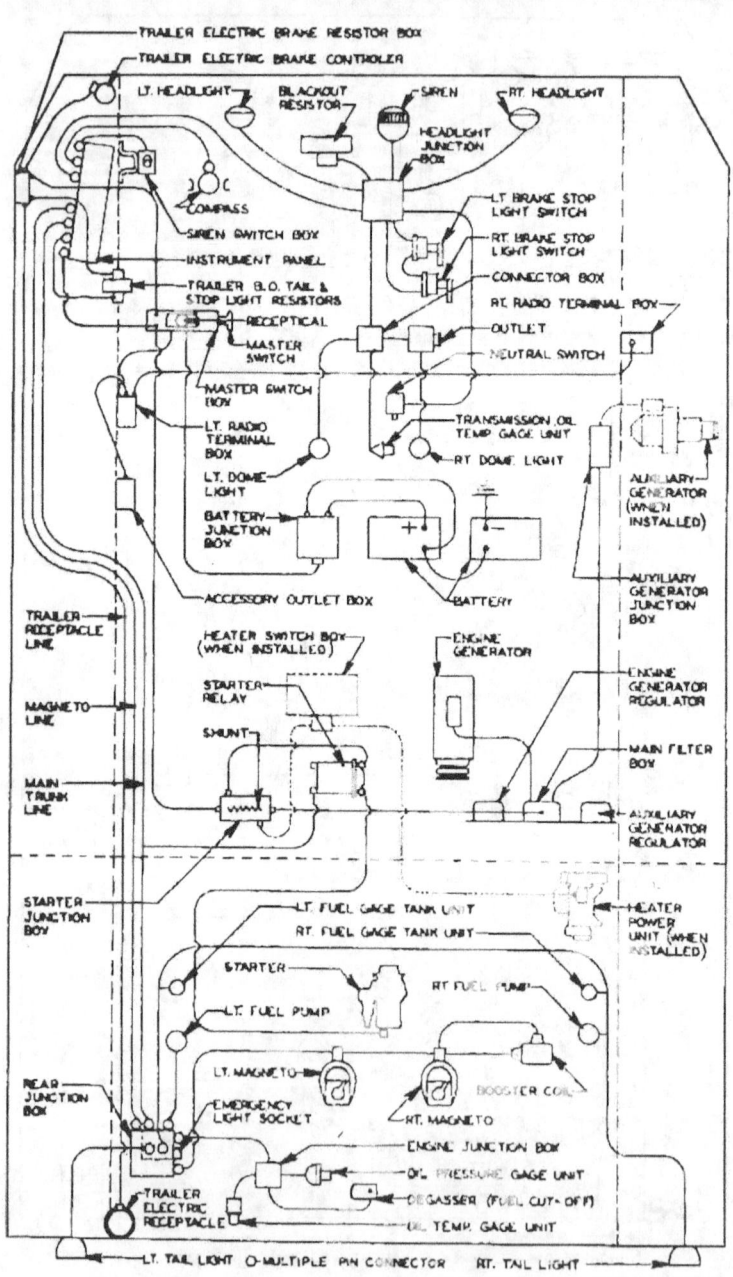

Figure 213—Electrical Units—M39

TM 9-755
164

Electrical Equipment

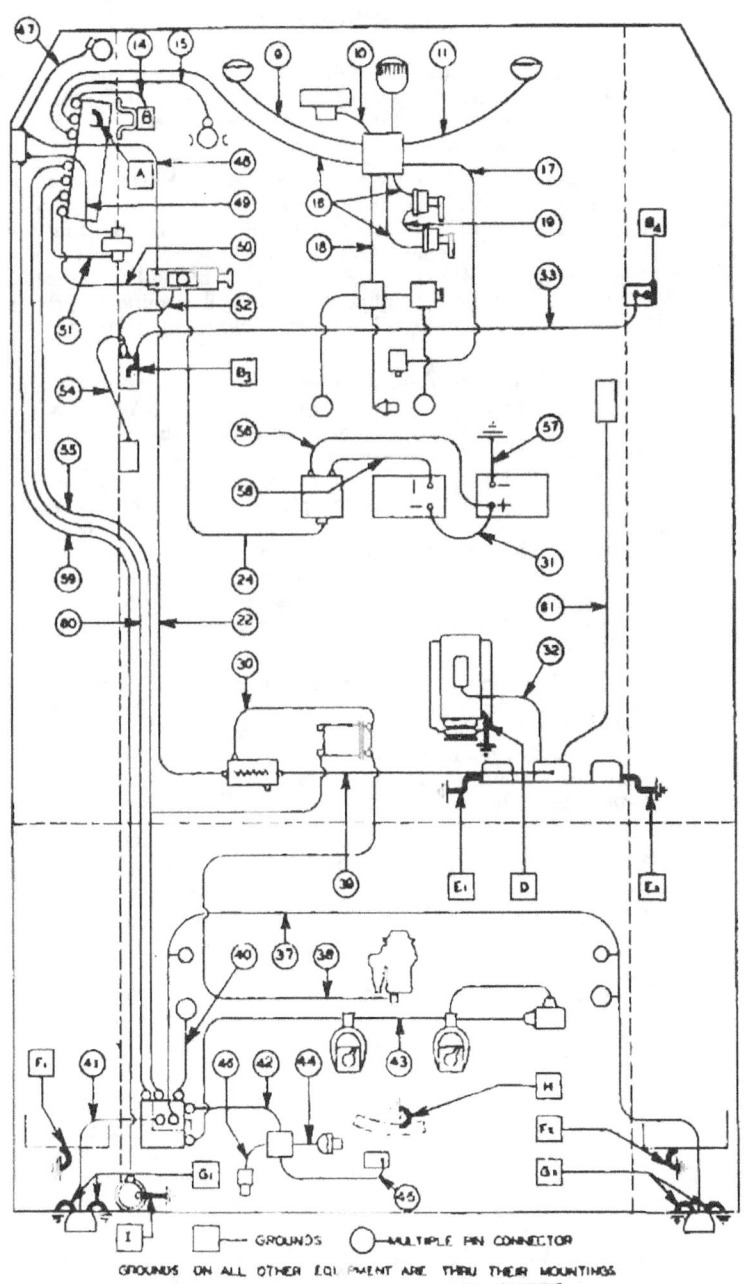

Figure 214—Wires, Cables, and Conduits—M39

TM 9-755

Part Three—Maintenance Instructions

plugs are not used, refer to paragraph pertaining to equipment involved for information regarding disconnection of wiring and conduits. Remove retaining clips, after taking out any stowage items or parts that interfere, and remove conduit from vehicle.

c. **Replacement of Wires in Conduits.** Where practicable, faulty wires may be replaced within conduit if conduit is in good condition, not broken or mashed. Remove terminal or unsolder faulty wire from connector plug pin and attach a new wire of proper size and length to old wire. Pull new wire into conduit by pulling out old one. Solder new wire to connector plug pins or install terminals as required. CAUTION: *Do not use liquid flux when soldering electrical connections.*

d. **Installation.** Position conduit in hull or turret and install all retaining clips and clamps in same manner as removed, with plated screws and lock washers to insure a positive ground bond. This is important in order to provide necessary bonding for radio interference suppression as well as for protection of conduits. Do not twist or kink conduit. When installing connector plugs in receptacles, make sure that letters on plugs and receptacles coincide and that tongue and groove line up. Do not force a plug; if plug is properly lined up it will slip into receptacle without forcing. Tighten all brass coupling nuts with pliers; tighten all die-cast coupling nuts finger-tight after connector pins are fully seated. Install all other parts and stowage items that were removed.

e. **Identification.** The wires, cables, and conduits are identified by key numbers, and ground straps and wires are identified by letters, in figure 210 (M18 turret), figure 212 (M18 hull), and figure 214 (M39). The following tabulation, arranged in key number order, gives the name, stock number and equipment connected for all parts identified in these illustrations.

Key No.	Name	Stock No.	Equipment Connected	Figure No.
1	Wire/conduit	G163-6325595	Turret wiring switch box to gun	210
2	Wire/conduit	G163-0128591	Turret wiring switch box to traversing electric motor	210
3	Wire/conduit	G163-04-47875	Turret wiring switch box to gun firing solenoid	210
4	Wire/conduit	G163-04-47880	Turret wiring switch box to turret front dome light	210
5	Wire	G163-01-28592	Slip ring box to turret wiring switch box (through drag link)	210
6	Wire/tube	G163-6325596	Gun firing relay to foot firing switch	210
7	Wires/conduit	G163-02-09452	Slip ring box to turret radio terminal box	210

Electrical Equipment

Key No.	Name	Stock No.	Equipment Connected	Figure No.
8	Wires/conduit	G163-02-09450	Turret wiring switch box to rear dome light and radio terminal box	210
9	Wires/conduit	G163-02-09404	Headlight junction box to left headlight	212, 214
10	Wire/conduit	G163-04-47796	Headlight junction box to blackout resistor	212, 214
11	Wires/conduit	G163-02-09405	Headlight junction box to right headlight	212, 214
12	Wires/conduit	G163-02-09457	Master switch box to instrument panel	212
13	Wire/conduit	G163-02-09462	Master switch box to hull radio terminal box	212
14	Wires/conduit	G163-02-09458	Instrument panel to siren switch box	212, 214
15	Wires/conduit	G163-02-09406	Instrument panel to compass	212, 214
16*	Wires/conduit	G163-02-09403	Instrument panel to headlight junction box	212, 214
17	Wires/conduit	G163-02-09402	Headlight junction box to starter neutral switch	212, 214
18	Wires/conduit	G163-02-09466	Headlight junction box to dome light and transmission oil temperature sending unit	212, 214
19	Wire	G163-04-47800	Jumper wire between stop light switches	212, 214
20	Wire/conduit	G163-01-39678	Master switch box to slip ring box	212
21	Wires/conduit	G163-02-09465	Hull radio terminal box to slip ring box	212
22	Wires/conduit	G163-02-09463	Master switch box to starter junction box	212, 214
23	Wires/conduit	G163-02-09455	Instrument panel magneto line to main trunkline junction box	212
24	Wires/conduit	G163-02-09461	Battery junction box to master switch box	212, 214
25	Wires/conduit	G163-7005646	Auxiliary generator control box to auxiliary generator junction box	212
26	Wires/conduit	G163-02-09456	Instrument panel to main trunkline junction box	212
27	Cable	G163-01-28568	Battery to battery junction box (24-volt)	212

TM 9-755

Part Three—Maintenance Instructions

Key No.	Name	Stock No.	Equipment Connected	Figure No.
28	Cable	G163-01-28569	Battery to battery junction box (12-volt)	212
29	Cable	G163-01-28567	Battery to ground	212
30	Wire/conduit	G163-6325626	Starter junction box to starter relay	212 214
31	Cable	B300267	Battery to battery	212 214
32	Wires/conduit	G163-02-09454	Engine generator to main filter box	212 214
33	Wires/conduit	G163-02-09401	Auxiliary generator junction box to main filter box	212
34	Wires/conduit	G163-02-09460	Main trunkline junction box to rear junction box (magnetos and coil)	212
35	Wires/conduit	G163-02-09459	Main trunkline junction box to rear junction box	212
36	Wire/conduit	G163-6325625	Main trunkline junction box to starter relay	212
37	Wires/conduit	G163-01-39676	Rear junction box to fuel gage tank units, right fuel pump and taillight	212 214
38	Wire/conduit	G163-5660271	Starter relay to starter	212 214
39	Wires/conduit	G163-02-09453	Main filter box to starter junction box	212 214
40	Wire/conduit	G163-01-39607	Rear junction box to left fuel tank pump	212 214
41	Wires/conduit	G163-01-39677	Rear junction box to left taillight	212 214
42	Wires/conduit	G163-02-09464	Rear junction box to engine junction box	212 214
43	Wires/conduit	G163-5652257	Rear junction box to magnetos and booster coil	212 214
44	Conduit	B163-01-39605	Engine junction box to oil pressure gage unit	212 214
45	Conduit	G163-01-39606	Engine junction box to degasser (fuel cut-off)	212 214
46	Conduit	G163-01-39604	Engine junction box to oil temperature gage unit	212 214
47	Wires/conduit	G163-7019315	Electric brake controller to electric brake resistor box	214

TM 9-755
164-165

Electrical Equipment

Key No.	Name	Stock No.	Equipment Connected	Figure No.
48	Wires/conduit	G163-7019379	Master switch box to electric brake resistor box	214
49	Wires/conduit	G163-7019306	Electric brake resistor box trailer blackout tail and stop light resistors	214
50	Wires/conduit	G163-7019302	Master switch box to instrument panel	214
51	Wires/conduit	G163-7019313	Instrument panel to trailer blackout tail and stop light resistors	214
52	Wires/conduit	G163-7019304	Master switch box to left radio terminal box	214
53	Wires/conduit	G163-7019305	Left to right radio terminal boxes	214
54	Wires/conduit	G163-7019455	Left radio terminal box to accessory outlet box	214
55	Wires/conduit	G163-7019301	Instrument panel to starter relay and rear junction box	214
56	Cable	G163-7019309	Battery to battery junction box (12-volt)	214
57	Cable	G163-7019310	Battery to ground	214
58	Cable	G163-7019308	Battery to battery junction box (24-volt)	214
59	Same as item 49		Electric brake resistor box to trailer electric receptacle	214
60	Wires/conduit	G163-7019300	Instrument panel to rear junction box	214
61	Wires/conduit	G163-7019303	Auxiliary generator junction box to main filter box	214

165. GROUND STRAPS AND WIRES.

  a. *Location.* The locations of all ground straps and wires are shown in figure 210 (M18 turret), figure 212 (M18 Hull), and figure 214 (M39). Individual views of these ground straps and wires are shown in figures 215 through 228.

  b. *Replacement.* Ground straps or wires that are frayed or broken must be replaced to insure positive ground and radio interference suppression. For replacement instructions, refer to paragraphs pertaining to replacement of equipment to which strap or wire is attached. When new parts are installed, thoroughly clean all contact surfaces to insure metal-to-metal contact, use plated screws and toothed lock washers where such parts were originally installed, and tighten attaching screws securely.

TM 9-755

Part Three—Maintenance Instructions

Figure 215—Instrument Panel Ground Strap

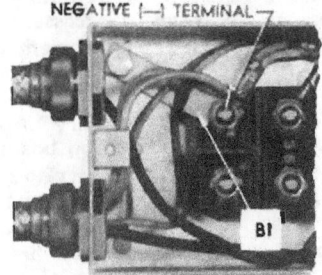

RA PD 340441

Figure 216—Hull Radio Terminal Box Ground Strap—M18

RA PD 340442

Figure 217—Turret Radio Terminal Box Ground Strap—M18

TM 9-755
165

Electrical Equipment

Figure 218—Left Radio Terminal Box Ground Strap—M39

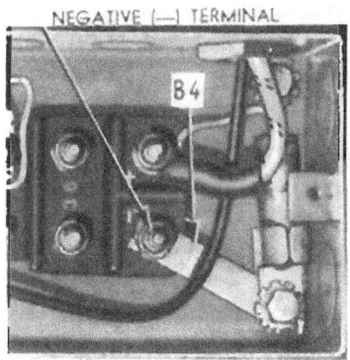

Figure 219—Right Radio Terminal Box Ground Strap—M39

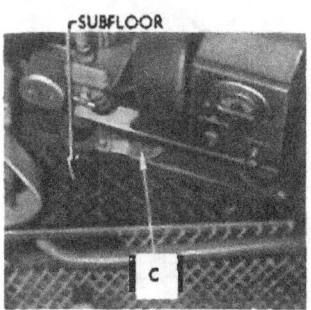

Figure 220—Auxiliary Generator Engine Ground Strap—M39

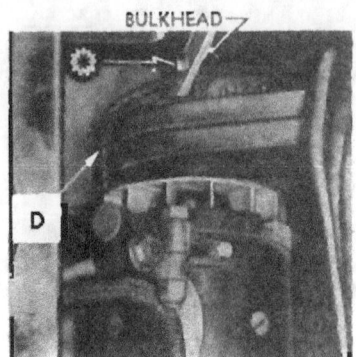

Figure 221—Engine Generator Ground Wire

Figure 222—Generator Regulator Ground Strap

Figure 223—Fuel Tank Ground Strap

TM 9-755

Electrical Equipment

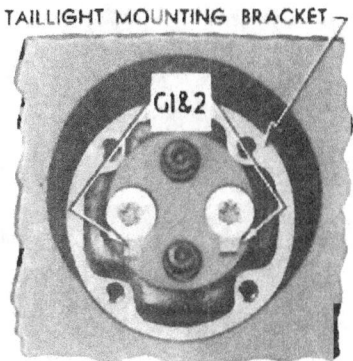

Figure 224—Taillight Ground Strap

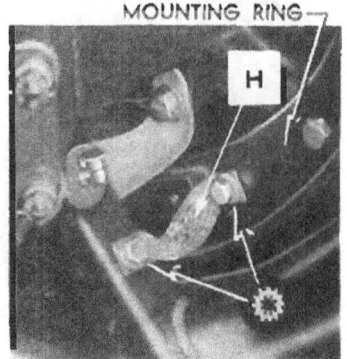

Figure 225—Engine Ground Strap

Figure 226—Trailer Electric Receptacle Ground Wire—M39

TM 9-755
166

Part Three—Maintenance Instructions

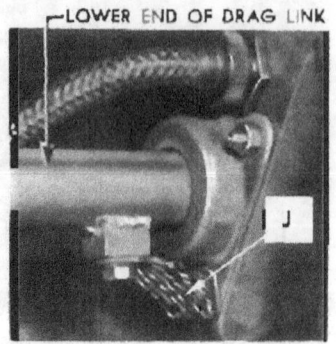

Figure 227—Drag Link to Slip Ring Box Ground Strap—M18

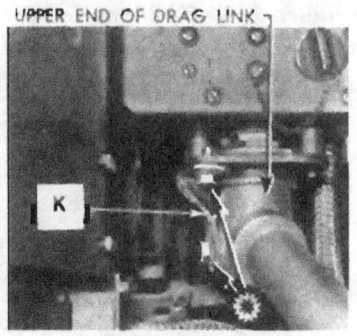

Figure 228—Drag Link to Race Ring Bracket Ground Strap—M18

166. JUNCTION AND TERMINAL BOXES.

a. **Location.** The locations of all junction and terminal boxes are shown in figure 210 (M18 turret), figure 211 (M18 hull), and figure 213 (M39).

b. **Unit Wiring Diagrams.** With few exceptions, junction and terminal boxes have diagrams on inside surfaces of the covers which show wiring connections within unit. These diagrams are reproduced in figures 229 through 234. Where diagrams are not contained in units, illustrations are used herein to show connections (fig. 235 through 239).

c. **Removal.** In general, removal of a junction or terminal box is effected by removing stowage items or vehicle parts which interfere, removing the cover and disconnecting wires from terminal parts, disconnecting conduits, and removing box attaching screws, and lock

TM 9-755

Electrical Equipment

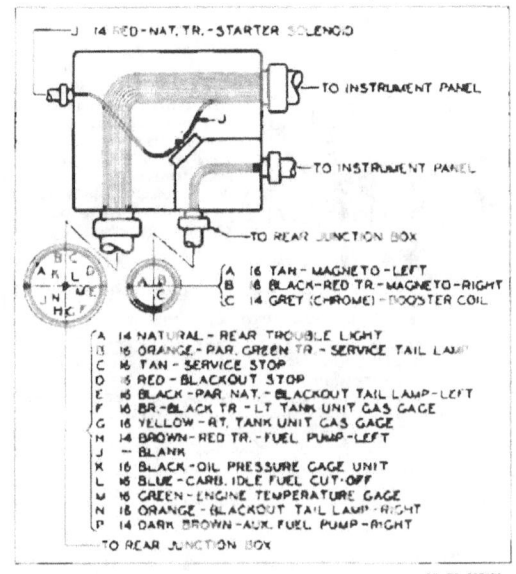

Figure 229—Main Trunk Line Junction Box Wiring Diagram—M18

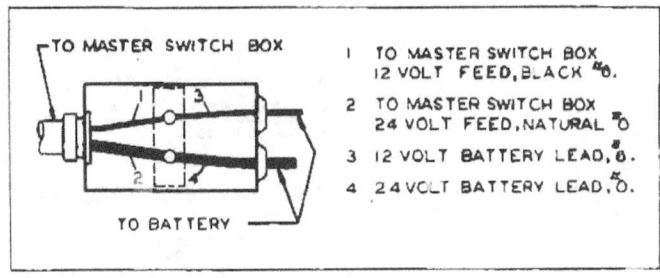

Figure 230—Battery Junction Box Wiring Diagram

Part Three—Maintenance Instructions

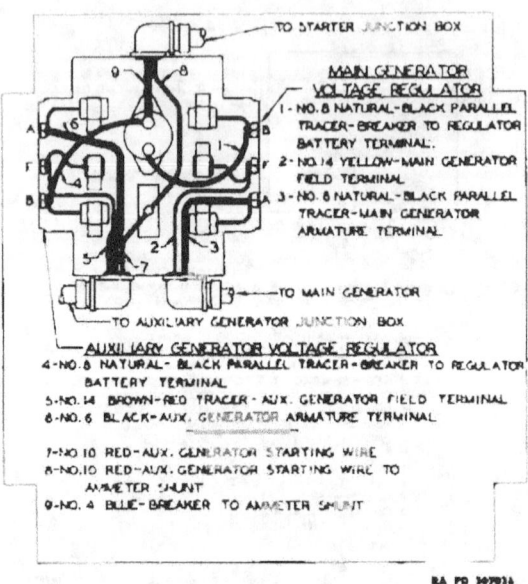

Figure 231—Main Filter Box Wiring Diagram

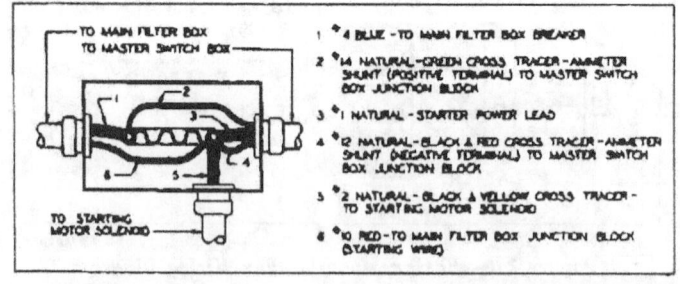

Figure 232—Starter Junction Box Wiring Diagram

Electrical Equipment

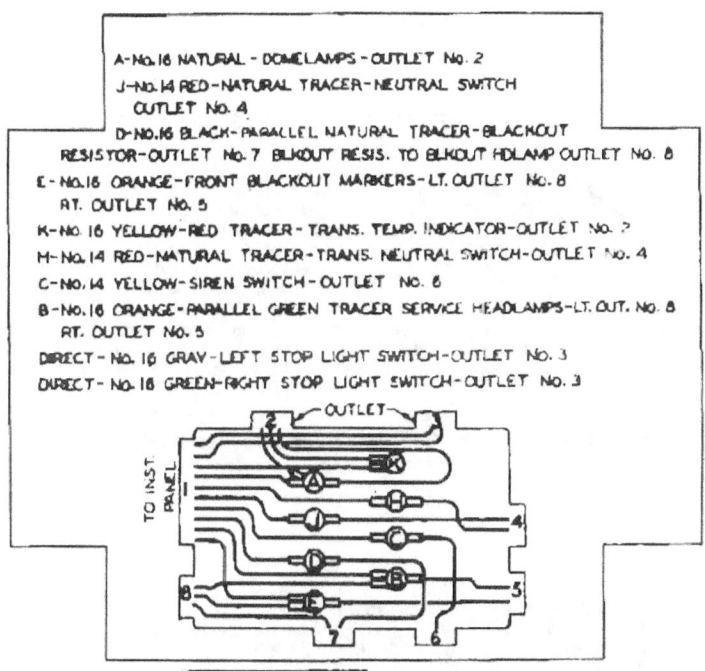

Figure 233—Headlight Junction Box Wiring Diagram

washers. If box is in good condition and securely mounted, it is not necessary to remove it in order to remove terminal blocks or other parts in interior which require replacement, as these parts can be removed with box in place.

d. **Installation.** Install all parts in interior of box and anchor box securely in place, using plated screws and toothed lock washers where originally used. Connect conduits to box and tighten hose coupling nuts with pliers; tighten die-cast coupling nuts finger-tight after connector pins are fully seated. Connect all wires as indicated by diagram in box or pertinent illustrations in this manual. Install cover, making sure cover gasket is in good condition. Install other vehicle parts or stowage items that were removed.

167. MASTER SWITCH BOX.

a. **Replacement of Internal Components.** It is not necessary to remove switch box assembly in order to replace internal components. Disconnect ground cable and 24-volt cable from battery (fig. 189). Remove switch box cover which is secured by four captive screws. Disconnect wires as required to remove and install faulty

TM 9-755
167

Part Three—Maintenance Instructions

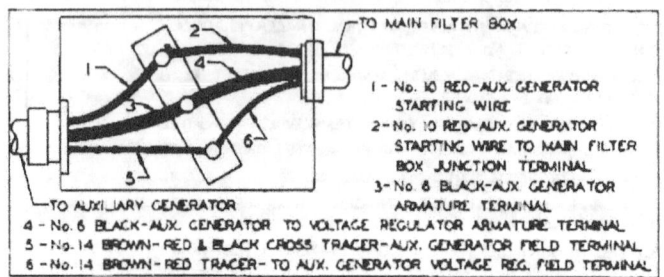

Figure 234—Auxiliary Generator Junction Box Wiring Diagram—M18

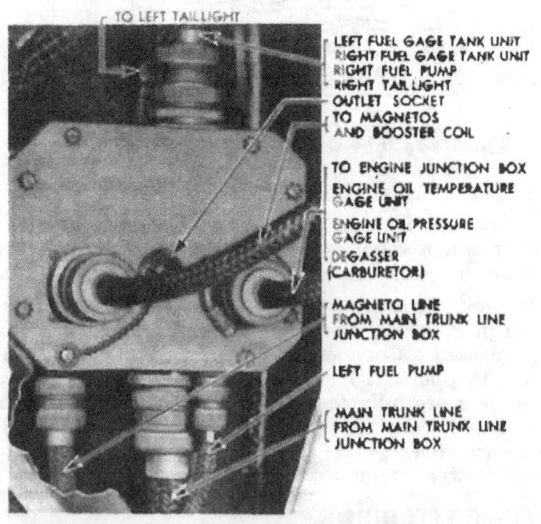

Figure 235—Rear Junction Box Conduit Connections

TM 9-755
167

Electrical Equipment

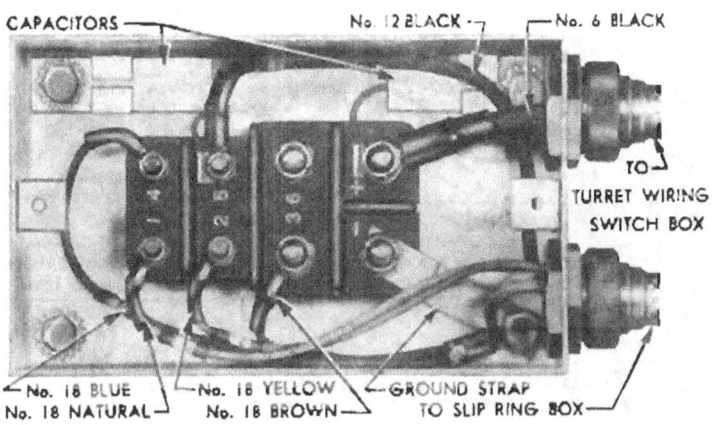

Figure 236—Turret Radio Terminal Box Wiring Connections—M18

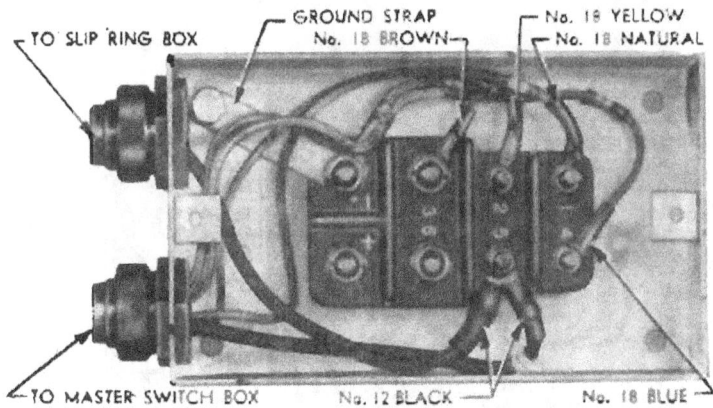

Figure 237—Hull Radio Terminal Box Wiring Connections—M18

component. When component has been installed connect wires as shown on diagram on box cover (fig. 241 or 242). When installing switches make sure that gasket is in good condition, that insulation sleeves are in position on connector straps, and that straps are well separated to prevent a short circuit (fig. 240). Install cover, making sure cover gasket is in good condition. Connect ground cable and 24-volt cable to battery.

Part Three—Maintenance Instructions

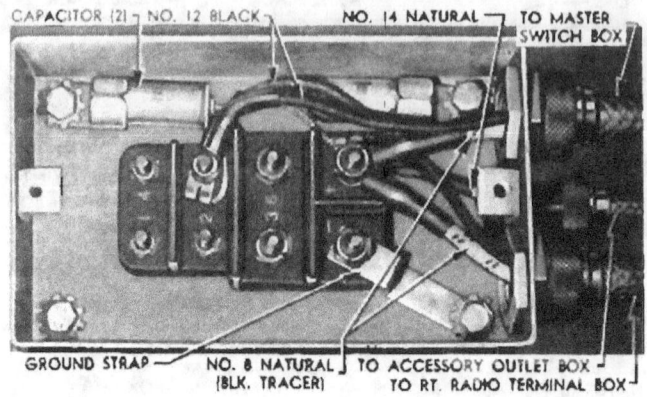

Figure 238—Left Radio Terminal Box Wiring Connections—M39

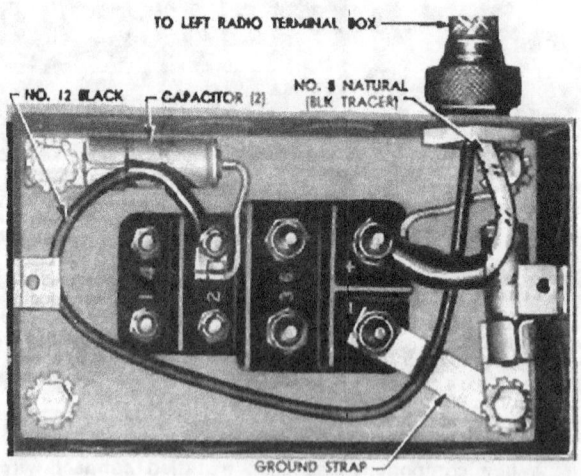

Figure 239—Right Radio Terminal Box Wiring Connections—M39

Electrical Equipment

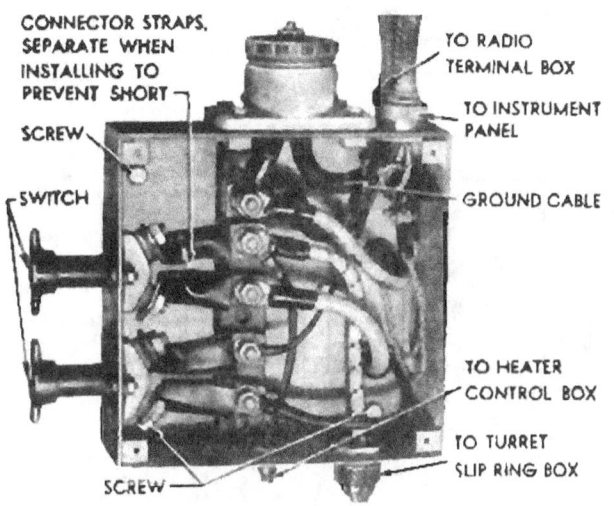

Figure 240—Master Switch Box—Cover Removed—M18

b. Removal (fig. 240). Disconnect ground cable and 24-volt cable from battery (fig. 189). Remove switch box cover which is secured by four captive screws. Disconnect all wires from terminal blocks, disconnect all conduits and pull wires out of box. Remove switch box which is secured to hull plate by four cap screws and lock washers.

c. Installation. Make certain that rubber seals are in position over the cables coming through oil cooler duct front plate, and that they are properly seated in key slots. The rubber seals must be compressed around holes in switch box when attaching screws are tightened, to prevent water leaking into box. Attach switch box to plate with four cadmium-plated cap screws and external-toothed lock washers. Push wires into proper openings in box, connect conduits to box, and connect all wires as shown on wiring diagram on box cover (fig. 241 or 242). Make sure that insulation sleeves are in position on switch connector straps and that straps are well separated to prevent a short circuit (fig. 240). Install cover, making sure that cover gasket is in good condition. Connect ground cable and 24-volt cable to battery (fig. 189).

168. **TURRET WIRING SWITCH BOX.**

a. General. All of components contained in turret wiring switch box are mounted on cover. Turn both master switches off (fig. 15). Remove eight round-head screws and detach switch box cover as-

Part Three—Maintenance Instructions

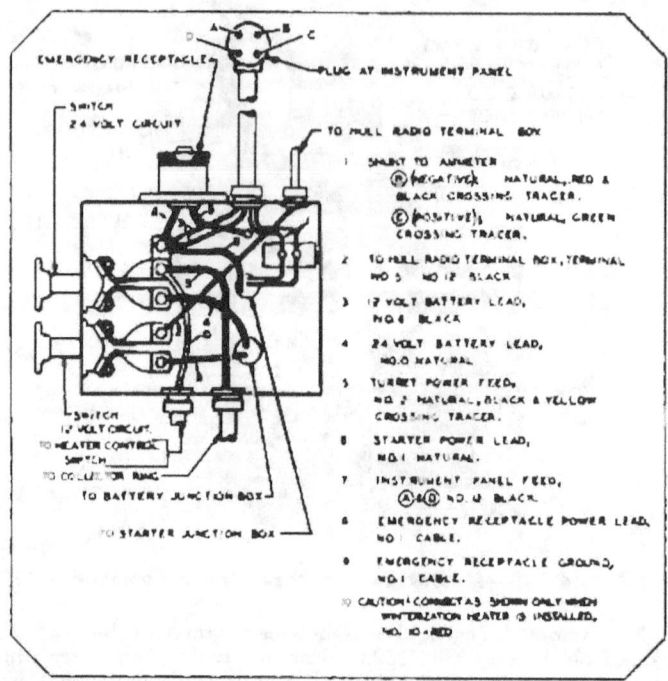

Figure 241—Master Switch Box Wiring Diagram—M18

sembly from switch box. Disconnect all attached wires and remove cover assembly. Remove faulty component and install new one as described in pertinent subparagraphs below, then install cover (subpar. g below).

b. **Replacement of Traverse Motor Master Switch** (fig. 243). Remove bus bar which connects traversing motor switch to circuit breaker, disconnect wire connected to firing circuit breaker, and remove condenser from bracket beside switch. Remove switch guard and boot which are attached to face of cover. Remove switch which is secured to cover by four flat-head screws, lock washers, and nuts. To install, place switch on cover with side marked "ON" towards upper edge of cover, and secure it with four flathead screws (8—32 x 42 in.), lock washers, and nuts. Install boot and guard with word "ON" toward upper edge of cover and secure guard with two cap screws (⅜ in.— 24 x ¾ in.), lock washers, and nuts. Attach condenser to bracket, and connect black wire to lower binding post on switch. Place condenser wire terminal over upper binding post on switch, place long end of bus

Electrical Equipment

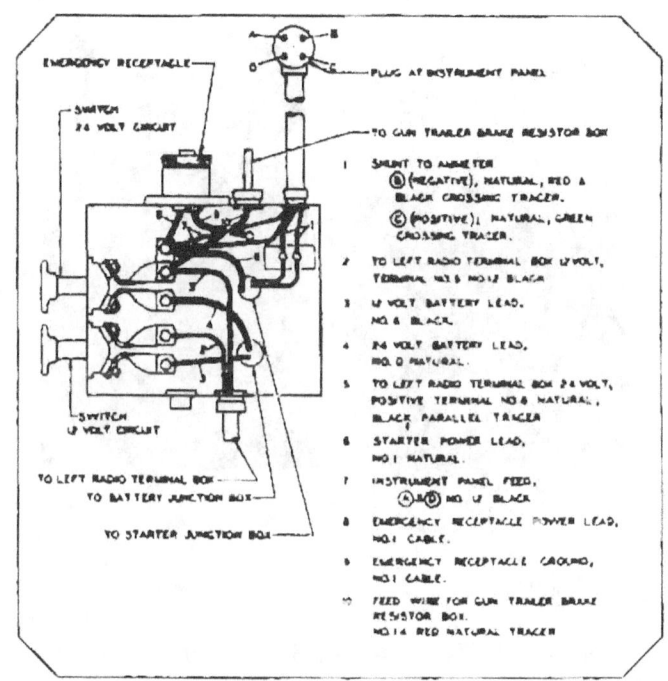

Figure 242—Master Switch Box Wiring Diagram—M39

bar over binding post, secure lower end of bus bar to circuit breaker with cap screw and lock washer, then secure upper end with plain washer, lock washer, and nut. CAUTION: *If short end of bus bar is connected to switch, bar will touch switch case and cause a short circuit. Check to make sure there is a minimum of 1/8-inch clearance between bar and edge of case* (fig. 244).

c. **Replacement of Traverse Motor Circuit Breaker** (fig. 243). Remove bus bar attaching screw, loosen nut at upper end of bar and swing bar to one side. Remove circuit breaker which is attached to switch box cover with two round-head screws. To install, attach breaker to cover using care not to tighten screws excessively. Connect bus bar to breaker with cap screw and lock washer and tighten nut at upper end.

d. **Replacement of Dome Lamp, Trouble Lamp or Firing Circuit Breaker** (fig. 243). Disconnect wires and bus bar and remove circuit breaker which is attached to cover with two round-head screws. To install, attach circuit breaker to cover using care not to

TM 9-755
168

Part Three—Maintenance Instructions

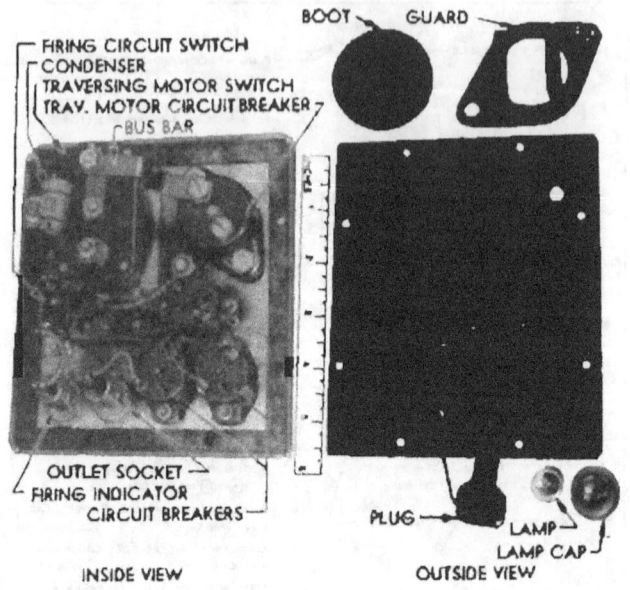

Figure 243—Turret Wiring Switch Box Components

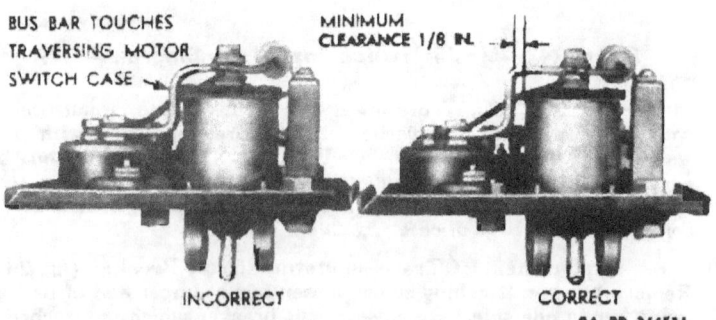

Figure 244—Incorrect and Correct Installation of Bus Bar

tighten attaching screws excessively. Connect bus bar and wires as shown on diagram in switch box (fig. 245).

e. Replacement of Firing Circuit Switch (fig. 243). Disconnect wires, and remove switch which is attached to switch box cover by two round-head screws. To install, place switch on cover with side

Electrical Equipment

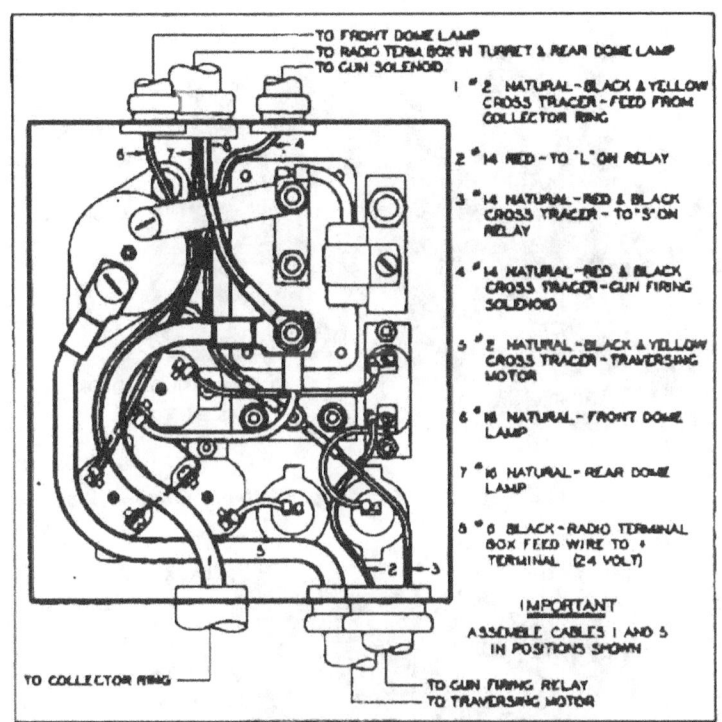

Figure 245—Turret Wiring Switch Box Wiring Diagram

marked "ON" in line with word "ON" stamped on face of cover, and secure it with two round-head screws, lock washers, and nuts. Attach two wires as shown on wiring diagram in switch box (fig. 245).

f. **Replacement of Firing Indicator Lamp** (fig. 243). Carefully pry firing indicator cap out of socket. Push in on lamp, turn counterclockwise, and pull out. Push new lamp into socket and turn clockwise. Push indicator cap into socket.

g. **Installation of Switch Box Cover.** Hold switch box cover in position and attach all wires as shown on wiring diagram in switch box (fig. 245). Attach cover to switch box with eight round head screws (12—24 x ½ in.) and external-tooth lock washers.

169. **TURRET SLIP RING BOX.**

a. **Cleaning and Replacement of Brushes.** Turn both master switches off (fig. 15). Remove arrow and cap which are attached by four cap screws and lock washers. Unscrew four captive knurled

TM 9-755

Part Three—Maintenance Instructions

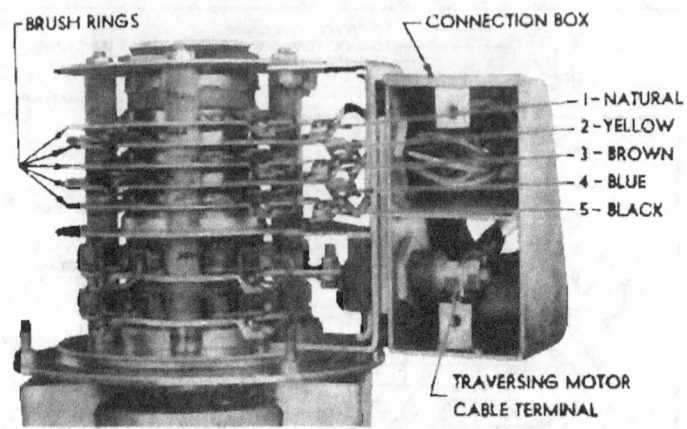

Figure 246—Slip Ring Box Upper Wiring Connections

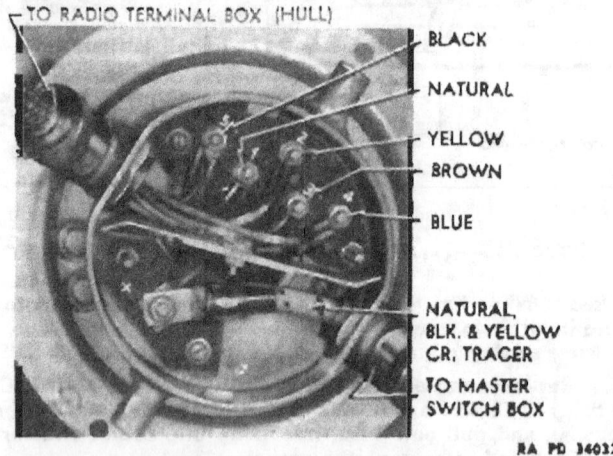

Figure 247—Slip Ring Box Lower Wiring Connections

nuts and lift off slip ring box cover. If brushes and collector rings are dirty, clean thoroughly with dry cleaning solvent. Examine all brushes and replace any that are worn excessively or which have broken or frayed lead wires. Inspect collector rings and, if dirty or burned, clean carefully with 2/0 flint paper and wipe off all traces of abrasive. CAUTION: *Do not use emery cloth or coarse flint paper.* Install cover, cap and arrow, with arrow pointing to front of vehicle.

Electrical Equipment

b. **Removal.**

(1) Turn both master switches off (fig. 15).

(2) Remove arrow and cap which are attached by four cap screws and lock washers. Unscrew four captive knurled nuts and lift off slip ring box cover. Remove connector box cover which is attached by two cap screws.

(3) Disconnect traversing motor cable from terminal stud and disconnect five wires from brush rings (fig. 246).

(4) Remove conduit guard from lower end of drag link, detach radio wiring conduit from connector box and move it out of way. Remove four $5/16$-inch safety nuts from studs which attach grommet retainer and ground strap to connector box and turn slip ring box around to free lower end of drag link.

(5) Remove four $3/8$-inch cap screws and lock washers which anchor slip ring box to floor plate, lift assembly up and remove bottom cover.

(6) Disconnect wires and cable coming from the two conduits (fig. 247) and disconnect conduits by unscrewing coupling nuts. Tie conduits so they will not drop through holes in floor plate.

c. **Installation.**

(1) Push wires and cables through proper openings, connect conduits to lower end of box, and connect wires and cable to terminal block as show in figure 247. Install bottom cover.

(2) Place slip ring box in position on floor plate and anchor it with four cadmium-plated cap screws ($3/8$ in.—24 x 1 in.) and external-tooth lock washers.

(3) Inspect rubber grommet at lower end of drag link to make sure it is in good condition and properly installed around ring on link.

(4) Remove slip ring cover, turn slip ring around to insert traverse motor cable (in drag link) into lower opening in connector box, and attach grommet retainer and ground strap to connector box with four safety nuts ($5/16$ in.—24) and external-toothed lock washers.

(5) Insert wires of radio conduit through upper opening in connector box and attach conduit securely to box. Install conduit guard and conduit clip to drag link with two cap screws ($5/16$ in.—24 x $1/2$ in.) and lock washers.

(6) Attach traversing motor cable to terminal stud in connector box with nut ($3/8$ in.—16, brass) and lock washer. Insert radio wires through opening in side of box and attach them to terminals on brush rings in order shown in figure 246. Each wire is numbered and also color coded for identification.

(7) Install connector box cover with two cap screws ($1/4$ in.—20 x $5/8$ in.) and lock washers. Install slip ring box cover and tighten knurled nuts. Place slip ring cover cap and arrow with arrow pointing to front of vehicle, and secure them with four cap screws ($1/4$ in.—24 x 1 in.) and lock washers.

TM 9-755
169

Part Three—Maintenance Instructions

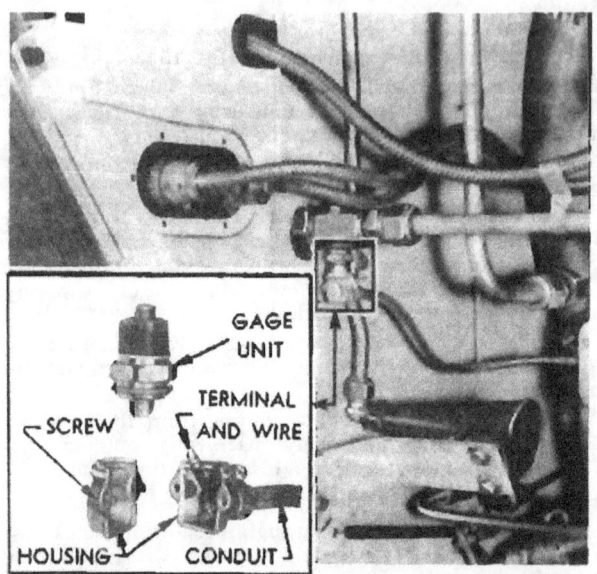

Figure 248—Engine Oil Temperature Gage Unit

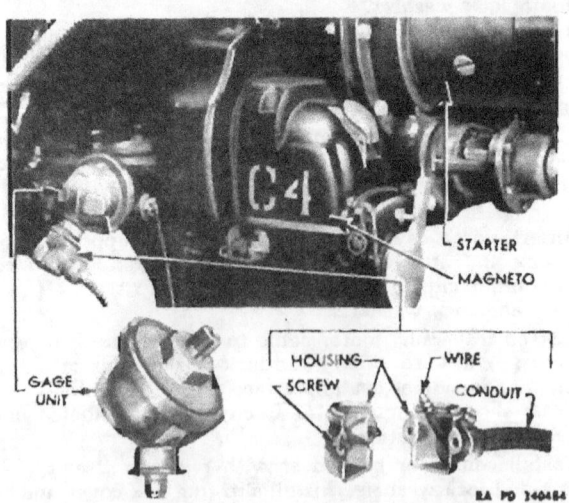

Figure 249—Engine Oil Pressure Gage Unit

Electrical Equipment

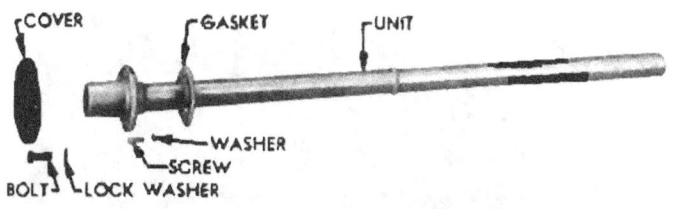

Figure 250—Fuel Gage Tank Unit and Cover

170. SIGNAL SENDING UNITS.

a. **Replacement of Transmission Oil Temperature Sending Unit.** Turn 24-volt master switch off (fig. 15). Remove drain plug in hull floor under torque converter (third plug from front), remove ¼-inch pipe plug from converter housing and drain oil; then install and tighten both plugs securely. Remove two screws which secure conduit housing to sending unit on elbow at right side of converter, remove conduit and wire, and unscrew sending unit from elbow. Clean oil from threads in elbow and on replacement unit, coat threads with anti-seize compound and screw new sending unit securely into elbow. Press wire terminal firmly into unit, and connect conduit housing to unit. Check and fill transmission (par. 38).

b. **Replacement of Engine Oil Temperature Gage Unit** (fig. 248). Turn 24-volt master switch off (fig. 15). Open hull rear door (par. 181 a). Remove two screws which secure conduit housing to gage unit on oil tank outlet tee, remove conduit and wire, and unscrew gage unit from elbow. Screw new gage unit securely into tee. Press wire terminal firmly into unit, and connect conduit housing to unit. Run engine and test operation of gage unit and gage. Close hull rear door (par. 181 b).

c. **Replacement of Engine Oil Pressure Gage Unit** (fig. 249). The gage unit is located in engine rear crankcase forward of the left magneto. To replace unit follow the same procedure as described for temperature gage unit in subparagraph b above.

d. **Replacement of Fuel Gage Tank Unit** (fig. 250). Turn 24-volt master switch off (fig. 15). Remove inspection hole cover which is attached to hull roof plate by four ⅜-inch bolts and lock washers. Disconnect wire from tank unit, remove six cap screws and lock washers which secure unit to tank, and remove unit from tank. To install, clean joint surface on tank, place new gasket (G137-0193886) over lower end of unit, install unit in tank, and secure it with six socket head screws (¼ in.—20 x ⅝ in.) and lock washers. Connect wire to terminal screw. Install cover with four bolts (⅜ in.—24 x ¾ in.) and lock washers.

Part Three—Maintenance Instructions

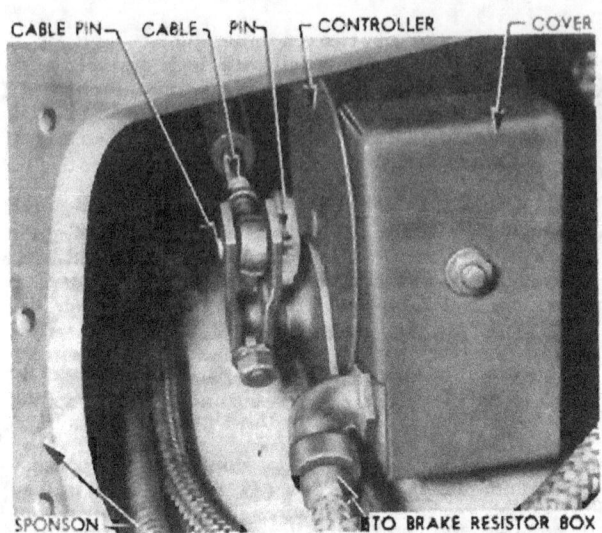

Figure 251—Trailer Electric Brake Controller—M39

171. SIREN AND SIREN SWITCH.

a. **Siren Removal.** Turn 24-volt master switch off (fig. 15). Remove headlight junction box cover and disconnect yellow wires from terminal block binder screw marked "C" (fig. 233). Remove hexagon nut, flat washer, and gasket from lower end of siren conduit sleeve. Remove two 1/4-inch cap screws and lock washers which anchor the siren to hull plate, lift siren and pull conduit sleeve out of opening in hull plate. Remove gasket and flat washer on top side of hull plate.

b. **Siren Installation.** Place flat washer and gasket over end of coupling sleeve, insert wire and sleeve through hole in hull plate as siren is placed in position, and anchor siren to hull plate with two cap screws (1/4 in.— 28 x 5/8 in.) and external-toothed lock washers. Place gasket, flat washer, and hexagon nut over lower end of conduit sleeve and tighten nut securely. Insert siren wire into headlight junction box and attach both yellow wires to terminal block binder screw marked "C" (fig. 233). Install headlight junction box cover.

c. **Siren Switch Removal.** Turn 24-volt master switch off (fig. 15). Remove siren foot switch junction box which is anchored to mounting bracket by four 3/16-inch cap screws and lock washers. Disconnect wires from switch and remove switch which is secured to box by two machine screws, lock washers and nuts.

Electrical Equipment

d. **Siren Switch Installation.** Place new gasket and switch in junction box and secure by two cadmium-plated machine screws (¼ in.—20 x ⅝ in.) with internal-tooth lock washers and nuts. Attach wires to switch terminal posts by lock washers and brass nuts. Place new gasket between switch junction box and mounting bracket and anchor in place with four cadmium-plated cap screws (5/16 in.—24 x ¾ in.) and external-tooth lock washers.

172. TRAILER ELECTRIC BRAKE CONTROLLER, M39.

a. **Removal** (fig. 251). Turn master switches off. Remove sponson side opening cover and gasket which are attached by twelve ½-inch bolts and lock washers. Disconnect cable from lever by removing cotter pin and cable pin. Remove terminal cover which is attached with one nut and lock washer. Disconnect wires from terminal studs, unscrew conduit coupling nut and pull wires out of elbow on controller. Remove controller assembly which is anchored to sponson plate by two ⅜-in. cap screws and external-toothed lock washers.

b. **Installation** (fig. 251). Anchor controller to sponson plate with two cadmium-plated cap screws (⅜ in.—24 x ½ in.) and external-toothed lock washers. Push wires through elbow on controller and connect conduit to elbow. Connect wire marked "BK" to terminal stud marked "BK". Connect wire marked "BA" to terminal stud marked "BA". Install terminal cover with one nut and internal-toothed lock washer. Connect cable to controller lever with cable pin secured by cotter pin (1/16 in. x 9/16 in.). Install sponson side opening cover and gasket with 12 bolts (½ in.—20 x 1¼ in.) and lock washers tightened to 75-85 foot-pounds tension.

173. TRAILER ELECTRIC BRAKE RESISTOR BOX, M39 VEHICLE.

a. **Removal** (fig. 252). Remove sponson side opening cover and disconnect conduits and shafts from instrument panel as described in steps (1) through (4) of paragraph 156 a. Remove six 5/16-inch cap screws and lock washers which anchor the instrument panel mounting supports to sponson, and remove instrument panel assembly through driving compartment. Remove covers from resistor box and trailer blackout tail and stop light resistor. Disconnect wires from resistor box and tail and stop light resistor. Disconnect three conduits at resistor box, and pull wires out of conduit leading to resistor. Remove resistor box assembly which is anchored by three cap screws (¼ in.) and lock washers.

b. **Installation** (fig. 252). Anchor resistor box assembly to mounting bracket with three cadmium-plated cap screws (¼ in.—20 x ½ in.) and external-tooth lock washers. Connect three conduits and push wires from rear conduit through conduit leading to tail and stop light resistor. Connect wires to terminal studs in resistor and install cover. Connect wire marked "CTR" to terminal stud marked "CTR" in resistor box. Connect wires marked "BRK" and "CTR"

TM 9-755
173-174

Part Three—Maintenance Instructions

Figure 252—Trailer Electric Brake Resistor Box and Blackout Tail and Stop Light Resistor

to terminal stud marked "BRK-CTR". Connect wire marked "MSTR-SW" to terminal stud marked "MSTR". Install resistor box cover. Install instrument panel assembly and anchor the mounting supports to sponson with six cap screws ($\frac{5}{16}$ in.—24 x $\frac{5}{8}$ in.) and internal-external-tooth lock washers. Connect shafts and conduits to instrument panel and install sponson side opening cover and gasket as described in steps (3) through (5) of paragraph 156 h.

174. TRAILER ELECTRIC RECEPTACLE, M39.

a. **Removal** (fig. 226). Turn master switches off. Open hull rear door (par. 181 a). Disconnect two conduits from rear junction box (fig. 235) and remove junction box inspection plate which is attached with five $\frac{5}{16}$-inch cap screws, special washers and two lock nuts. Disconnect upper and lower conduits and remove junction box which is anchored to hull by two $\frac{5}{16}$-inch bolts. Remove electric receptacle cover, lift conduit and terminal shield and disconnect wires. Remove nuts and lock washers from inner ends of the four attaching screws, remove receptacle and gasket from inner side of hull plate, and remove screws, dust shield, dust shield ring, and gasket from outer side of hull plate.

b. **Installation.** On outer side of opening in hull plate install a gasket (rubber) dust shield ring, dust shield with hinge up, and four cadmium-plated cap screws ($\frac{1}{4}$ in.—20 x 1$\frac{3}{8}$ in.) with external-tooth lock washers. Place gasket (composition) over screws on inside of hull and place receptacle over screws, with terminal marked "GR" to right (viewed from front) (fig. 226). Install cadmium-plated external-tooth lock washers and nuts on all screws except lower left (viewed from front). Place external-tooth lock washer, ground wire terminal and split washer on lower left screw and install nut. Tighten all nuts firmly. Connect wires by color to terminal studs marked as follows: black (ground) to "GR"; red to "SL"; natural to "BK"; and black (natural tracer) to "TL". Install terminal shield attached to conduit over the receptacle, with ground wire leading through notch in flange, and install cover with one nut and external-tooth lock washer. Install rear junction box with two bolts ($5/16$ in.—24 x 1¼ in.) and lock washers, and attach upper and lower conduits (fig. 235). Install junction box inspection plate with five cap screws ($5/16$ in.—24 x ⅝ in.), and square washers, with lock nuts on two top screws. Connect two conduits to junction box. Close hull rear door (par. 181 b).

Section XXXIV

RADIO INTERFERENCE SUPPRESSION

175. PURPOSE.

a. The purpose of radio interference suppression is to eliminate or minimize electrical disturbances within vehicle which would interfere with radio reception, or would disclose the location of vehicle to sensitive electrical detectors. It is important, therefore, that vehicles with, as well as vehicles without, radios be suppressed properly to prevent interference with radio reception of neighboring vehicles.

176. SUPPRESSION COMPONENTS.

a. General. Suppression is accomplished by use of capacitors (condensers), bonding jumpers (ground straps), and by tooth-type lock washers. Wiring, which may carry interfering electrical surges to a point where interference will affect radio reception, is shielded. The tooth-type lock washers and ground straps located throughout vehicle tend to bind entire unit together into solid shield. Location of suppression components is given in the following subparagraphs.

b. Ignition System. High tension wires are shielded between magnetos and spark plugs by inclosure in metal harness which is bonded to engine with tooth-type lock washers at attaching screws. The engine is bonded to hull through ground strap.

c. Charging System. Two capacitors are mounted inside engine generator housing, with one capacitor connected to each positive brush holder. Condensers are connected to each generator regulator terminal in main filter box. A suppressor is connected in high tension (spark plug) wire in magneto of the auxiliary generator engine (M18).

d. Fuel System. One capacitor is connected to battery terminal of each fuel tank pump, and is inclosed in Douglas housing which covers terminal on pump.

TM 9-755
176-177

Part Three—Maintenance Instructions

e. **Turret Traversing System.** One condenser is connected to traverse motor master switch in turret wiring switch box. Two capacitors are mounted inside the electric motor housing, with one capacitor connected to each positive brush holder.

f. **Radio Terminal Boxes.** One capacitor is connected to 12-volt positive terminal, and one capacitor is connected to 24-volt positive terminal in all radio terminal boxes except hull radio terminal box in M18 vehicle.

g. **Bonding by Ground Straps or Wires.** Ground straps or wires are used at following points:

   (1) Instrument panel to support bracket (fig. 215).
   (2) Radio terminal boxes (figs. 216, 217, 218 and 219).
   (3) Auxiliary generator to hull (fig. 220).
   (4) Engine generator to hull (fig. 221).
   (5) Generator regulators to hull (2 used) (fig. 222).
   (6) Right and left fuel tanks to hull (fig. 223).
   (7) Engine to engine mounting ring (fig. 225).
   (8) Drag link to race ring bracket—M18 (fig. 228).
   (9) Drag link to slip ring box—M18 (fig. 227).

h. **Bonding by Toothed Lock Washers.** The attaching screws of all clips or clamps which support conduits and pipes are provided with toothed lock washers except where clips are anchored to tapping blocks; these attaching screws have split lock washers. Toothed lock washers are used also to bond mechanical parts at the following points:

   (1) Center instrument panel support to sponson plate.
   (2) Fuel pump mounting plate to fuel tank.
   (3) Fuel pump terminal casting to mounting plate.
   (4) Fire extinguisher nozzles to brackets.
   (5) Fire extinguisher nozzle brackets to hull.
   (6) Primer pump line clamps to hull.
   (7) Subfloor center plate to floor support.
   (8) Attaching screws for all electrical units.

177. MAINTENANCE.

a. **Ignition System.** The attaching screws and all conduit coupling nuts of the ignition wiring harness must be kept securely tightened.

b. **Charging System.**

   (1) CAPACITORS. Capacitors in engine generator housing can be removed by removing generator (par. 143 e), removing the cover band, and disconnecting capacitors from brush holders and housing. When new capacitors are installed make sure that surfaces are clean and tight connections are made. Install cover band and install generator (par. 143 f).

(2) CONDENSERS. Condensers in the main filter box can be removed by removing filter box cover and disconnecting condensers from box and regulator terminals. When new condensers are installed make sure that condenser attached to regulator field terminal is 0.01 mfd capacity, and that condensers attached to armature and battery terminals are 0.1 mfd capacity. Surfaces must be clean, and condensers must be attached to box with machine screws (No. 10—32 x ⅜ in.) and internal-tooth lock washers. Connect condensers to terminals as shown in figure 231 and install filter box cover.

(3) SUPPRESSOR IN AUXILIARY GENERATOR. Remove magneto rotor by loosening rotor nut. Unscrew high tension wire from suppressor (fig. 192) and remove suppressor and rubber sleeve. Place rubber sleeve over new suppressor, and screw ends of high tension wire into ends of resistor. Make sure that rubber sleeve covers ends of suppressor, install magneto rotor and tighten nut securely.

c. Fuel System. Replace capacitor at fuel tank pump as follows: Disconnect battery wire and capacitor at fuel pump battery terminal by opening hull rear door (par. 181 a), closing fuel control valves (par. 14 a), disconnecting pipes and removing fuel pump sheet metal cover (par. 99 b (4) and (6)). Remove capacitor from housing and install replacement capacitor. Attach capacitor and wire to battery terminal of pump, install inspection cover, connect pipes to pump, and close hull rear door.

d. Turret Traversing System. The condenser attached to traverse motor master switch can be replaced by removing turret wiring switch box cover and disconnecting condenser from mounting bracket and master switch (par. 168 h). Replacement of the capacitors in the electric motor requires partial disassembly of motor. Remove motor (par. 198) for service by higher authority.

e. Radio Terminal Boxes. Remove terminal box cover and remove capacitors which are connected to positive terminals and anchored by box attaching screws. When new capacitors are installed, make sure that surfaces are clean, and that external-tooth lock washers are used on attaching screws. Connect capacitors to terminals as shown in figures 236, 238, or 239.

f. Wiring System. Conduit support clip attaching screws must be kept tight and secured by the proper type lock washers. Conduit coupling nuts must be kept properly tightened.

Section XXXV
HULL

178. DESCRIPTION AND DATA.

a. General Description. The hulls of the M18 and M39 vehicles are identical in construction except for differences in middle compartments and top structures described in subparagraphs b and c below.

(1) WELDED STRUCTURE. The hull is a completely welded structure except for certain sections which are removable for service

operations. All exterior plates are made of armor plate except sponson bottom plate which is made of mild steel. Thicknesses of these plates, which vary in accordance with structural requirements, are given in subparagraph d below. A transverse bulkhead, located between engine compartment and middle compartment, is welded and bolted to floor, side, sponson, and roof plates. This bulkhead supports roof, reinforces side plates and floor, and also provides a mounting for rear transfer case. Reinforcing plates for supporting vehicle on jacks or stands are welded to bottom surface of hull floor plate at four corners of hull.

(2) TOWING PINTLES AND SHACKLES. Heavy towing blocks are welded to front and rear hull plates on each side, to which towing shackles are attached by heavy shackle pins. An automatic towing pintle is attached to center of hull rear door by bolts.

(3) MOUNTING RAILS. Two rails are welded on top of No. 1 and No. 2 torsion bar housings on hull floor to provide a mounting for transmission and differential assembly. Two rails are welded to floor plate in rear compartment to provide a mounting for engine. Both sets of rails also serve as tracks upon which units may be moved in or out of hull.

(4) DRIVER'S DOORS. Two double section doors are hinged to roof of hull to provide entrances to driving compartment (fig. 6). Doors are made watertight with rubber seals and are anchored in closed position by two latch handles on each outer section. Each section is anchored in open position by a stop latch knob and pin located in bosses welded to hull roof. Periscope housing mounted in outer section of each door to support a removable periscope may be tilted or rotated to change line of vision as required.

(5) FRONT AND REAR DOORS. Opening provided in front hull plate for removal and installation of transmission and differential assembly is closed by a front door assembly which is bolted in place. The opening in rear hull plate for removal and installation of engine is closed by a rear door assembly, hinged at bottom and anchored to hull with bolts. Both front and rear doors are provided with rubber seals to exclude water. A cover and gasket bolted over an opening in rear door may be removed for inspection of carburetor. A spring-loaded cover closes a hole in rear door through which engine hand crank is inserted when cranking engine.

(6) ROOF DOOR AND GRILLES. A roof door, hinged to roof plate and anchored by bolts, permits access to engine from above. Grilles are bolted over openings in roof above the engine to provide an air inlet, air outlet, and escape of exhaust gas. A hinged section of the left outlet grille permits easy access to the engine oil tank filler opening and oil level indicator.

(7) HULL DRAIN VALVES. Two drain valves are located in floor of driving compartment and two also in engine compartment. Valves in both compartments are held closed by spring pressure. Driver's compartment drain valves are locked in open position by pushing

down on knob and turning it clockwise so that a pin in valve stem engages a bayonet type slot in cage. Engine compartment drain valves are connected by levers and cables to handles projecting through the roof. Pulling the handles up and turning them clockwise will lock these valves in open position.

b. **M18 Hull.** An opening, 77½ inches in diameter, in roof center plate provides a mounting for turret upon the roof while allowing inclosure of turret components and crew within hull. A turret base deflector is welded to roof plate around turret opening to protect joint between turret and hull against entrance of projectiles or other objects which would jam turret. A vertical tubular support, welded in place, supports roof at front edge of turret opening, directly under turret race in line with the 76-mm gun when in traveling position. Frames are welded to the floor and side plates within the middle compartment to support subfloor and plates. Doors covering stowage areas below subfloor are hinged and secured by flush type fastener handles. An escape door, which opens outward when released, is attached under an opening in hull floor by a quick-acting release lever. The escape door is reached through right front subfloor door.

c. **M39 Hull.** Middle or crew compartment of hull forms an open cockpit in which two transverse seats are provided for personnel. On each side of crew compartment, stowage boxes are welded to hull roof above the sponson. These stowage boxes are closed on top by hinged covers. Hinged covers on sponson side plates provide access to stowage space within sponson. Racks are welded to outer sides of sponson extensions for stowage of blankets, tarpaulins, etc. A caliber .50 machine gun is mounted over front of crew compartment. An escape door, which opens outward when released, is attached under an opening in hull floor by quick-acting release lever. Upper side of door forms part of crew compartment floor. Battery box is located across front of crew compartment.

d. Data.

(1) HULL PLATE MATERIAL AND THICKNESS.

Part	Material	Thickness
Floor plate	Armor plate	¼ in.
Side plates	Armor plate	½ in.
Sponson bottom plates	Mild steel	³⁄₁₆ in.
Sponson side plates	Armor plate	½ in.
Roof plates	Armor plate	³⁄₁₆ in.
Front plates, upper section	Armor plate	½ in.
Front plates, lower section	Armor plate	½ in.
Rear plates	Armor plate	½ in.
Front door, upper section	Armor plate	½ in.
Front door, lower section	Armor plate	⅝ in.
Drivers' doors	Armor plate	⅜ in.
Escape door	Armor plate	³⁄₁₆ in.

TM 9-755
178-179

Part Three—Maintenance Instructions

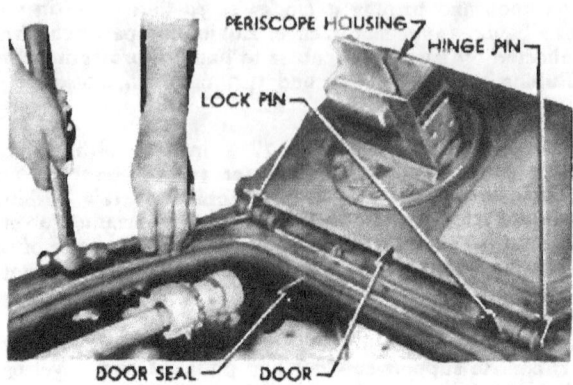

Figure 253—Removing Hinge Pin Lock Pin From Driver's Door

(2) DIMENSIONS.

Length, overall	208 in.
Width outside sponsons	110 in.
Width below sponsons	69½ in.
Ground clearance	14¼ in.

179. DRIVERS' DOORS AND SEALS.

a. **Adjustment of Door Latch Handle.** Two latch handles are installed on lower side of each outer section of door. A U-shaped spring attached to the eccentric latch handle by a pin engages a latch stud on the hull so that when handle is moved to locked position its leverage causes the spring to tighten and hold door securely against door seal. The spring tension is adjusted by screwing latch stud in or out as required, after which stud lock nut must be tightened.

b. **Replacement of Doors.**

(1) Drive out hinge pin lock pins with a ³⁄₃₂-inch blunt punch (fig. 253); then drive out hinge pins and remove door. Female section of each hinge is welded to roof plate and has flexible self-lubricating bushings pressed into eyes to provide bearings for hinge pins.

(2) Place door in position and aline hinge pin holes, drive hinge pins into position so that lock pin holes in pins and male section of hinges are alined; then anchor hinge pins by installing the lock pins.

c. **Replacement of Door Seals.** A rubber seal is attached to outboard edge of each inner door by a metal retainer and five cap screws (14 in.—28 x ½ in.) and lock washers. When installing a new seal, coat surface of door with non-vulcanizing rubber cement where seal is to be placed before installing seal and retainer. A rubber seal vulcanized to a metal retainer is attached to under side of roof plate to form a full seal around door opening. When installing a

Figure 254—Hull Rear Door Open and Supported by Turnbuckle Hooks 41-H-2742

new seal, coat it with non-vulcanizing rubber cement where it contacts roof plate, then attach seal and retainer to underside of roof plate with 17 flat head screws (¼ in.—28 x ⅜ in.) and lock washers.

180. HULL FRONT DOOR AND SEALS.

a. *Removal of Door.* Attach hoist to hooks in each upper corner of door, cut lock wires and remove the eighteen ½-inch bolts; then swing door forward to clear hull.

b. *Installation of Door.* Attach hoist to hooks in each upper corner of door and swing door into position on hull. Install 18 bolts (½ in.—20 x 1¾ in.) while door is supported by hoist, and tighten bolts evenly to 50-60 foot-pounds tension. Install lock wires between adjacent pairs of bolt heads.

c. *Replacement of Door Seals.* The joint between front door and hull is made water tight by a channel shaped rubber seal attached to hull by metal retainers and round head self-tapping screws. Eight curved retainers are used at corners and seven straight retainers are used in the straight sections. When new seal is installed, thoroughly coat flat side of seal and contact surface on hull plate with non-vulcanizing rubber cement, place seal smoothly on hull around door opening and install retainers and tapping screws.

TM 9-755
181

Part Three—Maintenance Instructions

Figure 255—Door Plate Lowered, Giving Access to Bottom of Engine

181. HULL REAR DOOR AND SEALS.

a. Procedure for Opening Rear Door. Cut and remove attaching bolt lock wires. Remove seventeen ¾-inch bolts and lock washers which anchor reinforcement rails and lower edge of door. Remove pilot bolt and lock washer from center of lower edge. NOTE: *M18 vehicles having serial numbers below 1242 do not have the pilot bolt.* Remove all top and side attaching bolts except the upper corner bolts. Support door to prevent falling while removing corner bolts, then lower door and support it with two turnbuckle hooks (41-H-2742) (fig. 254).

b. Procedure for Closing Rear Door. Coat floor-to-door brace and the mating surface of door with joint sealing compound (51-C-1616). Raise door and install two upper corner bolts (½ in.—

Figure 256—Hull Rear Door Hinge

20 x 2¼ in.) and lock washers, leaving bolts loose. Install pilot bolt and lock washer in center hole in lower edge of door (see note, subpar. a above). Install 12 remaining cap screws (½ in.) and lock washers around top and sides; install nine bolts (¾ in.—16 x 2 in.) and lock washers along lower edge; install 8 bolts (¾ in.—16 x 2¾ in.) through reinforcement rails. Tighten all bolts to 50-60 foot-pounds tension. Install lock wires through heads of two or more adjacent bolts.

c. **Hinged Door Plate.** In some vehicles, door plate is hinged separately from rails so that plate can be disconnected from the rails and lowered to provide access to bottom of engine. These doors are opened and closed in the manner described in subparagraphs a and b above. To lower door plate when door is open, dig a hole in ground for towing pintle to enter, disconnect door plate from rails by removing two bolts, and lower the plate as shown in figure 255. Before closing the door, raise the plate and attach it to rails with two bolts.

d. **Removal of Door Assembly.** Remove towing pintle (par. 187 a). Open door (subpar. a above) and place blocks to support front edge. Remove the ¼-inch cap screw which locks each hinge pin in place (fig. 256), install long cap screw (⁵⁄₁₆ in.—24) in threaded hole in hinge pin, and pull pin out of hinge. Lift door assembly from hull.

e. **Installation of Door Assembly.** Place door in horizontal position at door opening, supporting it upon blocks so that hinge pin holes line up. Drive each hinge pin into hinge until groove in pin is in line with hole for hinge pin lock screw (fig. 256). Install hinge pin lock screw (¼ in.—28 x 1½ in.) with lock washer in each hinge. Close the door (subpar. b above). Install towing pintle (par. 187 a).

f. **Replacement of Seals.** The joint between rear door and hull is made water tight by channel-shaped rubber seals. The seals along

TM 9-755
181-183

Part Three—Maintenance Instructions

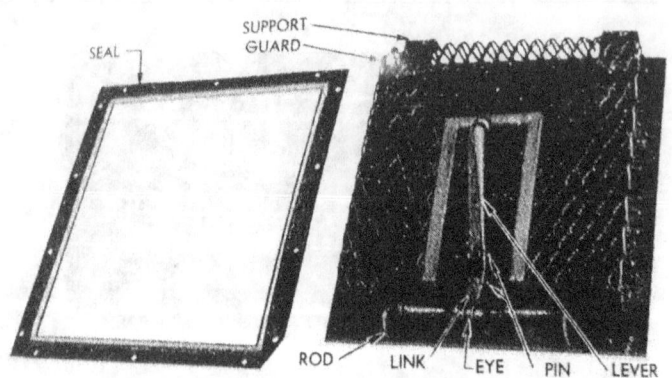

Figure 257—Escape Door and Seal—M18

the top and sides are held in place by metal retainers and round head self-tapping screws (10—32 x ¾ in.). The seal along the lower edge of door is vulcanized to metal retainer which is secured by round head self tapping screws (10—32 x ¾ in.). When new seals are installed, thoroughly coat both seal and metal contact surfaces with non-vulcanizing rubber cement, place seal smoothly on hull and install retainer and tapping screws. Be sure to make contact with adjoining seals at corner to avoid water leaks.

182. HULL FLOOR ESCAPE DOOR AND SEALS.

a. **Adjustment of Door Locking Lever** (fig. 257). Door must close tightly against seal to prevent entrance of water into hull. To adjust locking lever for proper pressure against seal, lower door by lifting lever; disconnect link from lever by removing cotter pin and clevis pin, and turn link on threaded eye bolt as required. Turning link in clockwise direction decreases pressure against seal; turning link counterclockwise increases pressure. When proper pressure is secured with clevis pin installed, secure pin with cotter pin.

b. **Replacement of Door Seal** (fig. 257). Escape door seal is made of fabric covered with neoprene sponge, and is secured to hull floor by metal retainer strips anchored by studs and nuts. When new seal is installed, thoroughly clean floor surface around door opening, coat surface of floor and surface of seal with non-vulcanizing rubber cement, and place seal over studs in floor with round section downward. Install retainers over studs, install nuts (⅜ in.—24) and tighten evenly.

183. COVERS AND GRILLES.

a. **Removal and Installation of Carburetor Inspection Hole Cover.** Remove the eight attaching cap screws (⅜ in.—24 x 1¼ in.)

Hull

A—ARMORED UTILITY VEHICLE—M-39

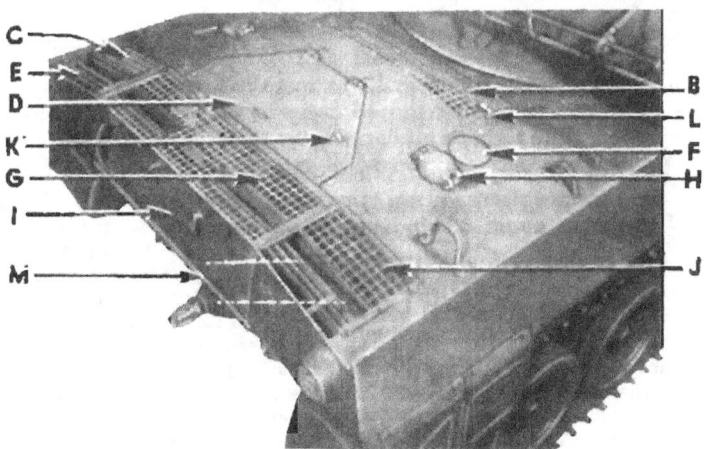

B—76-MM MOTOR GUN CARRIAGE—M18

A—WATER CAN RETAINER
B—HULL AIR INLET GRILLE
C—OIL TANK FILLER SCREEN DOOR
D—HULL REAR ROOF DOOR
E—HULL AIR OUTLET GRILLE, LEFT
F—FUEL TANK GAGE COVER
G—HULL AIR OUTLET GRILLE, CENTER
H—FUEL TANK CAP COVER
I—HULL REAR UPPER PLATE
J—HULL AIR OUTLET GRILLE, RIGHT
K—OIL FILTER HANDLE
L—DRAIN VALVE HANDLE
M—HULL REAR DOOR

RA PD 344528

Figure 258—Doors, Grilles, Covers, and Plates on Rear End of Hull

and lock washers. Before cover is installed, make certain that gasket is in good condition and securely cemented to cover. When cover is installed, tighten attaching screws and lock washers to 20-25 foot-pounds tension.

b. Removal and Installation of Fuel Tank Cap Cover (fig. 258). Remove cotter pin (⅛ in. x 1 in.) from one end of cover pin and remove cover pin. When cover is installed, make certain that cover pin is locked by a cotter pin in each end, and that locking pin is pushed fully into place.

c. Removal and Installation of Hull Hand Crank Hole Cover. Remove cotter pin, nut, spring, and washer from stud which secures cover to door. When cover is installed, place cover over stud, place washer and spring over stud, and install the nut. Screw nut up tight, then loosen to aline with nearest cotter pin hole and install cotter pin (⅛ in. x 1½ in.).

d. Removal and Installation of Air Inlet Grille (fig. 258). On M39 vehicle only, remove two water can retainers which are attached by four cap screws (⅜ in.—24 x ¾ in.), lock washers, and nuts. Remove the 15 cap screws (⅜ in.—24 x ⅝ in.), plain washers and lock washers, unhook tachometer shaft support clip from grille, and lift grille from hull. When grille is installed, hook tachometer shaft support clip to grille, start the 15 cap screws with plain washers and lock washers; then tighten all cap screws to 20-25 foot-pounds tension. Install water can retainers on M39 vehicle only.

e. Removal and Installation of Air Outlet Grilles (fig. 258). Three grille assemblies are installed over mufflers at rear end of hull roof. The center grille, which should be removed first, is anchored along the front edge and sides with seven cap screws (⅜ in.—24 x 1¼ in.) plain washers and lock washers, and along rear edge with three cap screws (⅜ in.—24 x 1¼ in.) and lock washers through hull rear upper plate. Right grille is anchored by center grille and by three cap screws (⅜ in.—24 x 1¼ in.) plain washers and lock washers on outer edge. Left grille has a hinged front section which is anchored to hull roof by a removable pin, and a rear section anchored to hull by center grille and two cap screws (⅜ in.—24 x 1¼ in.), plain washers and lock washers on outer edge. To install grilles, first place left and right grilles in place, then install center grille and all attaching cap screws with plain washers and lock washers. Tighten all cap screws to 20-25 foot-pounds tension.

184. HULL SUBFLOOR DOORS AND PLATES—M18 VEHICLE.

a. Removal and Installation of Subfloor Doors (fig. 260). Each subfloor door is attached by two hinges. Each battery door is locked by two cap screws (⅜ in.—24 x 1 in.) The four other doors are locked by flush-type door fasteners. To open, pull up on fastener handle. Any door is removed by opening door, removing cotter pin from one end of hinge pins, and removing hinge pins. When door is

Hull

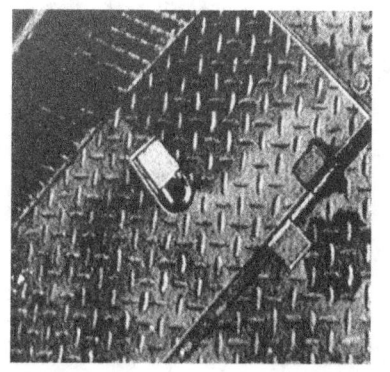 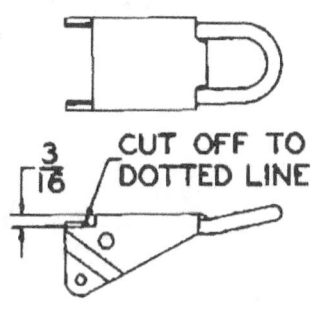

Figure 259—Modification of Subfloor Door Fastener Handle

installed, be sure that hinge pins are secured by a cotter pin in each end.

b. **Modification of Subfloor Door Fastener Handle, M18 Vehicle.** Starting with M18 vehicle serial number 482, subfloor door fastener handles were modified in production to provide more positive locking of doors. In vehicles of earlier production, doors sometimes become unlocked during operation of vehicle, and this causes interference when traversing turret. The four subfloor door fastener handles in each vehicle having serial numbers below 482 will be modified in field by filing off metal on both sides of handle body as shown in figure 259. The handle is removed from door by removing the retaining pin. When handle is installed, make certain that pin is secured with cotter pins ($3/16$ in. x 1½ in.) at each end.

c. **Removal and Installation of Subfloor Plates.** Subfloor plates may be removed individually by removing any stowage or installed item and then removing attaching screws. Some plates are attached by cap screws (⅜ in.—24 x 1 in.) only; others have plain washers under screw heads. Location of plates and attaching screws is shown in figure 260. Particular attention is called to center plate which is attached with cadmium-plated cap screws (⅜ in.—24 x 1 in.) and external-toothed lock washers. It is important that these screws and lock washers be used to attach plate to properly ground slip ring box and prevent radio interference. When plates are installed, tighten all attaching screws to 20-25 foot-pounds tension.

185. HULL SEATS.

a. **Driver's Seat Replacement** (fig. 8). Seat back may be removed from seat assembly by lifting it up out of sockets in seat frame. Early production seat backs are not removable but are hinged so they

TM 9-755
185

Part Three—Maintenance Instructions

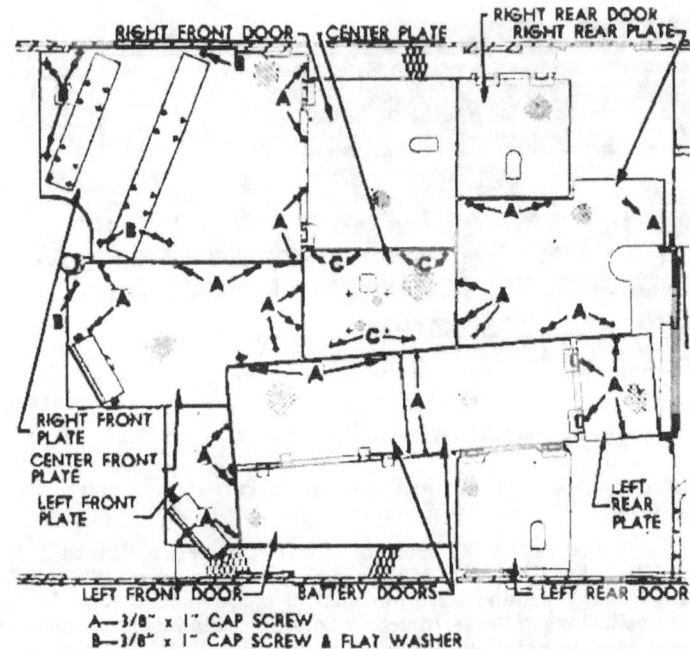

Figure 260—Hull Subfloor Doors and Plates—M18

may be folded down over seat. Seat back pads may be removed from frame by removing the attaching screws. Each seat assembly is anchored to a tapping plate on hull floor by four cap screws (3/8 in.—24 x 7/8 in.) and lock washers, which must be tightened to 20-25 foot-pounds tension when seat assembly is installed.

b. **Crew Seat Replacement, M39 Vehicle.** Front and rear seats in crew compartment are formed by hinged covers attached to supports welded to hull. The seats and seat backs are covered by pads which are attached with flat head screws (1/4 in.—20 x 3/4 in.) and finish washers. Each seat cover is locked by flush type fasteners on side opposite hinges, under pads. Seat covers are removed by detaching seat pads and removing hinge pins which are secured by cotter pins. When covers are installed, be sure that hinge pins are secured by a cotter pin on each end and that seat pads are attached with all screws and finish washers.

c. **Rear Seat Back Removal, M39 Vehicle.** Remove three 3/8-inch cap screws with lock washers and clamp which secures fire

TM 9-755
185-186

Hull

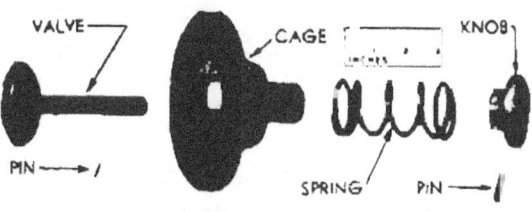

Figure 261—Hull Drain Valve—Disassembled

extinguisher dual-pull mechanism to mounting bracket on rear seat back (fig. 277). Disconnect both ends of rear seat back pad by removing flathead screws and finish washers. Remove ten ⅜-inch cap screws and lock washers which anchor rear seat back and lift out with pad attached. Remove pad from seat back.

d. **Rear Seat Back Installation, M39 Vehicle.** Place rear seat back in position and anchor it with 10 cap screws (⅜ in.—24 x 1 in.) and lock washers tightened to 20-25 foot-pounds tension. Attach seat back pad with flathead screws (¼ in.—28 x ¾ in.) and finish washers. Anchor fire extinguisher dual-pull mechanism to mounting bracket on seat back by means of clamp secured by three cap screws (⅜ in.—24 x 1 in.) and lock washers.

186. HULL DRAIN VALVES.

a. **Removal.** Remove engine (par. 75) if drain valve in engine compartment is to be removed. Remove six nuts, lock washers, and cap screws (5/16 in.—24 x ⅞ in.) which attach valve cage to floor plate and push cage down out of hole in floor plate. NOTE: *In M18 vehicle having serial numbers below 465, cage is welded to hull floor plate and it is necessary to remove valve parts from cage (subpar. c, below).*

b. **Repair (fig. 261).** Compress valve spring sufficiently to drive out upper pin which anchors knob to valve stem. Remove knob, spring and valve from cage. Lower pin in valve stem may be driven out if it is damaged. Before parts are installed in cage, examine valve face and seat in cage to make sure that valve will seat properly. Some valves have rubber-coated seats; these valves must be replaced if coating is damaged. If valve is not rubber-coated, it may be lapped into cage with valve grinding compound to secure a water tight seat. Drive lower pin into valve stem so that it projects on one side only. Place valve in cage, install spring and knob and anchor knob to valve stem by installing upper pin.

c. **Installation.** Thoroughly clean joint surfaces of valve cage and floor plate and coat surfaces with joint sealing compound (51-C-1616). Install valve assembly from underneath hull; then secure in place with six cap screws (5/16 in.—24 x ⅞ in.) lock washers

443

and nuts tightened to 10-12 foot-pounds tension. Each engine compartment drain valve is opened by a lever and cable which must be adjusted to permit valve to close tightly and open fully. Loosen lock nut on lower end of cable and screw end of cable in or out of threaded yoke to secure a slight clearance between lever and valve knob and then tighten lock nut. Install engine, if removed (par. 76).

187. TOWING PINTLES, HOOKS, AND SHACKLES.

a. **Replacement of Rear Towing Pintle.** A swivel-type automatic towing pintle is attached to hull rear door by four bolts (fig. 45). NOTE: *A non-swiveling type pintle, like front pintle on M39 vehicle, was installed on some M18 vehicles.* To remove swivel type pintle, cut lock wire and remove four bolts and lock washers while supporting pintle to keep it from falling. Before pintle is installed clean joint surface of door and pintle and coat surface with joint sealing compound (51-C-1616). Attach pintle to door with four bolts (1 in.—14 x 3 in.) and lock washers tightened to 200-250 foot-pounds tension. Install lock wires through heads of adjacent pairs of bolts.

b. **Replacement of Front Towing Pintle, M39.** A non-swiveling automatic towing pintle is attached to hull front door of M39 vehicle only. Remove four bolts (⅝ in.—18 x 1½ in.) and lock washers to remove pintle from door. When pintle is installed, tighten attaching bolts to 130-150 foot-pounds tension.

c. **Replacement of Towing Hooks and Shackles.** Heavy towing blocks are welded to front and rear corners of hull to which towing cable quick-attaching hooks are attached by heavy pins. A hook may be removed by pulling locking pin from grooved end of hook pin and removing pin. When hook is installed, make certain that hook pin is locked on both ends by locking pins (¼ in. x 2¼ in.). NOTE: *On M18 vehicles having serial numbers below 1601, a U-shaped shackle was installed instead of the hook. These shackles are attached by shackle pins which are locked by locking pin on one end and a cotter pin (¼ in. x 2¼ in.) on other end.*

188. BATTERY BOX, M39 VEHICLE.

a. Removal (fig. 262).

(1) Remove batteries (par. 142 g) and battery floor pads.

(2) Remove battery box left rear support which is attached with eight ⅜-inch cap screws having plain washers and lock washers.

(3) Remove one ⅜-inch cap screw and lock washer from each end of battery box floor plate flange and remove one ½-inch cap screw and lock washer which anchors flange to seat center support rail.

(4) Remove seven ⅜-inch cap screws and lock washers which anchor battery box front plate to hull plate and to left front support.

(5) Remove three ⅜-inch cap screws which attach ammunition box retainer rails to front plate.

TM 9-755
188

Hull

Figure 262—Battery Box Rear View With Seat Pads Removed—M39

(6) Remove one $5/_{16}$-inch self-tapping cap screw and plain washer which attaches oil pipe support to front plate and remove one $5/_{16}$-inch cap screw and external-toothed lock washer which attaches temperature sending unit conduit clip to front plate.

(7) Remove two $5/_{16}$-inch cap screws and special washers which attach the cooler grille to front plate and remove one $3/_{8}$-inch cap screw and lock washer which attaches blower belt guard.

(8) Remove two ¼-inch cap screws and lock washers which anchor junction box bracket to left front battery box support.

(9) Slide battery box to rear and remove three ¼-inch cap screws and external-toothed lock washers which attach junction box bracket to left end plate.

(10) Slide battery box to rear and remove it from vehicle.

b. **Installation** (fig. 262).

(1) Place battery box slightly to rear of final position and attach junction box bracket to left end plate with three cap screws (¼ in.—28 x ¾ in.) and external-toothed lock washers.

(2) Slide battery box forward and attach front plate to hull bracket with three cap screws (⅜ in.—24 x ¾ in.) and lock washers, and to left front support with four cap screws (⅜ in.—24 x 1 in.) and lock washers. Install two cap screws (¼ in.—28 x ¾ in.) and lock washers which anchor the junction box bracket to left front support. Leave all screws loose.

(3) Install one cap screw (⅜ in.—24 x ¾ in.) and lock washer at each end of floor plate flange, and install one cap screw (½ in.—

445

TM 9-755
188-189

Part Three—Maintenance Instructions

Figure 263—Periscope M6

20 x 1 in.) and lock washer through center of flange into center support rail. Leave screws loose.

(4) Install battery box left rear support, attaching it to left end plate with four cap screws (⅜ in.—24 x ¾ in.) and lock washers, and to hull brackets with four cap screws (⅜ in.—24 x 1 in.) and lock washers. Leave screws loose.

(5) Attach each ammunition box retainer rail to battery box front plate with one cap screw (⅜ in.—24 x 1 in.) with plain washer and lock washer. Attach oil cooler grille to front plate with two cap screws (5/16 in.—24 x ¾ in.) and special washers. Attach belt guard to front plate with one cap screw (⅜ in.—24 x ¾ in.) and lock washer.

(6) Attach temperature sending unit conduit clip to front plate with one cap screw (5/16 in.—24 x ¾ in.) and lock washer. Attach oil pipe bracket to front plate with one self-tapping cap screw (5/16 in.—24 x ¾ in.) and plain washer.

(7) Tighten all attaching screws installed in preceding steps, using care not to strip threads.

(8) Install battery floor pads, making sure inspection hole cover is in place. Install batteries (par. 142 h).

189. DRIVER'S PERISCOPE.

a. *Removal and Installation.* Open the sliding latch on periscope housing, loosen knurled locking nut and pull periscope out of housing from under side of door. To install, slide periscope into housing on door with guide and locking nut stud engaged in slot in housing. Close sliding latch and tighten locking nut.

b. **Replacement of Periscope Head** (fig. 263). Lift hinged head latch handles on each side of periscope body below head and turn handles until head clamp is completely disengaged from latches inside periscope body. Lift head from periscope body. When installing periscope head, position it on top of body with window facing same side as locking nut (front). Turn latch handles until reference arrow on each handle matches corresponding arrow on periscope body, then fold handles down flat with side of body.

Section XXXVI
TURRET—M18

190. DESCRIPTION AND DATA.

a. **Description.** Turret of M18 vehicle is made of armor plate with a heavy cast steel face plate welded across front to provide a firm base for mounting 76-mm gun. Turret is open at top and has an extended section in rear which provides space for housing radio. A steel box with hinged cover is bolted on rear end of extension to provide space for stowage of tools and spare parts. Brackets attached to rear end of this stowage box provide a mounting for spare track links. The turret extension and stowage box serve to balance weight of 76-mm gun. Stowage racks for tarpaulins and blanket rolls are welded to outer surface of turret walls. Two stowage boxes for driver's hoods are welded to front corners of turret wall. A turret roof and support plate is welded into right front corner of turret to support an ammunition stowage rack. A plate welded on rim of turret wall on left side, provides a support for caliber .50 machine gun mount (fig. 283), which is described in TM 9-308. Turret is supported on roof of hull by a ball-type race assembly (fig. 264) which permits traversing turret 360 degrees. Turret seats, turret lock, traversing mechanism, azimuth indicator, and gun-sighting periscope are described in paragraph 21.

b. **Data.**

Thickness of turret walls	½ in.
Turret wall material	Armor plate
Turret face plate material	Armor steel casting
Weight of turret, less guns and mounts	2400 lb
Diameter of turret race	81 in.
Turret traverse	360 degrees
Turret traversing mechanism	Manual and hydraulic

191. TURRET SEATS.

a. **Commander's Seat.** Commander's seat is provided by a cushion set in commander's platform. Cushion may be lifted out of platform for replacement or when it is necessary to stand on platform.

TM 9-755
191-192

Part Three—Maintenance Instructions

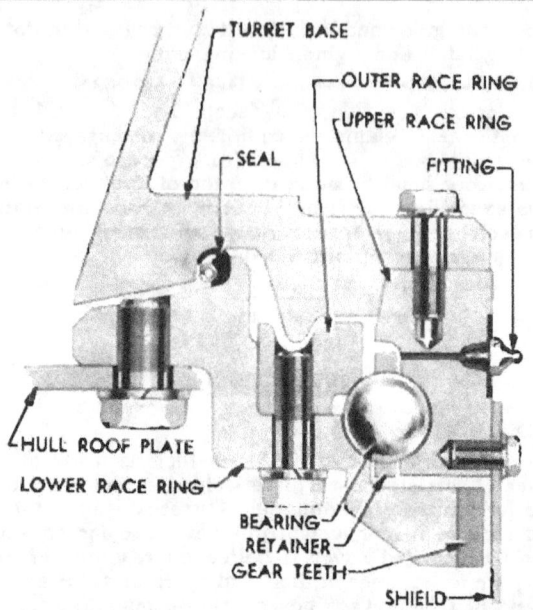

Figure 264—Turret Race Assembly—Sectional View

b. **Gunner's Seat** (A1, fig. 40). The gunner's seat stud is not removable from seat frame. The seat may be removed from mounting plate on upper end of stud by removing four nuts, lock washers, and plain washers. When seat is installed on mounting plate, secure it with plain washers, lock washers and $\frac{3}{8}$-inch nuts on front studs, and plain washers, lock washers and $\frac{5}{16}$-inch nuts on rear studs. Tighten all nuts securely.

c. **Gun Loader's Seat** (fig. 17). Gun loader's seat is removed from its support by pulling outward on seat lock handle and lifting seat out of wedge-shaped channels on support. The seat cushion is attached to seat body by two cap screws ($\frac{1}{4}$ in.—20 x $\frac{5}{8}$ in.) and lock washers. The seat back is attached to its support by three cap screws ($\frac{1}{4}$ in.—20 x $\frac{5}{8}$ in.) and lock washers. Removal of attaching screws permits replacement of cushion and back. The seat may be installed in upper or lower channel of support either in a stowed position or service position as shown in figure 17. Pull outward on lock handle when installing seat, and make certain that seat is firmly locked in place.

192. CALIBER .50 MACHINE GUN MOUNT, M18.

a. **Removal.** Remove four $\frac{1}{2}$-inch bolts which anchor machine gun mount to gun bearing support and to mounting plate on turret.

Turret—M18

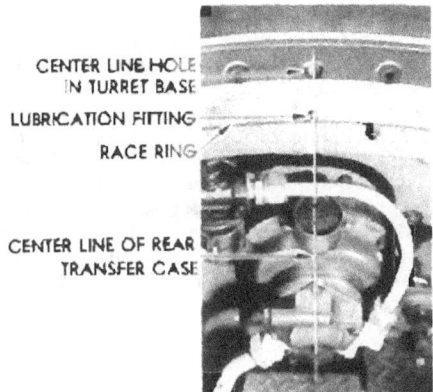

Figure 265—Locating Turret in Straight Ahead Position

Remove two ½-inch bolts with nuts which are installed through anchor bracket and mounting plate, attach chain hoist to mount and remove it from turret.

b. **Installation.** Hoist machine gun mount into position on mounting plate on left side of turret and anchor it with two bolts (½ in.—20 x 1¼ in.) and lock washers through mounting plate into mount. Secure anchor bracket on mount to mounting plate with two bolts (½ in.—20 x 1½ in.) with safety nuts. Attach gun bearing support to gun mount with two bolts (½ in.—20 x 1¼ in.) and lock washers. Tighten all bolts to 50-60 foot-pounds tension.

193. **AZIMUTH INDICATOR.**

a. **Removal.** Disconnect light wire from azimuth indicator by pulling out on knurled sleeve. Remove four ½-inch attaching bolts and lock washers, remove indicator and any spacers located between indicator and turret race ring.

b. **Installation.** Set turret in straight ahead position, with turret race rear lubrication fitting on centerline of rear transfer case (fig. 265) and anchor turret with lock. Turn azimuth indicator gear until bottom pointer sets at "0" on the inner dial (fig. 23). Attach indicator to turret race ring with four bolts (½ in.—20 x 1⅛ in.) and lock washers, placing two spacers 1/16 inch thick over each bolt between indicator and race ring, and turning indicator gear as required to mesh with gear teeth on lower race ring. Tighten bolts to 60-80 foot-pounds tension. When azimuth indicator is installed, bottom pointer must point to "0" on inner dial when turret and 76-mm gun are in straight ahead position. Attach light wire to indicator.

c. **Adjustment of Gear Lash.** Indicator gear is made in two sections. The upper section of gear is actuated by a spring so that its teeth are moved slightly out of alinement with teeth of lower section, thus acting as a cushion to take up slight lash in both directions. Unlock and rotate turret manually and check lash between indicator gear and race ring gear at equally spaced points. At each point, rock lower section of indicator gear back and forth to determine whether a very slight spring action exists without additional lash. There must be sufficient clearance between meshed gear teeth so that a very slight spring action can be felt at closest meshed point. There must not be so much clearance at any point that movement of indicator gear will move top pointer more than one-half mil on outside dial. Adjust lash, if necessary, by placing spacers of proper thickness between indicator and race ring, being careful to install spacers of same total thickness at each attaching bolt (subpar. h above). Spacers are furnished in thicknesses of $\frac{1}{32}$ inch, $\frac{1}{16}$ inch, and 0.015 inch.

194. TURRET LOCK.

a. **Removal.** Move turret traversing mechanism shifting lever to down position (fig. 20). Cut lock wires, remove the four ½-inch bolts which attach turret lock to turret race ring and remove lock and shims.

b. **Installation.** Install the turret lock and two shims on turret upper race ring with four bolts (½ in.—20 x 1 in.) tightened to 80-100 foot-pounds tension. Manually rotate turret and check engagement of lock at equally spaced points. The locking handle must fully engage the clip in the "LOCK" position without excessive lash between teeth of lock pawl and lower race ring. If teeth of pawl bottom in teeth of ring before handle fully engages clip, add one or more steel shims (0.010 in. thick) between lock and race ring. If lash is excessive with handle in "LOCK" position, remove one or more shims. With locking handle in "FREE" position, turret must rotate freely without contact between lock pawl and race ring at any point. Add shims if necessary to prevent contact. Install lock wire (0.0625 in. dia.) between pairs of attaching bolts.

195. TURRET HAND TRAVERSING MECHANISM.

a. **Description.** The hand traversing mechanism (fig. 20) consists of a train of reduction gears and a main drive shaft pinion enclosed in a traverse case and cover. The traverse case anchored on the turret, and the pinion engaged with the gear teeth on the turret lower race ring anchored to the hull, makes it possible to rotate the turret on its ball bearing mounting by applying power through gear train to main drive shaft pinion. Hand traversing mechanism also includes a traverse case extension which carries extension gears and shafts through which gear train can be operated manually by means of brake handle. This extension incorporates a brake which is applied by spring pressure to lock turret when the lever on brake handle is released. A sliding clutch gear, which is part of main gear train, connects extension parts to gear train for manual operation when external shifting lever is pushed down. When shifting lever is pushed

Turret—M18

up for power traversing, it moves clutch gear up to disconnect extension parts and connect gear train to gears operated by hydraulic motor and adapter (par. 196).

b. Removal.

(1) Place traverse motor switch in "OFF" position (fig. 18) and turn turret lock handle to "LOCK" position (fig. 19).

(2) Place a pan or a cloth in position to catch oil, and disconnect three oil tubes from the hydraulic motor (fig. 274). Wipe fittings and ends of tubes clean and cover with tape to prevent entrance of dirt or insects.

(3) Remove two ⅝-inch cap screws and lock washers which attach traverse case to turret. Remove two ½-inch safety nuts and pinion guard from studs which attach traverse case to turret upper race ring.

(4) Slide traversing mechanism off of studs, remove two spacers located between traverse case and turret base, remove shims located between traverse case and upper race ring, and remove locating key from keyway in race ring.

(5) Remove four safety nuts (¼ in.—18) from studs which attach adapter to traverse case cover and lift hydraulic motor and adapter from hand traversing mechanism. Remove gasket.

c. Installation.

(1) Place a new gasket (vellumoid, $\frac{1}{32}$ in. thick) over studs on traverse case cover. Install hydraulic motor and adapter on hand traversing mechanism, engaging splined adapter shaft with pinion in traverse case, and secure adapter with four safety nuts (¼ in.—18) on studs.

(2) Install locating key in keyway in turret upper race ring, and place special shim over studs and under key. Install traversing mechanism on studs, turn brake handle as required to mesh pinion with race ring gear teeth, and make certain that locating key enters keyway in traverse case.

(3) Place two round spacers between traverse case and turret base and install two cap screws (⅝ in.—18 x 3 in.) with lock washers through case and spacers into turret base. Place pinion guard over studs and install safety nuts (½ in.—20) on studs. Tighten nuts to 80-100 foot-pounds tension, and cap screws to 120-130 foot-pounds tension.

(4) Completely rotate turret manually, and if any bind exists between main drive shaft pinion and lower race ring, loosen nuts and cap screws and add steel shims (0.005 in. thick) between the special shim and upper race ring as required to eliminate the binding condition.

(5) Connect the three oil tubes to the hydraulic motor (fig. 274) and tighten connections securely. Test traversing mechanism (subpar. d, below).

TM 9-755
195
Part Three—Maintenance Instructions

Figure 266—Brake Adjustment

d. **Testing Traversing Mechanism.**

(1) Push traversing mechanism shifting lever up (fig. 20), turn turret lock handle to "FREE" position (fig. 19), and rotate turret a complete revolution in either direction by means of the hand traversing mechanism.

(2) At equally spaced points during rotation of turret, check for tight spots or binding between main drive shaft pinion on traversing mechanism and lower race ring gear. Check lash or play between pinion and gear teeth with feeler gages. Lash must not be less than 0.002 inch at tightest point, nor greater than 0.015 inch at point of greatest clearance.

(3) Adjust lash, if necessary, by loosening attaching screws and nuts and removing or installing steel shims between special shim and turret upper race ring. Be sure to again tighten nuts to 80-100 foot-pounds tension and cap screws to 120-130 foot-pounds tension.

(4) Turn traversing motor switch to "ON" position and allow motor and pump to run for 30 seconds; then turn switch to "OFF" position. Repeat this operation several times before operating pump continuously, to work any air out of hydraulic system.

(5) Turn switch to "ON" position, push traversing mechanism shifting lever up, turn turret lock handle to "FREE" position, and rotate turret several revolutions in both directions by means of hydraulic traversing mechanism.

(6) Turret must rotate smoothly without chatter at any point. A chattering condition indicates insufficient lash between main drive

Turret—M18

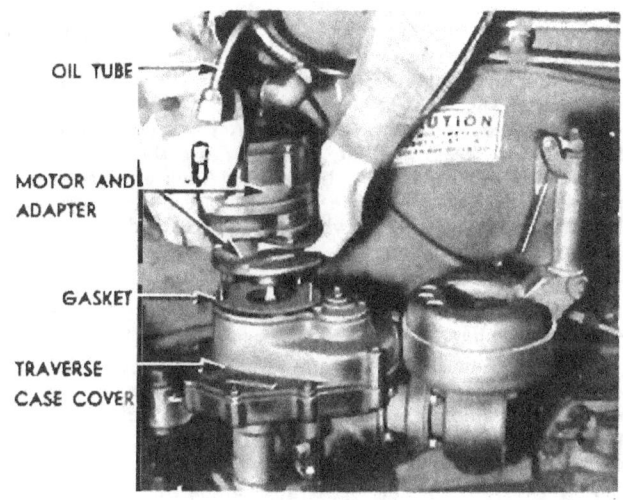

Figure 267—Removing Hydraulic Motor and Adapter

shaft pinion and turret race ring gear, and it must be corrected by adding steel shims as described above.

(7) Check all oil tube connections and correct any oil leaks. Check oil reservoir and add oil (par. 38) as required to bring level two-thirds of the way up on inspection window.

e. **Brake Adjustment.** A brake adjusting screw, with a $5/16$-inch square head, is located in an opening in brake cover opposite brake handle (fig. 266). Hold brake release lever against brake handle and turn adjusting screw clockwise until brake drags when brake cover is turned by means of brake handle; then turn adjusting screw counterclockwise one-quarter turn. The brake is properly adjusted when it does not drag while brake cover is turned with brake release lever squeezed against brake handle, but would drag if the adjusting screw were tightened one-quarter turn clockwise. When lever is released, the brake must grip firmly to stop and lock the turret securely.

196. TRAVERSING MECHANISM HYDRAULIC MOTOR AND ADAPTER.

a. **Description.** The hydraulic motor and adapter (fig. 20) is an assembled unit consisting of an oilgear type hydraulic motor and an adapter housing cover containing a pair of reduction gears. The hydraulic motor transmits controlled fluid power into rotary drive for traversing turret in either direction through adapter gears and gear train in hand traversing mechanism (par. 195). The controlled fluid power, in the form of oil under pressure, is supplied to the motor by the hydraulic pump (par. 197) through one of the two oil tubes which connect these units (fig. 274). One oil tube supplies oil pressure for traversing in one direction, and other tube supplies oil pressure for traversing in opposite direction, as controlled by the pump control handle (fig. 21). Oil is returned to pump for recirculation by tube which is not supplying pressure. A third oil tube connects the motor to the oil reservoir (par. 199) to drain the oil which leaks by the internal working parts of the motor. Adapter is an intermediate gear reducer to increase torque and decrease output speed of motor. It also serves as a coupling by which motor is joined to hand traversing mechanism.

b. Removal (fig. 267).

(1) Place traverse motor switch in "OFF" position (fig. 18) and turn turret lock handle to "LOCK" position (fig. 19).

(2) Place a cloth in position to catch oil, disconnect three oil tubes from hydraulic motor, disconnect two upper tubes at hydraulic pump, loosen support clip nut and swing tubes up out of way. Wipe clean fittings and ends of tubes and cover openings with tape to prevent entrance of dirt or insects.

(3) Remove two anchor straps which attach remote control bearing to turret wall above motor.

(4) Remove four ¼-inch safety nuts from studs which attach adapter to traverse case cover and lift motor and adapter assembly from hand traversing mechanism. Remove gasket.

c. Installation.

(1) Place a new gasket (vellumoid, $\frac{1}{32}$ in. thick) over studs on traverse case cover. Install hydraulic motor and adapter on hand traversing mechanism, engaging splined adapter shaft with pinion on traverse case, and secure adapter with four safety nuts (¼ in.—18) on studs.

(2) Connect three oil tubes to hydraulic motor and hydraulic pump (fig. 274) and tighten connections securely. Tighten oil tube support clip nut.

(3) Place anchor straps in groove around remote control bearing and secure control and straps to turret wall with two cap screws (¼ in.—20 x 1⅜ in.) and lock washers.

(4) Turn traversing motor switch to "ON" position, push traversing mechanism shifting lever up (fig. 20), unlock turret and traverse it several revolutions in both directions to check operation and control.

(5) Check oil tube connections and correct any oil leaks. Check oil reservoir and add oil (par. 38) as required to bring level two-thirds of way up on inspection window.

197. TRAVERSING MECHANISM HYDRAULIC PUMP.

a. Description. The Oilgear type hydraulic pump (fig. 21) is mounted on and directly connected to a vertical electric motor. The pump supplies fluid power, in form of oil under pressure, to hydraulic motor (par. 195 a) for the purpose of traversing turret. A spring

Turret—M18

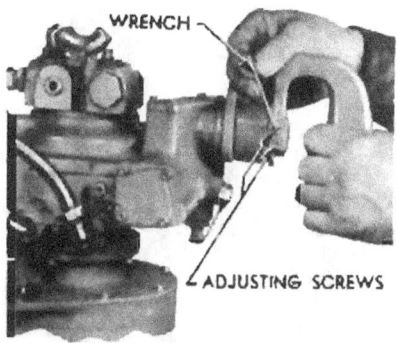

Figure 268—Adjusting Control Handle for Neutral Position

centered control handle, mounted integral with pump, controls speed and direction of turret rotation by regulating volume and direction of flow of oil supplied to hydraulic motor. Oil delivered by traverse pump is variable from zero to maximum in either direction and flows directly to and from traverse motor through oil tubes. Oil tubes also connect pump to oil reservoir (fig. 274). Control handle may be operated directly by hand, or through a remote control mechanism and lever mounted on turret wall just forward of caliber .50 machine gun mount.

b. **Adjusting Pump Control Handle and Remote Control.** Pump control handle must be adjusted so that turret will not creep in either direction with hydraulic pump running and control handle in neutral or vertical position. If adjustment is required, proceed as follows:

(1) Cut and remove lock wire between two socket head adjusting screws on left side of handle.

(2) If turret creeps clockwise, loosen upper adjusting screw about one-quarter turn and tighten lower screw same amount, using $\frac{5}{32}$-inch socket head screw wrench (fig. 268). If turret creeps counterclockwise, loosen lower adjusting screw and tighten upper screw same amount.

(3) Adjust screws until a positive neutral position is reached and so that turret will start to move when control handle is turned an equal distance from vertical in both directions. With both adjusting screws tightened securely, install lock wire.

(4) Check operation of control handle by means of the remote control lever (fig. 22) to make certain that it moves control handle to limit in both directions.

(5) Check remote control mechanism for binding which would prevent control handle from returning to neutral from either extreme position when the control lever is released.

TM 9-755
197

Part Three—Maintenance Instructions

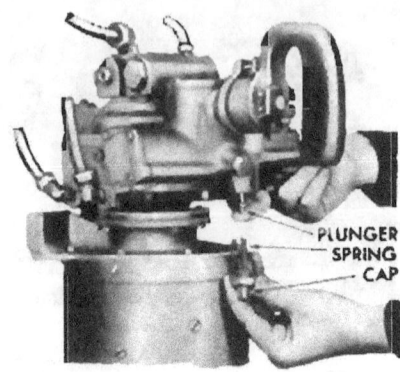

Figure 269—Cleaning Control Box Valve Plunger

(6) If remote control lever does not move control handle through full range, disconnect control rod end from control handle, adjust rod end on rod to secure full travel of handle, and attach rod end securely.

c. **Control Box Valve Plunger.** The control box valve plunger performs an important part in controlling turret rotation. If foreign matter collects around plunger causing it to stick, turret may not rotate when control handle is operated with pump running. If this occurs, turn traverse motor switch to "OFF" position (fig. 18) apply turret lock (fig. 19), and clean dirt from plunger in the following manner:

(1) Cut and remove lock wire, and remove plunger cap, gasket, and spring from pump case below control handle (fig. 269).

(2) Plunger should drop down, but not out, when parts are removed; if it does not, push upward lightly on plunger with a blunt punch 1/8 inch in diameter while moving control handle first to left then to right until plunger is free in case and drops down.

(3) Hold cloth over valve opening, then turn traverse motor switch to "ON" and "OFF" position as quickly as possible, which will cause the pump to force oil around valve and wash out foreign matter.

(4) Install spring, cap and gasket, tighten cap securely and install lock wire between cap and hexagon head screw directly above.

(5) Unlock turret and test traversing mechanism for oil leaks and proper control. Add hydraulic oil to reservoir, if necessary, to bring level two-thirds of way up on inspection window.

d. **Pump Check Valves.** Two spring loaded check valves are located in pump head. If these valves leak or stick because of foreign matter, turret traversing action may be slow or erratic in one or both directions or turret may traverse in one direction only. If any of these

Turret—M18

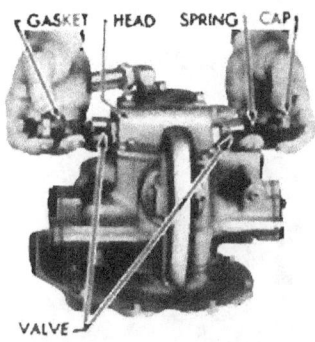

Figure 270—Removing Pump Check Valves

conditions exist, turn traverse motor switch to "OFF" position, apply turret lock, and clean check valves in the following manner:

(1) Cut and remove lock wire, place cloth to catch oil and remove check valve caps, gaskets, springs, and valves from pump head (fig. 270).

(2) Thoroughly clean parts and valve recesses in head with dry-cleaning solvent. Examine valves and replace if scored or deeply scratched. Replace broken or fatigued springs.

(3) Install valves, springs, gaskets and caps. Tighten caps securely and install lock wire through both caps and three pump head screws.

(4) Unlock turret and test traversing mechanism for proper control.

e. Removal of Pump.

(1) Place traverse motor switch in "OFF" position (fig. 18) and turn turret lock handle to "LOCK" position (fig. 19).

(2) Disconnect remote control rod from control handle by removing nut and lock washer from rod end screw.

(3) Loosen oil tube clamp attaching nuts on turret wall and above turret wiring switch box. Disconnect two tubes at upper fittings on hydraulic motor, and disconnect the seven oil tubes at hydraulic pump, using cloth to catch oil that drains from tubes. Wipe clean fittings and ends of tubes and cover openings with tape to exclude dirt or insects.

(4) Remove four ¼-inch cap screws and lock washers which attach pump to electric motor and lift pump from motor. Remove double-key coupling from end of motor armature shaft.

f. Installation of Pump.

(1) Place coupling centrally on end of motor armature shaft with key engaged in slot (fig. 271). Turn armature shaft so that upper

TM 9-755
197

Part Three—Maintenance Instructions

Figure 271—Installing Pump on Electric Motor

key on coupling will engage slot in end of pump shaft and install pump on electric motor with control handle parallel to 76-mm gun.

(2) Aline screw holes and attach pump to motor with four cap screws (¼ in.—20 x ¾ in.) and lock washers tightened securely.

(3) Attach the seven oil tubes to pump and two tubes to hydraulic motor as shown in figure 274 and tighten connections securely. Arrange oil tubes at bracket above turret wiring switch box so that they will be secured by the clamp and tighten clamp nut. Tighten nut at oil tube clip on turret wall.

(4) Attach remote control rod to bracket on control handle, with rod end screw placed in upper hole and secured by lock washer and nut.

(5) Turn traversing motor switch to "ON" position (fig. 18) and allow motor and pump to run for 30 seconds; then turn switch to "OFF" position. Repeat this operation several times before operating pump continuously, to fill hydraulic system and work out any air.

(6) Turn traversing motor switch to "ON" position, push traversing mechanism shifting lever up (fig. 20), unlock and traverse turret several revolutions in both directions to check operation and control.

(7) Check oil tube connections and correct any oil leaks. Check oil reservoir and add oil (par. 38) as required to bring level two-thirds of way up on inspection window.

Turret—M18

198. TURRET TRAVERSING ELECTRIC MOTOR.

a. **Description.** The electric motor is a direct current, 24-volt, 1¾ horsepower motor, operating at 2,000 revolutions per minute. It is mounted vertically on a bracket attached to turret, and drives hydraulic pump which is mounted on its upper end. Motor receives current from batteries through a switch in turret wiring switch box (fig. 18) which also contains a circuit breaker to protect wiring.

b. **Removal.**

(1) Place traversing motor switch in "OFF" position (fig. 18) and turn turret lock handle to "LOCK" position (fig. 19).

(2) Disconnect remote control rod from control handle by removing nut and lock washer from rod end screw.

(3) Loosen oil tube, clamp attaching nuts on turret wall and above turret wiring switch box. Disconnect two tubes at upper fittings on hydraulic motor, and disconnect seven oil tubes at hydraulic pump, using cloths to catch oil that drains from tubes. Wipe clean fittings and ends of tubes and cover opening with tape to exclude dirt or insects.

(4) Unscrew coupling nut on wire conduit at lower end of motor, remove terminal cover which is attached to motor with four machine screws (10-32) and lock washers, and disconnect wire which is attached to motor terminal stud with nut and lock washer.

(5) Remove two ½-inch lower cap screws and lock washers which attach the electric motor and gunner's platform support to motor bracket. Place block under front end of gunner's platform to support it.

(6) Place sling around and under hydraulic pump and attach hoist to support pump and motor assembly while removing two upper attaching cap screws; then lift assembly out of turret.

(7) Remove four ¼-inch cap screws and lock washers which attach pump to motor and lift pump from motor. Remove double-key coupling from end of motor armature shaft.

c. **Installation.**

(1) Set motor in vertical position, place coupling centrally on end of motor armature shaft with key engaged in slot (fig. 271). Turn armature shaft so that upper key on coupling will engage slot in end of pump shaft, and install pump on motor with control handle on opposite side from motor bracket flanges.

(2) Aline screw holes and attach pump to motor with four cap screws (¼ in.—20 x ¾ in.) and lock washers tightened securely.

(3) Place sling around and under pump, attach hoist and lift motor and pump assembly into position against motor bracket in turret.

(4) Anchor electric motor to upper end of bracket with two cap screws (½ in.—20 x 1¾ in.) and external-tooth lock washers. Install two lower attaching screws (½ in.—20 x 1¾ in.), with external-tooth lock washers, through bracket and gunner's platform support into

motor lower bracket. Tighten all cap screws to 80-100 foot-pounds tension. Remove block from under gunner's platform.

(5) Attach wire to terminal stud with lock washer and nut, attach terminal cover to motor with four machine screws (10—32 x ⅜ in.) and lock washers, and screw conduit coupling nut securely on terminal cover bushing.

(6) Attach the seven oil tubes to pump and two tubes to hydraulic motor as shown in figure 274 and tighten connections securely. Arrange oil tubes at bracket above turret wiring switch box so that they will be secured by the clamp and tighten clamp nut. Tighten nut at oil tube clip on turret wall.

(7) Attach remote control rod to bracket on control handle, with rod end screw placed in upper hole and secured by lock washer and nut.

(8) Turn traversing motor switch to "ON" position (fig. 18) and allow motor and pump to run for 30 seconds; then turn switch to "OFF" position. Repeat this operation several times before operating the pump continuously, to fill hydraulic system and work out any air.

(9) Turn traversing motor switch to "ON" position, push traversing mechanism shifting lever up (fig. 20), unlock and traverse the turret several revolutions in both directions to check operation and control.

(10) Check oil tube connections and correct any oil leaks. Check oil reservoir and add oil (par. 38) as required to bring level two-thirds of way up on inspection window.

199. TRAVERSING MECHANISM OIL RESERVOIR.

a. **Description.** A rectangular oil reservoir mounted on a bracket attached to turret to the right of hydraulic pump, is connected by oil tubes to the pump and the hydraulic motor (fig. 274). Oil from reservoir is used as fluid power medium in hydraulic system. The reservoir is filled through an opening in top which is closed by a combination filler and breather cap. A glass inspection window on side shows level of oil in reservoir. A high pressure relief valve (fig. 272) built into reservoir provides overload protection for high pressure section of pump, hydraulic motor, and turret traversing mechanism. Another relief valve (fig. 273) built into reservoir limits output pressure of a gear pump which is incorporated in pump assembly for purpose of supercharging high pressure section and for actuating pump control cam.

b. **High Pressure Relief Valve** (fig. 272). If the high pressure relief valve leaks, due to dirty or scored plunger or seat, or because of weak or broken spring, turret traversing speed will be low in either direction, or traversing action will be sluggish and unsteady. If these conditions exist, turn traverse motor switch to "OFF" position, apply turret lock, and clean relief valve parts in following manner:

(1) Remove hexagon cap (marked "I") and gasket nearest reservoir mounting lugs. Do not lose shims which will probably be in

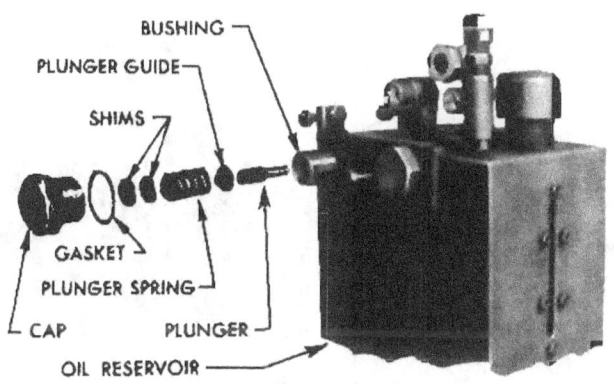

Figure 272—High Pressure Relief Valve Parts

counterbored hole in cap. Remove spring, guide, and spacer bushing from reservoir.

(2) Insert a No. 8-32 screw in end of relief valve plunger to pull it out. If plunger cannot be removed by this means, remove pipe plug on opposite side of reservoir and tap plunger out with a $\frac{3}{16}$-inch rod.

(3) Thoroughly clean plunger bushing and seat, and all parts with dry-cleaning solvent. Hold pump control handle to extreme right position, turn traverse motor switch to "ON" positions quickly, which will force oil out through seat and wash out foreign matter.

(4) Install plunger, small end first, into plunger bushing in reservoir. Install spacer bushing with small hole end inward. Place guide with rounded end against plunger and install spring. Place shims in counterbore in cap, place gasket on cap, screw cap into reservoir and tighten securely.

(5) Check oil level and add oil (par. 38) as required to bring level two-thirds of way up on inspection window. Test traversing mechanism for proper operation and control.

c. **Gear Pump Relief Valve** (fig. 273). If gear pump relief valve leaks, due to dirty or scored seat or plunger, or because of weak or broken springs, the turret traversing speed will be low in one or both directions. If relief valve plunger sticks open, turret may not turn with pump running, or traversing action will be sluggish and unsteady. If these conditions exist remove, clean, and install relief valve parts under hexagon cap (marked "2") farthest from reservoir mounting lugs, as described in subparagraph b above. CAUTION: *If both sets of relief valve parts are removed at same time, do not interchange springs or shims. The high pressure spring is much heavier than gear pump relief spring.*

TM 9-755

Part Three—Maintenance Instructions

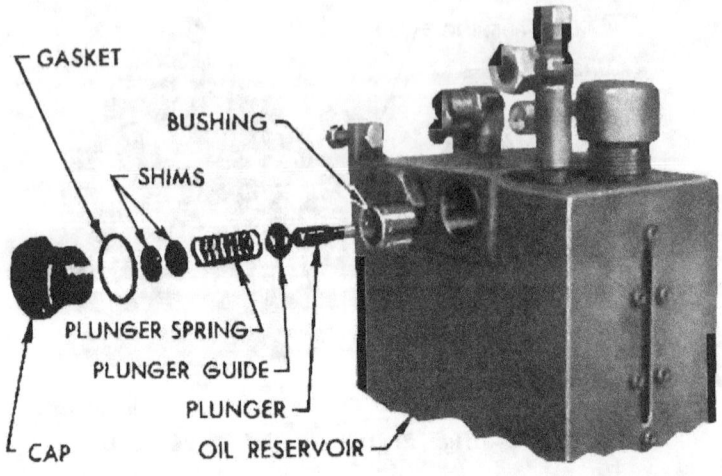

Figure 273—Gear Pump Relief Valve Parts

d. **Removal of Reservoir.**

(1) Disconnect oil tube from fitting at port No. 13 on bottom of oil reservoir (fig. 274) and drain oil from hydraulic system. NOTE: *If oil is to be used again, be sure that container is absolutely clean, and that it is covered to exclude dirt or insects.*

(2) Disconnect oil tubes from fittings on reservoir, wipe clean fittings and ends of tubes and cover opening with tape to exclude dirt or insects.

(3) Tag or otherwise mark oil tubes with the numbers stamped on reservoir at their respective fittings, to insure proper connections when reservoir is installed.

(4) Remove three ¼-inch cap screws and safety nuts which attach oil reservoir to mounting bracket and remove reservoir.

e. **Installation of Reservoir.**

(1) Attach oil reservoir to left side of mounting bracket with three cap screws (¼ in.—28 x 1¼ in.) and safety nuts tightened securely.

(2) Connect oil tubes to fittings on reservoir according to numbers stamped on reservoir and marked on oil tubes when they were disconnected (subpar. b, above).

(3) Fill reservoir with hydraulic oil (par. 38) until level is two-thirds of way up on inspection window.

(4) Turn traversing motor switch to "ON" position (fig. 18) and allow motor and pump to run for 30 seconds; then turn switch to "OFF"

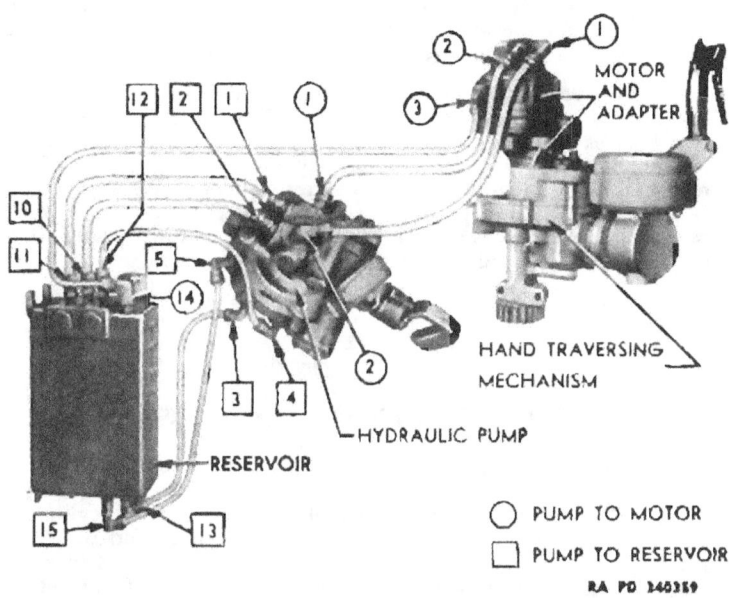

Figure 274—Hydraulic Traversing System and Tube Connections

position. Repeat this operation several times before operating pump continuously, to fill hydraulic system and work out any air.

(5) Test operation and control of hydraulic traversing mechanism by traversing turret several revolutions.

(6) Check oil tube connections and correct any oil leaks. Check oil reservoir and add oil (par. 38) as required to bring level two-thirds of way up on inspection window.

200. TRAVERSING MECHANISM OIL TUBES.

a. Description. Steel tubes are used to provide oil passages between three hydraulic units—reservoir, pump, and motor (fig. 274). Tubes are attached to these units by threaded connections which provide tight joints yet permit them to be readily disconnected. Tubes are supported between units by brackets and clips attached to turret.

b. Removal. Individual oil tubes may be removed by disconnecting them at both ends and removing supporting clips. When a number of tubes are removed, tag or mark the ends of each tube with same numbers as are stamped on hydraulic units adjacent to fittings to which tubes are connected; this will insure proper connections when tubes are installed. NOTE: *Each port on oil reservoir is stamped with a number, but ports on pump and motor are not always* so stamped. Number ends of tubes as shown in figure 274. Wipe clean the fittings and the ends of tubes which are to be used again, and cover openings with tape to exclude dirt or insects.

c. **Installation.** Before installation make certain that interior and ends of tubes are absolutely clean. Press ends of tubes squarely into fittings, start fittings with fingers to avoid crossed threads, and tighten nuts firmly. Anchor three tubes which lead to hydraulic motor with clip secured to stud on turret wall with one nut ($5/16$ in.—24) and lock washer. Anchor all tubes which run to reservoir by means of tube clamp secured to stud of bracket on turret base with one nut ($5/16$ in.—24) and lock washer.

Section XXXVII

FIXED FIRE EXTINGUISHER SYSTEM

201. DESCRIPTION (fig. 275).

a. The fixed fire extinguisher system consists of two cylinders of carbon-dioxide gas (10 lb. charge) which are connected by pipes to six horns in the engine compartment, and are discharged by pulling control handles located at accessible points on both the inside and outside of the hull. Location of the cylinders, their controls and operation are described and illustrated in paragraph 24.

b. The cylinders are rigidly supported side-by-side in mounting brackets welded to hull. They are connected together by pipes and a check valve which prevents one cylinder from discharging into other. From check valve, pipes connect to six horns located in engine compartment.

c. Each inside control handle connects to a separate fire extinguisher cylinder through a flexible steel cable attached to handle and to a cam in remote control assembly on cylinder discharge port. Each outside control handle cable is joined to a separate inside control handle cable by a clamp within barrel of a dual-pull mechanism, thus permitting discharge of a cylinder by pulling either inside or outside handle.

d. All cables are inclosed in conduits for support and protection, and pass over pulleys where required to make sharp turns. These pulleys are located on remote control mechanisms in the M18 vehicle, and on inside control handle mounting bracket in the M39 vehicle.

202. CYLINDERS.

a. *Handling.* Any cylinder containing gas under high pressure is as dangerous as a loaded shell. Fire extinguisher cylinders should never be dropped, struck, handled roughly, or exposed to unneces-

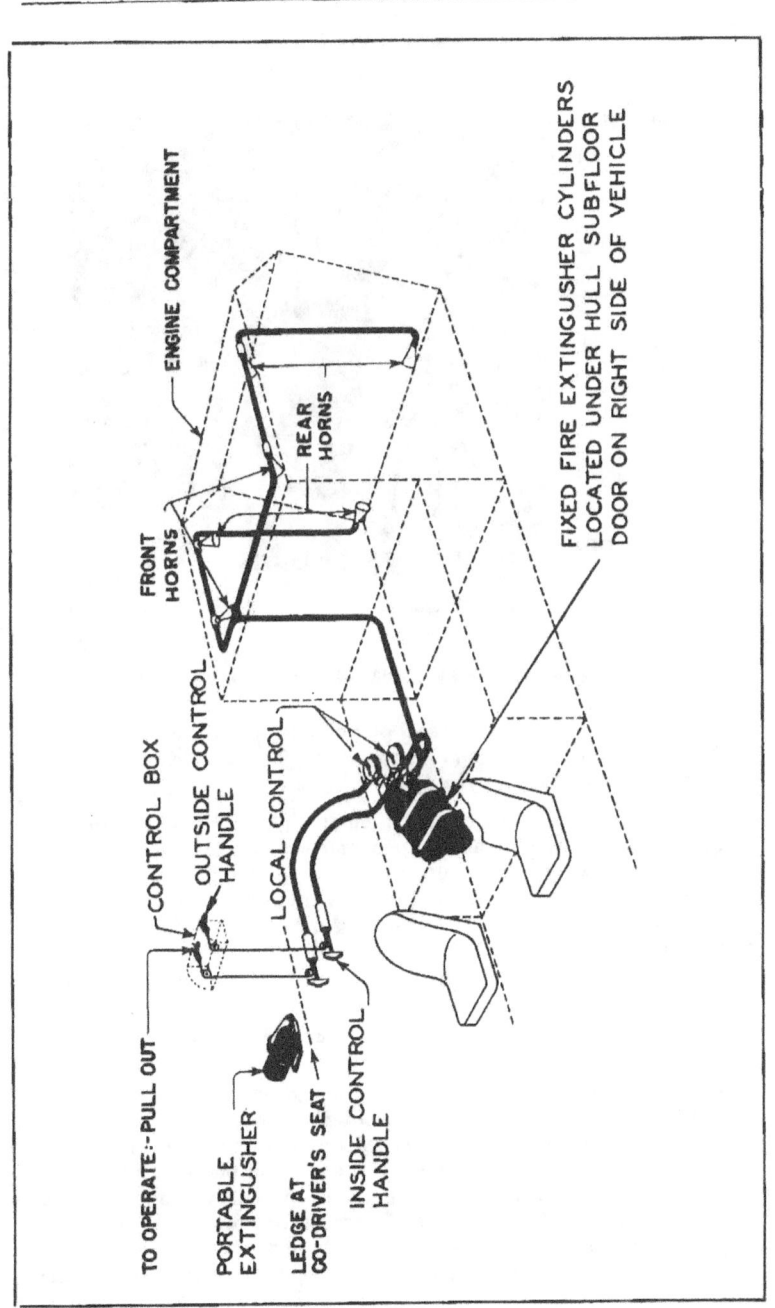

Figure 275—Fixed Fire Extinguisher System—M18

Part Three—Maintenance Instructions

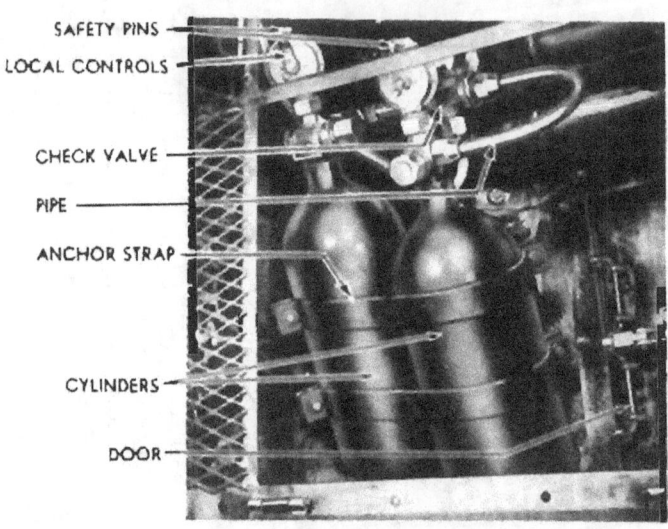

Figure 276—Fixed Fire Extinguisher Cylinders—M18

sary heat. Red safety blow-off seal on valve head indicates if cylinder has been discharged due to high temperature. This should be examined regularly; if it is missing, cylinder should be replaced.

b. **Removal** (fig. 276 or 277). Open right front and right rear subfloor doors on M18 vehicle; open rear seat center cover on M39 vehicle. Unscrew coupling nuts and remove remote controls from cylinders. Lay remote controls aside with pull cable conduits attached. Loosen discharge pipes attaching nuts at check valve and disconnect pipes at cylinders. Remove both anchor straps which are secured with cap screws (3/8 in.—16 x 2 in.) and plain washers, and lift out cylinders.

c. **Installation** (fig. 276 or 277). NOTE: *Before installation, weigh each replacement cylinder to make sure it contains proper charge of 10 pounds of gas plus weight of cylinder, which is stamped on head. If gas charge is less than 9 pounds, cylinder needs recharging.* Place cylinders in position and secure them with two anchor straps attached with cap screws (3/8 in.—16 x 2 in.) and plain washers tightened securely. Connect discharge pipes to cylinders and tighten attaching nuts at check valve. Install remote controls on cylinders so that cable conduits are free of sharp bends and tighten coupling nuts securely. Close subfloor doors in M18 vehicle; close rear seat center cover in M39 vehicle.

TM 9-755
203

Fixed Fire Extinguisher System

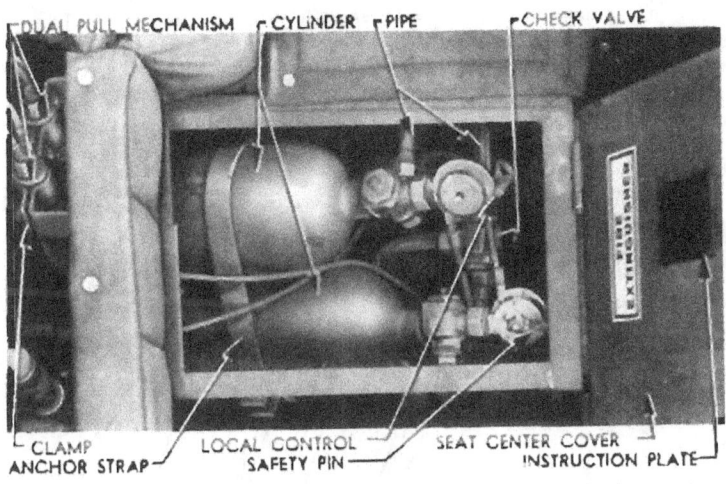

Figure 277—Fixed Fire Extinguisher Cylinders—M39

203. CONTROLS, M18 VEHICLE.

a. Removal of Control Handles and Cables. The following procedure covers removal of set of handles and cables attached to either cylinder:

(1) Open right rear subfloor door. Unscrew coupling nuts and remove remote control assemblies from both cylinders, to prevent accidental discharge of either cylinder. Lay to one side remote control that is not to be worked upon.

(2) Remove cover which is attached to other remote control by three screws secured with lock wire. Loosen set screws and pull cable from block (fig. 279). Lay control down so that there are no sharp bends in conduit.

(3) Remove clamp, attached by three $\frac{3}{8}$-inch cap screws and lock washers, which anchor both dual-pull mechanisms to mounting bracket on hull wall.

(4) Loosen coupling nut of rear conduit, and unscrew rear half of dual-pull mechanism barrel from front half. Pull cable out of rear conduit.

(5) Grasp both ends of cable clamp with pliers (step (4), fig. 278) and untwist clamp so that it can be removed from cables.

(6) Pull out inside handle and cable at support bracket. If outside handle and cable require replacement, disconnect conduit at pulley on dual-pull mechanism and pull out handle and cable at control box; otherwise, leave handle and cable in place.

467

TM 9-755

Part Three—Maintenance Instructions

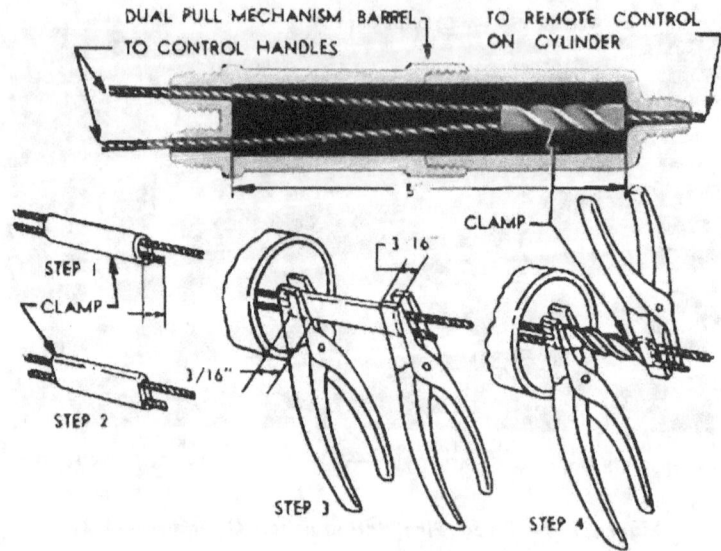

Figure 278—Installation of Cable Clamp

b. **Removal of Control Box.** Control box need not be removed unless it is damaged, as handles and cables can be removed and installed with box in place (subpar. a and c above).

(1) Remove cable clamps to disconnect both outside control handle cables from both inside control handle cables as described in steps (1) through (5) of subparagraph a above.

(2) Disconnect both outside control cable conduits at pulleys on dual-pull mechanisms and pull cables out of barrels and pulleys.

(3) Remove lock nut, jam nut, flat washers, and composition washer from control box extensions under hull roof plate, then lift control box and gasket from top side of roof plate.

(4) Disconnect conduits from control box.

c. **Installation of Control Handles and Cables.** Control handles are furnished with cables attached. Be sure that cables are the correct length.

(1) Apply a thin coat of graphite to cables before installation.

(2) Insert end of outside handle cable through opening in control box cover into groove in pulley, then gently push cable in, while turning it slightly, until cable enters hole in bottom of box leading to conduit. If difficulty is experienced, remove attaching screws and move control box cover out far enough to guide end of cable into hole;

Fixed Fire Extinguisher System

then install cover attaching screws. Push cable down through conduit, turn handle so that key on handle engages groove in box cover, and push handle in until firmly held by springs.

(3) Thread lower end of outside cable through pulley into dual-pull mechanism and connect conduit to pulley.

(4) Push end of inside control handle cable through socket and connector into dual-pull mechanism, turn handle so that key on handle engages groove in socket, and push handle in until firmly held by springs.

(5) Place cable clamp over ends of both cables and hold dual-pull mechanism so that outside cable pulley is bent at a right angle. Pull both cables to take up slack, position cable clamp so that its outer end is 5 inches from inside end of barrel, then squeeze cable clamp tightly against cables along its entire length, using pliers (steps (1) and (2), fig. 278).

(6) Grasp both ends of cable clamp with pliers placed $3/16$ inch from end, then rotate ends of clamp in opposite directions simultaneously until clamp has been twisted to $1\frac{1}{2}$ turns (steps (3) and (4), fig. 278). Cut off the shorter outside control cable $1/16$ inch from cable clamp.

(7) Push end of inside control cable through rear conduit until it comes out at remote control, then screw two halves of dual-pull mechanism barrel together and tighten rear conduit coupling nut.

(8) Anchor dual-pull mechanisms to mounting bracket on hull wall by means of clamp attached with three cap screws ($1/8$ in.— 24 x 1 in.) and lock washers.

(9) Push remote control lever against its stop, install safety pin and secure it with a very light, easily broken, wire. With open side of remote control facing up, turn the cam clockwise against its stop and hold it there.

(10) Place block over end of cable, pull on cable to take up slack, wind cable around cam in clockwise direction and place block in recess in cam (fig. 279). Push block against stops in open end of recess and tighten both set screws firmly. Cut off surplus end of cable and install cover with three screws secured by lock wire.

(11) Install remote controls on both cylinders and tighten coupling nuts securely. Remove safety pins. Close sub-floor door.

d. *Installation of Control Box.* The control box is furnished with handles and cables, gasket, and attaching parts.

(1) Remove nuts and washers from control box extensions. Apply a thin coat of graphite to cables, push cables through conduits and attach conduits to control box.

(2) Place control box with new gasket in position on hull roof and anchor it by installing a composition washer, flat washer, jam nut, and lock nut on each extension on under side of roof plate, in order named.

TM 9-755
203-204

Part Three—Maintenance Instructions

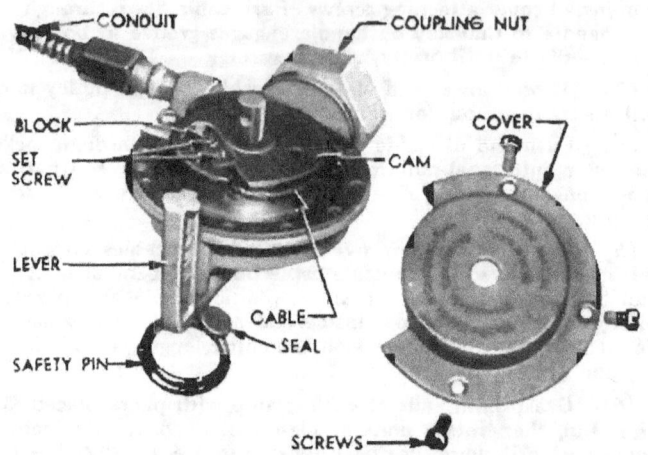

Figure 279—Remote Control Assembly—Cover Removed

(3) Turn control handles so that keys on handles engage grooves in box cover, and push handles in until firmly held by springs.

(4) Complete installation by performing steps (3) through (11) of subparagraph c above.

204. CONTROLS, M39 VEHICLE.

a. **Removal of Control Handles and Cables.** The following procedure covers removal of set of handles and cables attached to either cylinder:

(1) Open rear seat center cover. Unscrew coupling nuts and remove remote control assembly from cylinder from which cable is to be removed.

(2) Remove cover which is attached to remote control by three screws secured with lock wire. Loosen set screws and pull cable from block (fig. 279).

(3) Disconnect conduit from remote control and loosen attaching screw so that conduit can be freed from clamp which attaches it to seat support.

(4) Remove clamp, attached by three ⅜-inch cap screws and lock washers, which anchor both dual-pull mechanisms to mounting bracket on front side of rear seat back (fig. 277). Carefully move required dual-pull mechanism out over top of seat back.

(5) Unscrew lower half of dual-pull mechanism barrel from upper half and pull cable out of lower conduit.

(6) Grasp both ends of cable clamp with pliers (step (4), fig. 278) and untwist clamp so that it can be removed from cables.

(7) Disconnect inside cable conduit from pulley on inside control handle mounting bracket. Pull out handles and cables.

b. **Installation of Control Handles and Cables.** Control handles are furnished with cables attached. Outside control cable is 70 inches long; inside control cable is 52 inches long.

(1) Apply a thin coat of graphite to cables before installation.

(2) Straighten pulley on inside control handle mounting bracket and push end of inside control cable through socket and pulley. Push cable through conduit and attach conduit to pulley.

(3) Push end of outside control cable through socket on outside control handle mounting bracket, and down through conduit to dual-pull mechanism.

(4) Turn both handles so that keys on handles engage grooves in sockets and push handles in until firmly held by springs.

(5) Place cable clamp over ends of both cables. Pull cables evenly to take up slack, position cable clamp so that its outer end is 5 inches from inside end of barrel, then squeeze cable clamp tightly against cables along its entire length, using pliers (steps (1) and (2), fig. 278).

(6) Grasp both ends of cable clamp with pliers placed $3/16$ inch from end, then rotate ends of clamp in opposite directions simultaneously until clamp has been twisted to $1\frac{1}{2}$ turns (steps (3) and (4), fig. 278). Cut off shorter outside control cable $1/16$ inch from cable clamp.

(7) Push end of inside control cable through lower half of barrel and through conduit, then screw two halves of dual-pull mechanism barrel together.

(8) Anchor dual-pull mechanisms to mounting bracket on rear seat back by means of clamp attached with three cap screws ($3/8$ in.— 24 x 1 in.) and lock washers.

(9) Place conduit under clamp on seat support and tighten clamp screw. Push end of cable through fitting on remote control and attach conduit to remote control.

(10) Push remote control lever against its stop, install the safety pin and secure it with a very light, easily broken, wire. With open side of remote control facing up, turn cam clockwise against its stop and hold it there.

(11) Place block over end of cable, pull on cable to take up slack, wind cable around cam in clockwise direction and place block in recess in cam (fig. 279). Push block against stops in open end of recess and tighten both set screws firmly. Cut off surplus end of cable and install cover with three screws secured by lock wire.

(12) Install remote control on cylinder and tighten coupling nut securely. Remove safety pins. Close rear seat center cover.

TM 9-755
205-207

Part Four—Auxiliary Equipment

PART FOUR—AUXILIARY EQUIPMENT

Section XXXVIII
GENERAL

205. SCOPE.

a. Part four contains information for guidance of personnel responsible for operation of this equipment. It contains only the information necessary to using personnel to properly identify, connect, and protect such auxiliary equipment while being used or transported with the main equipment. Detailed instructions on this equipment are contained in separate technical manuals.

Section XXXIX
ARMAMENT

206. SCOPE.

a. This section contains instructions for the operation of the 76-mm Gun M1A1, M1A1C, or M1A2 in the Gun Mount M1 for the gun Motor Carriage M18 (figs. 1 and 280). Refer to paragraphs 37 and 38 for lubrication instructions on the armament. Sighting and fire control equipment is covered in section XXXX; Ammunition is covered in section XXXXI.

b. The other armament in M18 gun motor carriage is a caliber .50 HB, M2 machine gun. It is mounted as a flexible gun on a revolving ring in top of turret (fig. 281). This gun can be removed from the vehicle and used with a tripod mount M3 which is an accessory.

c. The caliber .50 HB, M2 machine gun is the only weapon mounted in M39 vehicle (fig. 3). This is the same gun and mount as used in M18 gun motor carriage. Refer to FM 23-65 for complete information on its operation.

207. CHARACTERISTICS.

a. Armament is employed chiefly against enemy tanks and other ground objectives. The turret can be traversed 360 degrees manually or by a hydraulic traversing mechanism. The 76-mm gun can be elevated 20 degrees and depressed 10 degrees.

b. Caliber .50 machine gun M2 is employed chiefly as a defense against enemy aircraft. However, it may also be used against ground

Armament

TM 9-755
207

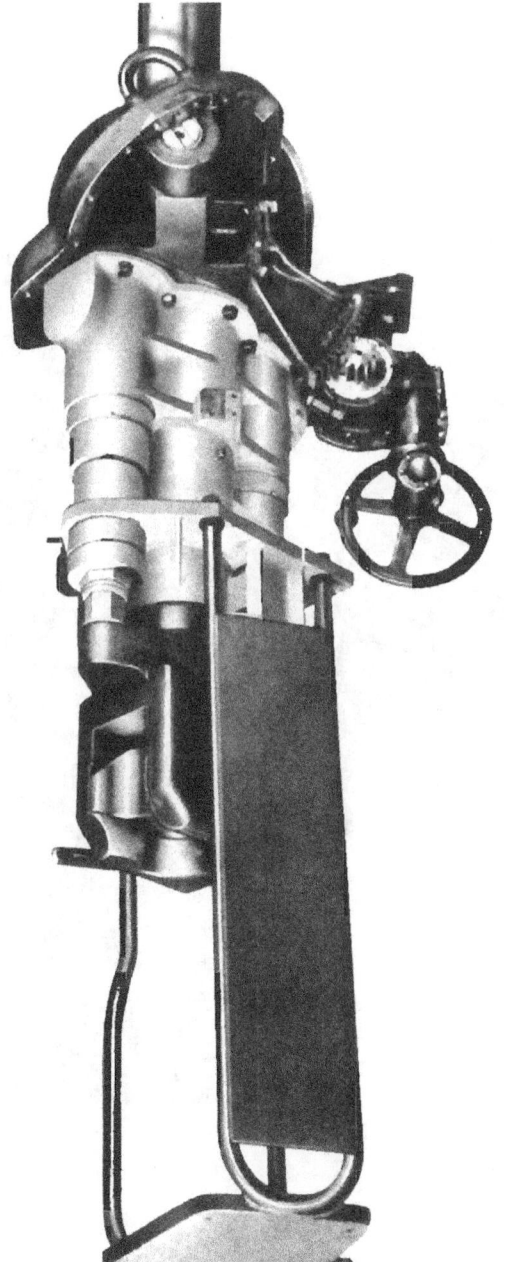

RA PD 92244

Figure 280—76-mm Gun and Mount M1—Right Rear View

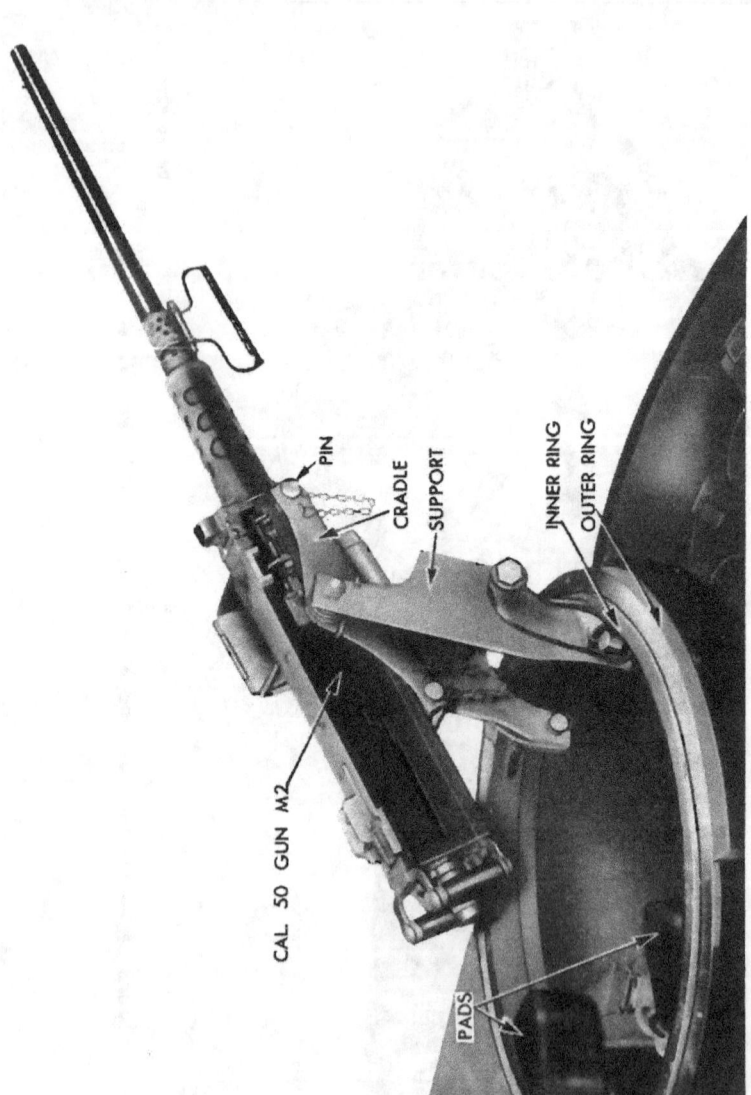

Figure 281—Browning Machine Gun—Caliber .50 HB, M2, and Mount

Figure 282—Close-up of Muzzle Brake

Figure 283—Close-up of Thread Protecting Ring

forces. This machine gun and ring mount, in both M18 gun motor carriage and M39 armored utility carriage, can be traversed 360 degrees independently of turret.

208. DIFFERENCES AMONG MODELS.

a. There are three models of 76-mm guns in gun mount M1, in gun motor carriage M18. They are M1A1, M1A1C, and M1A2. The M1A1 is rifled with uniform right-hand twist, one turn in 40 calibers. The M1A1C is modified by addition of threads at muzzle end to take a muzzle brake (fig. 282). When muzzle brake is not installed,

TM 9-755
208-209

Part Four—Auxiliary Equipment

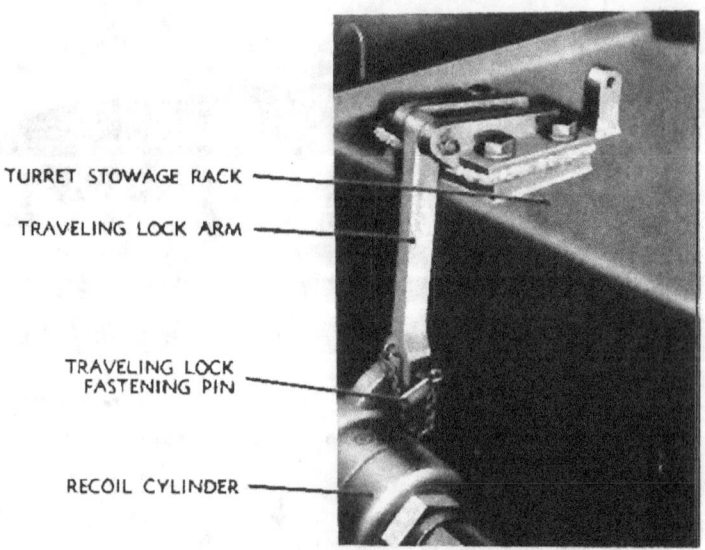

Figure 284—Gun Traveling Lock Engaged—Early Model

Threads will be protected by a ring (fig. 283). The M1A2 model has the rifling changed to a uniform right-hand twist, one turn in 32 calibers with the muzzle end threaded to take the muzzle brake. If the muzzle brake is not installed, a ring is used to protect threads. Guns are so marked that proper identification can be made.

b. Two types of gun traveling locks are in use with the gun mount M2, in the gun motor carriage M18. The earlier vehicles were equipped with the traveling lock arm as shown in figure 284. Beginning with vehicle No. 1858, the traveling lock shown in figure 285 has been installed.

209. PLACING THE 76-MM GUN IN FIRING POSITION.

a. Remove gun covers and store them. Likewise store other equipment not required for operation of armament.

b. Before disengaging traveling lock make sure elevating mechanism shifter lever is in engaged or left position (fig. 293).

c. *Disengage Traveling Lock, Early Model.* Remove traveling lock fastening pin and place traveling bar in its stowed position (fig. 284).

Armament

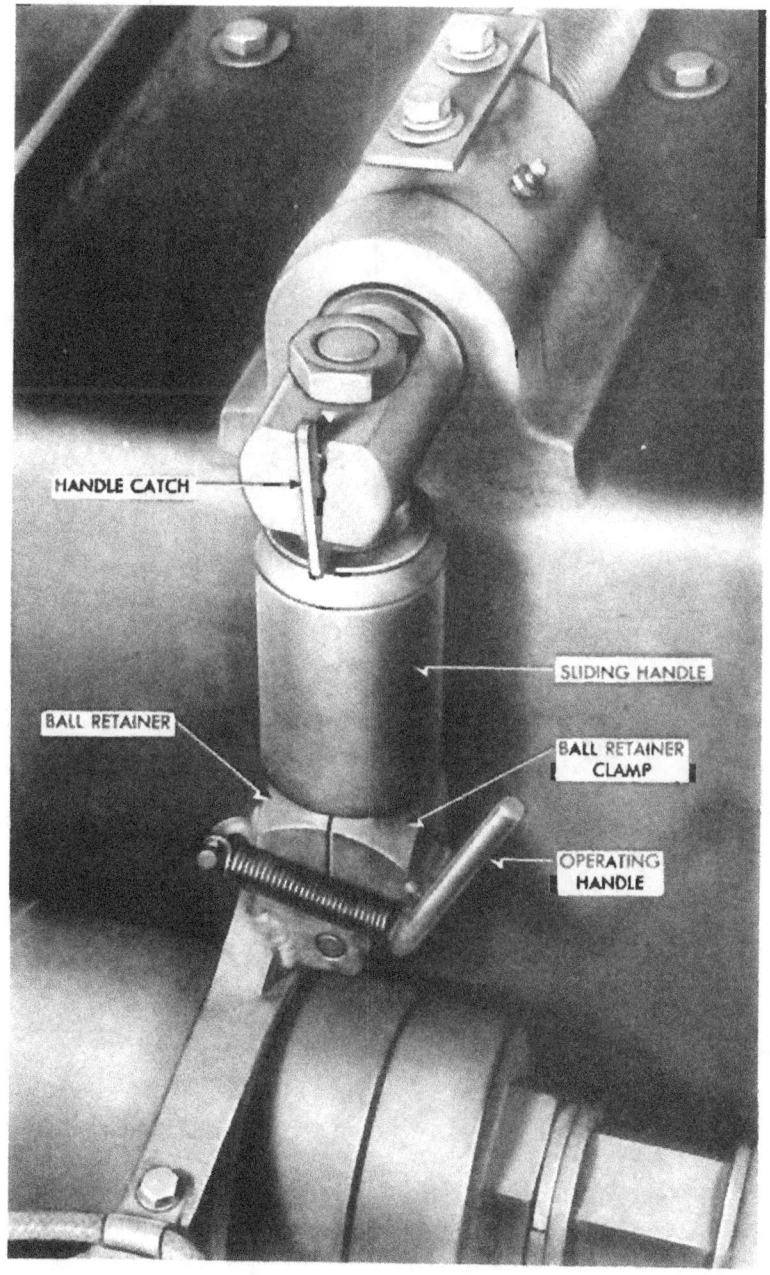

Figure 285—Gun Traveling Lock Engaged

TM 9-755
209

Part Four—Auxiliary Equipment

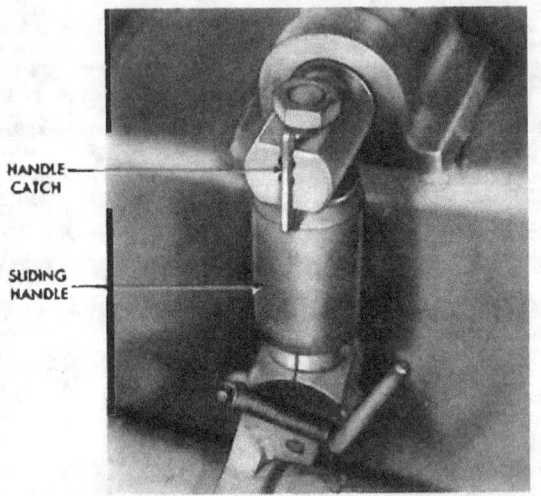

Figure 286—Sliding Handle in "UP" Position

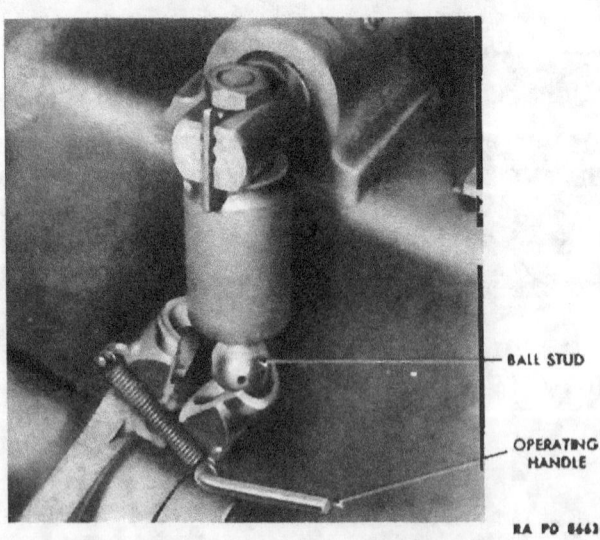

Figure 287—Ball Stud Released

TM 9-755
209

Armament

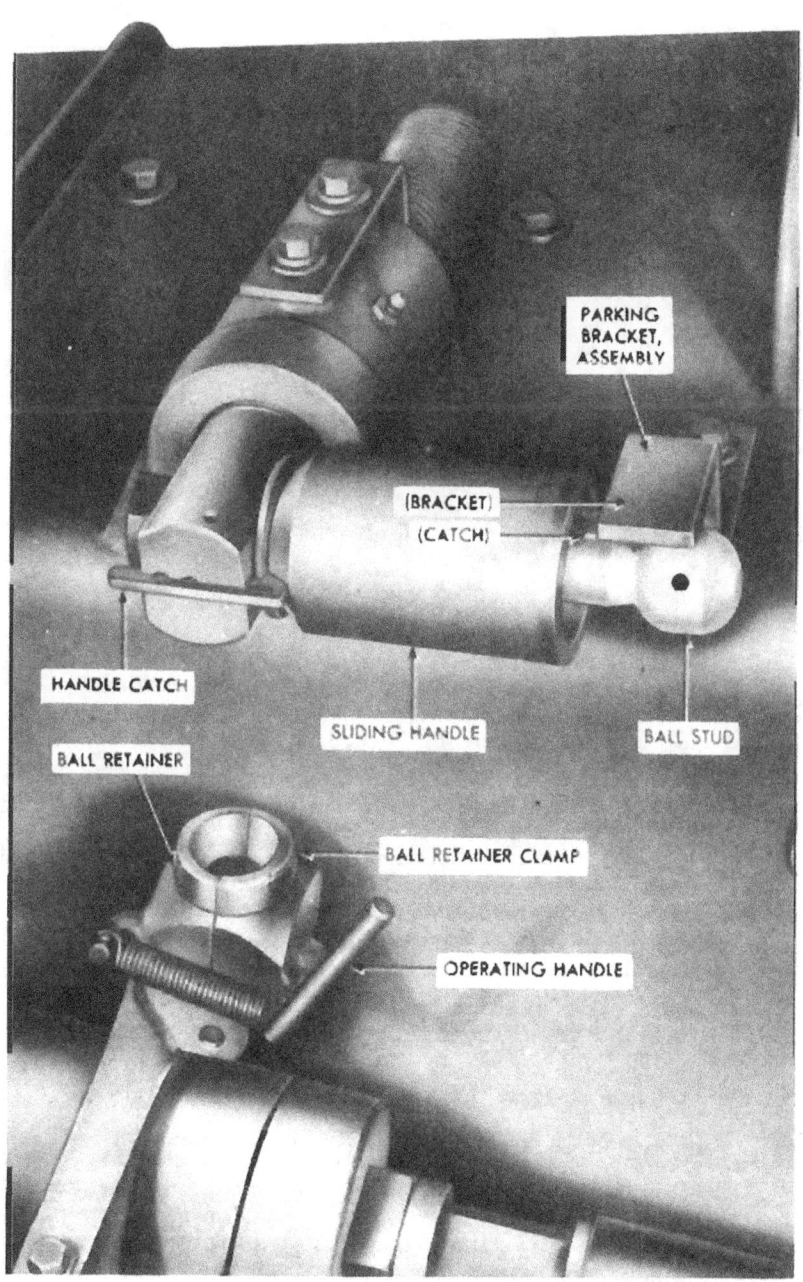

Figure 288—Ball Stud in "UP" Position

TM 9-755
209-210

Part Four—Auxiliary Equipment

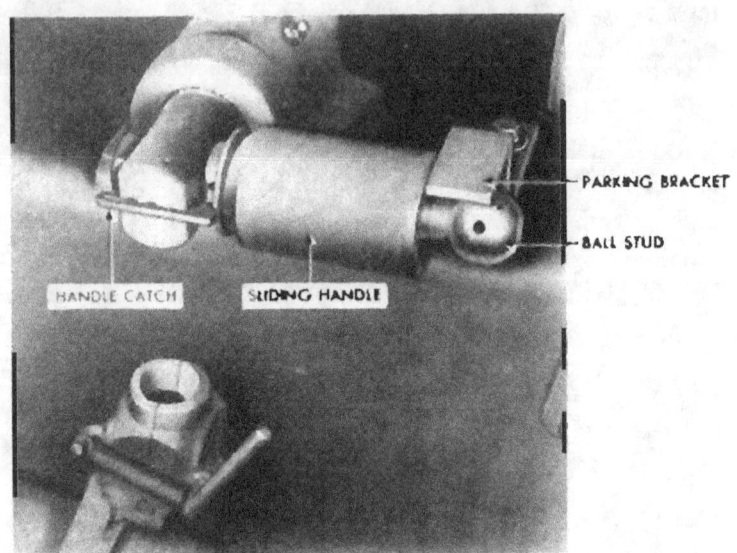

Figure 289—Ball Stud Locked In "UP" Position

d. Disengage Traveling Lock, Late Model.

(1) Press in on top of handle catch (fig. 285). Slide the sliding handle up and engage in the release position with handle catch (fig. 286).

(2) Press down on operating handle while elevating gun. This will release ball stud from ball retainer and retainer clamp (fig. 287).

(3) Swing ball stud and sliding handle up with ball stud under parking bracket (fig. 288).

(4) Lock in this position by pressing handle catch and moving sliding handle into engagement with parking bracket catch (fig. 289).

e. Disengage the turret lock by turning lock handle to its forward position which is marked "FREE" (figs. 290, 291, and 292).

210. INSPECTION BEFORE FIRING.

a. To check oil level in recoil cylinders, depress gun one degree and remove top plug at rear of each cylinder (AP, fig. 41). The oil level in cylinder should reach bottom of hole. If oil level does not reach this level, fill as instructed in paragraph 38.

b. Check path of recoil to make sure that it is free from all obstructions.

Armament

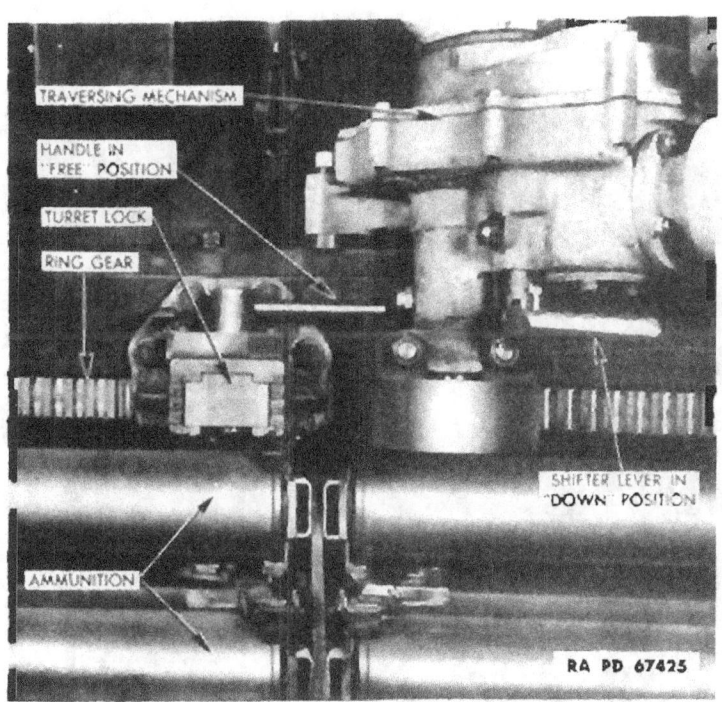

Figure 290—Turret Lock in "FREE" Position—Early Model

c. Open breech and inspect bore of gun for cleanliness.

d. Refer to paragraph 221 for bore sighting instructions.

211. TRAVERSING TURRET.

a. *General.* The turret may be traversed manually or hydraulically for the full 360 degrees in either direction. Before traversing turret make sure all doors are closed and that equipment and personnel are not in a position to be damaged or injured. Also elevate gun sufficiently to clear rear deck equipment. Turret lock must be disengaged from turret ring gear by turning lock handle to its forward position which is marked, "FREE" (figs. 290, 291, 292). CAUTION: *It is very important to have lock completely disengaged from ring gear teeth before traversing turret.* Turret lock may be engaged by turning handle to its rearward position marked "LOCK" and making sure handle is held by spring clip (figs. 290 and 291). It may sometimes be necessary to rotate turret slightly to left or right to permit gear teeth and lock teeth to mesh when locking turret.

TM 9-755

Part Four—Auxiliary Equipment

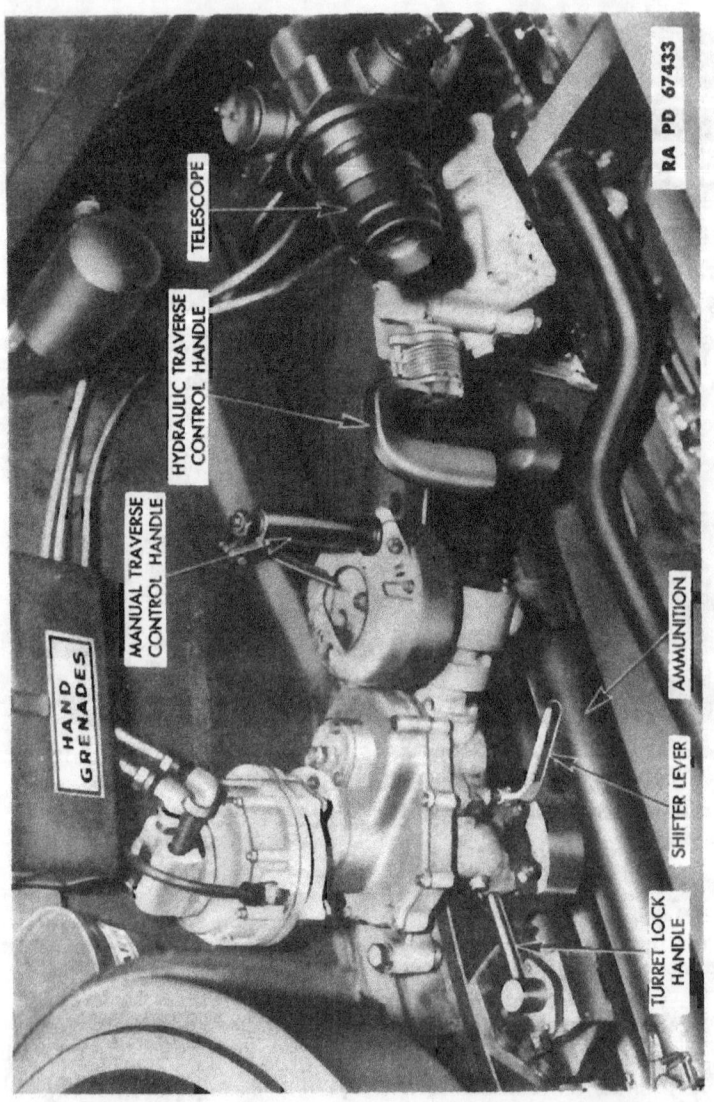

Figure 291—Turret Controls—Early Model

Figure 292—Turret Controls—Late Model

TM 9-755

Part Four—Auxiliary Equipment

b. **Traversing Turret Hydraulically.**

(1) Turn vehicle master switch located directly behind driver's seat to on position. Two identical switches will be found here; vehicle switch is the upper one; lower one being radio switch. CAUTION: *Be sure traverse pump control handle is in neutral or vertical position, and gear box shifting lever is in down position.*

(2) Turn traverse motor master switch to "ON" position (fig. 293) to start electric motor and traverse pump.

(3) Move shifting lever under gear box to up position (fig. 292). It may be necessary to turn brake handle on top of gear box slightly to engage gears. Power operated gears are thus engaged with turret ring gear and pinion.

(4) To traverse turret clockwise, or to right, turn top of pump control handle to right. To traverse turret to left, turn handle to left. Speed of traverse may be increased by turning this handle further to right or left, depending upon direction of traverse.

(5) Turret traverse may be stopped by releasing control handle or bringing it to upright or neutral position.

(6) Turret rotation can be reversed instantly without damage to fluid power mechanism. Automatic hydrodynamic braking is provided.

(7) When power turret traversing is completed, point gun in traveling position, move traversing shifting lever under gear box to down position, and engage turret lock with ring gear (fig. 292).

c. **Traversing Turret Manually.**

(1) Move shifting lever under gear box to down position (figs. 290 and 291). It may be necessary to turn vertical brake handle on top of gear box (figs. 290 and 291) slightly to engage gears. Manually operated gears are thus engaged with turret ring gear and pinion.

(2) Grip vertical brake handle and its associate lever on top of gear box, to release turret brake.

(3) Rotate control handle to the left or counterclockwise to traverse turret counterclockwise. Turning control handle to right or clockwise will traverse turret clockwise.

(4) If this control handle turns freely and fails to traverse turret, shifter lever under gear box is in the up or power traverse position and should be moved to down or manual traverse position.

(5) Always traverse turret 360 degrees in both directions to make sure turret is free.

d. **Remote Control Lever.** Mounted on a bracket on the inner left turret wall is remote control lever (fig. 292). If commander so desires, he may take traverse control of turret away from gunner, since operation of lever automatically operates gunner's hydraulic pump control handle.

TM 9-755

Armament

Figure 293—Turret Switch Box, Elevating Controls, and Foot Firing Switch

TM 9-755
212-213

Part Four—Auxiliary Equipment

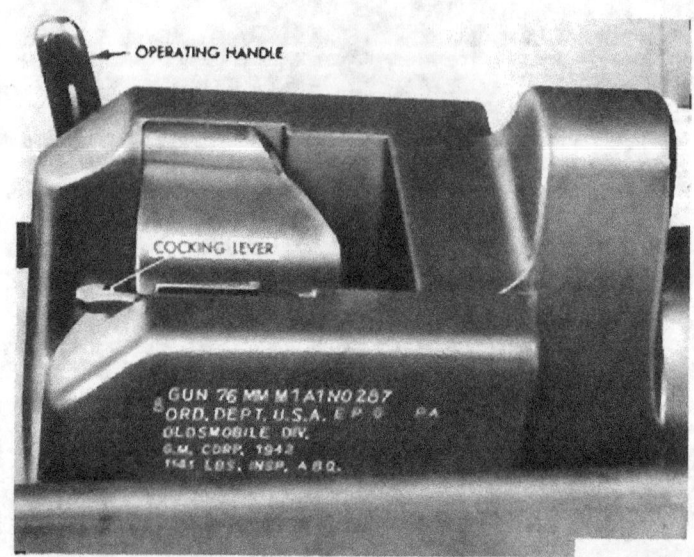

Figure 294—Breech Operating Handle and Cocking Lever

212. ELEVATING OR DEPRESSING GUN.

a. Make certain elevating gears are engaged. They are engaged by releasing the trigger on the elevating shifter lever and moving this lever to left (fig. 293). It may be necessary to turn elevating handwheel slightly to permit meshing of the gears when moving shifting lever.

b. The gun is elevated or depressed by turning elevating handwheel (fig. 293).

213. OPERATING THE BREECH MECHANISM.

a. Opening Breech (figs. 294 and 295). To open breech, grasp grip portion of breech operating handle. Release latch on grip and pull breech operating handle rearward and through opening in recoil guard. This moves breechblock down and thus opens breech. CAUTION: *Keep hands out of gun when breech is open.* Return operating handle to closed position and make sure it is latched immediately after opening breech in order to avoid injury to personnel and mechanism.

Armament

Figure 295 — Breech Operating Handle and Firing Mechanisms

b. Closing Breech (figs. 294 and 295). Close breech by unlatching operating handle and pulling it rearward through opening in recoil guard. Bear sufficient weight on the handle to overcome tension of closing spring and release breechblock extractors from their locked position by pressing them forward with base of an empty cartridge case. The breechblock is then free to be eased into its closed position by means of operating handle which should finally be latched in place. CAUTION: *Never use hands to release the extractors.*

c. Semiautomatic Operation. Under ordinary circumstances, it will be necessary to open breech only at start of firing operations. The counter recoil of gun, by means of a cam on mount, actuates operating crank which is keyed to spline shaft. This shaft is in turn keyed to breechblock crank which opens and closes breechblock.

214. LOADING THE GUN.

a. Before loading, open breech and examine bore of gun to see that it is clear and free of foreign material. Be sure to return operating handle to its closed position. Refer to paragraphs 222 and 223 for information on authorized ammunition and preparation of ammunition for firing.

b. To load gun, place a round in breech, with the nose entering bore, then impel round into chamber with sufficient force so that flange of cartridge case will drive extractors forward and automatically close breech. Loader's hand should be moving toward the right as he shoves projectile home in order to clear breechblock as it automatically closes.

215. FIRING THE GUN.

a. The 76-mm gun can be fired electrically, or manually in case of failure of electrical system. A firing switch is located in turret switch box and controls current to foot firing switch convenient to gunner's foot (fig. 293). The foot firing switch energizes solenoid which actuates firing lever (fig. 295) to fire 76-mm gun. The hand-firing lever is used to fire 76-mm gun manually.

b. To fire 76-mm gun electrically, throw firing switch in turret switch box to "ON" position. Place firing shaft release lever toward rear in firing position (figs. 295 and 296). To fire, depress foot firing switch on gunner's platform. If gun fails to fire using foot firing switch, gun may be fired manually by pulling rearward on hand firing lever mounted on top of cradle (fig. 297).

TM 9-755
215

Armament

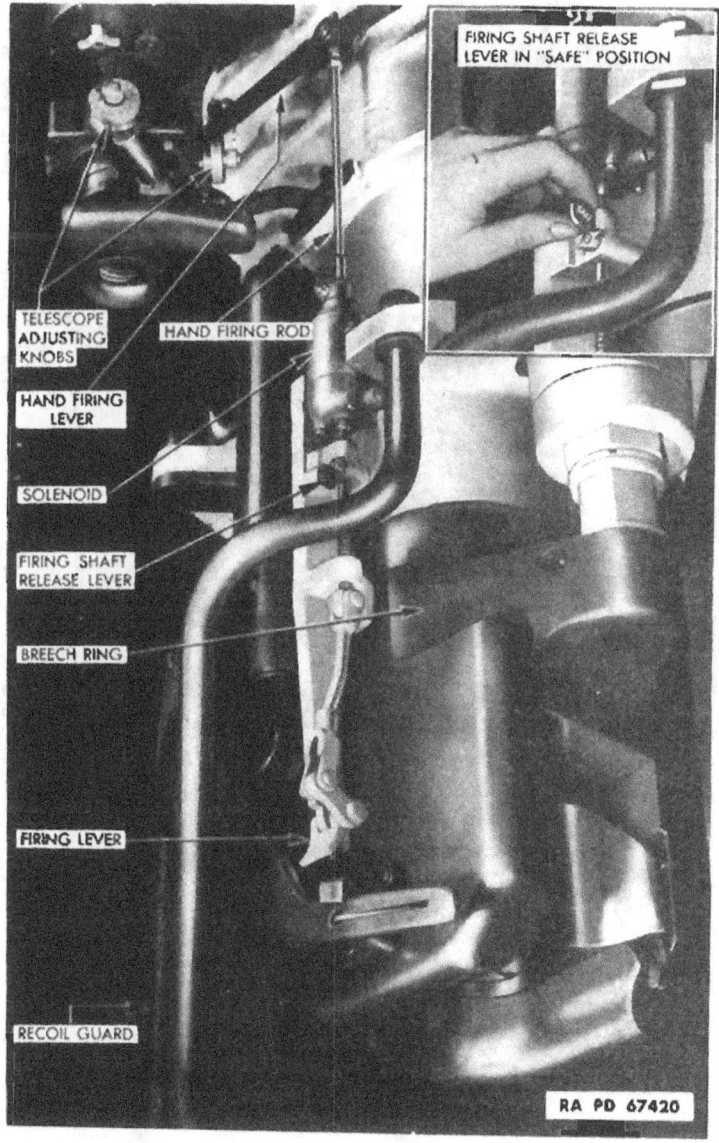

Figure 296—Firing Shaft Release Lever

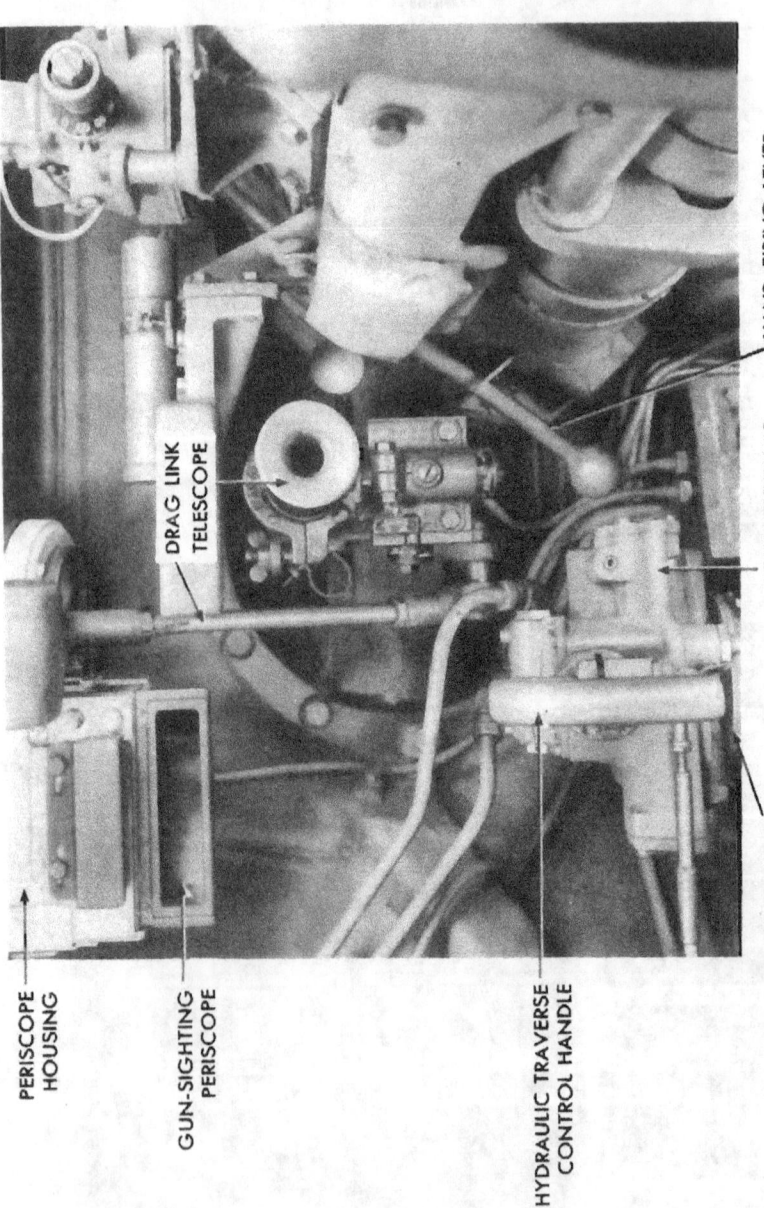

Figure 297—Hand Firing Lever

c. If gun fails to fire using either electrical or manual means, position of firing shaft release lever must be checked. Failure to fire may be due to gun staying out of battery, failure of firing mechanism, failure of breech to close, or because of defective ammunition. If gun is in battery, recock by means of cocking lever (fig. 294) located in breech ring, and attempt to fire again. CAUTION: *In case of misfire, open firing switch immediately before recocking.* If gun still fails to fire after three attempts, wait 30 seconds before opening breech; then remove round, reload, and attempt to fire again. After need for firing is completed, throw firing switch to "OFF" position and place release lever in "SAFE" position (fig. 296).

216. PLACING GUN IN TRAVELING POSITION.

a. Make sure release lever is in "SAFE" position.

b. Clean and lubricate gun, and install covers.

c. Traverse turret until gun points forward and attach traveling lock arm to yoke on cradle (fig. 284).

d. In case vehicle is equipped with new type gun lock engage in traveling position as follows:

(1) Release from locked position in parking bracket catch (fig. 289) by pressing handle catch and moving sliding handle out.

(2) Press down on operating handle (fig. 287) and swing sliding handle with ball stud down and engage it in retainer clamp. It will probably be necessary to elevate or depress gun in order to facilitate engagement of ball stud in retainer clamp. Release operating handle which locks ball stud in position (fig. 286).

e. Engage turret lock by turning handle to left (fig. 292), and disengage elevating gears by moving elevating shifter lever to right.

Section XXXX

SIGHTING AND FIRE CONTROL EQUIPMENT

217. CHARACTERISTICS.

a. Sighting and fire control equipment for 76-mm Gun Motor Carriage M18 includes Telescope M72C or M76C with Telescope Mount M55, Periscope M4A1 with Telescope M47A2, and Elevation Quadrant M9.

TM 9-755

Part Four—Auxiliary Equipment

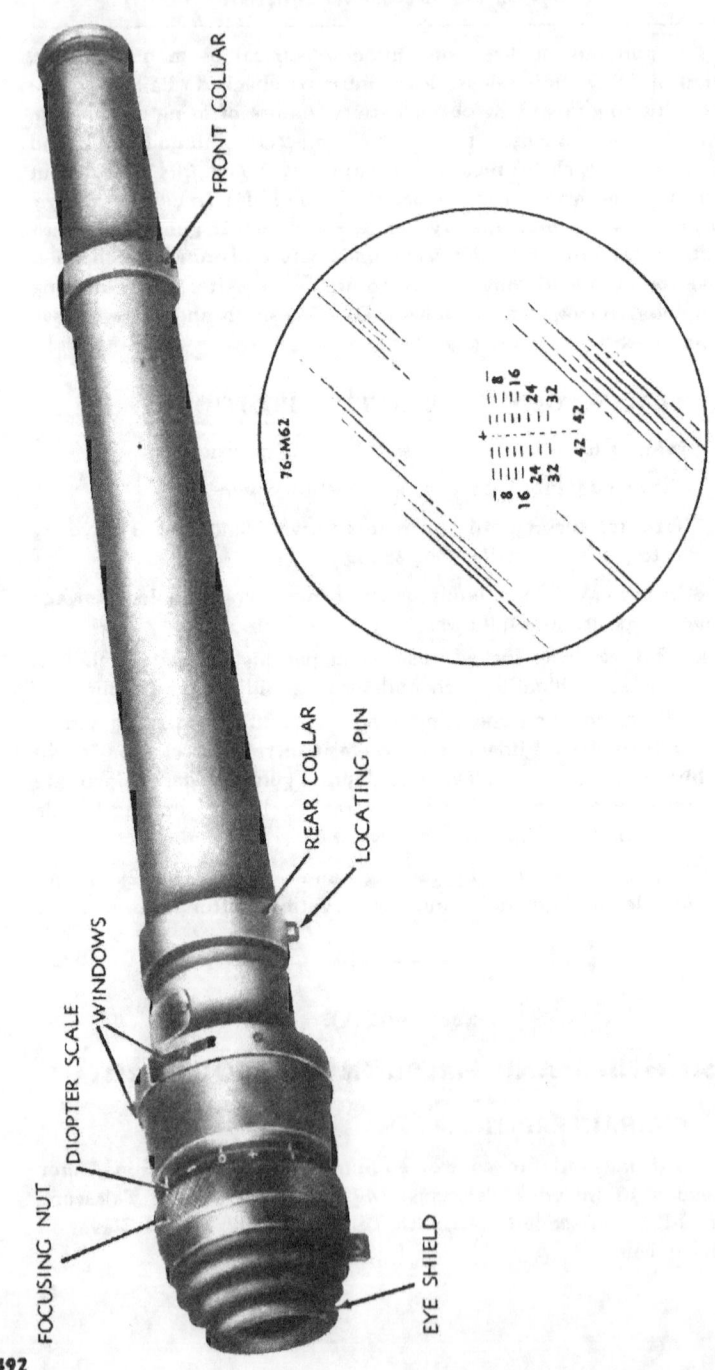

Figure 298—Telescope M72C or M76C and Reticle Pattern

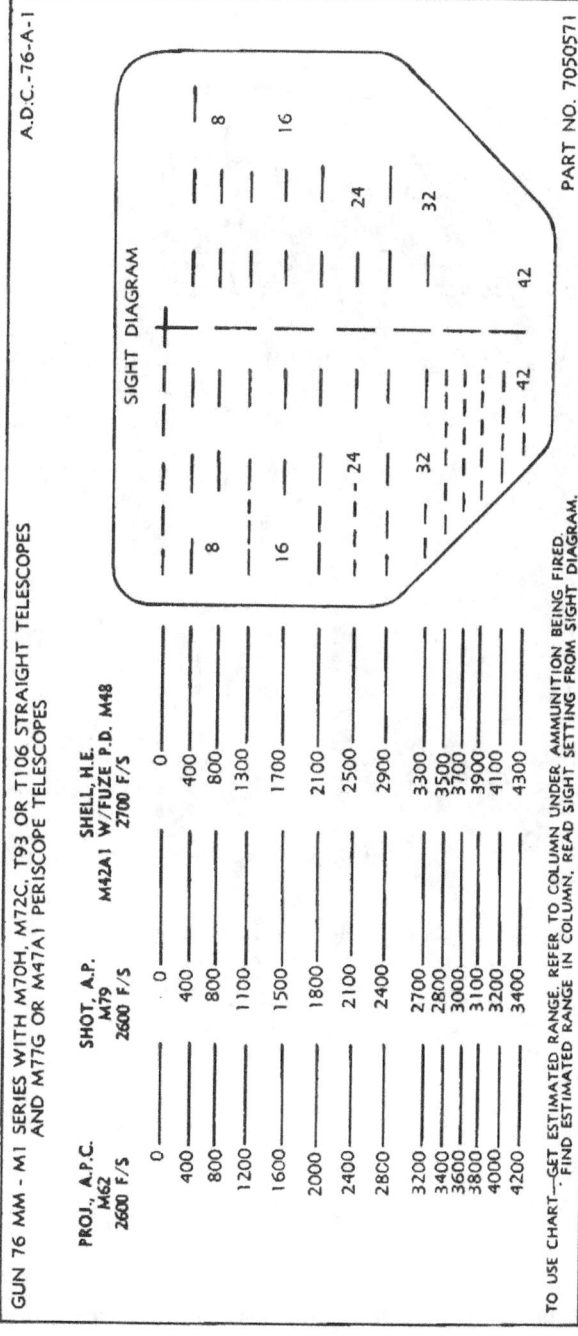

Figure 299—Decalcomania Sign on Turret Wall—Conversion of Reticle Range Markings for Use With Different Ammunition

TM 9-755
217

Part Four—Auxiliary Equipment

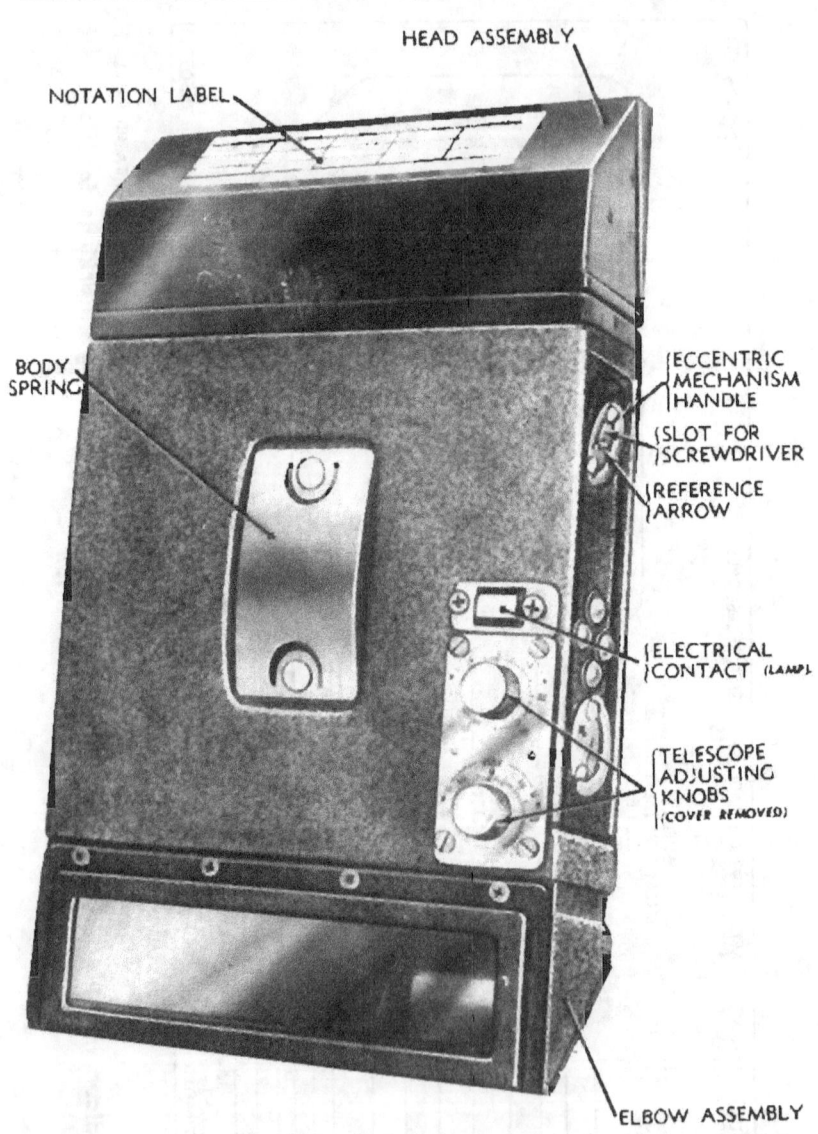

Figure 300—Periscope M4A1—Rear View

TM 9-755

Sighting and Fire Control Equipment

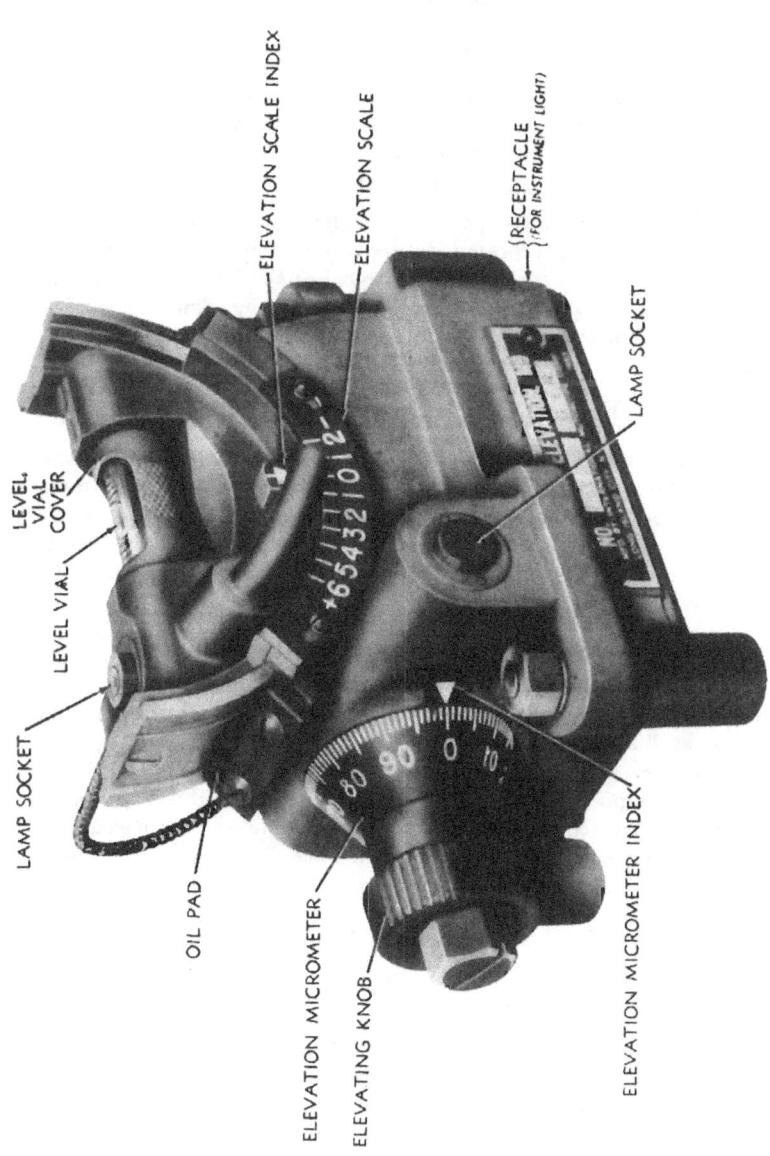

Figure 301—Elevation Quadrant M9

218. TELESCOPE M72C OR M76C.

a. Telescope M72C or M76C (fig. 298) is used for direct laying of 76-mm Gun. It is mounted in Telescope Mount M55 on left hand side of gun cradle, and moves with gun. Windows above reticle provide for illumination of reticle pattern.

b. To operate telescope for direct laying, bring image of target to point on reticle representing required range and deflection by rotating elevating and traversing handwheels. Gunner's eye should be approximately 1¼ inches from telescope eye lens. For night observing, turn rheostat knob on instrument light until reticle pattern is seen clearly.

219. PERISCOPE M4A1.

a. Periscope M4A1 (fig. 300) contains Telescope M47A2, and is used for direct laying against moving targets when firing 76-mm Armor-Piercing Capped Projectile M62. Head of periscope is constructed of plastic material so it will shatter into small pieces if struck by a projectile. Head is readily replaced with spare heads which are provided.

b. To operate periscope, observe through telescope, and bring image of target to point on reticle representing required range and deflection, by rotating elevating and traversing handwheels. Reticle may be illuminated for night operation. By moving eye to left of telescope eyepiece, periscope may be used for observation. If periscope is equipped with window wiper, window in head can be cleaned by pulling periscope down to retracted position and then pushing it back into viewing position. Repeat this operation several times until window is clean.

220. ELEVATION QUADRANT M9.

a. Elevation Quadrant M9 (fig. 301) is used to lay 76-mm gun in elevation for indirect fire.

b. To lay gun in elevation, set off elevation angle on coarse scale (100-mil intervals) and on micrometer (1-mil intervals). The quadrant has two scales and two micrometer indexes. Use micrometer index on side corresponding to scale in use.

221. BORE SIGHTING.

a. The purpose of bore sighting is to test alinement of sighting equipment for parallelism with gun bore. For expediency it may be performed by sighting on a well defined fixed object at least 1,000 yards distant.

b. Open breech of gun, and while looking through barrel, aline gun on distant object. With telescope in position in its mounting, observe through telescope and note position of cross on reticle representing zero range and zero deflection, with respect to aiming point. If they do not coincide, move line of sighting of telescope, by first loosening clamping nuts, and then turning adjusting knobs on rear of telescope mount until coincidence is obtained.

c. Check Periscope M4A1 in a similar manner, and if coincidence is not observed, move line of sighting by turning adjusting knobs on periscope until coincidence is obtained. Record adjustment and serial number of periscope on notation label on head of periscope.

Section XXXXI

AMMUNITION

222. AUTHORIZED AMMUNITION.

a. Ammunition authorized for 76-mm gun, M1A1, M1A1C and M1A2, is listed in Table I below. Standard nomenclature which completely identifies ammunition is used in listing. Identification is provided for by painting and marking on rounds themselves and on all packing.

Part Four—Auxiliary Equipment

TABLE 1. AUTHORIZED AMMUNITION[1]

Nomenclature	Action of Fuze	Approx. Weight of Projectile as Fired (lb)
Service Ammunition		
PROJECTILE, APC-T, M62A1, w/FUZE, B.D., M66A1, 76-mm guns, M1, M1A1, and M1A2	Delay	15.44
PROJECTILE, fixed, APC-T, M62A1, NH, w/FUZE, B.D., M66A1, 76-mm guns, M1, M1A1, and M1A2	Delay	15.44
PROJECTILE, fixed, A.P.C., M62, w/FUZE, B.D., M66A1, and TRACER, 76-mm guns, M1, M1A1, and M1 and M1A2	Delay	15.44
PROJECTILE, fixed, A.P.C., M62, NH, w/FUZE, B.D., M66A1, and TRACER, 76-mm guns, M1, M1A1, and M1A2	Delay	15.44
PROJECTILE, fixed, A.P.C., M62, w/TRACER, 76-mm guns, M1, M1A1, and M1A2	....	15.11
SHELL, fixed, H.E., M42A1, w/FUZE, P.D., M48A1, 76-mm guns, M1, M1A1, and M1A2[2]	SQ & 0.15-sec. delay[3]	12.87
SHELL, fixed, H.E., M42A1, w/FUZE, P.D., M48A2 SQ & 0.05-sec. delay, 76-mm guns, M1, M1A1, and M1A2[2]	SQ & 0.05-sec. delay[3]	12.87
SHELL, fixed, H.E., M42A1, NH, w/FUZE, P.D., M48A1, 76-mm guns, M1, M1A1, and M1A2[2]	SQ & 0.15-sec. delay[3]	12.87
SHELL, fixed, H. E., M42A1, NH, w/FUZE, P.D., M48A2, SQ & 0.05-sec. delay, 76-mm guns, M1, M1A1, and M1A2[2]	SQ & 0.05-sec. delay[3]	12.87
SHELL, fixed, H.E., M42A1, w/FUZE, P.D., M48, 76-mm guns, M1, M1A1, and M1A2[2]	SQ & 0.05-sec. delay[3]	12.87
SHELL, fixed, H.E., M42A1, NH, w/FUZE, P.D., M48, 76-mm guns, M1, M1A1, and M1A2[2]	SQ & 0.05-sec. delay[3]	12.87

Ammunition

SHELL, fixed, H.E., M42, w/FUZE, P.D., M48A1, 76-mm guns, M1, M1A1, and M1A2[2]	SQ & 0.15-sec. delay[3]	12.81
SHELL, fixed, H.E., M42A1, reduced charge, w/FUZE, P.D., M48A2, SQ & 0.15-sec. delay, 76-mm guns, M1, M1A1, and M1A2[2]	SQ & 0.15-sec. delay[3]	12.87
SHELL, fixed, illuminating, Mk. 24-Mod. 1 (Navy), w/FUZE, TSQ, M54, 76-mm guns, M1, M1A1, and M1A2	Time & SQ	......
SHELL, fixed, smoke, HC, B.I., M88, NH, 76-mm guns, M1, M1A1, and M1A2		7.60
SHELL, fixed, smoke, HC, B.I., M88, 76-mm guns, M1, M1A1, and M1A2		7.60
SHOT, fixed, A.P., M79, w/TRACER, 76-mm guns, M1, M1A1, and M1A2		15.00

Blank Ammunition

AMMUNITION, blank (double pellet), 76-mm guns, M1, M1A1, and M1A2	......	......

Drill (Dummy) Ammunition

CARTRIDGE, drill, M20, w/FUZE, dummy, M59, 76-mm guns, M1 and M1A1	Inert	......

A.P.—armor-piercing
A.P.C.—armor-piercing-capped
APC-T—armor-piercing-tracer

B.D.—base detonating
B.I.—base-ignition
H.E.—high-explosive

P.D.—point detonating
SQ—superquick
TSQ—Time and SQ

1.—(a) Service rounds which do not have "NH" in the nomenclature contain flashless (FNH) propellent powder.

(b) Nomenclature listed refers to brass-cased rounds. Nomenclature of steel-case rounds includes the words "steel case" immediately following model designation. The words are also stenciled on packing boxes and crates, when applicable. Steel cartridge cases are signified in stamping on base of case by addition of suffix "B" and arabic numeral to the model designation of case. The ammunition lot number has an "X" suffix when the round is steel cased.

2.—In future manufacture, 76-mm rounds requiring M48 series fuze, other than reduced charge rounds, will be fuzed with M48 or M48A2 with 0.05-second delay. Reduced-charge rounds will be fuzed with M48A2 fuze with 0.15-second delay.

3.—The delay of FUZE, P.D., M48 is 0.05 second; of FUZE, P.D., M48A1, 0.15 second. M48A2 fuze may have either 0.05 second or 0.15 second delay, depending on the lot. Provision is made for identification by stamping length of delay in seconds on fuze, immediately following model number. Thus, M48A2 fuzes with 0.05 second delay element will have stamped on body, "FUZE, P.D., M48A2 (.05 SEC.)."

TM 9-755
223-224

Part Four—Auxiliary Equipment

223. PREPARATION FOR FIRING.

a. 76-mm rounds are ready for firing as removed from packing, except for setting M48 and M54 fuzes for required action.

(1) M48, M48A1, and M48A2 fuzes. As shipped these fuzes are set for superquick action (SQ), that is, the slot in setting screw is parallel to axis of fuze and in line with "SQ". To adjust for delay action, setting screw should be turned by means of screwdriver end of fuze wrench, M7A1, or similar instrument, so that slot is alined with "DELAY," that is, at right angles to axis of fuze. Delay action is provided for in fuze by a delay pellet. The setting may be made or changed at will, and can be done in the dark by noting position of slot in setting sleeve.

(2) M54 FUZE.—Prior to firing, with either superquick or time setting, safety pull wire securing time plunger during shipment must be withdrawn from fuze. (Pull lower end of wire from hole and slide wire off end of fuze.) To obtain superquick action, fuze may be left at safe (S) setting, as shipped, or may be set for a time longer than the expected time of flight. Since superquick action is always operative, it will function on impact unless prior functioning has been caused by time action. If time action is required, graduated time-ring (graduated to 25 seconds) is set for required time of burning by means of a fuze setter. NOTE: *If, after setting fuze preparatory to firing, round is not fired, fuze should be reset "safe" and safety pull wire replaced before returning round to its packing.*

Section XXXXII

RADIO AND INTERPHONE EQUIPMENT

224. RADIO SETS AND INTERPHONE EQUIPMENT.

a. General. 76-mm Gun Motor Carriage M18 and Armored Utility Vehicle M39 are equipped with radio set SCR-610 and interphone equipment RC-99. This radio set obtains power from the 24-volt vehicular electrical system if provided with power unit TE-120. The 12-volt tap must be used if power unit PE-117 is provided. The interphone equipment operates off the 24-volt vehicular electrical system. Power connections for radio and interphone equipment in 76-mm Gun Motor Carriage M18, are made in turret radio terminal box which is mounted below radio shelf located in rear of turret (fig. 304). Power connections for Armored Utility Vehicle M39, are made in radio terminal box which is mounted in right sponson opposite co-driver (fig. 306). Radio Set SCR-610 is frequency modulated and voice operated only. For 76-mm Gun Motor Carriage M18 a reel assembly RL-106/VI is available. This reel mounts in front left quarter of turret and provides commander with a 25-yard extension cord from his control box (fig. 309). Signal Corps

Radio and Interphone Equipment

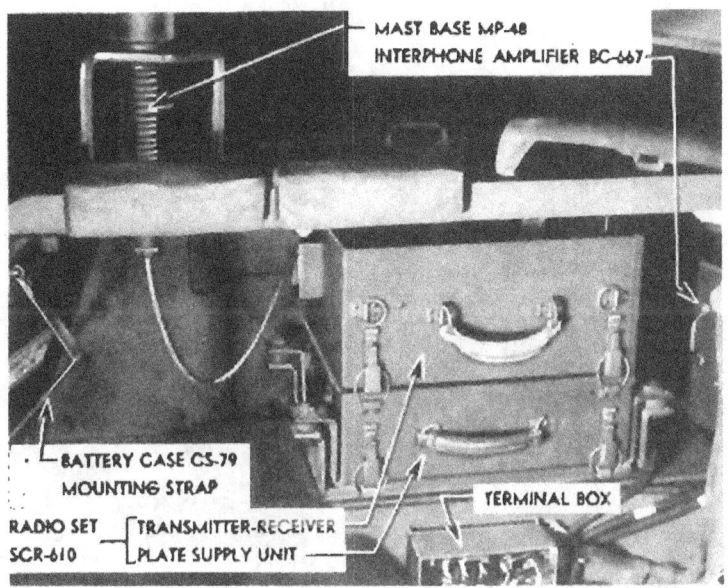

Figure 302—Radio Set SCR-610 and Antenna Mast Base MP-48 Installed—M18

drawing and installation instructions have been prepared for radio and interphone installations referred to above. If needed, copies can be obtained through organization signal officer.

(1) RADIO SET SCR-610 (figs. 302 and 303). Major components of this set consist of a radio receiver and transmitter BC-659 and plate supply unit PE-117 or PE-120 mounted on mounting base FT-250. Also supplied is battery case CS-79 which is stowed on 76-mm Gun Motor Carriage M18 in right hand side of turret extension, and on Armored Utility Vehicle M39, on right end of engine compartment bulkhead.

(2) INTERPHONE EQUIPMENT BC-99. Components of this equipment are Interphone Amplifier BC-667 (figs. 307 and 308) and control boxes BC-606 and BC-739 (figs. 303 and 305). Interphone Amplifier on M18 (fig. 307) is located in turret extension and Interphone Amplifier on M39 (fig. 308) is located on ceiling of vehicle directly above and in front of co-driver. Control boxes are provided for driver and co-driver in each vehicle; commander, gunner and loader in 76-mm Gun Motor Carriage M18, and one on each side of crew compartment in Armored Utility Vehicle M39.

(3) REEL ASSEMBLY RL-106/VI (fig. 309). Major components

TM 9-755
224-226

Part Four—Auxiliary Equipment

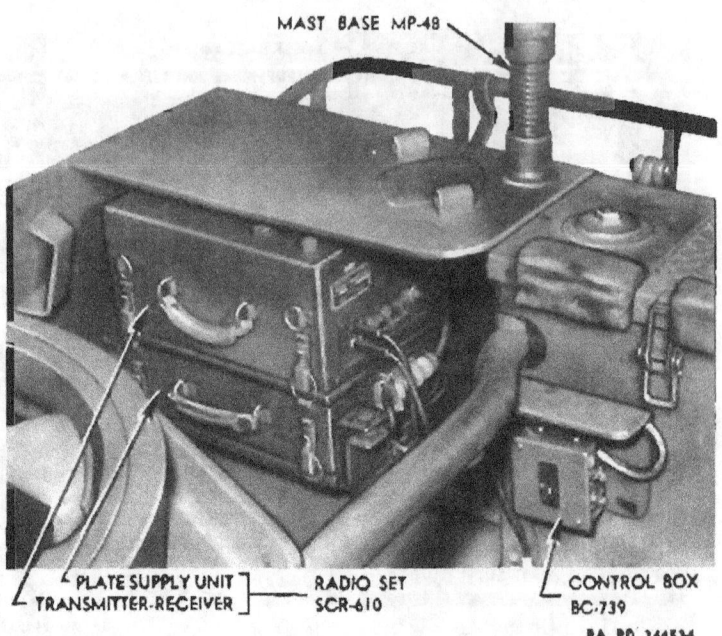

Figure 303—Radio Set SCR-610, Control Box BC-739 and Mast Base MP-48 Installed—M39

of the reel assembly consist of Reel RL-108/VI, cord CD-264 and CD-265, and guide ring MX-171/VI.

225. MOUNTINGS.

a. Mounting Base FT-250 (fig. 304). This mounting base is used with Radio Set SCR-610 described in paragraph 224 a (1). Base is made up of two sections which are connected together through four rubber grommets. Lower section is fastened to vehicle brackets with four cap screws. Plate supply unit fastens to top portion of mounting base by use of four snap fasteners. Transmitter-receiver is located on top of plate supply unit and is fastened thereon by use of snap fasteners.

226. ANTENNA.

a. Mast Base MP-48 (figs. 302 and 303). This base is equipped with large helical springs for flexing, and a small porcelain insulator incorporated above spring. This base is secured to its mounting bracket (or surface) by clamping action with use of a large hexagonal nut on lower end of base. This base is used with Radio Set SCR-610.

Figure 304—Mounting Base FT-250, Radio Terminal Box, and Interphone Amplifier BC-667 Installed—M18

b. **Mast Sections.** This frequency modulated radio set uses three mast sections, numbers MS-51, 52 and 53. Sections are made of high tensile steel and are secured together so that ends with like color enamel are joining. Body of the mast bears type number. Clamps are provided to keep the mast sections from loosening while in use. These sections and spare sections are stowed in Roll Bag BC-56 when not in use.

c. **Light Weight Antenna.** Light weight antenna have been designed to replace components described in subparagraphs a and b above. This mast base is designated AB-15/GR and mast sections are MS-116, 117 and 118.

227. INSPECTIONS.

a. Antenna.

(1) MAST SECTIONS. Inspect antenna mast sections to be sure that they are securely screwed and clamped together and are not damaged.

(2) HELICAL SPRING. Inspect helical spring on base; be sure that it maintains a vertical position and is not damaged to prevent flexibility.

TM 9-755
227

Part Four—Auxiliary Equipment

Figure 305—Commander's and Gunner's Control Box BC-739 Installed—M18

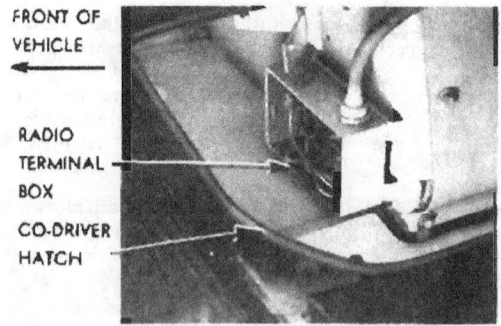

Figure 306—Radio Terminal Box Installed—M39

(3) MAST BASE. See that mast base is secured to its bracket or mounting surfaces and that insulator is not cracked or chipped.

(4) LEADS TO SET. Check leads to set and be sure that there is no interference that may damage cords and that any stand-off insulators are not cracked or chipped.

Radio and Interphone Equipment

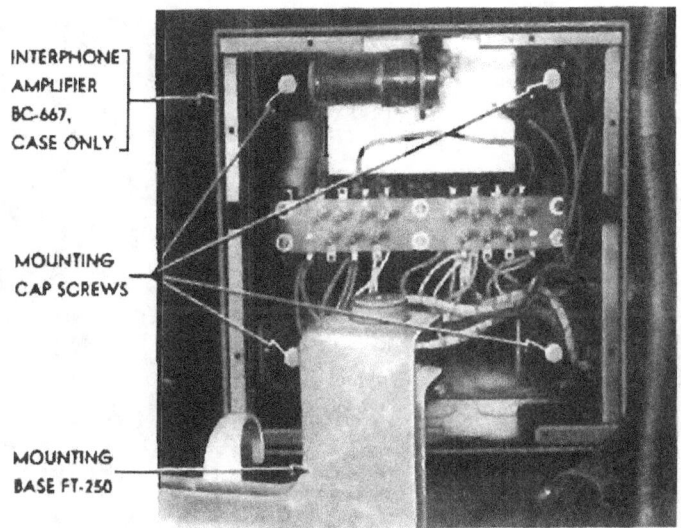

Figure 307—Interphone Amplifier Mounting BC-667 Installed—M18

b. **Mountings.**

(1) SNAP FASTENERS. Be sure that radio components are securely fastened onto mounting base.

(2) LOCK WASHERS. When re-installing radio or interphone equipment, make sure that all toothed lock washers are replaced in locations where they were originally used.

(3) SHOCK MOUNTS. Inspect mounting screws to see that they are tight and that shock mountings are in good condition. Rock set and interphone amplifier to determine if they bump any other equipment, and observe whether shock absorbers are deteriorated and permit excessive movement.

c. **Cords and Connections.** Inspect all cords which connect radio or interphone equipment to see that they are not damaged; make sure they are properly secured in clips. Report any damage to proper authority.

d. **Microphones and Headsets.** Handle microphones and headsets with care to see that they are hung on hooks provided for this purpose when not in use. Be sure that cords are not twisted or knotted to prevent movement of wearer. Inspect jack plugs on microphones and headsets to see that they are not damaged.

TM 9-755
227-228

Part Four—Auxiliary Equipment

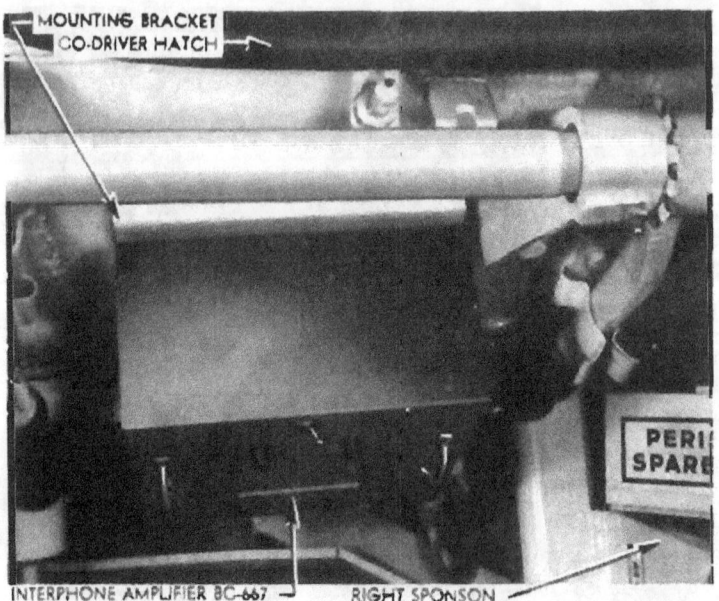

Figure 308—Interphone Amplifier BC-667 Installed—M39

e. Radio Terminal Box (figs. 304 and 306). Remove cover and check tightness of all terminal nuts. Tighten nuts, if necessary, to prevent any movement of wire on terminal stud, thereby eliminating possibility of radio interference from this source. At this time also check presence and fastening of condenser in terminal box.

f. Covers. Be sure that cover for protection of radio set is available in vehicle and that cover is installed when equipment is not in use. See that all fasteners and zippers are in good condition. Cover BG-153 is used with radio set SCR-610.

228. PRECAUTIONS.

a. Antenna. Tie antenna down securely when vehicle is in motion and radio is not in use to prevent damage to antenna. Be sure antenna is vertical and not touching anything when radio is in use.

b. Radio.

(1) Keep radio covered when vehicle is not in use to prevent dust

Figure 309—Reel Assembly RL-108/VI Installed—M18

and moisture from entering set. Keep all cover plates closed and securely fastened.

(2) Turn off all radio and interphone switches when not in use. Do not turn off master battery switch with radio and interphone on.

(3) Do not store equipment behind radio where it can prevent movement on mountings or damage to connections.

c. **Batteries and Charging System** (fig. 186).

(1) Be sure batteries are charged at all times to insure satisfactory operation of set. Low batteries will cause set to be weak and unstable resulting in poor reception, and may make it difficult to start vehicle.

(2) See that all battery cables and terminals are in good condition and tight.

(3) Test operation of generator and regulator (par. 59). Excessive charging rate may cause damage to radio set and interphone amplifier.

APPENDIX

Section XXXXIII

SHIPMENT AND LIMITED STORAGE

229. GENERAL INSTRUCTIONS.

a. Preparation for domestic shipment of vehicle is the same as preparation for limited storage. Preparation for shipment by rail includes instructions for loading and unloading vehicle, blocking necessary to secure vehicle on freight cars, clearance, weight, and other information necessary to properly prepare vehicle for rail shipment. For more detailed information and for preparation for indefinite storage refer to AR 850-18 and FM 9-25.

230. PREPARATION FOR LIMITED STORAGE OR DOMESTIC SHIPMENT.

a. A vehicle to be prepared for limited storage or domestic shipment is one temporarily out of service for less than 30 days, or a vehicle that must be ready for operation on call. If vehicle is to be indefinitely stored after shipment by rail, it will be prepared for such storage at its destination.

b. If vehicle is to be placed in limited storage, take the following precautions.

(1) LUBRICATION. Completely lubricate entire vehicle, except engine (par. 38). For preparation of engine, see step (8) below.

(2) BATTERIES. Check batteries and terminals for corrosion and if necessary, clean and thoroughly service batteries (par. 142).

(3) ROAD TEST. Preparation for limited storage will include a road test of at least 5 miles, after battery and lubrication services, to check on general condition of vehicle. Correct any defects noted in vehicle operation, before vehicle is stored or note on a tag attached to steering levers, stating repairs needed or describing condition present. A written report of these items will then be made to officer in charge.

(4) FUEL IN TANKS. It is not necessary to remove fuel from tanks during temporary storage or shipment within the United States, nor to label tanks under Interstate Commerce Commission Regulations. Leave fuel in tanks except when storing in locations where fire ordinances or other local regulations require removal of all gasoline before storage. If vehicles are to be maintained ready for operation on call in excess of 30 days, the following precautions against gum formation must be taken:

(a) Fuel system must be free from accumulated gum. Unless vehicle is entering its first storage and has never been issued for use, inspect and clean fuel pump; carburetor accelerator pump plunger,

Shipment and Limited Storage

venturi tube, choke and throttle valves, float mechanism; fuel lines; fuel tanks; fuel filters; fuel shut-off valves; and screens.

(b) If gum is present in the above parts, it can best be removed by benzol, acetone, alcohol, or a mixture of these solvents. Deposited gum is not readily soluble in fresh gasoline. When gum has dried, it may be necessary to resort to mechanical means to remove it.

(c) Parts which cannot be thoroughly cleaned and freed from gum deposit without damage should be replaced.

(d) After cleaning and reassembling, fill fuel tank half full of fresh gasoline which has not been long in storage.

(e) Add three containers (12 oz) of gum-preventive compound to each set of fuel tanks.

(f) Fill fuel tank to capacity and operate vehicle for at least 5 minutes.

(5) BREECH MECHANISM (M18). When possible, partially disassemble breech mechanism and dip, spray, or brush parts with light rust preventive compound. Assemble breech mechanism.

(6) GUN TUBE (M18). Clean bore of 76-mm gun with dry-cleaning solvent and thoroughly dry. Swab bore with light rust preventive compound. Seal muzzle with non-hygroscopic adhesive tape. Install the muzzle cover, if available, and seal with non-hygroscopic adhesive tape. If a muzzle cover is not available, wrap waterproof barrier wrapping paper over tape and seal with non-hygroscopic adhesive tape.

(7) EXTERIOR OF VEHICLES. If practicable, remove rust appearing on vehicle exterior with flint paper. Repaint painted surfaces whenever necessary to protect wood or metal. Coat exposed polished metal surfaces susceptible to rust with light rust preventive compound. Close firmly all doors, hatches, and vision slots. Make sure paulins are in place and firmly secured. Leave rubber mats, such as floor mats, where provided, in an unrolled position on floor, and not rolled or curled up. Equipment such as pioneer tools and fire extinguishers will remain in place in vehicle. For treatment of small arms carried on or within vehicles, refer to pertinent technical manuals.

(8) ENGINE.

(a) Remove spark plugs and spray into tops of cylinders with preservative engine oil, SAE 30, grade II, while slowly rotating engine. Replace spark plugs.

(b) If spark plugs cannot be removed, spray preservative oil into air intake with engine running at a fast idle until smoke comes from exhaust pipe. CAUTION: *Preservative oil should never be poured through carburetor.* After spraying preservative oil into air intake, shut off engine and allow to cool for about 15 minutes. Start engine and again spray preservative oil into air intake for several minutes only. Second spraying is necessary in order to coat exhaust valves. Do not run engine for more than several minutes as exhaust valves will become so hot that preservative oil will not adhere properly.

TM 9-755
230-231

Appendix

Perform this treatment when further running of the engine is not necessary.

(c) If it becomes necessary to run engine after treatment, it should not be operated at over 1,600 revolutions per minute. Hold operation to a minimum, and spray cylinders again after operation.

(9) INSPECTION. Make a systematic inspection, just before shipment or temporary storage, to insure all above steps have been covered and that vehicle is ready for operation on call. Make a list of all missing or damaged items and attach it to one of steering levers. Refer to Before-operation Service (par. 41).

(10) BRAKES. Release brakes and chock tracks.

c. Inspections in Limited Storage. When vehicle is placed in limited storage, inspect batteries weekly. If water is added to batteries when freezing weather is anticipated, recharge batteries with a portable charger or remove them for charging. Do not attempt to charge batteries by running auxiliary generator. Remove any rust from vehicle with flint paper.

231. LOADING AND BLOCKING FOR RAIL SHIPMENT.

a. **Preparation.** In addition to preparation described in paragraph 230, when Ordnance vehicles are prepared for domestic shipment, the following preparation and precautions will be taken:

(1) EXTERIOR. Cover body of vehicle with a canvas cover ordinarily supplied as an accessory.

(2) BATTERIES. Disconnect batteries to prevent their discharge by vandalism or accident. This may be accomplished by disconnecting the positive lead, taping end of lead, and tying it back away from battery.

(3) MARKING CARS. All cars containing Ordnance vehicles must be placarded "DO NOT HUMP."

b. **Placing Vehicle on Car.**

(1) TYPES OF CARS. Ordnance vehicles may be shipped on flat cars, end door box cars, side door cars, or drop end gondola cars, whichever type is most convenient.

(2) FACILITIES FOR LOADING. Whenever possible, load and unload vehicle from open cars, using permanent end ramps and spanning platform. Movement from one flat car to another along length of train is made possible by crossover plates or spanning platforms. If no permanent end ramp is available, an improvised ramp can be made from railroad ties. Vehicle may be loaded in a gondola car without drop ends by using a crane.

(3) BRAKE WHEEL CLEARANCE. If a flat car is used, position vehicle with a railroad brake wheel clearance of at least 6 inches (A, fig. 310). When more than one vehicle is loaded on car, locate vehicle on car in such a manner as to prevent car from carrying an unbalanced load. Apply brakes.

TM 9-755
231

Shipment and Limited Storage

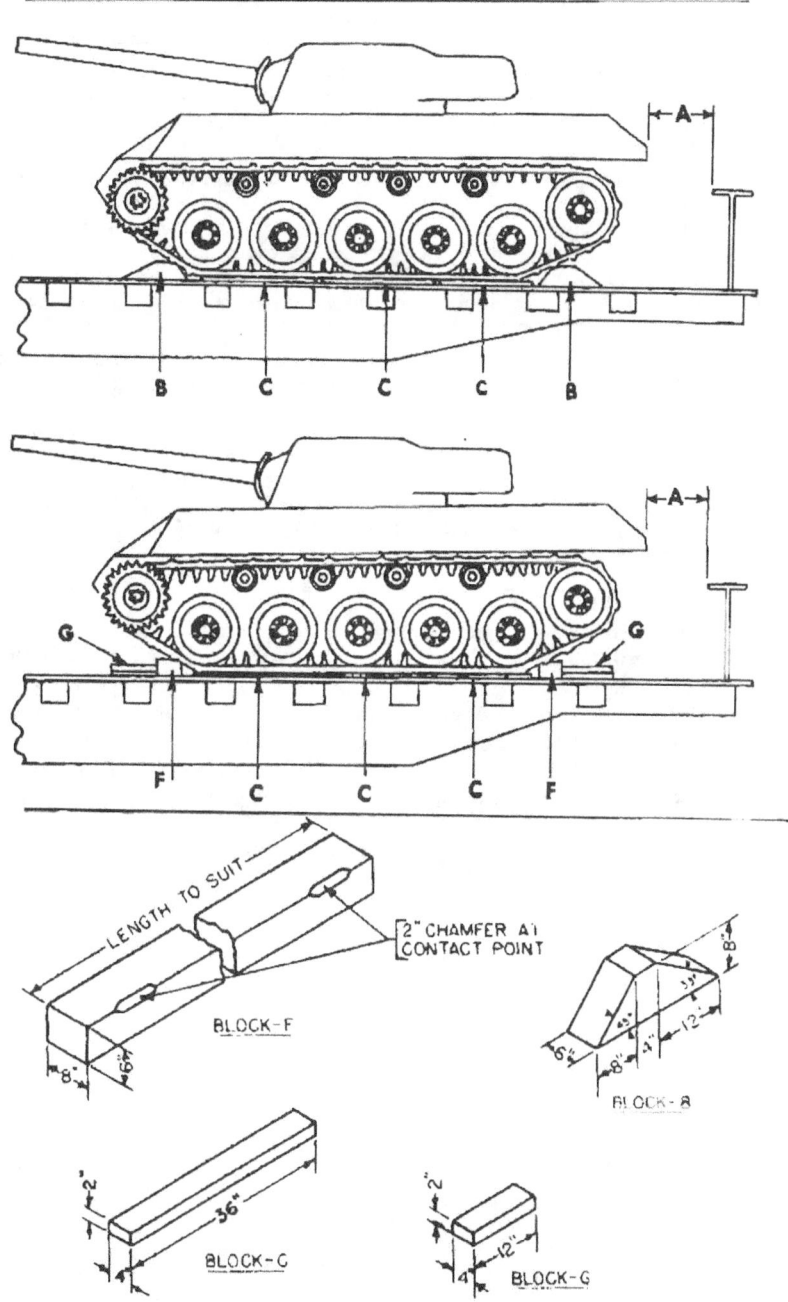

Figure 310—Blocking Requirements for Rail Shipment

TM 9-755
231

Appendix

---

c. **Securing Vehicles.** In securing or blocking a vehicle, three motions, lengthwise, sidewise, and bouncing must be prevented. There are two approved methods of blocking vehicles on freight cars, as described below.

(1) METHOD ONE. Place four blocks (B, fig. 310), one to front and one to rear of each track. Nail heel of each block to car floor with five 40-penny nails. Nail portion of each block which is under track to car floor with two 40-penny nails. Locate three blocks (C, fig. 310) on each side of vehicle on outside of each track. Nail each block to car floor with three 40-penny nails. These blocks may be located on the inside of tracks if conditions warrant.

(2) METHOD TWO. Place two blocks (F, fig. 310), one to front and one to rear of tracks. These blocks are to be at least as long as over-all width of vehicle at car floor. Locate eight blocks (G) against blocks (F) to front and to rear of each track. Nail lower block to floor with three 40-penny nails and top block to lower block with three 40-penny nails. Locate three blocks (C, fig. 310) on each side of vehicle on outside of each track. Nail each block to car floor with three 40-penny nails. These blocks may be located on inside of tracks if conditions warrant.

d. Shipping Data.

	M18	M39
Length, over-all, (gun in traveling position)	21 ft. 2 in.	17 ft. 10 in.
Width, over-all	9 ft. 5 in.	9 ft. 5 in.
Height, over-all, A.A. gun removed	8 ft. 5 in.	6 ft. 9 in.
Area of car floor occupied per vehicle	199 sq ft	168 sq ft
Volume occupied per vehicle	1,678 cu ft	1,133 cu ft
Shipping weight per vehicle	37,557 lb	35,500 lb

References

Section XXXXIV
REFERENCES

232. PUBLICATIONS INDEXES.

The following publications indexes should be consulted frequently for latest changes to or revisions of the publications given in this list of references and for new publications relating to materiel covered in this manual:

Introduction to Ordnance Catalog (explains SNL system)	ASF Cat. ORD-1 IOC
Ordnance publications for supply index (index to SNL's)	ASF Cat. ORD-2 OPSI
Ordnance major items and combinations, and pertinent publications (alphabetical listing of ordnance major items with available publications pertaining thereto, including TM's, OFSTB's, WDTB's, MWO's, and ASF catalogs)	SB 9-1
List of publications for training (lists MTP's, TR's, TC's, FM's, TM's, WDTB's, Firing Tables and Charts and Lubrication Orders)	FM 21-6
List of miscellaneous publications (lists MP's, MWO's, SB's, RR's, and War Department Pamphlets)	WD Pamphlet 12-6
List of training films, film strips and film bulletins (lists TF's, FS's, and FB's by serial number and subject)	FM 21-7
Military training aids (lists graphic training aids, models, devices, and displays)	FM 21-8

233. STANDARD NOMENCLATURE LISTS.

    a. **Ammunition.**

Ammunition, fixed and semifixed, including subcaliber, for pack, light and medium field, aircraft, tank and antitank artillery, including complete round data	SNL R-1
Ammunition, blank, for pack, light and medium field artillery	SNL R-5
Ammunition instruction material for pack, light and medium field, aircraft, tank, and antitank artillery	SNL R-6
Ammunition, rifle, carbine, and automatic gun	SNL T-1
Grenades, hand and rifle, and fusing components	SNL S-4

Appendix

        Service fuzes and primers, for pack, light and medium field artillery .... SNL R-3

b. **Armament.**

        Carbine, cal. .30, M1, M1A1, M1A3, and M2.... SNL B-28

        Gun, 76-mm, M1A1 mount, gun, 76-mm, M1 ... SNL C-58

        Gun, 76-mm, M1A1 and M1A2 mount, combination gun, M62 (T80) .... SNL C-64

        Gun, machine, cal. .50 Browning, M2, heavy barrel turret type.................. SNL A-59

c. Carriage, motor, 76-mm gun, M18 (T70) vehicle, utility armored, T41 (M39) vehicle, utility armored, T41E1 .... SNL G-163

d. **Maintenance.**

        Antifriction bearings and related items ORD-5.. SNL H-12

        Cleaning, preserving and lubricating materials: recoil fluids, special oils, and miscellaneous related items, ORD-5................... SNL K-1

        Elements, oil filter, ORD-5 .... SNL K-4

        Lubricating equipment, accessories, and related dispensers, ORD-5 .... SNL K-3

        Soldering, brazing and welding materials, gases and related items, ORD-5 .... SNL K-2

        Standard hardware, ORD-5 .... SNL H-1

        Tool-sets for maintenance of sighting and fire control equipment .... SNL F-272

        Tools, maintenance, for repair of automatic guns, automatic gun, antiaircraft materiel, automatic and semiautomatic cannon, and mortars, ORD-6 .... SNL A-35

        Tools, maintenance, for repair of automotive and semiautomotive vehicles:

            ORD-6, Tool-sets (special) automotive and semiautomotive .... SNL G-27 (Section 1)

            ORD-6, Tool-sets (common) specialists' and organizational .... SNL G-27 (Section 2)

e. **Sighting Equipment.**

        Lights, instrument .... SNL F-205

TM 9-755
233-234

References

Periscopes, telescopes for periscopes, and direct sighting telescopes for use in tanks	SNL F-235
Quadrant, elevation, M9 (T10) for gun motor carriages, tanks and combination gun mounts M34 and M34A1	SNL F-281
Quadrant, gunner's, M1 (mils)	SNL F-140

234. EXPLANATORY PUBLICATIONS.

a. Fundamental Principles.

Ammunition, general	TM 9-1900
Automotive electricity	TM 10-580
Auxiliary fire-control instruments (field glasses, eyeglasses, telescopes and watches)	TM 9-575
Basic maintenance manual	TM 37-250
Browning Machine Gun, Cal. .50, HB, M2 (mounted in combat vehicles)	FM 23-65
Care and maintenance of ball and roller bearings	TM 37-265
Driver selection and training	TM 21-300
Driver's manual	TM 21-305
Electrical fundamentals	TM 1-455
Field artillery and field motor ammunition	OFSB 3-3
Firing Tables for: Gun, 76-mm M1A1 and M1A2 Shell, fixed, H.E., M42A1 w/fuze, P.D., M48 and Mod's. Fuze C. P. T105 (nose) Projectile, fixed, A.P.C., M62, and M62A1 w/fuze B. D., M66A1 and tracer Shot, fixed, A. P., M79 w/tracer Shell, fixed illuminating MK. 24-Mod. 1 (Navy)	FT 76-A-5
Fuels and carburetion	TM 10-550
Instruction guide, small arms data	TM 9-2200
Military motor vehicles	AR 850-15
Motor vehicle inspections and preventive maintenance service	TM 9-2810
Ordnance service in the field	FM 9-5
Precautions in handling gasoline	AR 850-20
Qualifications in arms and ammunition training allowances	AR 775-10

515

TM 9-755
234

Appendix

Radio fundamentals	TM 11-455
Radio Set SCR 610	TM 11-615
Range regulations for firing ammunition for training and target practice	AR 750-10
76-mm Gun, Materiel M1 (combat vehicle)	TM 9-308
Small arms ammunition	TM 9-1990
Small arms ammunition	OFSB 3-5
Standard military motor vehicles	TM 9-2800
Targets, target materials, and rifle range construction	TM 9-855
The radio operator	TM 11-454
U. S. Carbines, cal. .30, M1 and M1A1	FM 23-7

b. **Maintenance and Repair.**

Cleaning, preserving, sealing, lubricating and related materials issued for ordnance materiel	TM 9-850
Maintenance and care of pneumatic tires and rubber treads	TM 31-200
Ordnance maintenance: Accessories for Wright R975-EC2 engines for medium tanks M3 and M4 (Scintilla Magnetos)	TM 9-1750D
Ordnance maintenance: Auxiliary generator (Homelite Model HRUH-28) for medium tanks M4 and modifications	TM 9-1731K
Ordnance maintenance: Carburetors (Stromberg)	TM 9-1826B
Ordnance maintenance: Electrical equipment (Delco-Remy)	TM 9-1825A
Ordnance maintenance: Fuel pumps	TM 9-1828A
Ordnance maintenance: Hydraulic traversing mechanism for medium tank M4 and modifications (Oilgear)	TM 9-1731G
Ordnance maintenance: 9-cylinder, radial gasoline engine (Continental Model R975-C1)	TM 9-1751
Ordnance maintenance: Ordnance engine Model R975-C4 (Continental)	TM 9-1725
Ordnance maintenance: Power train for 76-mm gun motor carriage T70 (M18)	TM 9-1755A
Ordnance maintenance: Speedometers, tachometers, and recorders	TM 9-1829A

References

	Ordnance maintenance: Tracks, suspension, hull, turret, and related components for 76-mm gun motor carriage T70 (M18)	TM 9-1755B
c.	**Protection of Materiel.**	
	Camouflage	FM 5-20
	Decontamination	TM 3-220
	Decontamination of armored force vehicle	FM 17-59
	Defense against chemical attack	FM 21-40
	Explosives and demolitions	FM 5-25
d.	**Storage and Shipment.**	
	Ordnance company, depot	FM 9-25
	Ordnance packing and shipping (posts, camps, and stations)	TM 9-2854
	Ordnance storage and shipment chart, group G—Major items	SB 9-OSSC-G
	Protection of ordnance materiel in open storage	SB 9-47
	Registration of motor vehicles	AR 850-10
	Rules governing the loading of mechanized and motorized army equipment also major caliber guns, for the United States Army and Navy, on open top equipment published by Operations and Maintenance Department of Association of American Railroads.	
	Storage of motor vehicle equipment	AR 850-18

TM 9-755

INDEX

A

	Page
A.A. Guns and mounts (See Guns and Gun mounts)	
Accelerator linkage	119
Accelerator pedals	38
Accessories	
after-operation service	104
before-operation service	98
cold weather	72
during-operation service	103
gun and carriage	16
run-in test	27
Accessory drives	116
Accessory equipment	60
Adapter	365
After-operation and weekly service	45, 103
Aircleaners	
after-operation service	105
description	237
during-operation service	103
maintenance chart	116
operation under dusty conditions	73
removal, cleaning, inspection, and installation	240
run-in test	27
Air duct seals	197
Air filter, carburetor	368
Air intake system	
data	237
description	
pipes	183
system	231, 235
Ammeter	
before-operation service	99
inoperative	155
replacement (auxiliary generator)	368
road test	109
run-in test	29
shows discharge	149
Ammunition	
authorized	
general	497
preparation for firing	500
tables	498
data	11
M18 vehicle	15
Armament	472
Armor plate, run-in test	28
Auxiliary equipment	472
Auxiliary fuel pump circuit breaker	394
Auxiliary fuel pumps switch	41
Auxiliary generator, M18 vehicle	
cleaning	
carburetor air filter and commutator	368
spark plug and adapter	365
data	357
description and operation	60
installation	370

	Page
maintenance operations	122, 123
removal	369
trouble shooting	149
Axle shaft housing	
description	349
installation	351
removal	349
Azimuth indicator	
description and operation	58
removal and installation	449

B

	Page
Batteries	
after-operation service	104
cable replacement, removal, and installation	360
care of in torrid zones	74
checking specific gravity	358
corroded terminals	359
data	357
description of batteries and circuits	354
flashlight	20
maintenance	119
operation in sub-zero weather	71
run-in test	27
trouble shooting	146
Battery ammeter	
installation and removal	388
operation	42
Battery box (M39 vehicle)	
installation	445
removal	444
Battery charging circuits	355
Before-operation service	97
Belts	
adjustment	
blower belts	306
generator drive belt	362
installation	
blower belts	307
generator drive belt	363
removal	
blower belts	306
generator drive belt	362
run-in test	27
Blackout lights	
description	376
inoperative	152, 158
operation of switch	42
Blackout marker lights	
inoperative	152
replacement of lamp	382
Blackout resistor	
description	
headlight	376
trailer taillight and stop light	377
replacement	382
Blackout taillight, trailer	380

518

Index

B—Contd.

Blower (battery air tube)... 67
Blower belts (transmission and differential oil cooler)
  adjustment and removal... 306
  installation... 307
Bonding (radio interference)... 430
Booster coil
  maintenance operation... 115
  removal and installation... 230
  replacement... 394
  trouble shooting... 132
Booster switch... 41
Bore sighting... 496
Brake drums... 289
Brake hand lever locking pawl... 290
Brake shoes... 289
Brakes
  adjustment... 453
  do not hold... 142
  maintenance... 120
  road test... 110
  run-in test... 30
  use of... 49
  (See also Electric brake controls and Steering brake controls)
Breaker contact point... 227
Breather caps
  after-operation service... 105
  maintenance chart... 115
  run-in test... 27
Breathers
  after-operation service... 105
  cleaning, installation, and removal... 181
  oil flows out of... 137
  operation under dusty conditions... 73
Breech mechanism
  operating... 486
  shipment and storage... 509

C

Cable, attaching towing... 51
Carbine (cal. .30), accessories... 16
Carburetor
  adjustment... 347, 366
  description of assembly... 237, 240
  draining carburetor... 242
  identification... 241
  idle adjustment... 241
  improper adjustment... 128
  installation... 245
  maintenance... 116, 123
  non-interchangeability... 241
  removal... 244
  water drawn in through... 51
Carriage M18 (See Vehicles)
Cartridges, pyrotechnic... 18
Cautions
  road test... 111
  turret traversing lock... 98

Charging system, radio interference suppression... 429, 430
Chassis lubricants... 70
Circuit breakers
  description... 55, 355, 377, 395
  open or defective... 155
  operation... 41
  replacement... 384, 393, 394, 419
Circuits
  description
    battery charging... 354
    generator... 357
    lighting system... 376, 377, 379
  (See also Fuel cut-off wiring circuit)
Conduits
  description... 355, 377
  identification and installation... 402
  removal... 396
  replacement of wires... 402
Clutch pedal... 35
Clutches
  fail to apply... 140
  road test for unusual noise... 111
Collector (slip) ring (M18 vehicle)... 122
Communications
  data... 11
  general... 9
  gun accessories... 18
Commutator
  cleaning... 368
  description... 363
Compass, maintenance operations... 121
Compass light, inoperative... 153
Compensating idler wheels... 113
Compensating links
  data... 325
  description, replacement, and staking oil seals... 335
Compensating wheels
  data... 324
  description and removal... 337
  installation... 338
  lubrication points... 94
Condensers, radio (See Radio condensers)
Connectors
  installation... 262
  removal... 261
Control box (M18 vehicle)... 469
Controls (engine)
  after-operation service... 105
  during-operation service... 101
  engine... 105
  M18 vehicle... 467
  M39 vehicle... 470
  maintenance... 121
  operation... 31, 43
  steering brake... 290
  testing engine controls... 201
Converter, oil drains out of... 140

Index

C—Contd.

Converter oil temperature warning light
- before-operation service 99
- driving precautions 48
- inoperative switch 160
- road test 110
- test switch 41

Cooling systems 73
Covers, stowage location 21
Crankcase, road test 112
Crew seat, replacement (M39 vehicle) 442
Crews of carriage and utility vehicle 7
Cylinders (fire extinguisher system) 116, 464

D

Decontaminating apparatus
- affected by gas 74
- after-operation service 104
- before-operation service 100
- description and operation 64
- maintenance 125

Definition of terms 96
Degasser
- adjustment and replacement 244
- testing fuel cut-off 136

Demolition to prevent enemy use 74
Differential
- description and data 286
- installation 284
- lubrication 70
- maintenance operation 112, 120
- overheating 143
- removal 281
- replacement 294
- trouble shooting 142

Differential oil cooler
- data 298
- description 296

Differential oil cooler blower installation
- M18 vehicle 310
- M39 vehicle 311
- removal
  - M18 vehicle 308
  - M39 vehicle 310

Differential oil cooler cores installation
- M18 vehicle 305
- M39 vehicle 306
- removal
  - M18 vehicle 303
  - M39 vehicle 304

Differential lubrication system
- data 298
- description 296

Dome lights
- description of circuit 380
- replacement 383

Doors (hull)
- adjustment and replacement of driver's 434
- closing 201, 436
- controls (driver's) 37
- description 432
- floor escape 438
- opening 192, 436
- removal and installation
  - front 435
  - rear 437

Drain plugs, inspection 112
Drain valves, hull 443
Drive belts (engine generator)
- adjustment 362
- installation 363
- removal 362

Drive pulley, replacement 365
Driver's periscope
- removal and installation 446
- replacement of head 447

Driver's Permit and Form No. 26 100
Driver's seat
- adjustment 31
- replacement 441

Drives
- before-operation service 98
- during-operation service 103

Driving compartments, lubrication points 86
Driving precautions 48
During-operation service 101

E

Echelon maintenance
- first 97
- second 107

Electric brake controls 396
Electric brake resistor box, M39 vehicle 427
Electric motor, description and operation 56
Electrical equipment
- description 394
- testing equipment 163
- trouble shooting 161
Electrical systems 71
Electrical wiring (See Wiring)
Elevating mechanism
- maintenance 123, 124
- road test 111
- run-in test 30

Elevation quadrant
- description (M9) 496
- stowage 20

Engine
- before-operation service 44
- check alinement 198
- construction 20
- data 22, 178, 180
- description 20, 177

520

Index

E—Contd.

Engine—Cont.
- during-operation service ... 101
- failure to idle before stopping ... 50
- installation ... 196
- lubrication ... 180
- maintenance ... 116, 122, 123, 177
- observation of idle ... 119
- operation in sub-zero weather ... 72
- removal ... 114, 192
- road test ... 110
- run-in test
  - controls ... 29
  - engine ... 30
  - warm-up ... 28
- shipment and storage ... 509
- stopping ... 47
- trouble shooting ... 126, 149

Engine compartment
- lubrication points ... 88
- maintenance ... 117

Engine mountings
- description ... 184
- removal and installation
  - front ... 186
  - rear ... 185

Engine oiling system
- data ... 204
- description ... 201
- trouble shooting ... 129

Engine operation
- after-operation service ... 104
- before-operation service ... 100

Engine speed, driving precautions ... 48
Engine warm-up, before-operation service ... 98

Equipment
- after-operation service ... 107
- auxiliary ... 472
- before-operation service ... 100
- electrical ... 161, 163, 394
- interphone ... 500
- maintenance ... 124
- on-vehicle ... 15
- run-in test ... 29
- sighting and fire control ... 491
- special organizational ... 76

Escape door ... 112
Exhaust pipes ... 118
Exhaust system ... 237
Extension, firing pin assembly ... 24

F

Fenders
- after-operation service ... 105
- before-operation service ... 100
- during-operation service ... 103
- lubrication points ... 87
- maintenance chart ... 120

Filler caps ... 114
Filters
- oil ... 117, 313
- fuel ... 105, 123

Final drive assembly
- installation ... 321
- removal ... 319
- replacement record ... 321

Final drives
- description and data ... 315
- maintenance ... 112
- removal of sprockets ... 315
- trouble shooting ... 143

Final road test ... 125

Fire extinguisher system, fixed
- description ... 63, 464
- install nozzles and supports ... 200
- maintenance ... 118
- operation of controls ... 63

Fire extinguishers
- after-operation service ... 104
- before-operation service ... 98
- description (portable) ... 19, 61
- maintenance (portable) ... 125
- operation (portable) ... 62
- precautions in handling ... 63
- removal of horns and supports ... 192
- run-in test ... 27

Firing controls ... 123
Firing indicator lamp ... 421
Flashlight batteries ... 20
Float needle valve ... 51

Fuel
- after-operation service ... 104
- before-operation service ... 98
- run-in test ... 27

Fuel-air control system ... 235
Fuel cut-off wiring circuit ... 137

Fuel filter
- after-operation service ... 105
- maintenance operations (auxiliary generator) ... 123

Fuel gage
- before-operation service ... 99
- driving precautions ... 48
- inoperative ... 157
- operation ... 43
- road test ... 110
- run-in test ... 29

Fuel intake system
- data ... 237
- description ... 231

Fuel pipes
- connect ... 199
- disconnect ... 195

Fuel pump (engine)
- inoperative ... 136
- installation ... 252

Fuel strainer
- cleaning ... 368
- run-in test ... 27

Fuel supply valve ... 66
Fuel system
- data ... 179
- description ... 231
- radio interference suppression ... 429, 431

TM 9-755

521

TM 9-755

Index

F—Contd.

Fuel system—Contd.
  trouble shooting ... 134
Fuel tank pumps
  cleaning and installation ... 255
  description ... 253
  inoperative ... 135
  maintenance ... 117
  removal ... 251, 254
  road test ... 112
Fuel tanks
  description ... 233
  installation ... 259
  removal ... 258
Fuel valve control handles ... 32

G

Gages
  driver's attention to ... 48
  run-in test ... 29
  (See also under specific items)
Gasoline, grades, storage, and handling in sub-zero temperatures ... 68
Gear pump relief valve ... 461
Generator, auxiliary (See Auxiliary generator, M18 vehicle)
Generator, engine
  description ... 354
  drive belt adjustment and removal ... 362
  maintenance ... 114
  polarizing ... 365
  removal and installation ... 364
  trouble shooting ... 147
Generator circuits ... 357
Generator drive belts (engine) ... 362
Generator regulators
  data ... 358
  description of assemblies ... 355
  improperly adjusted ... 148
  installation ... 371
  removal ... 371
  (See also Regulator units)
Glass
  after-operation service ... 104
  before-operation service ... 99
  during-operation service ... 103
Governor (engine)
  adjustment ... 250
  description ... 248
  installation ... 250
  removal ... 249
Gun barrels, spare ... 124
Gun loader's seat ... 448
Gun mount
  lubrication points ... 91
  maintenance operations ... 123, 124
Gunner's quadrant ... 20
Gunner's seat ... 448
Guns
  accessories ... 16

after-operation service ... 106
characteristics ... 472
destruction (76-mm) ... 75
differences among models ... 475
elevating or depressing ... 486
firing ... 488
inspection before firing ... 480
loading ... 488
lubrication points ... 91, 92
maintenance operations ... 123, 124
on-vehicle equipment ... 15
placing in firing position ... 476
placing in traveling position ... 491
run-in test ... 30
tools
  (See also Machine gun)
Gun-sighting periscope ... 58

H

Hand throttle control ... 38
Headlight circuit ... 379
Headlight switch ... 42
Headlights
  adjustment ... 382
  description ... 376
  inoperative ... 152, 158
  installation and removal ... 380
  replacement of blackout resistor ... 382
Heater, operation ... 67
Heater control box ... 64, 396
Heater power unit ... 64
High pressure relief valve ... 460
Hoods, driver's ... 31
Hot air distribution tubes ... 66
Hubs
  data ... 324
  description
    compensating wheels ... 337
    track wheels ... 339
  installation
    compensating wheels ... 338
    track wheels ... 342
  removal
    compensating wheels ... 338
    track wheels ... 340
Hull dome lights ... 153
Hull drain valves ... 443
Hull seats ... 441
Hull subfloor doors and plates (M18 vehicle) ... 440
Hulls
  after-operation service ... 106
  before-operation service ... 100
  data ... 431, 433
  description ... 431, 433
  during-operation service ... 103
Hydraulic motor (turret)
  abnormal noise ... 76
  description ... 57, 453
  operation ... 57

Index

H—Contd.

	Page
Hydraulic pump, traversing mechanism adjusting control handle and remote control	455
description	56, 454
operation	56
removal and installation	457
Hydrostatic lock	50

I

Idle fuel cut-off	136
Ignition system	
data	179, 222
description	220
maintenance of wiring and harness	115
suppression components	429, 430
timing	222
trouble shooting	132
Instrument panel assembly	384
Instrument panel face plate	
installation	388
removal	385
Instrument panel gages	
installation	391
removal	388
Instrument panel lights	377
Instrument panel outlet	42
Instruments	
after-operation service	48
before-operation service	43, 99
driver's attention to	48
during-operation service	101
road test	109
run-in test	29
trouble shooting	154
Intake pipes	
description	183
installation	184
removal	183
Interphone equipment	
accessories	118
components of BC-659	501
general discussion	500
inspection	503
precautions	506

J

Junction boxes	
description	355, 377
installation	413
removal	410

L

Lamps (flashlights)	21
Lamps (lights)	
after-operation service	104
before-operation service	99
maintenance operations	122

	Page
run-in test	29
Leaks	
after-operation service	105
before-operation service	98
maintenance chart (oil and fuel)	119
oil leaks into differential	142
road test	112
run-in test	29, 30
Lighting system	
description	376
trouble shooting	152
Lights	
adjustment (headlights)	382
description	
dome lights	377
instrument panel	377
taillight and stop light	377
inoperative	152
installation	383
operation of switches	42
removal	382
replacement	383
Loading and blocking for rail shipment	510
Lubrication	
after-operation service	106
detailed instructions	79
for operation in sub-zero temperature	70
run-in test	28
Lubrication order	79

M

Machine gun	
accessories	17
on-vehicle spare parts	23
Machine gun mount, cal. .50, M18	
installation	449
removal	448
Magneto ground receptacle	226
Magneto switch	41
Magnetos	
correct timing	223
identification	226
ignition circuit	220
inoperative	133, 159
maintenance	96, 115, 123
non-interchangeability	225
removal and installation	228
testing, inspection, and adjustment	365
timing	119
Maintenance	
general	76
preventive	
crew (first echelon)	97
organizational (second echelon)	107
Maintenance operations	112
Manifolds	
description	181
maintenance chart	116
removal and installation	182

523

M—Contd.

Manual shift lever (transmission) 279
Markings 112
Master switch box
   installation and removal 417
   replacement of components 413
Modification (MWO completed) 125
Mufflers
   installation 200, 262
   maintenance 118
   removal 192, 261

N

Noises, unusual 49, 111

O

Odometer, road test 109
Oil, engine
   after-operation service 104
   before-operation service 98
   changing 239
   keeping fluid in sub-zero temperatures 69
   leaks at oil seals and gaskets 138
   run-in test 27
Oil cooler blower (C4 engine) 217
Oil cooler oil pump
   description 298
   installation and removal 299
   priming 300
Oil coolers
   C1 engine 210
   C4 engine 211
Oil dilution valve, engine
   description 67
   operation 68
Oil filter
   cleaning 209
   description 207
   differential oil cooler 313
   removal, assembly, and installation 209
Oil pipe coupling hoses 131
Oil pipes
   connecting 199
   description 313
   disconnecting 195
   inspection and replacement 219
Oil pressure adjustment 204
Oil pressure gage (See Pressure gage, oil)
Oil pumps
   description
      air cooler 298
      pressure and scavenge 204
      transfer case 301
   installation
      air cooler 299
      pressure and scavenge 205
      transfer case 302

   removal
      air cooler 299
      pressure and scavenge 204
      transfer case 301
Oil reservoir
   description 460
   removal and installation 462
Oil strainers
   cleaning
      oil suction 206
      oil sump
         engine C1 206
         engine C4 207
   description 206
Oil sump drain plug 206
Oil tank (engine)
   cleaning 196, 218
   description 217
   leaking check valve 51
   removal and installation 218
Oil temperature gage 157
Oil temperature warning light 394
Oiling system (See Engine oiling system)
Operating instructions 26
Operation under:
   unusual conditions 68
   usual conditions 43
Outlet box, accessory 396
Outlet check valve
   assembly 219
   disassembly and cleaning 218
Outlet sockets, description
   circuit 380
   socket 55, 377

P

Packing, replacement 183
Panel lights 394
Parking brakes (See Brakes)
Periscope, driver's
   controls and instruments 31
   removal and installation 446
   replacement 447
Periscope, gun-sighting 58
Periscope M4A1
   checking 497
   description 496
   maintenance 120
Periscope M6, stowage location 20
Periscopes, maintenance 120
Pipes
   connect (oil and fuel) 199
   disconnect (oil and fuel) 195
   description (fuel) 234
Pistol, pyrotechnic, M2 18
Power train, description 7
Pressure gage, oil
   before-operation service 99
   driver's attention to 48
   inoperative 157

Index

P—Contd.

	Page
Pressure gage, oil—Contd.	
operation	43
road test	109
Primer pump	
before-operation service	99
description	253
operation	35
run-in test	29
Priming system (fuel)	235
Propeller shaft (M18 vehicle)	
description	268
installation	
M18 vehicle	272
M39 vehicle	273
maintenance	121, 138
removal	
M18 vehicle	268
M39 vehicle	270
road test	111
run-in test	28
weekly service	105
Publications and Form No. 26	125
Pulleys (transmission and differential oil cooler)	
adjustment and removal	306
installation	307
Pumps, description	
fuel	234
fuel tank	233
oil cooler	298
oil pressure and scavenge	204
primer	253

Q

Quadrants, stowage location	20

R

Radio bonding	122
Radio interference suppression	
components	429
maintenance	430
trouble shooting	164
Radio sets	
accessories	18
antenna	502
before-operation service	99
general discussion	500
inspections	503
major components	501
mounting base	502
mountings	505
precautions	506
Radio terminal boxes	430
Rations	19
Receptacle, trailer electric	
description	396
removal	428
Recoil control, maintenance	124
Reflectors, run-in test	29
Regulator units, maintenance	119

	Page
Regulators (See Generator regulators)	
Remote control lever	57
Replacement records	201, 286, 321, 335, 352
Reports and records	1, 30, 201
Resistor, blackout (See Blackout resistor)	
Road test	109
Rocker assemblies (See Valve rocker assemblies)	
Rocker box covers	190
Roof support	196
Run-in test procedures	27

S

Sand shields	
before-operation service	100
during-operation service	103
Screen, transmission oil	313
Seats	
replacement	441, 443
turret seats	447
(See also Driver's seats)	
Services	
after-operation	48
before-operation	43
preventive maintenance	96
upon receipt of equipment	26
Shackles	
maintenance operations	112
replacement (towing)	444
Shipping data	512
Shipment and storage	508
Shock absorbers	
data	324
removal and installation	353
Sighting and fire control equipment	
accessories	19, 20
characteristics	491
Signal sending units	
description	395
replacement	425
trouble shooting	154
Siren	
after-operation service	104
before-operation service	99
description	396
inoperative	161
removal and installation	426
road test	110, 112
run-in test	29
Spare parts	
on-vehicle	23
organizational	22
Spark plugs	
installation	191, 230
maintenance	114, 122
removal	229
test and replace	71
Special services (second echelon)	108

S—Contd.

Speedometer
- driving precautions ... 48
- inoperative ... 161
- operation ... 43
- removal and installation ... 391
- road test ... 109

Springs, before-operation service ... 100

Sprockets
- installation (final drive) ... 317
- maintenance operations ... 113
- removal (final drive) ... 315
- run-in test ... 27

Starter circuit, description ... 372
Starter control circuit breaker ... 394
Starter neutral safety switch
- installation and checking switch timing ... 376
- removal ... 375

Starter relay (auxiliary generator) ... 374

Starter system
- data ... 373
- description ... 371
- trouble shooting ... 131

Starting motor ... 114

Steering brake
- adjustment ... 287
- do not hold ... 142
- during-operation service ... 101
- run-in test ... 28, 30

Steering brake controls
- assembly and disassembly ... 292
- description and adjustment ... 290
- installation ... 294
- removal ... 291

Steering brake hand levers ... 38

Steering brake linkage
- after-operation service ... 105
- before-operation service ... 100

Steering brake shoe ... 287

Stop lights
- description
  - circuits ... 379, 380
  - switch ... 376
- inoperative (trailer) ... 153
- installation of assembly ... 383
- removal of assembly ... 382

Strainer, oil (See Oil strainers)
Support arm spring bumpers ... 324
Support arms, data ... 324
Support rollers, lubrication ... 94
Suspension arms ... 113

Suspensions
- after-operation service ... 105
- before-operation service ... 100
- trouble shooting ... 144

Switches
- description
  - converter oil temperature warning light and test switch ... 41
  - lighting system ... 376
  - starter neutral safety switch ... 376
  - turret ... 55

- inoperative
  - master ... 161
  - starter ... 159
- installation ... 427
- operation
  - fuel gage ... 43
  - instrument panel ... 41
  - lighting system ... 376
  - outside of instrument panel ... 39
- removal ... 426
- replacement
  - carburetor idle fuel cut-off ... 393
  - firing circuit ... 420
  - fuel gage and auxiliary fuel pump ... 392
  - instrument panel ... 391
  - instrument panel lights ... 384, 392
  - light testing ... 393
  - stop lights ... 384
  - starter neutral safety switch ... 375
  - trouble shooting ... 159
  - turn off master ... 192

T

Tachometer
- before-operation service ... 97
- disconnect shaft ... 194
- driving precautions ... 48
- inoperative ... 161
- operation ... 43
- road test ... 109
- run-in test ... 29

Taillights
- description ... 377
- inoperative ... 153, 158
- installation of assembly ... 383
- operation of switch ... 42
- removal of assembly ... 382

Taillight circuit ... 379
Tampering and damage ... 98

Tarpaulin
- after-operation service ... 106
- before-operation service ... 100
- during-operation service ... 103

Telescopes, M72C and M76C ... 20, 496

Temperature gage
- before-operation service ... 99
- driver's attention to ... 48
- operation ... 43
- run-in test ... 29

Temperatures
- road test ... 109, 111
- run-in test ... 30

Terminal boxes
- installation ... 413
- removal ... 410

Throttle
- description of linkage ... 237
- disconnect rod ... 195

Tires
- after-operation service ... 105
- before-operation service ... 100
- rapid wear of track wheel ... 145

Index

T—Contd.

Tools
- after-operation service 107
- before-operation service 100
- gun 14
- maintenance 124
- pioneer 11
- run-in test 29
- special organizational 76
- vehicular 12

Torqmatic transmission assembly
- description and data 276
- (See also Transmission)

Torque converter oil cooler cores
- installation
  - M18 vehicle 304
  - M39 vehicle 305
- removal
  - M18 vehicle 302
  - M39 vehicle 303

Torsion bars
- description and identification 342
- removal and installation 345

Towing connections
- after-operation service 106
- before-operation service 100
- during-operation service 103
- run-in test 27

Towing hooks 444

Towing pintle
- lubrication points 94
- replacement 444

Towing the vehicle 51

Track guards
- description 325
- run-in test 27

Track link guide lugs 145
Track links, replacement 331

Track support rollers
- data 325
- description 324, 336
- inoperative 145
- installation 337
- removal 336

Track suspension system
- care of under dusty conditions 73
- lubrication points 93
- run-in test 28, 30

Track tension
- check with spacer 114
- road test 111

Track wheels
- data 324
- lubrication points 94

Trailer electric brake controller 427
Trailer electric brake resistor box 396

Tracks
- adjustment 329
- after-operation service 105
- before-operation service 100
- data 324
- description 322, 327
- installation 331

maintenance operations 113
run-in test 27
thrown 145
trouble shooting 144
Trailer cable connector plug 153
Trailer electric receptacle 154

Transfer case
- description and data (rear assembly) 262
- installation 267
- lubrication in sub-zero temperatures 70
- removal 265
- road test 110
- trouble shooting rear 137

Transfer case lubrication system
- data 298
- description 296

Transfer case oil pump 301
Transfer case shifter lever 39, 264
Transfer unit 118, 121

Transmission
- band adjustment 279
- description 276
- during-operation service torqmatic 101
- installation 284
- lubrication in sub-zero temperatures 70
- maintenance operations 121
- overheats 141
- removal 281
- road test torqmatic 110
- run-in test 30
- trouble shooting torqmatic 138

Transmission lubrication system 294
Transmission manual shift lever 35, 46

Transmission oil cooler
- data 298
- description 296

Transmission oil cooler blower
- installation
  - M18 vehicle 310
  - M39 vehicle 311
- removal
  - M18 308
  - M39 310

Traveling lock (late model) 480
Traverse motor master switch 418

Traversing mechanism
- description
  - hand 55
  - hydraulic 56
- maintenance 123, 124
- operation of turret
  - hand 58
  - hydraulic 59
- road test 111
- run-in test 30
- trouble shooting (turret) 172
- (See also Turret hand traversing mechanism)

527

RESTRICTED

TM 9-755

Index

T—Contd.

	Page
Traversing mechanism hydraulic pump	454
Traversing mechanism oil reservoirs and tubes	463
Turret	
after-operation service	106
controls and operation (M18)	52
description and data	447
lubrication points	90
operation	58
traversing	484
traversing troubles	172
Turret hand traversing mechanism	
brake adjustment	453
description	450
removal and installation	451
testing	452
Turret lock (M18)	
maintenance	120
removal and installation	450
Turret platforms	52
Turret seats	52, 447
Turret slip ring box	
cleaning and replacement of brushes	421
description	395
removal and installation	423
Turret traversing electric motor	459
Turret traversing gear	94
Turret traversing system, radio interference suppression	430, 431
Turret wiring switch box	55, 417

U

	Page
Universal joints	
description	268, 317
installation	
final drive	318
vehicle	274, 275
maintenance	120, 138
removal	
final drive	318
vehicle	270, 272
run-in test	28
trouble shooting	143

V

	Page
Valve clearance adjustment	189
Valve mechanism	114
Valve rocker assemblies	
description and removal	187
installation	189
Valves, fuel	234
Vehicle lubrication	125
Vehicle publications and reports	30
Vehicles	
accessories	
M18	16
M39	18

	Page
care of after fording	74
cleaning	106
data	
M18	9
M39	10
description	7
destruction (M18)	75
inspection and parking	72
loading and blocking	510
moves forward with shift lever in neutral	141
operation	
under dusty conditions	73
under ordinary conditions	43
sags to one side	145
serial numbers and data	9
shipment and storage	509
steering	45
stopping	47
tools	12
towing	51
Vents, after-operation service	105
Vision devices	28, 106

W

	Page
W.D. Form No. 48	97
Wheel support arm	
description	346
installation	351
removal	349
replacement of spring bumper	354
Wheels	
before-operation service	99
data	324
description	
compensating	337
track	339
installation	
compensating	338
track	341
maintenance operations	113
removal	
compensating	337
track	339
Windshield wiper	
after-operation service	104
before-operation service	99
road test	110
run-in test	29
Winterizing equipment	64
Wiring	
after-operation service	104
check	71
disconnect	194
faulty	133, 155
maintenance operations	122
(See also Circuits)	
Wiring diagrams	410
Work sheet	108

IN HIGH DEFINITION NOW AVAILABLE!

COMPLETE LINE OF WWII AIRCRAFT FLIGHT MANUALS

WWW.PERISCOPEFILM.COM

©2013 Periscope Film LLC
All Rights Reserved
ISBN#978-1-937684-46-4
www.PeriscopeFilm.com

www.ingramcontent.com/pod-product-compliance
Lightning Source LLC
Chambersburg PA
CBHW070157240426
43671CB00007B/478